www.wadsworth.com

www.wadsworth.com is the World Wide Web site for Wadsworth and is your direct source to dozens of online resources.

At *www.wadsworth.com* you can find out about supplements, demonstration software, and student resources. You can also send email to many of our authors and preview new publications and exciting new technologies.

www.wadsworth.com
Changing the way the world learns®

RACE, CLASS, AND GENDER

An Anthology
Fifth Edition

Margaret L. Andersen
University of Delaware

Patricia Hill Collins
University of Cincinnati

THOMSON
™
WADSWORTH

Australia • Canada • Mexico • Singapore • Spain
United Kingdom • United States

THOMSON
WADSWORTH
™

Acquisitions Editor: Bob Jucha
Development Editor: Natalie Cornelison
Assistant Editor: Stephanie Monzon
Editorial Assistant: Melissa Walter
Technology Project Manager: Dee Dee Zobian
Marketing Manager: Matthew Wright
Marketing Assistant: Michael Silverstein
Advertising Project Manager: Linda Yip
Project Manager, Editorial Production: Ritchie Durdin
Print/Media Buyer: Karen Hunt
Permissions Editor: Joohee Lee
Production Service: G & S Typesetters, Inc.
Text Designer: Cynthia Schultz

Copy Editor: Kathleen Deselle
Illustrator: Glenda Barlow
Cover Designer: Gopa & Ted2, Inc.
Cover Image: Alma Woodsey Thomas (1891–1978), *Atmospheric Effects II*, 1971. Gift of Vincent Melzac. Smithsonian American Art Museum, Washington, D.C. Photo credit: Smithsonian American Art Museum, Washington, D.C./Art Resource, NY.
Cover Printer: Coral Graphic Services, Inc.
Compositor: G & S Typesetters, Inc.
Printer: Maple-Vail Book Manufacturing Group/Binghamton

For more information about
our products, contact us at:
**Thomson Learning Academic
Resource Center
1-800-423-0563**

For permission to use material from this text,
contact us by: **Phone:** 1-800-730-2214
Fax: 1-800-730-2215
Web: http://www.thomsonrights.com

Library of Congress Control Number: 2003100217

ISBN 0-534-60903-1
Instructor's Edition: ISBN 0-534-62443-X

**Wadsworth/Thomson Learning
10 Davis Drive
Belmont, CA 94002-3098
USA**

Asia
Thomson Learning
5 Shenton Way #01-01
UIC Building
Singapore 068808

Australia
Nelson Thomson Learning
102 Dodds Street
South Melbourne, Victoria 3205
Australia

Canada
Nelson Thomson Learning
1120 Birchmount Road
Toronto, Ontario M1K 5G4
Canada

Europe/Middle East/Africa
Thomson Learning
High Holborn House
50/51 Bedford Row
London WC1R 4LR
United Kingdom

Latin America
Thomson Learning
Seneca, 53
Colonia Polanco
11560 Mexico D.F.
Mexico

Spain
Paraninfo Thomson Learning
Calle/Magallanes, 25
28015 Madrid, Spain

Contents

II Conceptualizing Race, Class, and Gender   75

III Rethinking Institutions 215

V Making a Difference   511

Preface

Race, Class, and Gender is an anthology that introduces students to how race, class, and gender shape the experiences of diverse groups in the United States. We want the book to help students see how the lives of different groups develop in the context of their race, class, and gender location in society. Central to the book is the idea that race, class, and gender are interconnected. As a result, the particular configuration of these factors in a given life is critical to understanding a person's or group's experience.

Since the publication of the first edition of this book, the study of race, class, and gender has become much more present in people's thinking and teaching. Yet, as our introductory essay will argue, not all of the work on race, class, and gender is centered in the intersectional framework that is central to this book. Often, people continue to treat one or the other of these social factors in isolation. Many studies of race, class, and gender also treat them as if they are equivalent. And, with the growth of race, class, and gender studies, many have added other social factors, such as sexuality, nationality, or disability—to name a few—to the list of social differences that are sources of oppression and inequality. We are glad to see such a profusion of work, although we continue to think that a framework that emphasizes the structural intersections of race, class, and gender is critical.

Thus, in this edition we have expanded the analysis of race, class, and gender to further develop the framework of intersectionality. We elaborate this framework in our revised and expanded Introduction to the new edition. We have developed the new edition with several other things in mind. We have significantly revised all of the introductions to the five sections of the book. This means that our anthology is more than a collection of readings—it is strongly centered in an analytical framework about the interconnections between race, class, and gender. The part introductions explain the importance of such an analysis ("Shifting the Center"); provide a strong conceptual foundation for understanding race, class, and gender ("Conceptualizing Race, Class, and Gender"); illustrate how you can apply such an analysis to understanding both social institutions ("Rethinking Institutions") and important social issues ("Applying the Framework"); and show how you can use such an understanding to shape social change ("Making a Difference").

This book is grounded in a sociological perspective; several articles also provide a historical foundation for understanding how race, class, and gender have emerged. Because of the focus on change, the new edition also includes more on how people have mobilized for change—through social movements, community activism, in their teaching, and in other ways that challenge oppressive social systems. We have also added materials that bring a more global dimension to the study of race, class, and gender—not just by looking comparatively at other cultures but also by analyzing how the process of globalization is shaping life in the United States. We have also added more materials— and analysis in our introductions—of how sexuality is linked to the structures of race, class, and gender in society.

As in earlier editions, we have selected articles based primarily on two criteria: (1) accessibility to undergraduate readers and the general public, not just highly trained specialists; and (2) articles that are grounded in race *and* class *and* gender—in other words, intersectionality. Not all articles accomplish this as much as we would like, but we try not to select articles that focus exclusively on one while ignoring the others. In this regard, our book differs significantly from other anthologies on race, class, and gender that include a lot of articles on each, but do less to show how they are connected. We also distinguish our book from those that are centered in a multicultural perspective. Although multiculturalism is important, we do not think that the appreciation of cultural differences is the only thing that race, class, and gender are about. Rather, we see race, class, and gender as embedded in the structure of society and significantly influencing group cultures. But race, class, and gender are also structures of group opportunity, power, and privilege. Certainly, multiculturalism is increasingly part of contemporary society, and we want students to appreciate and understand cultural differences, but, as we argue in the Introduction, multiculturalism is only one dimension of the complexities created by race, class, and gender. "Understanding multiculturalism" can be just another source of group privilege. We want students to understand differences that mark group experience in society, but we think that a focus on difference alone neither reveals the workings of race, class, and gender in society nor provides an inclusive vision for progressive social change.

Organization of the Book

We open the book with an introductory essay that establishes the framework with which we approach the study of race, class, and gender. Following that introduction, we have organized the book into five parts that follow from the logic that we see in studying these complex social systems. We introduce each of these sections with an essay by us that analyzes the issues raised by the reading selections. These essays are an important part of this book, because they establish the framework that we use to think about race, class, and gender.

Part I, "Shifting the Center," includes several classic articles that help students expand their thinking by seeing the complex interconnections between race, class, gender, sexuality, and other sources of group experience. Because they are grounded in personal experiences, these essays capture student interest, but we also use them analytically to think beyond any one person's life.

Part II, "Conceptualizing Race, Class, and Gender," provides an introduction to all three concepts, but emphasizes the linkage between them. The introductory essay provides working definitions for each and also presents some of the contemporary data through which students can see how race, class, and gender stratify contemporary society. Each concept—Race and Racism, Class and Inequality, and Gender and Sexism—is then given a separate section of readings. Together, these articles show that no one can be subsumed under the others, not studied in isolation.

Part III, "Rethinking Institutions," examines how race, class, and gender structure various social institutions. Here we focus on Work and Economic Transformation, Families, Cultural Institutions and the Production of Ideas, and State Institutions and Social Policy. We show how race, class, and gender are manifested in institutional forms and also how institutions shape these dimensions of human life. These readings also link issues of group identity to broader structures, such as the influence of cultural institutions on gender, race, and class formation.

In Part IV, "Applying the Framework," we use the race/class/gender framework we have developed to understand three important social issues: ethnicity and migration, sexuality, and violence. Although we could have selected any number of social issues, we think these are particularly relevant in contemporary society and are important to understand in the context of the linkages between race, class, and gender that this book explores. We want students to be able to apply the study of race, class, and gender to any of the social issues that concern them and think this section provides a model for doing so.

Finally, Part V, "Making a Difference," examines some of the ways that people try to create change in society so as to create a more just system of race, class, and gender relations. The theme of activism runs throughout the anthology, but here we focus on some of the arenas in which people have worked to make a difference. We want to communicate to students that activism means different things to different people and can be done in a variety of social contexts.

New to the Fifth Edition

We have made several changes in the fifth edition that strengthen and refresh *Race, Class, and Gender.* These include

- 22 *new readings.*
- 6 *new and expanded introductions* by us. These articulate the analytical model we use to understand the intersections between race, class, and gender. See especially the comparison of a "diversity model" and an "intersectional model" developed in the introductory essay to the book.
- More material on *sexuality.* We use a race/class/gender framework to understand sexuality and its relationship to these other systems. Our introductions point out some of the similarities in how sexuality patterns group experiences, but we also discuss ways that the analysis of sexuality differs from the analysis of race, class, and gender in society.
- A more *global perspective.* We are conscious of how globalization is shaping race, class, and gender relations in the United States at the same time that we are increasingly aware of how we would understand race, class, and gender differently in different global settings. We focus here on how the process of globalization reaches into race, class, and gender relations in the United States, reminding readers that it is no longer possible to understand social relations in the United States without a global perspective.
- A *new section on ethnicity and migration.* The United States is currently being transformed by processes of ethnicity and migration, as are other parts of the world. This new section of readings is fundamentally rooted in global social change.
- New articles that discuss the impact of *welfare reform.* Scholars and activists are now able to see some of the effects of welfare reform on group well-being, and we have changed the articles that discuss welfare and policy to include these recent developments.
- *Reproductive rights and health care issues.* These topics take on new meanings when viewed through a race/class/gender lens. Some of the new articles in this edition explore issues related to reproductive rights and health care, particularly as they involve the health of women of color.
- A stronger presentation of *human activism and agency.* We have added more articles that include discussion of the many ways that people have organized to work toward social justice. This also brings a perspective on *social movements* to some of the material in the book.
- The *aftermath of 9/11,* which brings race, class, and gender into new focus. We have included a new article ("How Safe Is America?") to raise some of the questions that emerge from thinking about issues of national safety.
- Several *new pedagogical features,* which we detail in the following section.

New Pedagogical Features

We realize that the context in which you teach matters. If you teach in an institution where students are more likely to be working class, perhaps how the class system works will be more obvious to them than it is for students in a more privileged college environment. Many of those who use this book will be teaching in segregated environments, given the high degree of race and class—and even gender—segregation in education. Thus, how one teaches this book should reflect the different environments in which faculty are working. Ideally, the material in this book should be discussed in a multiracial, multicultural atmosphere, but we realize that is not always the case. We hope that the content of the book and the pedagogical features that enhance it will help bring a more inclusive analysis to educational settings than might be there to start with.

We see this book as more than just a collection of readings. The book has an analytical logic to its organization and content, and we think it can be used to format a course. Of course, some faculty will use it in an order different than the way we imagine it, but we hope the five part frames will help people develop the presentation of their course. We also hope that the various pedagogical tools will help people expand their teaching and learning beyond the pages of the book.

We have added some new features to this edition that provide faculty with additional teaching tools. They include

- ***Thinking Further:*** This feature is included at the end of each major section and is meant to push people's thinking beyond what is in our introductions and the articles. We realize that the study of race, class, and gender is evolving—and, in some cases, incomplete. We use this feature to ask readers to think further about the issues raised in each section. Sometimes this means asking what is missing from the discussion; other times, it means developing an intersectional framework by pursuing unanswered questions, even if specific readings are not available on a given subject. We expect this feature to be a good basis for class discussion and for further work in this area of study.
- ***Bonus Reading:*** Wadsworth's InfoTrac® College Edition feature enables us to provide an additional recommended reading at the end of every section of the book. These are articles that we have selected because they further examine the issues in each major section. Via InfoTrac College Edition, students and faculty can retrieve an online, full-text copy of articles not included in the text per se.

We have retained the following pedagogical tools from the prior edition:

- ***Suggested Readings:*** In addition, we have included lists of suggested readings at the conclusion of each section for students and faculty who want to explore these topics further. We have selected these books based on their currency and their accessibility to undergraduate students.

- *Search Terms:* Students who purchase the book are given a password to Wadsworth's InfoTrac College Edition feature. We have included terms at the end of each section that students can use on InfoTrac College Edition to find additional information on subjects that interest them. These terms can also be used as the basis for in-class discussion, exercises, exams, or research projects.
- *Instructor's Manual:* This edition includes an Instructor's Manual with suggestions for class exercises, discussion and examination questions, and course assignments.
- *Index:* This will help students and faculty locate particular topics in the book quickly and easily.
- *Web Site:* Online support is available via the Wadsworth Sociology Resource Center. We encourage people to browse the *Race, Class, and Gender* page at the Wadsworth Web site or related resources. The address is http://sociology.wadsworth.com

A Note on Language

Reconstructing existing ways of thinking to be more inclusive requires many transformations. One transformation needed involves the language we use when referring to different groups. Language reflects many assumptions about race, class, and gender; and for that reason, language changes and evolves as knowledge changes. The term *minority*, for example, marginalizes groups, making them seem somehow outside the mainstream or dominant culture. Even worse, the phrase *non-White*, routinely used by social scientists, defines groups in terms of what they are not and assumes that Whites have the universal experiences against which the experiences of all other groups are measured. We have consciously avoided using both of these terms throughout this book, although this is sometimes unavoidable.

We have capitalized Black in our writing because of the specific historical experience, varied as it is, of African Americans in the United States. We also capitalize White when referring to a particular group experience; however, we recognize that White American is no more a uniform experience than is African American. We realize that these are arguable points, but we want to make our decision apparent and explicit. For the benefit of purists who like to follow the rules, we note that the *Chicago Manual of Style* recognizes that these grammatical questions are also political ones; this edition suggests that writers might want to capitalize Black and White to reflect the fact that they are referring to proper names of groups. We use *Hispanic* and *Latino* interchangeably, though we recognize that is not how groups necessarily define themselves.

When citing data from other sources (typically government documents), we use "Hispanic," because that is usually how such data are reported.

Language becomes especially problematic when we want to talk about features of experience that different groups share. Using shortcut terms like Hispanic, Latino, Native American, and women of color homogenizes distinct historical experiences. Even the term *White* falsely unifies experiences across such factors as ethnicity, region, class, and gender, to name a few. At times, though, we want to talk of common experiences across different groups, so we have used labels like Latino, Asian American, Native American, and women of color to do so. Unfortunately, describing groups in this way reinforces basic categories of oppression. We do not know how to resolve this problem but want readers to be aware of the limitations and significance of language as they try to think more inclusively about diverse group experiences.

Acknowledgments

An anthology rests on the efforts of more people than the editors alone. This book has been inspired by our work with scholars and teachers from around the country who are working to make their teaching and writing more inclusive and sensitive to the experiences of all groups. Over the years of our own collaboration, we have each been enriched by the work of those trying to make higher education a more equitable and fair institution. In that time, our work has grown from many networks that have generated new race, class, and gender scholars. These associations continue to sustain us. We especially thank Amalia Amaki, Maxine Baca Zinn, Patrice Dickerson, Bonnie Thornton Dill, Valerie Hans, Elizabeth Higginbotham, Carole Marks, and Howard Taylor for inspiring our work and discussing various parts of this project with us. We also thank them for the encouragement and vision that sparks our work.

Many other people contributed to the development of this book. We especially thank Tina Diablo at the University of Delaware and Rachel Clark at the University of Kentucky for their expert research assistance. We appreciate the support given by our institutions, with special thanks to Joel Best, chair of the Department of Sociology (University of Delaware); Gerald Smith, head of the African American Studies Program (University of Kentucky); and Debra Harley, acting head of the Women's Studies Program (University of Kentucky). We also thank Vicky Baynes, Linda Keen, and Judy Watson (University of Delaware), Josephine Wynne (University of Cincinnati), and Betty Pasley (University of Kentucky) for their invaluable secretarial support. We especially thank Linda for taping down all those cut pages!

Bob Jucha and Natalie Cornelison have joined this project with enthusiasm, fresh ideas, and expert advice. We sincerely appreciate their hard work and vision for the book and look forward to continuing our collaboration with them. We also thank the reviewers, who provided valuable commentary on the prior edition and thus helped enormously in the development of the fifth edition: Pat Washington (San Diego State University), Susan Rose (Dickinson College), Ralph Pyle (Michigan State University), Eleanor Hubbard (University of Colorado), Magalene Taylor (University of Arkansas), Margaret Wilder (University of Georgia), M. Bahati Kuumba (Spelman College), and Melinda Miceli (University of Wisconsin).

Developing this book has been an experience based on friendship, hard work, travel, and fun. We especially thank Valerie, Roger, and Richard for giving us the support, love, and time we needed to do this work well. This book has deepened our friendship, as we have grown more committed to transformed ways of thinking and being. We are lucky to be working on a project that continues to enrich our work and our friendship.

About the Editors

Margaret L. Andersen (B.A., Georgia State University; M.A., Ph.D., University of Massachusetts, Amherst) is Professor of Sociology and Women's Studies at the University of Delaware. She is the author of *Thinking about Women: Sociological Perspectives on Sex and Gender*, 6th ed. (Allyn and Bacon, 2003); *Sociology: Understanding a Diverse Society*, 3d ed. (Wadsworth, 2004; co-authored with Howard F. Taylor); *Sociology: The Essentials*, 2d ed. (Wadsworth, 2003; also co-authored with Howard F. Taylor); and *Understanding Society: An Introductory Reader* (Wadsworth, 2001; co-edited with Kim Logio and Howard F. Taylor). She is a recipient of the University of Delaware's Excellence-in-Teaching Award, former President of the Eastern Sociological Society, and chair of the National Advisory Board for the Center for Comparative Studies in Race and Ethnicity at Stanford University, where she has been a Visiting Professor.

Patricia Hill Collins (B.A., Brandeis; M.A.T, Harvard University; Ph.D., Brandeis University) is Professor of African American Studies and Sociology at the University of Cincinnati, and former chair of the Department of African American Studies. She is the author of numerous articles and books, including *Fighting Words: Black Women and the Search for Social Justice* (University of Minnesota Press, 1998) and *Black Feminist Thought: Knowledge, Consciousness, and the Politics of Empowerment* (Unwin Hyman, 1990; Routledge, 2000), which won the C. Wright Mills Award of the Society for the Study of Social Problems. She was the Bryan University Chair in African American Studies and Women's Studies at the University of Kentucky while completing this edition. She has a forthcoming book entitled *Black Sexual Politics*.

About the Contributors

Paula Gunn Allen is a poet, novelist, and critic. Her heritage is Laguna, Sioux, and Lebanese. A major Native American poet, writer, and scholar, she has published numerous volumes of poetry, a novel, a collection of essays, and several anthologies. She is now professor of English at the University of California, Los Angeles.

Julia Álvarez is the author of several novels, including *How the Garcia Girls Lost Their Accents, In the Time of Butterflies,* and *¡Yo!* She is originally from the Dominican Republic but migrated to the United States when she was ten years old.

Teresa Amott is on the faculty in economics at Bucknell University. She is the author of *Race, Gender, and Work: A Multicultural Economic History of Women in the United States* (with Julie Matthaei) and *Caught in the Crisis: Women in the U.S. Economy Today,* along with numerous articles. She is an editorial associate with *Dollars and Sense* magazine and is committed to sharing economic analysis with unions, welfare rights organizations, women's organizations, and other progressive groups.

John Anner is the founder and executive director of the Independent Press Association. He is the editor of *Beyond Identity Politics: Emerging Social Justice Movements in Communities of Color* and has worked in the social justice movement, primarily with South African and Central American solidarity movements.

Charon Asetoyer is the founder and executive director of the Native American Women's Health Education Resource Center, based on the Yankton Sioux Reservation in Lake Andes, South Dakota. She has been involved in community organizing and activism on women's health issues for the past twenty years.

Maxine Baca Zinn is professor of sociology at Michigan State University. She is the coauthor of *Diversity in Families, In Conflict and Order*, and *Social Problems* (with D. Stanley Eitzen), and the co-editor of *Gender Through the Prism of Difference: A Sex and Gender Reader* (with Pierrette Hondagneu-Sotelo and Michael A. Messner).

Kenneth W. Brown is a contributor to *Essence* magazine. While unemployed, he completed an associate's degree; later, he worked in financial services while pursuing a bachelor's degree.

Sarah L. Brownlee is an expert in community organizing and gay and lesbian issues. She is the former director of the National Center for Human Rights Education and a massage therapist.

Linda Burnham is cofounder and executive directer of the Women of Color Resource Center. She is the author of *Women's Education in the Global Economy* and co-author of *Working Hard, Staying Poor*.

Ward Churchill (enrolled Keetoowah Band Cherokee) is a long-time native rights activist and professor of ethnic studies and coordinator of American Indian studies for the University of Colorado. His books include *Agents of Repression, Fantasies of the Master Race, From a Native Son*, and *A Little Matter of Genocide: Holocaust and Denial in the Americas*.

Judith Ortiz Cofer is the Franklin Professor of English and Creative Writing at the University of Georgia. She is the author of numerous books of poetry and essays, and a novel, *The Line of the Sun*. She has won the PEN/Martha Albrand Special Citation in nonfiction for her book *Silent Dancing*.

Chuck Collins is the cofounder and director of United for a Fair Economy —an organization that draws attention to the income and wealth inequality in the United States. He is the author of several books, including *Economic Apartheid in America: A Primer on Economic Inequality and Insecurity*, and he writes regularly for *Dollars and Sense*.

Gary David Comstock is University Protestant Chaplain and Visiting Associate professor at Wesleyan University. He has published numerous books, including *Gay Theology Without Apology; Unrepentant, Self-Affirming, Practicing: Lesbian/Bisexual/Gay People Within Organized Religion; Que(e)rying Religion: A Critical Anthology;* and *Violence against Lesbians and Gay Men*.

Dalton Conley is associate professor of sociology and director of the Center for Advanced Social Science Research at New York University. He is author of *Being Black, Living in the Red: Race, Wealth and Social Policy in America*, which received the award for best in the field from the American Sociological Association. He is also author of *Honky*, a memoir of growing up White in a predominantly minority, inner-city housing project.

Deirdre E. Davis received her B.A. from Wesleyan University and J.D. from Boalt Hall School of Law at the University of California, Berkeley. While in law school, she edited the *Berkeley Women's Law Journal* and is now in private practice.

Dazon Dixon Diallo is the founder and president of SisterLove Women's AIDS project and a public health expert and human rights activist.

Bonnie Thornton Dill is professor of women's studies and director of the Research Consortium on Gender, Race, and Ethnicity at the University of Maryland, College Park. Her books include *Women of Color in U.S. Society*, co-edited with Maxine Baca Zinn, and *Across the Boundaries of Race and Class: Work and Family Among Black Female Domestic Servants*.

Yen Le Espiritu is professor in the department of ethnic studies at the University of California, San Diego. She is the author of *Asian American Pan-ethnicity, Filipino American Lives*, and *Asian American Women and Men: Labor, Laws, and Love*, which received a book award from the American Sociological Association.

Abby L. Ferber is associate professor and co-assistant Vice Chancellor at the University of Colorado at Colorado Springs. She is the author of *White Man Falling: Race, Gender and White Supremacy; Hate Crime in America: What Do We Know?*; and *Making a Difference: University Students of Color Speak Out*.

Marilyn Frye teaches philosophy and women's studies at Michigan State University. She is the author of numerous books and essays, including *The Politics of Reality* and *Willful Virgin*. With her partner Carolyn Shafer, she created and manages Bare Bones Studios for Women's Art—space and facilities for art-making in a wide range of media.

Rose Campbell Gibson is professor emerita at the University of Michigan. She is the author of several books, including *Different Worlds: Inequality in the Aging Experience* (with Eleanor Palo Stoller), *Blacks in an Aging Society*, and *Health in Black America* (with James Jackson).

Amy Gluckman is a social studies teacher in an alternative high school in Lowell, Massachusetts. She is part of an editorial collective that publishes the progressive economics magazine *Dollars and Sense* and, with Betsy Reed, is the editor of a book, *Homo Economics: Capitalism, Community, and Lesbian and Gay Life.*

Pierrette Hondagneu-Sotelo is associate professor of sociology, American studies, and ethnicity at the University of Southern California. She is author of *Gendered Transitions: Mexican Experiences of Immigration* and *Doméstica: Immigrant Workers Cleaning and Caring in the Shadow of Affluence*, which won the Society for Social Problems C. Wright Mills Award.

Shani Jamila is a poet and activist and a graduate of Spelman College with a master's degree in African Diaspora Cultural Studies from the University of California, Los Angeles. She has traveled extensively throughout Africa and the Caribbean and is known for her work on hip-hop and feminism.

James Jennings is professor of urban and environmental policy and planning at Tufts University and senior fellow at the Trotter Institute, University of Massachusetts, Boston. He is the author of *The Politics of Black Empowerment* and *Understanding the Nature of Poverty*, among other books and articles.

June Jordan, deceased in 2002, was a prolific poet, essayist, and activist. Among her many books are *Some of Us Did Not Die; Soldier: A Poet's Childhood;* and *Affirmative Acts.*

Robin D. G. Kelley is a professor of history and Africana studies at New York University. He is the author of numerous books, including *Freedom Dreams: The Black Radical Imagination; Three Strikes: The Fighting Spirit of Labor's Last Century; Race Rebels: Culture Politics and the Black Working Class;* and *Yo' Mama's DisFunktional!*

Kamala Kempadoo is assistant professor of women's studies and sociology at the University of Colorado, Boulder.

Nazli Kibria is associate professor of sociology at Boston University. She is the author of *Becoming Asian American: Identities of Second Generation Chinese and Korean Americans* and *Family Tightrope: The Changing Lives of Vietnamese Americans.*

Bruce Kokopeli is the author of *Leadership for Change: Toward a Feminist Model* (with George Lakey).

Celene Krauss is professor of sociology and women's studies at Kean University. She has written extensively on women's issues and toxic waste protests, with a focus on race, ethnicity, and class. Her work has appeared in *Sociological Forum, Qualitative Sociology*, and other journals and books.

Louis Kushnick teaches at the University of Manchester (England), and is director of the Ahmen Iqbal Ullah Race Relations Archive. He is also vice chair of the Institute of Race Relations and editor of *Sage Race Relations Abstracts*. He is author of *Race, Class, and Struggle: Essays on Racism and Inequality in Britain, the U.S., and Western Europe* and co-editor with James Jennings of *A New Introduction to Poverty: The Role of Race, Power, and Politics.*

George Lakey is the director of Training for Change—an organization promoting nonviolent social change. He is the author of six books, including *Grassroots and Nonprofit Leadership: A Guide for Organizations in Changing Times.* An openly gay grandfather of four, he frequently lectures at major colleges and universities.

Donna Langston is chair of ethnic studies at California Polytechnic State University and author of articles in *Race, Class, and Gender: Nature, Society, and Thought;* and other journals. She is completing a biography of Wilma Mankiller, the first woman chief of Cherokee Nation.

Audre Lorde, who passed away in 1992, grew up in the West Indian community of Harlem in the 1930s, the daughter of immigrants from Grenada. She was professor of English at Hunter College and a major figure in the lesbian and feminist movements. Among her works are *Sister Outsider; Zami: A New Spelling of My Name; Uses of the Erotic;* and numerous other books and essays.

Arturo Madrid is Murchison Distinguished Professor of the Humanities at Trinity University and the recipient of the Charles Frankel Prize in the Humanities, National Endowment for the Humanities, in 1996. From 1984 until 1993 he served as founding president of The Tomás Rivera Center, a national institute for policy studies on Latino issues.

Julianne Malveaux holds a Ph.D. in economics and is a syndicated columnist who writes regularly for *Essence* and also appears regularly on CNN, BET, Fox News, MSNBC, and other television news shows. She is the author of *Wall Street, Main Street, and the Side Street: A Mad Economist Takes a Stroll.* She is president and CEO of her own multimedia company.

Gregory Mantsios is the director of worker education at Queens College, the City University of New York.

Elizabeth Martinez is a Chicana scholar and activist. She is the cofounder of the Institute for MultiRacial Justice. Her latest book is *De Colores Means All of Us: Latina Views for a Multi-Colored Century.*

Julie Matthaei is a professor of economics at Wellesley College. Her books include *An Economic History of Women in America: Women's Work, The Sexual Division of Labor, and the Development of Capitalism* and (with Teresa Amott) *Race, Gender, and Work: A Multicultural Economic History of Women in the United States.*

Peggy McIntosh is associate director of the Wellesley College Center for Research on Women. She is founder and codirector of the National SEED Project on Inclusive Curriculum—a project that helps teachers make school climates fair and equitable. She is cofounder of the Rocky Mountain Women's Institute.

Michael A. Messner is associate professor of sociology and gender studies at the University of Southern California. His books include *Masculinities, Gender Relations, and Sport; Power at Play: Sports and the Problem of Masculinity; Politics of Masculinities: Men in Movements;* and others.

Roslyn Arlin Mickelson is professor of sociology and women's studies at the University of North Carolina at Charlotte. She is the author of *Children on the Streets of the Americas: Globalization, Homelessness, and Education in the United States, Brazil, and Cuba.*

Barbara Miner is co-author of *Selling Out Our Schools: Vouchers, Markets, and the Future of Public Education* and *False Choice: Why School Vouchers Threaten our Children's Future.* She is also managing editor of *Rethinking Schools,* a journal providing a grassroots look at urban educational reform.

Robert B. Moore writes frequently on feminist issues and is the author of the classic "Racist Stereotyping in the English Language" included in this volume.

Cherríe Moraga is a poet, playwright, and essayist. She is the co-editor of *This Bridge Called My Back: Writing by Radical Women of Color* and the author of numerous plays including *Shadow of a Man,* winner of the 1990 Fund for New American Plays Award, and *Heroes and Saints,* winner of the Will Glickman Prize and the Pen West Award. She is also a recipient of the National Endowment for the Arts' Theatre Playwrights' Fellowship.

Philip Moss is a professor in the Department of Regional Economic and Social Development at the University of Massachusetts, Lowell. He studies the impacts of structural change on economic opportunity. With Chris Tilly he has published *Stories Employers Tell: Race, Skill, and Hiring in America*.

Katherine S. Newman is Malcolm Wiener Professor of Urban Studies at the John F. Kennedy School of Government, Harvard University and Dean of Social Science at the Radcliffe Institute for Advanced Study. She is the author of *No Shame in My Game: The Working Poor in the Inner City* and *Falling from Grace: Downward Mobility in the Age of Affluence*.

Mary Pattillo-McCoy is associate professor of sociology at Northwestern University and the author of *Black Picket Fences: Privilege and Peril among the Black Middle Class* and a forthcoming book (with William Julius Wilson and Richard Taub), *The Roots of Racial Tensions: Urban Ethnic Neighborhoods*.

Betsy Reed, a former editor of *Dollars and Sense*, is a senior editor at *The Nation*. She is co-editor with Amy Gluckman of the anthology *Homo Economics: Capitalism, Community, and Lesbian and Gay Life*.

Luz Rodriquez is an expert on Latina health. She is the former executive director of Casa Atabex Ache, a health organization for women of color.

Loretta J. Ross is the executive director of the National Center for Human Rights Education, which focuses on domestic human rights violations.

Lillian Rubin is a senior research associate at the Institute for the Study of Social Change at the University of California, Berkeley. She has written numerous books, including *Families on the Fault Line*, *Worlds of Pain*, and *The Transcendent Child*.

Virginia Rutter is finishing her Ph.D. in sociology at the University of Washington, where she teaches on topics related to gender, sexuality, families, women in the social structure, research methodology, and writing. Prior to teaching she worked as a journalist and in public affairs and Congress in Washington, D.C. She is the author of two books, *The Gender of Sexuality* and *The Love Test*.

Almas Sayeed majored in philosophy, women's studies, and international studies at the University of Kansas where she was also a columnist for *Kansan*. She received a Fulbright grant to study the strategies of conflict resolution by six Israeli and Palestinian women's peace organizations and was affiliated with the Truman Institute for the Advancement of Peace at the Hebrew University in Jerusalem in 2002–2003.

Jason Schultz is the director of special projects for Stir Fry Productions, an organization specializing in diversity training and racial conflict resolution. He was also the coordinator of the Duke Men's Project, a group for college men working on issues of gender and sexuality.

Pepper Schwartz is a professor of sociology at the University of Washington in Seattle. She is past president of the Society for the Scientific Study of Sexuality and a member of the International Academy of Sex Research. She is the author of many books, among them *The Gender of Sexuality* (with Virginia Rutter), *The Great Sex Weekend, Everything You Know about Love and Sex is Wrong*, and *Ten Talks Parents Must Have with Their Children about Sex and Character* (with Dominic Cappello).

Jael Silliman is assistant professor of women's studies at the University of Iowa. She is the co-editor of *Dangerous Intersections: Feminist Perspectives on Population, Environment, and Development* and the author of *Jewish Portraits, Indian Frames: Women's Narratives from a Diaspora of Hope*.

Robert Smith is assistant professor of sociology at Barnard College. He is the author of *Migration, Settlement, and Transnational Life*, as well as numerous articles on migration and transnational communities.

Stephen Samuel Smith is associate professor of political science at Winthrop University. His interests include the politics of education, urban political economy, social movements, and program evaluation.

C. Matthew Snipp is a professor of sociology at Stanford University. He is the author of *American Indians: The First of This Land* and *Public Policy Impacts on American Indian Economic Development*. His tribal heritage is Oklahoma Cherokee and Choctaw.

Brent Staples is a journalist who writes for the *New York Times* among other sources. He has published his memoir, *Parallel Time: Growing up in Black and White*.

Eleanor Palo Stoller is the Selah Chamberlain Professor of Sociology at Case Western Reserve University. She is the author of numerous articles and co-author (with Rose Campbell Gibson) of *Worlds of Difference: Inequality in the Aging Experience*.

Ronald T. Takaki is professor of ethnic studies at the University of California, Berkeley. He is the author of several books, including *Iron Cages:*

Race and Culture in 19th Century America; Strangers from a Different Shore: A History of Asian Americans; and *A Different Mirror: A History of Multicultural America.*

Desiree Taylor is a mixed-race woman whose work has been published in *Reconstructing Gender: A Multicultural Anthology.*

Chris Tilly is University Professor in the Department of Regional Economic and Social Development at the University of Massachusetts, Lowell. He is the author of *Half a Job: Bad and Good Part-Time Jobs in a Changing Labor Market; Glass Ceilings and Bottomless Pits: Women's Work, Women's Poverty* (with Randy Albelda); and *Stories Employers Tell: Race, Skill, and Hiring in America* (with Philip Moss).

Lynet Uttal is an assistant professor in the Department of Child and Family Studies at the University of Wisconsin, Madison. She uses in-depth interviewing to study employed mothers and their child care arrangements and racial ethnic families.

Felice Veskel teaches in the Social Justice Education Program at the University of Massachusetts, Boston.

Mary C. Waters is professor of sociology and Harvard College Professor at Harvard University. She is the author of *Black Identities: West Indian Immigrant Dreams and American Realities; Ethnic Options: Choosing Identities in America;* and numerous articles on race, ethnicity, and immigration.

Cornel West is professor of Afro-American studies and the philosophy of religion at Harvard University. He is the author of numerous books and essays, including *Restoring Hope: Conversations on the Future of Black America* (with Shawn Sealey) and *Race Matters.*

Kath Weston is a sociocultural anthropologist who has written several books, including *Families We Choose: Lesbians, Gays, and Kinship; Render Me, Gender Me; Long Slow Burn;* and *Gender in Real Time.*

Patricia J. Williams is the James L. Dohr Professor of Law at Columbia University and the author of *The Alchemy of Race and Rights* and *The Diary of a Law Professor.* She is a regular contributor to *The Nation.*

Gloria Yamato has worked with numerous social change agencies in the Pacific Northwest, including The Women's Funding Alliance, the American

Friends Service Committee, and Seattle Rape Relief, among others. She has contributed essays to *Women's Lives: Multicultural Perspectives; Sinister Wisdom;* and *Making Face, Making Soul/Haciendo Caras.*

Helen Zia is a former executive editor of *Ms.* magazine and is the author of *Asian American Dreams: The Emergence of an American People.*

Introduction

DEVELOPING A RACE, CLASS, AND GENDER FRAMEWORK

Anyone who takes even a cursory look at the contemporary United States has to acknowledge that it is one of the most diverse nations in the world. In schools, students speak a variety of languages. At work, although different groups are still segregated into specific jobs, the presence of women, new immigrants, and racial minorities is growing. African Americans, Latinos, Asian Americans, and Native Americans together account for an increasing proportion of the total population—so much so that even the term *minority* is misleading, at least in the numerical sense. Furthermore, as shown in Figure 1, the population of the United States will become even more diverse in years to come, as African Americans, Latinos, Asian Americans, and American Indians become a larger proportion of the total population.

What does diversity mean? Because the American public has become more conscious of these trends, *diversity* has become a buzzword—popularly used, but loosely defined. People use diversity to mean cultural variety, numerical representation, changing social norms, and the inequalities that characterize the status of different groups. In thinking about diversity, people have recognized that gender, age, race, sexual orientation, and ethnicity matter; thus, groups who have previously been invisible, including people of color, gays, lesbians, and bisexuals, older people, and immigrants, are now in some ways more visible. At the same time that diversity is more commonly recognized, however, these same groups continue to be defined as "other"; that is, they are perceived through dominant group values, treated in exclusionary ways, and subjected to social injustice and economic inequality. Moreover, dominant ways of thinking remain centered in the experiences of a few. As a result, groups who in total are the majority are still rendered invisible.

Studying diversity is not simply a matter of learning about other people's cultures, values, and ways of being; it involves discovering how race, class, and gender—along with factors like age, ethnicity, sexual orientation, and religion—frame people's lives. The point is not just that people are diverse, as if that were an interesting yet benign fact of life, but that race, class, and

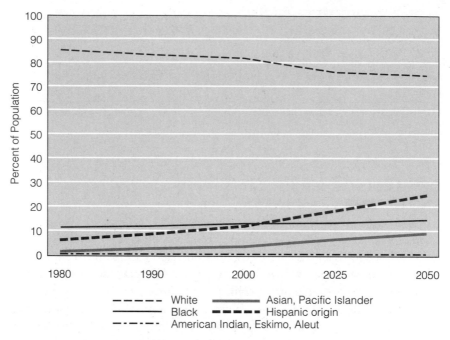

Note: The data total more than 100% because Hispanics may be of any race.

Source: U.S. Census Bureau. 2002. *Statistical Abstract of the United States: 2001.* Washington, DC: U.S. Census Bureau, p. 13.

FIGURE 1 Diversity in the U.S. Population

gender are fundamental axes of society and, as such, are critical to understanding people's lives, institutional systems, contemporary social issues, and the possibilities for social change. Thus, while diversity is widely acknowledged, there is still a need to reconstruct how we think—and how we behave—if we are to create a more just society.

Understanding the social dynamics of race, class, and gender is central to this new vision. Studying race, class, and gender brings new facts and new perspectives to light, and it illuminates familiar things in new ways. For example, college basketball can be seen as pure entertainment, but it can also be seen as a social system in which race, class, and gender play a significant part. Also, places that appear to be all-White or all-men can be understood as products of a social system that is premised on certain race, class, and gender arrangements. Understanding the significance of race, class, and gender puts the experience of all groups in the United States into a broader context. Furthermore, knowing how race, class, and gender operate within U.S. national borders helps you see beyond those borders, developing an awareness of how the increasingly global basis of society influences the configuration of race, class, and gender relationships in the United States.

This book is part of the new effort to transform thinking by seeing how race, class, and gender operate together to create the experiences of all groups in society. We want readers to understand that race, class, and gender are linked experiences, no one of which is more important than the others; the three are interrelated and together configure the structure of this society.

FRAMEWORKS OF DIVERSITY, DIFFERENCE, AND MULTICULTURALISM

Because we approach the study of race, class, and gender with an eye toward transforming thinking, we see our work as differing somewhat from the concepts implicit in the language of diversity, difference, or multiculturalism. Thinkers and activists who use these concepts have made important contributions in working for social justice. Clearly, our work is part of these movements, but the framework we develop differs from those provided by diversity, difference, and multiculturalism.

Diversity has become one catchword for trying to understand the complexities of race, class, and gender in the United States. Diversity initiatives hold that the diversity created by race, class, and gender differences are pleasing and important, both to individuals and to society as a whole—so important, in fact, that diversity should be celebrated. Under diversity initiatives, ethnic foods, costumes, customs, and festivals are celebrated, and students and employees receive diversity training to heighten their multicultural awareness. Diversity initiatives also advance a notion that, despite their differences, "people are really the same." Under this view, the diversity created by race, class, and gender, although worthy of appreciation, does not and should not affect how society functions.

The movement to "understand diversity" has made people more sensitive and aware of the intersections of race, class, and gender. Thinking about diversity has also encouraged students and activists to see linkages to other categories of analysis, including sexuality, age, religion, physical disability, national identity, and ethnicity. But appreciating diversity is not the only point. The very term *diversity* implies that understanding race, class, and gender is simply a matter of recognizing the plurality of views and experiences in society—as if race, class, and gender were benign categories that foster diverse experiences instead of systems of power that produce social inequalities.

Certainly, opening our awareness of distinct group experiences is important, but some approaches to diversity can erase the very real differences in power that race, class, and gender create. For example, some people in recent years have tended to think of race, class, and gender in terms of a "voices" metaphor—a metaphor that appropriately arose from challenging the silence surrounding many group experiences. In this framework, people think about diversity as "listening to the voices" of a multitude of previously silenced groups. This is an important part of coming to understand race, class, and gender, but it is not enough.

One problem is that people may begin hearing the voices as if they were disembodied from particular historical and social conditions—a framework that has been aggravated by postmodernist social theory. This perspective sees experience as a matter of competing discourses, personifying "voice" as if the voice or discourse itself constituted lived experience. Second, the "voices" framework suggests that any analysis is incomplete unless every voice is heard. In a sense, of course, this is true, because inclusion of silenced people is one of the goals of multicultural work. But in a situation where it is impossible to hear every voice, how does one decide which "voices" are more important than others? One might ask, who are the privileged listeners within these "voice" metaphors?

We want readers to be more analytical. Analyzing race, class, and gender involves more than appreciating cultural diversity. It requires analysis and criticism of existing systems of power and privilege; otherwise, understanding diversity becomes just one more privilege for those with the greatest access to education—something that has always been a mark of the elite class. Analyzing race, class, and gender as they shape different group experiences also involves addressing issues of power, privilege, and equity. This means more than just knowing the cultures of an array of human groups. It means recognizing and analyzing the hierarchies and systems of domination that permeate society and that limit our ability to achieve true democracy and social justice.

Difference constitutes another catchword for trying to understand the complexities of race, class, and gender. Diversity and difference both foster comparative thinking, where individuals are encouraged to compare their experiences with those supposedly unlike them. Like diversity, this comparative process can foster greater understanding and tolerance. But comparative thinking alone can also leave intact the power relations that create race, class, and gender relations. Because the concept of difference contains the unspoken question "different from what?" this framework can privilege those who are deemed to be "normal" and stigmatize people who are labeled as "different." Also, because it is based on comparison, the very concept of difference fosters dichotomous, "either/or" thinking. Some difference models place people in "either/or" categories, as if one is either Black or White, oppressed or oppressor, powerful or powerless, normal or "different." Those who are "different" routinely form the subordinate side of the dichotomy and are often seen as victims.

Thinking about race, class, and gender is not just a matter of studying victims. Relying too heavily on ideas about difference overemphasizes the impact of race, class, and gender on poor people, women, and people of color, rather than seeing race, class, and gender as an integral part of social structures that shape the experiences of everyone. We remind students that race, class, and gender have affected the experiences of all individuals and groups. As a result, we do not think we should talk only about women when talking about gender, or only about people of color when talking about race. Because race, class, and gender affect the experience of all, it is important to study men when analyzing gender, to study Whites when analyzing race, and to study the experiences of the affluent when analyzing class. Furthermore, we should not forget women when studying race or think only about Whites when studying gender.

Ideas about difference can also foster "additive models" of thinking that do little to challenge existing social inequalities. The additive model is reflected in terms like *double* and *triple jeopardy*. But social inequality cannot be quantified in this fashion: Within this logic, poor African American women seemingly experience the triple oppression of race, gender, and class, whereas poor Latina lesbians encounter quadruple oppression, and so on. Adding together "differences" (thought to lie in the difference from the norm) produces a hierarchy of difference that ironically reinstalls those who are additively privileged at the top while relegating those who are additively oppressed to the bottom. We do not think of race and gender oppression in the simple additive terms implied by phrases like *double* and *triple jeopardy*. The effects of race, class, and gender do "add up," both over time and in intensity of impact, but seeing race, class, and gender only in additive terms misses the social structural connections between them and the particular ways in which different configurations of race, class, and gender affect group experience.

When combined with the concept of difference, additive models of thinking can foster another troubling outcome. One can begin with the concepts of race, class, and gender and continue to "add on" additional types of difference. Ethnicity, sexuality, religion, age, and ability all can be "added on" to race, class, and gender in ways that suggest that any of these forms of difference can substitute for others. This use of difference fosters a view of oppressions as equivalent and as being the same. Recognizing that difference encompasses more than race, class, and gender is a step in the right direction. But continuing to "add on" many distinctive forms of difference can be a never-ending process. After all, there are as many forms of difference as there are individuals. Ironically, this form of recognizing difference can erase the workings of power just as effectively as diversity initiatives.

Finally, we see limits to the concepts of diversity and difference because we think it is important to challenge the idea that they are important only at the level of culture—an implication of the catchword *multiculturalism*. Culture is traditionally defined as the "total way of life" of a group of people. It encompasses both material and symbolic components and is an important dimension of understanding human life. Analysis of culture per se, however, tends to look at the group itself, rather than at the broader conditions within which the group lives. Of course, as anthropologists know, a sound analysis of culture situates group experience within these external conditions. Nonetheless, a narrow focus on culture tends to ignore social conditions of power, privilege, and prestige. The result is that multicultural studies often seem tangled up with notions of cultural pluralism—as if knowing a culture other than one's own is the goal of a multicultural education.

We think it is important to see the diversity and plurality of different cultural forms, but in our view this perspective, taken in and of itself, misses the broader point of understanding how racism, sexism, homophobia, and class relations have shaped the experience of groups. Imagine, for example, trying to study the oppression of gays and lesbians in terms of gay culture only. Obviously, doing this turns attention onto the group itself and away from the dominant society. Likewise, studying race only in terms of Latino culture or Asian American culture or African American culture, or studying women's oppression only by looking at women's culture, encourages thinking that blames the victims for their own oppression. For all of these reasons, the focus of this book is on the institutional, or structural, bases for race, class, and gender relations.

The languages of diversity, difference, and multiculturalism all encourage comparative thinking. People think comparatively when they learn about experiences other than their own and begin comparing and contrasting the experiences of different groups. This is a step beyond centering one's thinking in a single group (typically one's own), but it is nonetheless limited. For example, when students encounter studies of race, class, and gender for the first time, they often ask, "How is this group's experience like or not like my own?" This is an important question and a necessary first step, but it is not enough. For one thing, it tends to promote groups ranking one anothers' oppression, as if the important thing were to determine who is most victimized. Furthermore, it frames one's understanding of different groups only within the context of other groups' experiences; thus, it can assume an artificial norm against which different groups are judged. Thinking comparatively also assumes that race, class, and gender constitute separate and independent components of human experience that can be compared for their similarities and differences. Although thinking comparatively is an important step toward thinking more inclusively, we must go further.

In this sense, we distinguish "thinking comparatively" from "thinking relationally." Relational thinking involves seeing the interrelationships among diverse group experiences as well as among race, class, and gender themselves. When you think relationally, you see the social structures that simultaneously generate unique group histories and link them together in society. This does not mean that one group's experience is the same as another's, although finding commonalities is an important step toward more inclusive thinking. In thinking relationally, you untangle the workings of social systems that shape the experiences of different people and groups, and you move beyond just comparing (for example) gender oppression with race oppression, or the oppression of gays and lesbians with that of racial groups. When you recognize the systems of power that mark different groups' experiences, you possess the conceptual apparatus to think about changing the system, not just about documenting the effects of that system on different people.

DEVELOPING A NEW FRAMEWORK: INTERSECTIONS OF RACE, CLASS, AND GENDER

Race, class, and gender shape the experiences of all people in the United States. This fact has been widely documented in research and, to some extent, is commonly understood. Thus, for years, social scientists have studied the consequences of race, class, and gender inequality for diverse groups in society. The new framework of race, class, and gender studies presented here, however, explores how race, class, and gender operate *together* in people's lives. Fundamentally, race, class, and gender are intersecting categories of experience that affect all aspects of human life; thus, they simultaneously structure the experiences of all people in this society. At any moment, race, class, or gender may feel more salient or meaningful in a given person's life, but they are overlapping and cumulative in their effect on people's experience.

In this volume, we emphasize several core features of this intersectional framework for studying race, class, and gender. First, we emphasize social structures in our efforts to conceptualize intersections of race, class, and gender. We use the approach of a *matrix of domination* to analyze race, class, and gender. A matrix of domination posits multiple, interlocking levels of domination that stem from the societal configuration of race, class, and gender relations. This structural pattern affects individual consciousness, group interaction, and group access to institutional power and privileges (Collins 2000). Within this structural framework, we focus less on the similarities and differences among race, class, and gender than on patterns of connection that join them. Because of their simultaneity in people's lives, intersections of race, class, and gender can be seen in individual stories and personal experience. In fact, much exciting work on the intersections of race, class, and gender appears in autobiographies, fiction, and personal essays. We do recognize the significance of these individual narratives and include many here, but we emphasize social structures that provide a context for such stories.

Second, studying interconnections among race, class, and gender within a context of structural power helps us understand group experiences. Race, class, and gender are manifested differently, depending on their configuration with the others. Thus, although from the perspective of an additive model, one might say Black men are privileged as men, this makes no sense when their race and gender and class are taken into account. Otherwise, how can we possibly explain the particular disadvantages African American men experience as men—in the criminal justice system, in education, and in the labor market? For that matter, how can we explain the experiences that Native American women undergo as Native American women—disadvantaged by the unique experiences that they have based on race, class, *and* gender—none of which is isolated from the effects of the others?

Frameworks that study the connections among race, class, and gender interpret the division between powerful and powerless, or between oppressor and oppressed, as not being clear-cut. White women, for example, may be disadvantaged because of gender, but privileged by race and perhaps (but not necessarily) by class. Increasing class differentiation within racial-ethnic groups reminds us that race is not a monolithic category. Rather, race and class shape the experiences of, for example, Asian Americans, African Americans, and White Americans. For that matter, White Americans do not fit in "either/or" categories—either oppressor or oppressed. Isolating White Americans as if their experience stands alone ignores how White experience is intertwined with that of other groups. Moreover, thinking relationally about race, class, and gender need not entail thinking about all groups at one time. Difference models often aim to include the experiences of all groups, precisely because they aim for coverage. Because comprehensive coverage is impossible, difference models can never be complete. Rather, thinking relationally can mean understanding the experience of any group in context. An analysis grounded in race, class, and gender can be complete even if centered on the experience of a single group—as long as that group's experience is situated within a framework that recognizes the combined influence of race, class, and gender on the groups' experience.

Third, our approach to race, class, and gender studies is historically grounded. We have chosen to emphasize the intersections of race, class, and gender as institutional systems that have had a special impact in the United States. Race, class, gender, sexuality, ethnicity, age, ability, religion, and nationality are all important categories of analysis. Theoretically, they all interconnect. But historically, they have taken varying forms from one society to the next; and within any given society, the connections among them shift. Thus, race is not inherently more important than gender, just as sexuality is not inherently more significant than class and ethnicity. Given the complex and changing relationships among these categories of analysis, in this volume, why do we focus on race, class, and gender?

Rather than treating these concepts as free-floating, abstract ideas, we ground our analysis in the historical, institutional context of the United States.

Doing so means that race, class, and gender emerge as fundamental categories of analysis in the American setting, so significant that in many ways, they shape all other categories. For example, the capitalist class relations that have characterized all phases of U.S. history have routinely privileged or penalized groups organized by gender and by race. American social institutions have reproduced economic equalities for poor people, people of color, and for women from one generation to the next. Thus, in the United States, race, class, and gender demonstrate visible, longstanding, material effects that in many ways foreshadow more recently visible categories of ethnicity, religion, age, ability, and/or sexuality. These categories, among others, have been important sources of prejudice that also have structural forms. At the same time, systems of race, class, and gender have been consistently and deeply codified in U.S. laws, have profoundly affected U.S. customs, and have had intergenerational material effects within the U.S. capitalist class system. Social activism such as the abolitionist movement, trade unionism, women's suffrage, and the civil rights movement addressed social inequalities of race, class, and gender and created intellectual and political space for more recent social movements for more inclusive notions of social justice. In essence, in this volume, we study all categories of analysis by focusing on the historically interconnected categories of race, class, and gender. We recognize the importance of all categories of analysis and feel that our goal of developing inclusive thinking can be achieved through focusing on race, class, and gender.

Fourth, we ground this volume in the lived experiences of people whose individual and group experiences do not fit comfortably, if at all, within the confines of traditional knowledge. Thirty years ago, the experiences and work of historically marginalized groups were not considered knowledge. Today they are. Our goal is not to replace one set of truths with another. Rather, intersectional frameworks aim to transform knowledge itself, so that it is more inclusive, comprehensive, complex, and fair.

Finally, one important component of an intersectional framework is that it fosters social justice. One develops an intersectional framework by exploring the connections among individuals, groups, social institutions, and social issues with an eye toward developing more just social relations. Just as social justice is a goal to strive for, the transformative thinking that is needed to move in this direction requires an intersectional framework.

ORGANIZATION OF THE BOOK

We have organized the book to follow this thinking. Part I, "Shifting the Center," contains personal reflections on how race, class, and gender shape individual and collective experiences. These articles provide a fresh beginning point for social thought by putting those who have traditionally been excluded at the center of our thought. These different accounts are not meant to represent everyone who has been excluded, but to ground readers in the practice of thinking about the lives of people who rarely occupy the center of dominant group thinking. When excluded groups are placed in the forefront of thought, they are typically stereotyped and distorted. When you look at society through the different perspectives of persons reflecting on their location within systems of race, class, and gender oppression, you begin to see just how much race, class, and gender really matter. The accounts also model relational thinking of new race, class, and gender studies.

In Part II, "Conceptualizing Race, Class, and Gender," we provide a basic conceptual grounding for these three major concepts. Although we do not consider them separate systems, we treat them separately here for analytical purposes, so students will understand how each operates and then be better able to see their interlocking nature. We also want students to see the continuing effects of intersecting systems of race, class, and gender on people's experiences: Our introduction to this section and the articles within document those effects and point to the interrelationships between them.

In Part III, "Rethinking Institutions," we examine how intersecting systems of race, class, and gender shape the organization of social institutions and how, as a result, these institutions affect group experience. Social scientists routinely document how Latinos, African Americans, women, workers, and other distinctive groups are affected by institutional structures. We know this is true but want to go beyond these analyses to scrutinize how institutions are constructed through race, class, and gender relations. The articles in Part III show how we should not relegate the study of racial-ethnic groups, the working class, and women only to subjects marked explicitly as race, class, or gender studies. As categories of social experience, race, class, and gender shape all social institutions and systems of meaning; thus, it is important to think about people of color, different class experiences, and women in analyzing all social institutions and belief systems.

Part IV, "Applying the Framework," uses the inclusive perspective gained by understanding intersections of race, class, and gender to analyze contemporary social issues. We examine immigration, sexuality, and violence—social issues that are hotly contested in public discourse and that we think can be properly understood only if they are located within a race, class, and gender analysis. Indeed, these social issues are hotly contested because they evolve from contemporary race, class, and gender politics. Understanding the social issues on the public agenda requires an analysis of race, class, and gender relations.

Finally, in Part V, "Making a Difference," we look at social change. When studying the effects of race, class, and gender, a person can all too easily think only in terms of how people are victimized and oppressed by these intersecting systems. Throughout the book, we have avoided a "social problems approach" because we want readers to move away from thinking exclusively in a problem-centered framework. Race, class, and gender do indeed form the basis of many social problems, but a problems-based approach has three serious limitations: It tends to portray people primarily as victims, while ignoring their independent views of themselves and their society; it tends to see oppressed groups only from the perspective of the more privileged, relegating those who most suffer under race, class, and gender oppression to the status of "others," thereby reproducing the hierarchical viewpoints that have permeated traditional thinking; and, it tends to leave students feeling that they have no power to make changes in the society.

People are not just victims; they are creative and visionary. As a result, people organize to resist oppression and to work for social justice. In fact, oppression generates resistance. In Part V we examine the meaning of activism and its connection to the conditions in which people live. We see these articles as providing a basis for understanding how people can make a difference— something that we find often puzzles students. These articles suggest that change takes place through new actions and new social structures, and sometimes through existing channels; but unless one has an inclusive framework from which to make change, one's action is limited. Resistance to oppression on behalf of one's own group is not enough. Achieving social justice can only take place when people and groups build coalitions with others. Change also takes place over the long run. Short-term actions are needed, but the dedication needed to achieve long-term, institutional changes must be sustained by new visions. We hope the articles in this volume provide a visionary approach to the future.

In selecting articles for the anthology, we tried to find pieces that would show the intersections of race, class, and gender. Not every article does this, but we consider it equally important for students to learn to recognize connections between groups and social structural conditions even when these are not obvious. We searched for articles that are conceptually and theoretically informed and at the same time accessible to undergraduate readers. Although it is important to think of race, class, and gender as analytical categories, we do not want to lose sight of how they affect human experiences and feelings; thus, we have included personal narratives that are reflective and analytical. We think that personal accounts generate empathy and also help us connect personal experiences to social structural conditions.

We have tried to be as inclusive of all groups as possible, representing the richness of social experience within the United States. It is impossible to include all historically and presently marginalized groups in one book, so we encourage readers to explore the many other works available. We have selected materials that explain the relationships among race, class, and gender and that illuminate the experiences of many groups, not just the groups about whom an article is specifically written. As we begin to untangle the structure of race, class, and gender relations, we can better see both the commonalities and differences that various historical and contemporary experiences have generated.

We also caution students that, even though writing exclusively about groups in the United States can be seen as writing from a singular place, understanding the United States in a global context means seeing the factors that shape any given experience more completely. The United States is a diverse nation that is embedded in a diverse world. Thus, the process of relational thinking explored in this volume, one focused on the specifics of race, class, and gender in the United States, constitutes a process with wider implications.

Developing truly inclusive thinking and teaching is a long-term process that involves personal, intellectual, and political change. We do not claim to be models of perfection in this regard. We have been pleased by the strong response to the first four editions of this book, and we are fascinated by how race, class, and gender studies have developed in the short period of time since the publication of the first edition. We know further work is needed. Our own teaching and thinking has been transformed by developing this book. We imagine many changes still to come.

Reference

Collins, Patricia Hill. 2000. *Black Feminist Thought: Knowledge, Consciousness, and the Politics of Empowerment*, 2d ed. New York: Routledge.

Suggested Readings

Anzaldua, Gloria, ed. 1990. *Making Face, Making Soul / Haciendo Caras: Creative and Critical Perspectives by Women of Color.* San Francisco: Aunt Lute Books.

Baca Zinn, Maxine, Pierrette Hondagneu-Sotelo, and Michael A. Messner, eds. 1999. *Gender through the Prism of Difference*, 2d ed. Needham Heights, MA: Allyn & Bacon.

Chow, Esther Ngan-Ling, Doris Wilkinson, and Maxine Baca Zinn, eds. 1996. *Race, Class, & Gender: Common Bonds, Different Voices.* Thousand Oaks, CA: Sage.

Davis, Angela. 1981. *Women, Race, and Class.* New York: Random House.

Delgado, Richard, and Jean Stefancic, eds. 1998. *The Latino/a Condition: A Critical Reader.* New York: New York University Press.

Dill, Bonnie Thornton, and Maxine Baca Zinn, eds. 1994. *Women of Color in U.S. Society.* Philadelphia: Temple University Press.

Guy-Sheftall, Beverly, ed. 1995. *Words of Fire: An Anthology of African-American Feminist Thought.* New York: The New Press.

Schneider, Beth E., and Peter M. Nardi, eds. 1998. *Social Perspectives in Lesbian and Gay Studies: A Reader.* New York: Routledge.

Segrest, Mab. 1994. *Memoir of a Race Traitor.* Boston: South End Press.

 ## InfoTrac College Edition: Search Terms

Use your access to the online library, InfoTrac College Edition, to learn more about subjects covered in this section. The following are some suggested terms for searches:

African American studies
anti-Semitism
assimilation
Chicano studies (and Latino studies)
critical race theory
de jure/de facto segregation
labor studies
lesbian, gay, bisexual, transgendered (LGBT) identities
marginality
multiculturalism
Native American studies
segregation
social justice
women's studies

InfoTrac College Edition: Bonus Reading

You can use the InfoTrac College Edition feature to find an additional reading pertinent to this section and can also search on the author's name or keywords to locate the following reading:

Markus, Hazel Rose, Claude M. Steele, and Dorothy M. Steele. 2000. "Colorblindness as a Barrier to Inclusion: Assimilation and Nonimmigrant Minorities." *Daedalus* 129 (Fall): 233–59.

Difference and diversity are common themes in the discussion of race, class, and gender. How do Markus, Steele, and Steele see the connection between diversity and inclusion? What do they mean by *identity threat*, and how can teachers engage in different practices so that all students have an opportunity to learn? As you read this article, you can also think about how the belief in color blindness is an obstacle to building a more inclusive and just society.

Wadsworth Sociology Resource Center: Virtual Society

For additional links, quizzes, and learning tools related to this section, see the Wadsworth Sociology Resource Center, Virtual Society, at www.wadsworth .com/sociology_d/

I

Shifting the Center

We begin this book by asking, How might we see the world differently if we were to shift our vision of society from one that is typically centered in the voices and experiences of dominant groups to the lives and thoughts of those who have been devalued, marginalized, and excluded? This shift is central to thinking about race, class, and gender in ways that transform, rather than buttress, existing social arrangements. Shifting one's center of thinking so as to include previously silenced voices is a starting point for analyzing the complex interrelationship of race, class, and gender in society and for thinking about social relationships, actions, experiences, and institutions in new ways. Part of this process is recognizing that knowledge has been constructed largely from the experiences of the most powerful groups—because they have had the most access to systems of education and communication. Thus, shifting the center is fundamentally about reconstructing what you know to include those whose perspectives and life experiences have not been heard.

Think of the process of taking a photograph. For years, women and people of color—and especially women of color—were often totally outside people's frame of vision. But, as you move your angle of sight to include those who have been overlooked, new subjects come into sight. This is more than a matter of sharpening one's focus (although that is required for clarity). Instead, it means actually seeing things differently, perhaps even changing the lens you look through—thereby removing the filters (or stereotypes and misconceptions) that groups bring to what they see.

The articles in Part I are classics in that they are among the early writings that questioned how a transformation of knowledge might begin. Arturo Madrid, for example, shows how his experience as a young Latino student was silenced throughout his educational curriculum, leaving him to feel like an "other" in a society where he seemingly had no place, no history, no culture. For him, "shifting the center" means acknowledging the diverse histories, cultures, and experiences of groups who have been defined as marginal in society—what we have come to think of as "valuing diversity."

But shifting the center is more than just valuing the diverse histories and cultures of the different groups who constitute society. It is also recognizing how groups whose experiences have been vital in the formation of society and culture have been silenced in the construction of knowledge about this society. The result is that what we know—about the experiences of both these silenced groups and the dominant culture—is distorted and incomplete. How much did you learn about the history of group oppression in your formal education? You probably touched briefly on topics like slavery, women's suffrage, perhaps even the Holocaust, but most likely these were brief excursions from an otherwise dominant narrative that ignored people of color and women, along with others. For that matter, how much of what you study now is centered in the experiences of the most dominant groups in society? Think about the large number of social science studies that routinely make general conclusions about the population when they have been based on research done only about and by men. Or, how much of the literature you read and artistic creations that you study are the work of Asian Americans, Latinos/as, African Americans, Native Americans, gays, lesbians, or women?

By minimizing the experiences and creations of these different groups, we communicate that they have no history or that their work and creativity and is less important and less central to the development of culture than is the history of White American men. What false or incomplete conclusions does this exclusionary thinking generate? When you learn, for example, that democracy and egalitarianism were central cultural beliefs in the early history of the United States, how do you explain the enslavement of millions of African Americans? the genocide of Native Americans? the laws against intermarriage between Asian Americans and White Americans?

Shifting the center asks you to think more inclusively. Without doing so you are prone to understand society, your own life within it, and the experiences of others through stereotypes and misleading information. If this is the basis for our knowledge about each other, then we have little ground for building more just and liberating relationships and social institutions.

Thus, "shifting the center" means putting at the center of our thinking the experiences of groups that have formerly been excluded. Otherwise, many groups simply remain invisible. When they are seen, they are typically judged through the experiences of White people, rather than understood on their own terms; this establishes a false norm through which all groups are judged. This is well expressed by Scott Kurashige, who writes, "The public view of Asian Americans is a lot like that of Casper the Ghost: we're either white or we're invisible" (Kurashige 1994).

Shifting the center is not just about illuminating the experiences of oppressed groups, however. It also changes how we understand the dominant culture and groups who have more power and privilege than others. For example, the development of women's studies has changed what we know and how we think about women; at the same time, it has changed what we know and how we think about men. This does not mean that women's studies is about "male-bashing." It means that we take the experiences of women and men seriously and analyze how gender, race, and class shape the experiences of both men and women—in different, but interrelated, ways. Likewise, the study of racial-ethnic groups begins by learning the diverse histories and experiences of these groups, but in doing so, we transform our understanding of White experiences, too. The exclusionary thinking we have relied upon in the past simply does not reveal the intricate interconnections that exist between the different groups composing U.S. society.

Exclusionary thinking is increasingly being challenged by scholars and teachers who want to include the diversity of human experience in the construction and transmission of knowledge. Thinking more inclusively opens up the way the world is viewed, making the experience of previously excluded groups more visible and central in the construction of knowledge. By shifting our perspective, we can better understand the intersections of race, class, and gender in the experiences of all groups, including those with privilege and power.

Why would we want to do this? Does reconstructing knowledge matter? To begin with, knowledge is not just some abstract thing—good to have, but not all that important. There are real consequences to having partial or distorted knowledge. First, learning about other groups helps you realize the partiality of your own perspective; furthermore, this is true for both dominant and subordinate groups. Knowing only the history of African Americans, for example, or seeing that history only in single-minded terms will not reveal the historical linkages between the oppression of African Americans and the exclusionary and exploitative treatment of other groups. This is well pointed out by Ronald Takaki in his essay on the multicultural history of American society included here.

Second, having misleading and incorrect knowledge leads to the formation of bad social policy—policy that then reproduces, rather than solves, social problems. Finally, knowledge is not just about content and information; it provides an orientation to the world. What you know frames how you behave and how you think about yourself and others. If what you know is wrong because it is based on exclusionary thought, you are likely to act in exclusionary ways, thereby reproducing the racism, anti-Semitism, sexism, class oppression, and homophobia of society. This may not be because you are intentionally racist, anti-Semitic, sexist, elitist, or homophobic; it may simply be because you do not know any better. Challenging oppressive race, class, and gender relations in society requires reconstructing what we know so that we have some basis from which to change these damaging and dehumanizing systems of oppression.

As we said in the Introduction, developing inclusive thinking is more than just "understanding diversity" or valuing cultural pluralism. Inclusive thinking begins with the recognition that the United States is a multicultural and diverse society. Population data and even casual observations reveal that obvious truth, but inclusive thinking is more than recognizing the plurality of experiences in this society. Understanding race, class, and gender means coming to see the systematic exclusion and exploitation of various groups. This is more than just adding in different group experiences to already established frameworks of thought. It means constructing new analyses that are focused on the centrality of race, class, and gender in the experiences of us all.

You can begin to develop a more inclusive perspective by asking, How does the world look different if we put the experiences of those who have been excluded at the center of our thinking? At first, people might be tempted to simply assert the perspective and experience of their own group. Nationalist movements of different kinds have done this. They are valuable because they recognize the contributions and achievements of oppressed groups, but ultimately they are unidimensional frameworks because they are centered in a single group experience and they encourage, rather than discourage, exclusionary thought.

Seeing inclusively is more than just seeing the world through the perspective of any one group whose views have been distorted or ignored. Remember that group membership cuts across race, class, and gender categories. For example, one may be a working-class, Asian American woman or a Black, middle-class man or a gay, White, working-class woman. Inclusive thinking means seeing the interconnections between these experiences and not reducing a given person's or group's life to a single factor. In addition, developing inclusive thinking is more than just summing up the experiences of individual groups. Race, class, and gender are social structural categories. This means that they are embedded in the institutional structure of society. Understanding them requires a social structural analysis—by which we mean revealing the race, class, and gender patterns and processes that form the very framework of society.

We believe that shifting our perspective by thinking about the experiences of those who have been excluded from knowledge changes how we think about society, history, and culture. No longer do different groups seem just "different," "deviant," or exotic. Rather, specific patterns of the intersections of race, class, and gender are revealed, as are the connections that exist between groups. We then learn how our different experiences are linked, both historically and now.

Once we understand that race, class, and gender are *simultaneous* and *intersecting* systems of relationship and meaning, we come to see the different ways that other categories of experience intersect in society. Age, religion, sexual orientation, nationality, physical ability, region, and ethnicity also shape systems of privilege and inequality. We have tried to integrate these different experiences throughout the book, although we could not include as much as we would have liked.

Seeing these connections is what we think is most important about analyzing race, class, and gender. It is the reason for taking the perspective of people other than those like you. The purpose of the articles in "Shifting the Center" is not to appropriate others' experiences but to begin uncovering the structures of race, class, and gender that are embedded in the experiences of all people.

By providing personal accounts of what exclusion means and how it feels, the articles in this section show the limits of existing knowledge. We have selected personal accounts that reflect the diverse experiences of race, class, gender, and/or sexual orientation. We intend for the personal nature of these accounts to build empathy between groups—an emotional stance that we think is critical to seeing linkages and connections. By describing how exclusion shapes individual and collective consciousness, these accounts also document the diversity of experience within the United States. As a result, we begin to see how diverse experiences can shape the concepts and theories we develop in the academy. Together, they suggest new possibilities for thinking inclusively and imagining a society that would be more enabling for everyone.

In "Missing People and Others," Arturo Madrid tells how exclusion and marginalization in the educational curriculum have affected us all. Madrid describes his schooling as a process of denial of the specific experience of Latinos, Asian Americans, Native Americans, and all other groups together considered to be "other." He asks us to consider what it feels like to be the missing person. As we read his account, we see how education has contributed to exclusion by marginalizing the history of Chicanos and groups considered to be "other."

In "La Güera," Cherríe Moraga writes about her developing awareness of her class, race, and sexual identity. Like the other authors here, Moraga tells us what exclusion has meant in her experience and how it feels. Her essay also reveals the intersections of race, class, and gender with the system of compulsory heterosexuality—that is, the institutionalized structures and beliefs that define and enforce heterosexual behavior as the only natural and permissible form of sexual expression. For Moraga, acknowledging her lesbian identity deepened her understanding of her mother's oppression as a poor Chicana. Moraga reminds us of the importance of breaking the silences that protect oppression.

Readers should keep the accounts of Madrid and Moraga in mind as they read the remainder of this book. What new experiences, understandings, theories, histories, and analyses do these readings inspire? What does it take for a member of one group (say a White male) to be willing to learn from and value the experiences of another (for example, a Chicana lesbian)? June Jordan's essay, "Report from the Bahamas," is a model of this inclusive thinking. In it she begins from her own experience as a Black, middle-class, well-educated woman and reflects on her connections with other women in different locations in the race/class/gender system. In doing so, she finds connections in surprising places while also understanding the differences that distinguish our experiences. She also explains how one student, a White Catholic Irish woman, finds common ground with her South African classmate. There are no simple or singular answers to these questions, but recentering our frame of reference can reshape what we know, both formally and in everyday life. These essays show that, although we are caught in multiple systems, we can learn to see our connection to others. This is not just an intellectual exercise. As Paula Gunn Allen shows in her article on Native American women, "Angry Women Are Building," it is a matter of survival. Resisting systems of oppression means revising the ideas about ourselves and others that have been created as a part of a system of social control.

Reading these classic articles shows us the radical shift that thinking inclusively requires. Such a shift begins with valuing the experiences of those who have been excluded and questioning the assumptions made about all groups. When we begin to take a more inclusive perspective, divisions between privileged and underprivileged, oppressor and oppressed are challenged because we start to see race, class, and gender as intersecting systems of experience. For example, White women and women of color share some common experiences based on their gender, but their racial experiences are quite distinct; moreover, experiences within the race/gender system are further conditioned by one's class.

Engaging oneself at the personal level is critical to this process of thinking inclusively. Changing one's mind is not just a matter of assessing facts and data, though that is important; it also requires examining one's feelings. That is why we begin with several personal narratives. Unlike more conventional forms of sociological data (such as surveys, interviews, and even direct observations), personal accounts are more likely to elicit emotional responses. Traditionally, social science has defined emotional engagement as an impediment to objectivity. Sociology, for example, has emphasized rational thought as the basis for social action and has often discouraged more personalized reflection, but the capacity to reflect on one's experience makes us distinctly human. Personal documents tap the private, reflective dimension of life, enabling us to see the inner lives of others and, in the process, revealing our own lives more completely.

The idea that objectivity is best reached only through rational thought is a specifically Western and masculine way of thinking—one that we challenge throughout this book. Including personal narratives is not meant to limit our level of understanding only to individuals. As sociologists, we study individuals in groups as a way of revealing the social structures shaping collective experiences. Through doing so, we discover our common experiences and see the impact of the social structures of race, class, and gender on our experiences. Marilyn Frye's article, "Oppression," introduces the structural perspective that lies at the heart of this book. She distinguishes oppression from suffering, pointing out that many individuals suffer in society, but that oppression is structured into the fabric of social institutions. Using the metaphor of a birdcage, Frye artfully explains the concept of social structure. Looking only at an individual wire in a cage does not reveal the network of wires that forms a cage; likewise, social structure refers to the patterns of behavior, belief, resource distribution, and social control that constitute society.

Analysis of the multicultural basis of society is critical to understanding who we are as a society and a culture. The new focus on multiculturalism is, as Ronald Takaki says in "A Different Mirror," a debate over our national identity. Takaki makes a point of showing the common connections in the histories of African Americans, Chicanos, Irish Americans, Jews, and Native Americans. He argues that only when we understand this multidimensional history will we see ourselves in the full complexity of our humanity. Likewise, Audre Lorde writes that various forms of difference—race, sexuality, age, gender, among others—have separated and oppressed people. In a society marked by difference, she argues, we learn views of others and of ourselves that separate and divide people. For Lorde, relating across differences is critical to human survival.

As all of the authors in this section show, the strength and richness of our society lies in its diversity, but the potential of this diversity will be realized only through social structures that value and protect all human experiences.

Further Resources

For additional materials relating to this section, see the features on pages 72–73 at the conclusion of this section.

Reference

Kurashige, Scott. 1994. Cited in Karin Aguilar-San Juan, "Linking the Issues: From Identity to Activism." In *The State of Asian America: Activism and Resistance in the 1990s*, edited by Karin Aguilar-San Juan, 1–15. Boston: South End Press.

Shifting the Center

MISSING PEOPLE AND OTHERS

Joining Together to Expand the Circle

Arturo Madrid

1

I am a citizen of the United States, as are my parents and as were their parents, grandparents, and great-grandparents. My ancestors' presence in what is now the United States antedates Plymouth Rock, even without taking into account any American Indian heritage I might have.

I do not, however, fit those mental sets that define America and Americans. My physical appearance, my speech patterns, my name, my profession (a professor of Spanish) create a text that confuses the reader. My normal experience is to be asked, "And where are *you* from?"

My response depends on my mood. Passive-aggressive, I answer, "From here." Aggressive-passive, I ask, "Do you mean where am I originally from?" But ultimately my answer to those follow-up questions that ask about origins will be that we have always been from here.

Overcoming my resentment I will try to educate, knowing that nine times out of ten my words fall on inattentive ears. I have spent most of my adult life explaining who I am not. I am exotic, but—as Richard Rodriguez of *Hunger of Memory* fame so painfully found out—not exotic enough . . . not Peruvian, or Pakistani, or Persian, or whatever.

I am, however, very clearly *the other*, if only your everyday, garden-variety, domestic *other*. I've always known that I was *the other*, even before I knew the vocabulary or understood the significance of being *the other*.

I grew up in an isolated and historically marginal part of the United States, a small mountain village in the state of New Mexico, the eldest child of

From *Change* 20 (May/June 1988): 55–59. Reprinted by permission.

parents native to that region and whose ancestors had always lived there. In those vast and empty spaces, people who look like me, speak as I do, and have names like mine predominate. But the *americanos* lived among us: the descendants of those nineteenth-century immigrants who dispossessed us of our lands; missionaries who came to convert us and stayed to live among us; artists who became enchanted with our land and humanscape and went native; refugees from unhealthy climes, crowded spaces, unpleasant circumstances; and, of course, the inhabitants of Los Alamos, whose socio-cultural distance from us was moreover accentuated by the fact that they occupied a space removed from and proscribed to us. More importantly, however, they—*los americanos*—were omnipresent (and almost exclusively so) in newspapers, newsmagazines, books, on radio, in movies and, ultimately, on television.

Despite the operating myth of the day, school did not erase my otherness. It did try to deny it, and in doing so only accentuated it. To this day, schooling is more socialization than education, but when I was in elementary school—and given where I was—socialization was everything. School was where one became an American. Because there was a pervasive and systematic denial by the society that surrounded us that we were Americans. That denial was both explicit and implicit. My earliest memory of the former was that there were two kinds of churches: theirs and ours. The more usual was the implicit denial, our absence from the larger cultural, economic, political and social spaces—the one that reminded us constantly that we were *the other*. And school was where we felt it most acutely.

Quite beyond saluting the flag and pledging allegiance to it (a very intense and meaningful action, given that the U.S. was involved in a war and our brothers, cousins, uncles, and fathers were on the front lines) becoming American was learning English and its corollary—not speaking Spanish. Until very recently ours was a proscribed language—either *de jure* (by rule, by policy, by law) or *de facto* (by practice, implicitly if not explicitly; through social and political and economic pressure). I do not argue that learning English was not appropriate. On the contrary. Like it or not, and we had no basis to make any judgments on that matter, we were Americans by virtue of having been born Americans, and English was the common language of Americans. And there was a myth, a pervasive myth, that said that if we only learned to speak English well—and particularly without an accent—we would be welcomed into the American fellowship.

Senator Sam Hayakawa notwithstanding, the true text was not our speech, but rather our names and our appearance, for we would always have an accent, however perfect our pronunciation, however excellent our enunciation, however divine our diction. That accent would be heard in our pigmentation, our physiognomy, our names. We were, in short, *the other*.

Being *the other* means feeling different; is awareness of being distinct; is consciousness of being dissimilar. It means being outside the game, outside the circle, outside the set. It means being on the edges, on the margins, on the periphery. Otherness means feeling excluded, closed out, precluded, even disdained and scorned. It produces a sense of isolation, of apartness, of disconnectedness, of alienation.

Being *the other* involves a contradictory phenomenon. On the one hand being *the other* frequently means being invisible. Ralph Ellison wrote eloquently about that experience in his magisterial novel *The Invisible Man.* On the other hand, being *the other* sometimes involves sticking out like a sore thumb. What is she/he doing here?

If one is *the other*, one will inevitably be perceived unidimensionally; will be seen stereotypically; will be defined and delimited by mental sets that may not bear much relation to existing realities. There is a darker side to otherness as well. *The other* disturbs, disquiets, discomforts. It provokes distrust and suspicion. *The other* makes people feel anxious, nervous, apprehensive, even fearful. *The other* frightens, scares.

For some of us being *the other* is only annoying; for others it is debilitating; for still others it is damning. Many try to flee otherness by taking on protective colorations that provide invisibility, whether of dress or speech or manner or name. Only a fortunate few succeed. For the majority, otherness is permanently sealed by physical appearance. For the rest, otherness is betrayed by ways of being, speaking or of doing.

I spent the first half of my life downplaying the significance and consequences of otherness. The second half has seen me wrestling to understand its complex and deeply ingrained realities; striving to fathom why otherness denies us a voice or visibility or validity in American society and its institutions; struggling to make otherness familiar, reasonable, even normal to my fellow Americans.

I am also a missing person. Growing up in northern New Mexico I had only a slight sense of our being missing persons. *Hispanos,* as we called (and call) ourselves in New Mexico, were very much a part of the fabric of the society and there were Hispano professionals everywhere about me: doctors, lawyers, school teachers, and administrators. My people owned businesses, ran organizations and were both appointed and elected public officials.

To be sure, we did not own the larger businesses, nor at the time were we permitted to be part of the banking world. Other than that, however, people who looked like me, spoke like me, and had names like mine, predominated. There was, to be sure, Los Alamos, but as I have said, it was removed from our realities.

My awareness of our absence from the larger institutional life of society became sharper when I went off to college, but even then it was attenuated by the circumstances of history and geography. The demography of Albuquerque still strongly reflected its historical and cultural origins, despite the influx of Midwesterners and Easterners. Moreover, many of my classmates at the

University of New Mexico in Albuquerque were Hispanos, and even some of my professors were.

I thought that would obtain at UCLA, where I began graduate studies in 1960. Los Angeles already had a very large Mexican population, and that population was visible even in and around Westwood and on the campus. Many of the groundskeepers and food-service personnel at UCLA were Mexican. But Mexican-American students were few and mostly invisible, and I do not recall seeing or knowing a single Mexican-American (or, for that matter, black, Asian, or American Indian) professional on the staff or faculty of that institution during the five years I was there.

Needless to say, persons like me were not present in any capacity at Dartmouth College—the site of my first teaching appointment—and, of course, were not even part of the institutional or individual mind-set. I knew then that we—a "we" that had come to encompass American Indians, Asian Americans, black Americans, Puerto Ricans, and women—were truly missing persons in American institutional life.

Over the past three decades, the *de jure* and *de facto* segregations that have historically characterized American institutions have been under assault. As a consequence, minorities and women have become part of American institutional life, and although there are still many areas where we are not to be found, the missing persons phenomenon is not as pervasive as it once was.

However, the presence of *the other*, particularly minorities, in institutions and in institutional life, is, as we say in Spanish, *a flor de tierra;* spare plants whose roots do not go deep, a surface phenomenon, vulnerable to inclemencies of an economic, political, or social nature.

Our entrance into and our status in institutional life is not unlike a scenario set forth by my grandmother's pastor when she informed him that she and her family were leaving their mountain village to relocate in the Rio Grande Valley. When he asked her to promise that she would remain true to the faith and continue to involve herself in the life of the church, she assured him that she would and asked him why he thought she would do otherwise.

"Doña Trinidad," he told her, "in the Valley there is no Spanish church. There is only an American church." "But," she protested, "I read and speak English and would be able to worship there." Her pastor's response was: "It is possible that they will not admit you, and even if they do, they might not accept you. And that is why I want you to promise me that you are going to go to church. Because if they don't let you in through the front door, I want you to go in through the back door. And if you can't get in through the back door, go in the side door. And if you are unable to enter through the side door I want you to go in through the window. What is important is that you enter and that you stay."

Some of us entered institutional life through the front door; others through the back door; and still others through side doors. Many, if not most of us, came in through windows and continue to come in through windows. Of those who entered through the front door, some never made it past the lobby; others were ushered into corners and niches. Those who entered through back and side doors inevitably have remained in back and side rooms. And those who entered through windows found enclosures built around them. For despite the lip service given to the goal of the integration of minorities into institutional life, what has occurred instead is ghettoization, marginalization, isolation.

Not only have the entry points been limited: in addition, the dynamics have been singularly conflictive. Gaining entry and its corollary—gaining space—have frequently come as a consequence of demands made on institutions and institutional officers. Rather than entering institutions more or less passively, minorities have, of necessity, entered them actively, even aggressively. Rather than taking, they have demanded. Institutional relations have thus been adversarial, infused with specific and generalized tensions.

The nature of the entrance and the nature of the space occupied have greatly influenced the view and attitudes of the majority population within those institutions. All of us are put into the same box; that is, no matter what the individual reality, the assessment of the individual is inevitably conditioned by a perception that is held of the class. Whatever our history, whatever our record, whatever our validations, whatever our accomplishments, by and large we are perceived unidimensionally and are dealt with accordingly.

My most recent experience in this regard is atypical only in its explicitness. A few years ago I allowed myself to be persuaded to seek the presidency of a large and prestigious state university. I was invited for an interview and presented myself before the selection committee, which included members of the board of trustees. The opening question of the brief but memorable interview was directed at me by a member of that august body. "Dr. Madrid," he asked, "why does a one-dimensional person like you think he can be the president of a multi-dimensional institution like ours?"

If, as I happen to believe, the well-being of a society is directly related to the degree and extent to which all of its citizens participate in its institutions, we have a challenge before us. One of the strengths of our society—perhaps its main strength—has been a tradition of struggle against clubbishness, exclusivity, and restriction.

Today, more than ever, given the extraordinary changes that are taking place in our society, we need to take up that struggle again—irritating, grating, troublesome, unfashionable, unpleasant as it is. As educated and educator members of this society, we have a special responsibility for leading the struggle against marginalization, exclusion, and alienation.

Let us work together to assure that all American institutions, not just its precollegiate educational and penal institutions, reflect the diversity of our society. Not to do so is to risk greater alienation on the part of a growing segment of our society. It is to risk increased social tension in an already conflictive world. And ultimately it is to risk the survival of a range of institutions that, for all their defects and deficiencies, permit us the space, the opportunity, and the freedom to improve our individual and collective lot; to guide the course of our government, and to redress whatever grievances we have. Let us join together to expand, not to close the circle.

LA GÜERA

2

Cherríe Moraga

It requires something more than personal experience to gain a philosophy or point of view from any specific event. It is the quality of our response to the event and our capacity to enter into the lives of others that help us to make their lives and experiences our own.

Emma Goldman [1]

I am the very well-educated daughter of a woman who, by the standards in this country, would be considered largely illiterate. My mother was born in Santa Paula, Southern California, at a time when much of the central valley there was still farm land. Nearly thirty-five years later, in 1948, she was the only daughter of six to marry an anglo, my father.

I remember all of my mother's stories, probably much better than she realizes. She is a fine story-teller, recalling every event of her life with the vividness of the present, noting each detail right down to the cut and color of her dress. I remember stories of her being pulled out of school at the ages of five, seven, nine, and eleven to work in the fields, along with her brothers and sisters; stories of her father drinking away whatever small profit she was able to make for the family; of her going the long way home to avoid meeting him on the street, staggering toward the same destination. I remember stories of my

---

1. Alix Kates Shulman, "Was My Life Worth Living?" *Red Emma Speaks* (New York: Random House, 1972), p. 388.

mother lying about her age in order to get a job as a hat-check girl at Agua Caliente Racetrack in Tijuana. At fourteen, she was the main support of the family. I can still see her walking home alone at 3 a.m., only to turn all of her salary and tips over to her mother, who was pregnant again.

The stories continue through the war years and on: walnut-cracking factories, the Voit Rubber factory, and then the computer boom. I remember my mother doing piecework for the electronics plant in our neighborhood. In the late evening, she would sit in front of the T.V. set, wrapping copper wires into the backs of circuit boards, talking about "keeping up with the younger girls." By that time, she was already in her mid-fifties.

Meanwhile, I was college-prep in school. After classes, I would go with my mother to fill out job applications for her, or write checks for her at the supermarket. We would have the scenario all worked out ahead of time. My mother would sign the check before we'd get to the store. Then, as we'd approach the checkstand, she would say—within earshot of the cashier—"oh honey, you go 'head and make out the check," as if she couldn't be bothered with such an insignificant detail. No one asked any questions.

I was educated, and wore it with a keen sense of pride and satisfaction, my head propped up with the knowledge, from my mother, that my life would be easier than hers. I was educated; but more than this, I was "la güera": fair-skinned. Born with the features of my Chicana mother, but the skin of my anglo father, I had it made.

No one ever quite told me this (that light was right), but I knew that being light was something valued in my family (who were all Chicano, with the exception of my father). In fact, everything about my upbringing (at least what occurred on a conscious level) attempted to bleach me of what color I did have. Although my mother was fluent in it, I was never taught much Spanish at home. I picked up what I did learn from school and from overheard snatches of conversation among my relatives and mother. She often called other lower-income Mexicans "braceros," or "wet-backs," referring to herself and her family as "a different class of people." And yet, the real story was that my family, too, had been poor (some still are) and farmworkers. My mother can remember this in her blood as if it were yesterday. But this is something she would like to forget (and rightfully), for to her, on a basic economic level, being Chicana meant being "less." It was through my mother's desire to protect her children from poverty and illiteracy that we became "anglocized"; the more effectively we could pass in the white world, the better guaranteed our future.

From all of this, I experience, daily, a huge disparity between what I was born into and what I was to grow up to become. Because, (as Goldman suggests) these stories my mother told me crept under my "güera" skin. I had no choice but to enter into the life of my mother. *I had no choice.* I took her life into my heart, but managed to keep a lid on it as long as I feigned being the happy, upwardly mobile heterosexual.

When I finally lifted the lid to my lesbianism, a profound connection with my mother reawakened in me. It wasn't until I acknowledged and confronted my own lesbianism in the flesh, that my heartfelt identification with and empathy for my mother's oppression—due to being poor, uneducated, and Chicana—was realized. My lesbianism is the avenue through which I have learned the most about silence and oppression, and it continues to be the most tactile reminder to me that we are not free human beings.

You see, one follows the other. I had known for years that I was a lesbian, had felt it in my bones, had ached with the knowledge, gone crazed with the knowledge, wallowed in the silence of it. Silence *is* like starvation. Don't be fooled. It's nothing short of that, and felt most sharply when one has had a full belly most of her life. When we are not physically starving, we have the luxury to realize psychic and emotional starvation. It is from this starvation that other starvations can be recognized—if one is willing to take the risk of making the connection—if one is willing to be responsible to the result of the connection. For me, the connection is an inevitable one.

What I am saying is that the joys of looking like a white girl ain't so great since I realized I could be beaten on the street for being a dyke. If my sister's being beaten because she's Black, it's pretty much the same principle. We're both getting beaten any way you look at it. The connection is blatant; and in the case of my own family, the difference in the privileges attached to looking white instead of brown are merely a generation apart.

In this country, lesbianism is a poverty—as is being brown, as is being a woman, as is being just plain poor. The danger lies in ranking the oppressions. *The danger lies in failing to acknowledge the specificity of the oppression.* The danger lies in attempting to deal with oppression purely from a theoretical base. Without an emotional, heartfelt grappling with the source of our own oppression, without naming the enemy within ourselves and outside of us, no authentic, nonhierarchical connection among oppressed groups can take place.

When the going gets rough, will we abandon our so-called comrades in a flurry of racist/heterosexist/what-have-you panic? To whose camp, then, should the lesbian of color retreat? Her very presence violates the ranking and abstraction of oppression. Do we merely live hand to mouth? Do we merely struggle with the "ism" that's sitting on top of our own heads?

The answer is: yes, I think first we do; and we must do so thoroughly and deeply. But to fail to move out from there will only isolate us in our own oppression—will only insulate, rather than radicalize us.

To illustrate: a gay male friend of mine once confided to me that he continued to feel that, on some level, I didn't trust him because he was male; that he felt, really, if it ever came down to a "battle of the sexes," I might kill him. I admitted that I might very well. He wanted to understand the source of my distrust. I responded, "You're not a woman. Be a woman for a day. Imagine being a woman." He confessed that the thought terrified him because, to him, being a woman meant being raped by men. He *had* felt raped by men; he wanted to forget what that meant. What grew from that discussion was the realization that in order for him to create an authentic alliance with me, he must deal with the primary source of his own sense of oppression. He must, first, emotionally come to terms with what it feels like to be a victim. If he—or anyone—were to truly do this, it would be impossible to discount the oppression of others, except by again forgetting how we have been hurt.

And yet, oppressed groups are forgetting all the time. There are instances of this in the rising Black middle class, and certainly an obvious trend of such "unconsciousness" among white gay men. Because to remember may mean giving up whatever privileges we have managed to squeeze out of this society by virtue of our gender, race, class, or sexuality.

Within the women's movement, the connections among women of different backgrounds and sexual orientations have been fragile, at best. I think this phenomenon is indicative of our failure to seriously address ourselves to some very frightening questions: How have I internalized my own oppression? How have I oppressed? Instead, we have let rhetoric do the job of poetry. Even the word "oppression" has lost its power. We need a new language, better words that can more closely describe women's fear of and resistance to one another; words that will not always come out sounding like dogma.

What prompted me in the first place to work on an anthology by radical women of color was a deep sense that I had a valuable insight to contribute, by virtue of my birthright and background. And yet, I don't really understand first-hand what it feels like being shitted on for being brown. I understand much more about the joys of it—being Chicana and having family are synonymous for me. What I know about loving, singing, crying, telling stories, speaking with my heart and hands, even having a sense of my own soul comes from the love of my mother, aunts, cousins. . . .

But at the age of twenty-seven, it is frightening to acknowledge that I have internalized a racism and classism, where the object of oppression is not only someone outside of my skin, but the someone inside my skin. In fact, to a large degree, the real battle with such oppression, for all of us, begins under the skin. I have had to confront the fact that much of what I value about being Chicana, about my family, has been subverted by anglo culture and my own cooperation with it. This realization did not occur to me overnight. For example, it wasn't until long after my graduation from the private college I'd attended in Los Angeles, that I realized the major reason for my total alienation from and fear of my classmates was rooted in class and culture. CLICK.

Three years after graduation, in an apple-orchard in Sonoma, a friend of mine (who comes from an Italian Irish working-class family) says to me, "Cherríe, no wonder you felt like such a nut in school. Most of the people there were white and rich." It was true. All along I had felt the difference, but not until I had put the words "class" and "color" to the experience, did my feelings make any sense. For years, I had berated myself for not being as "free" as my classmates. I completely bought that they simply had more guts than I did—to rebel against their parents and run around the country hitch-hiking, reading books and studying "art." They had enough privilege to be atheists, for chrissake. There was no one around filling in the disparity for me between their parents, who were Hollywood filmmakers, and my parents, who wouldn't know the name of a filmmaker if their lives depended on it (and precisely because their lives didn't depend on it, they couldn't be bothered). But I knew nothing about "privilege" then. White was right. Period. I could pass. If I got educated enough, there would never be any telling.

Three years after that, another CLICK. In a letter to Barbara Smith, I wrote:

> I went to a concert where Ntosake Shange was reading. There, everything exploded for me. She was speaking a language that I knew—in the deepest parts of me—existed, and that I had ignored in my own feminist studies and even in my own writing. What Ntosake caught in me is the realization that in my development as a poet, I have, in many ways, denied the voice of my brown mother—the brown in me. I have acclimated to the sound of a white language which, as my father represents it, does not speak to the emotions in my poems—emotions which stem from the love of my mother.
>
> The reading was agitating. Made me uncomfortable. Threw me into a week-long terror of how deeply I was affected. I felt that I had to start all over again. That I turned only to the perceptions of white middle-class women to speak for me and all women. I am shocked by my own ignorance.

Sitting in that auditorium chair was the first time I had realized to the core of me that for years I had disowned the language I knew best—ignored the words and rhythms that were the closest to me. The sounds of my mother and aunts gossiping—half in English, half in Spanish—while drinking cerveza in the kitchen. And the hands—I had cut off the hands in my poems. But not in conversation; still the hands could not be kept down. Still they insisted on moving.

The reading had forced me to remember that I knew things from my roots. But to remember puts me up against what I don't know. Shange's reading agitated me because she spoke with power about a world that is both alien and common to me: "the capacity to enter into the lives of others." But you can't just take the goods and run. I knew that then, sitting in the Oakland auditorium (as I know in my poetry), that the only thing worth writing about is what seems to be unknown and, therefore, fearful.

The "unknown" is often depicted in racist literature as the "darkness" within a person. Similarly, sexist writers will refer to fear in the form of the vagina, calling it "the orifice of death." In contrast, it is a pleasure to read works such as Maxine Hong Kingston's *Woman Warrior*, where fear and alienation are described as "the white ghosts." And yet, the bulk of literature in this country reinforces the myth that what is dark and female is evil. Consequently, each of us—whether dark, female, or both—has in some way *internalized* this oppressive imagery. What the oppressor often succeeds in doing is simply *externalizing* his fears, projecting them into the bodies of women, Asians, gays, disabled folks, whoever seems most "other."

> call me
> roach and presumptuous
> nightmare on your white pillow
> your itch to destroy
> the indestructible
> part of yourself
>
> Audre Lorde[2]

But it is not really difference the oppressor fears so much as similarity. He fears he will discover in himself the same aches, the same longings as those of the people he has shitted on. He fears the immobilization threatened by his own incipient guilt. He fears he will have to change his life once he has seen himself in the bodies of the people he has called different. He fears the hatred, anger, and vengeance of those he has hurt.

This is the oppressor's nightmare, but it is not exclusive to him. We women have a similar nightmare, for each of us in some way has been both oppressed and the oppressor. We are afraid to look at how we have failed each other. We are afraid to see how we have taken the values of our oppressor into our hearts and turned them against ourselves and one another. We are afraid to admit how deeply "the man's" words have been ingrained in us.

To assess the damage is a dangerous act. I think of how, even as a feminist lesbian, I have so wanted to ignore my own homophobia, my own hatred of myself for being queer. I have not wanted to admit that my deepest personal sense of myself has not quite "caught up" with my "woman-identified" politics. I have been afraid to criticize lesbian writers who choose to "skip over" these issues in the name of feminism. In 1979, we talk of "old gay" and "butch and femme" roles as if they were ancient history. We toss them aside as merely patriarchal notions. And yet, the truth of the matter is that I have sometimes taken society's fear and hatred of lesbians to bed with me. I have sometimes hated my lover for loving me. I have sometimes felt "not woman enough" for her. I have sometimes felt "not man enough." For a lesbian trying to survive in a heterosexist society, there is no easy way around these emotions. Similarly, in a white-dominated world, there is little getting around racism and our own internalization of it. It's always there, embodied in someone we least expect to rub up against.

---

2. From "The Brown Menace or Poem to the Survival of Roaches," *The New York Head Shop and Museum* (Detroit: Broadside, 1974), p. 48.

When we do rub up against this person, *there* then is the challenge. *There* then is the opportunity to look at the nightmare within us. But we usually shrink from such a challenge.

Time and time again, I have observed that the usual response among white women's groups when the "racism issue" comes up is to deny the difference. I have heard comments like, "Well, we're open to *all* women; why don't they (women of color) come? You can only do so much. . . ." But there is seldom any analysis of how the very nature and structure of the group itself may be founded on racist or classist assumptions. More importantly, so often the women seem to feel no loss, no lack, no absence when women of color are not involved; therefore, there is little desire to change the situation. This has hurt me deeply. I have come to believe that the only reason women of a privileged class will dare to look at *how* it is that *they* oppress, is when they've come to know the meaning of their own oppression. And understand that the oppression of others hurts them personally.

The other side of the story is that women of color and working-class women often shrink from challenging white middle-class women. It is much easier to rank oppressions and set up a hierarchy, rather than take responsibility for changing our own lives. We have failed to demand that white women, particularly those who claim to be speaking for all women, be accountable for their racism.

The dialogue has simply not gone deep enough.

I have many times questioned my right to even work on an anthology which is to be written "exclusively by Third World women." I have had to look critically at my claim to color, at a time when, among white feminist ranks, it is a "politically correct" (and sometimes peripherally advantageous) assertion to make. I must acknowledge the fact that, physically, I have had a *choice* about making that claim, in contrast to women who have not had such a choice, and have been abused for their color. I must reckon with the fact that for most of my life, by virtue of the very fact that I am white-looking, I identified with and aspired toward white values, and that I rode the wave of that Southern California privilege as far as conscience would let me.

Well, now I feel both bleached and beached. I feel angry about this—the years when I refused to recognize privilege, both when it worked against me, and when I worked it, ignorantly, at the expense of others. These are not settled issues. That is why this work feels so risky to me. It continues to be discovery. It has brought me into contact with women who invariably know a hell of a lot more than I do about racism, as experienced in the flesh, as revealed in the flesh of their writing.

I think: what is my responsibility to my roots—both white and brown, Spanish-speaking and English? I am a woman with a foot in both worlds; and I refuse the split. I feel the necessity for dialogue. Sometimes I feel it urgently.

But one voice is not enough, nor two, although this is where dialogue begins. It is essential that radical feminists confront their fear of and resistance to each other, because without this, there *will* be no bread on the table. Simply, we will not survive. If we could make this connection in our heart of hearts, that if we are serious about a revolution—better—if we seriously believe there should be joy in our lives (real joy, not just "good times"), then we need one another. We women need each other. Because my/your solitary, self-asserting "go-for-the-throat-of-fear" power is not enough. The real power, as you and I well know, is collective. I can't afford to be afraid of you, nor you of me. If it takes head-on collisions, let's do it: this polite timidity is killing us.

As Lorde suggests in the passage I cited earlier, it is in looking to the nightmare that the dream is found. There, the survivor emerges to insist on a future, a vision, yes, born out of what is dark and female. The feminist movement must be a movement of such survivors, a movement with a future.

<hr>

REPORT FROM THE BAHAMAS   3

June Jordan

I am staying in a hotel that calls itself The Sheraton British Colonial. One of the photographs advertising the place displays a middle-aged Black man in a waiter's tuxedo, smiling. What intrigues me most about the picture is just this: while the Black man bears a tray full of "colorful" drinks above his left shoulder, both of his feet, shoes and trouserlegs, up to ten inches above his ankles, stand in the also "colorful" Caribbean salt water. He is so delighted to serve you he will wade into the water to bring you Banana Daquiris while you float! More precisely, he will wade into the water, fully clothed, oblivious to the ruin of his shoes, his trousers, his health, and he will do it with a smile.

I am in the Bahamas. On the phone in my room, a spinning complement of plastic pages offers handy index clues such as CAR RENTAL and CASINOS. A message from the Ministry of Tourism appears among these

From June Jordan, *On Call: Political Essays* (Boston: South End Press, 1985), pp. 39–49. Reprinted by permission of the author.

travelers' tips. Opening with a paragraph of "WELCOME," the message then proceeds to "A PAGE OF HISTORY," which reads as follows:

> New World History begins on the same day that modern Bahamian history begins—October 12, 1492. That's when Columbus stepped ashore— British influence came first with the Eleutherian Adventurers of 1647—After the Revolutions, American Loyalists fled from the newly independent states and settled in the Bahamas. Confederate blockade-runners used the island as a haven during the War between the States, and after the War, a number of Southerners moved to the Bahamas . . .

There it is again. Something proclaims itself a legitimate history and all it does is track white Mr. Columbus to the British Eleutherians through the Confederate Southerners as they barge into New World surf, land on New World turf, and nobody saying one word about the Bahamian people, the Black peoples, to whom the only thing new in their island world was this weird succession of crude intruders and its colonial consequences.

This is my consciousness of race as I unpack my bathing suit in the Sheraton British Colonial. Neither this hotel nor the British nor the long ago Italians nor the white Delta airline pilots belong here, of course. And every time I look at the photograph of that fool standing in the water with his shoes on I'm about to have a West Indian fit, even though I know he's no fool; he's a middle-aged Black man who needs a job and this is his job—pretending himself a servile ancillary to the pleasures of the rich. (Compared to his options in life, I am a rich woman. Compared to most of the Black Americans arriving for this Easter weekend on a three nights four days' deal of bargain rates, the middle-aged waiter is a poor Black man.)

We will jostle along with the other (white) visitors and join them in the tee shirt shops or, laughing together, learn ruthless rules of negotiation as we, Black Americans as well as white, argue down the price of handwoven goods at the nearby straw market while the merchants, frequently toothless Black women seated on the concrete in their only presentable dress, humble themselves to our careless games:

"Yes? You like it? Eight dollar."

"Five."

"I give it to you. Seven."

And so it continues, this weird succession of crude intruders that, now, includes me and my brothers and my sisters from the North.

This is my consciousness of class as I try to decide how much money I can spend on Bahamian gifts for my family back in Brooklyn. No matter that these other Black women incessantly weave words and flowers into the straw hats and bags piled beside them on the burning dusty street. No matter that these other Black women must work their sense of beauty into these things that we will take away as cheaply as we dare, or they will do without food.

We are not white, after all. The budget is limited. And we are harmlessly killing time between the poolside rum punch and "The Native Show on the Patio" that will play tonight outside the hotel restaurant.

This is my consciousness of race and class and gender identity as I notice the fixed relations between these other Black women and myself. They sell and I buy or I don't. They risk not eating. I risk going broke on my first vacation afternoon.

We are not particularly women anymore; we are parties to a transaction designed to set us against each other.

"Olive" is the name of the Black woman who cleans my hotel room. On my way to the beach I am wondering what "Olive" would say if I told her why I chose The Sheraton British Colonial; if I told her I wanted to swim. I wanted to sleep. I did not want to be harassed by the middle-aged waiter, or his nephew. I did not want to be raped by anybody (white or Black) at all and I calculated that my safety as a Black woman alone would best be assured by a multinational hotel corporation. In my experience, the big guys take customer complaints more seriously than the little ones. I would suppose that's one reason why they're big; they don't like to lose money anymore than I like to be bothered when I'm trying to read a goddamned book underneath a palm tree I paid $264 to get next to. A Black woman seeking refuge in a multinational corporation may seem like a contradiction to some, but there you are. In this case it's a coincidence of entirely different self-interests: Sheraton/cash = June Jordan's short run safety.

Anyway, I'm pretty sure "Olive" would look at me as though I came from someplace as far away as Brooklyn. Then she'd probably allow herself one indignant query before righteously removing her vacuum cleaner from my room; "and why in the first place you come down you without your husband?"

I cannot imagine how I would begin to answer her.

My "rights" and my "freedom" and my "desire" and a slew of other New World values; what would they sound like to this Black woman described on the card atop my hotel bureau as "Olive the Maid"? "Olive" is older than I am and I may smoke a cigarette while she changes the sheets on my bed. Whose rights? Whose freedom? Whose desire?

And why should she give a shit about mine unless I do something, for real, about hers?

It happens that the book that I finished reading under a palm tree earlier today was the novel, *The Bread Givers*, by Anzia Yezierska. Definitely autobiographical, Yezierska lays out the difficulties of being both female and "a person" inside a traditional Jewish family at the start of the 20th century. That any Jewish woman became anything more than the abused servant of her father or her husband is really an improbable piece of news. Yet Yezierska managed such an unlikely outcome for her own life. In *The Bread Givers*, the heroine also manages an important, although partial, escape from traditional

Jewish female destiny. And in the unpardonable, despotic father, the Talmudic scholar of that Jewish family, did I not see my own and hate him twice, again? When the heroine, the young Jewish child, wanders the streets with a filthy pail she borrows to sell herring in order to raise the ghetto rent and when she cries, "Nothing was before me but the hunger in our house, and no bread for the next meal if I didn't sell the herring. No longer like a fire engine, but like a houseful of hungry mouths my heart cried, 'herring—herring! Two cents apiece!'" who would doubt the ease, the sisterhood of conversation possible between that white girl and the Black women selling straw bags on the streets of paradise because they do not want to die? And is it not obvious that the wife of that Talmudic scholar and "Olive," who cleans my room here at the hotel, have more in common than I can claim with either one of them?

This is my consciousness of race and class and gender identity as I collect wet towels, sunglasses, wristwatch, and head towards a shower.

I am thinking about the boy who loaned this novel to me. He's white and he's Jewish and he's pursuing an independent study project with me, at the State University where I teach whether or not I feel like it, where I teach without stint because, like the waiter, I am no fool. It's my job and either I work or I do without everything you need money to buy. The boy loaned me the novel because he thought I'd be interested to know how a Jewish-American writer used English so that the syntax, and therefore the cultural habits of mind expressed by the Yiddish language, could survive translation. He did this because he wanted to create another connection between us on the basis of language, between his knowledge/his love of Yiddish and my knowledge/my love of Black English.

He has been right about the forceful survival of the Yiddish. And I had become excited by this further evidence of the written voice of spoken language protected from the monodrone of "standard" English, and so we had grown closer on this account. But then our talk shifted to student affairs more generally, and I had learned that this student does not care one way or the other about currently jeopardized Federal Student Loan Programs because, as he explained it to me, they do not affect him. He does not need financial help outside his family. My own son, however, is Black. And I am the only family help available to him and that means, if Reagan succeeds in eliminating Federal programs to aid minority students, he will have to forget about furthering his studies, or he or I or both of us will have to hit the numbers pretty big. For these reasons of difference, the student and I had moved away from each other, even while we continued to talk.

My consciousness turned to race, again, and class.

Sitting in the same chair as the boy, several weeks ago, a graduate student came to discuss her grade. I praised the excellence of her final paper; indeed it had seemed to me an extraordinary pulling together of recent left brain/right brain research with the themes of transcendental poetry.

She told me that, for her part, she'd completed her reading of my political essays. "You are so lucky!" she exclaimed.

"What do you mean by that?"

"You have a cause. You have a purpose to your life."

I looked carefully at this white woman; what was she really saying to me?

"What do you mean?" I repeated.

"Poverty. Police violence. Discrimination in general."

(Jesus Christ, I thought: Is that her idea of lucky?)

"And how about you?" I asked.

"Me?"

"Yeah, you. Don't you have a cause?"

"Me? I'm just a middle-aged woman: a housewife and a mother. I'm a nobody."

For a while, I made no response.

First of all, speaking of race and class and gender in one breath, what she said meant that those lucky preoccupations of mine, from police violence to nuclear wipe-out, were not shared. They were mine and not hers. But here she sat, friendly as an old stuffed animal, beaming good will or more "luck" in my direction.

In the second place, what this white woman said to me meant that she did not believe she was "a person" precisely because she had fulfilled the traditional female functions revered by the father of that Jewish immigrant, Anzia Yezierska. And the woman in front of me was not a Jew. That was not the connection. The link was strictly female. Nevertheless, how should that woman and I, another female, connect, beyond this bizarre exchange?

If she believed me lucky to have regular hurdles of discrimination then why shouldn't I insist that she's lucky to be a middle-class white Wasp female who lives in such well-sanctioned and normative comfort that she even has the luxury to deny the power of the privileges that paralyze her life?

If she deserts me and "my cause" where we differ, if, for example, she abandons me to "my" problems of race, then why should I support her in "her" problems of housewifely oblivion?

Recollection of this peculiar moment brings me to the shower in the bathroom cleaned by "Olive." She reminds me of the usual Women's Studies curriculum because it has nothing to do with her or her job: you won't find "Olive" listed anywhere on the reading list. You will likewise seldom hear of Anzia Yezierska. But yes, you will find, from Florence Nightingale to Adrienne Rich, a white procession of independently well-to-do women writers. (Gertrude Stein/Virginia Woolf/Hilda Doolittle are standard names among the "essential" women writers.)

In other words, most of the women of the world—Black and First World and white who work because we must—most of the women of the world persist far from the heart of the usual Women's Studies syllabus.

Similarly, the typical Black history course will slide by the majority experience it pretends to represent. For example, Mary McLeod Bethune will scarcely receive as much attention as Nat Turner, even though Black women who bravely and efficiently provided for the education of Black people hugely outnumber those few Black men who led successful or doomed rebellions against slavery. In fact, Mary McLeod Bethune may not receive even honorable mention because Black history too often apes those ridiculous white history courses which produce such dangerous gibberish as The Sheraton British Colonial "history" of the Bahamas. Both Black and white history courses exclude from their central consideration those people who neither killed nor conquered anyone as the means to new identity, those people who took care of every one of the people who wanted to become "a person," those people who still take care of the life at issue: the ones who wash and who feed and who teach and who diligently decorate straw hats and bags with all of their historically unrequired gentle love—the women.

> *Oh the old rugged cross*
> *on a hill far away*
> *Well I cherish the old rugged cross*

It's Good Friday in the Bahamas. Seventy-eight degrees in the shade. Except for Sheraton territory, everything's closed.

It so happens that for truly secular reasons I've been fasting for three days. My hunger has now reached nearly violent proportions. In the hotel sandwich shop, the Black woman handling the counter complains about the tourists; why isn't the shop closed and why don't the tourists stop eating for once in their lives. I'm famished and I order chicken salad and cottage cheese and lettuce and tomato and a hard boiled egg and a hot cross bun and apple juice.

She eyes me with disgust.

To be sure, the timing of my stomach offends her serious religious practices. Neither one of us apologizes to the other. She seasons the chicken salad to the peppery max while I listen to the loud radio gospel she plays to console herself. It's a country Black version of "The Old Rugged Cross."

As I heave much chicken into my mouth, tears start. It's not the pepper. I am, after all, a West Indian daughter. It's the Good Friday music that dominates the humid atmosphere.

> *Well I cherish the old rugged cross*

And I am back, faster than a 747, in Brooklyn, in the home of my parents where we are wondering, as we do every year, if the sky will darken until Christ has been buried in the tomb. The sky should darken if God is in His heavens. And then, around 3 p.m., at the conclusion of our mournful church service at the neighborhood St. Phillips, and even while we dumbly stare at the black cloth covering the gold altar and the slender unlit candles, the sun should return through the high gothic windows and vindicate our waiting faith that the Lord will rise again, on Easter.

How I used to bow my head at the very name of Jesus: ecstatic to abase myself in deference to His majesty.

My mouth is full of salad. I can't seem to eat quickly enough. I can't think how I should lessen the offense of my appetite. The other Black woman on the premises, the one who disapprovingly prepared this very tasty break from my fast, makes no remark. She is no fool. This is a job that she needs. I suppose she notices that at least I included a hot cross bun among my edibles. That's something in my favor. I decide that's enough.

I am suddenly eager to walk off the food. Up a fairly steep hill I walk without hurrying. Through the pastel desolation of the little town, the road brings me to a confectionary pink and white plantation house. At the gates, an unnecessarily large statue of Christopher Columbus faces me down, or tries to. His hand is fisted to one hip. I look back at him, laugh without deference, and turn left.

It's time to pack it up. Catch my plane. I scan the hotel room for things not to forget. There's that white report card on the bureau.

"Dear Guests:" it says, under the name "Olive." "I am your maid for the day. Please rate me: Excellent. Good. Average. Poor. Thank you."

I tuck this memento from the Sheraton British Colonial into my notebook. How would "Olive" rate *me?* What would it mean for us to seem "good" to each other? What would that rating require?

But I am hastening to leave. Neither turtle soup nor kidney pie nor any conch shell delight shall delay my departure. I have rested, here, in the Bahamas, and I'm ready to return to my usual job, my usual work. But the skin on my body has changed and so has my mind. On the Delta flight home I realize I am burning up, indeed.

So far as I can see, the usual race and class concepts of connection, or gender assumptions of unity, do not apply very well. I doubt that they ever did. Otherwise why would Black folks forever bemoan our lack of solidarity when the deal turns real. And if unity on the basis of sexual oppression is something natural, then why do we women, the majority people on the planet, still have a problem?

The plane's ready for takeoff. I fasten my seatbelt and let the tumult inside my head run free. Yes: race and class and gender remain as real as the weather. But what they must mean about the contact between two individuals is less obvious and, like the weather, not predictable.

And when these factors of race and class and gender absolutely collapse is whenever you try to use them as automatic concepts of connection. They may serve well as indicators of commonly felt conflict, but as elements of connection they seem about as reliable as precipitation probability for the day after the night before the day.

It occurs to me that much organizational grief could be avoided if people understood that partnership in misery does not necessarily provide for partnership for change: *When we get the monsters off our backs all of us may want to run in very different directions.*

And not only that: even though both "Olive" and "I" live inside a conflict neither one of us created, and even though both of us therefore hurt inside that conflict, I may be one of the monsters she needs to eliminate from her universe and, in a sense, she may be one of the monsters in mine.

I am reaching for the words to describe the difference between a common identity that has been imposed and the individual identity any one of us will choose, once she gains that chance.

That difference is the one that keeps us stupid in the face of new, specific information about somebody else with whom we are supposed to have a connection because a third party, hostile to both of us, has worked it so that the two of us, like it or not, share a common enemy. *What happens beyond the idea of that enemy and beyond the consequences of that enemy?*

I am saying that the ultimate connection cannot be the enemy. The ultimate connection must be the need that we find between us. It is not only who you are, in other words, but what we can do for each other that will determine the connection.

I am flying back to my job. I have been teaching contemporary women's poetry this semester. One quandary I have set myself to explore with my students is the one of taking responsibility without power. We had been wrestling ideas to the floor for several sessions when a young Black woman, a South African, asked me for help, after class.

Sokutu told me she was "in a trance" and that she'd been unable to eat for two weeks.

"What's going on?" I asked her, even as my eyes startled at her trembling and emaciated appearance.

"My husband. He drinks all the time. He beats me up. I go to the hospital. I can't eat. I don't know what/anything."

In my office, she described her situation. I did not dare to let her sense my fear and horror. She was dragging about, hour by hour, in dread. Her husband, a young Black South African, was drinking himself into more and more deadly violence against her.

Sokutu told me how she could keep nothing down. She weighed 90 lbs. at the outside, as she spoke to me. She'd already been hospitalized as a result of her husband's battering rage.

I knew both of them because I had organized a campus group to aid the liberation struggles of Southern Africa.

Nausea rose in my throat. What about this presumable connection: this husband and this wife fled from that homeland of hatred against them, and now what? He was destroying himself. If not stopped, he would certainly murder his wife.

She needed a doctor, right away. It was a medical emergency. She needed protection. It was a security crisis. She needed refuge for battered wives and personal therapy and legal counsel. She needed a friend.

I got on the phone and called every number in the campus directory that I could imagine might prove helpful. Nothing worked. There were no institutional resources designed to meet her enormous, multifaceted, and ordinary woman's need.

I called various students. I asked the Chairperson of the English Department for advice. I asked everyone for help.

Finally, another one of my students, Cathy, a young Irish woman active in campus IRA activities, responded. She asked for further details. I gave them to her.

"Her husband," Cathy told me, "is an alcoholic. You have to understand about alcoholics. It's not the same as anything else. And it's a disease you can't treat any old way."

I listened, fearfully. Did this mean there was nothing we could do?

"That's not what I'm saying," she said. "But you have to keep the alcoholic part of the thing central in everybody's mind, otherwise her husband will kill her. Or he'll kill himself."

She spoke calmly. I felt there was nothing to do but to assume she knew what she was talking about.

"Will you come with me?" I asked her, after a silence. "Will you come with me and help us figure out what to do next?"

Cathy said she would but that she felt shy: Sokutu comes from South Africa. What would she think about Cathy?

"I don't know," I said. "But let's go."

We left to find a dormitory room for the young battered wife.

It was late, now, and dark outside.

On Cathy's VW that I followed behind with my own car, was the sticker that reads BOBBY SANDS FREE AT LAST. My eyes blurred as I read and reread the words. This was another connection: Bobby Sands and Martin Luther King Jr. and who would believe it? I would not have believed it; I grew up terrorized by Irish kids who introduced me to the word "nigga."

And here I was following an Irish woman to the room of a Black South African. We were going to that room to try to save a life together.

When we reached the little room, we found ourselves awkward and large. Sokutu attempted to treat us with utmost courtesy, as though we were honored guests. She seemed surprised by Cathy, but mostly Sokutu was flushed with relief and joy because we were there, with her.

I did not know how we should ever terminate her heartfelt courtesies and address, directly, the reason for our visit: her starvation and her extreme physical danger.

Finally, Cathy sat on the floor and reached out her hands to Sokutu.

"I'm here," she said quietly, "because June has told me what has happened to you. And I know what it is. Your husband is an alcoholic. He has a disease. I know what it is. My father was an alcoholic. He killed himself. He almost killed my mother. I want to be your friend."

"Oh," was the only small sound that escaped from Sokutu's mouth. And then she embraced the other student. And then everything changed and I watched all of this happen so I know that this happened: this connection.

And after we called the police and exchanged phone numbers and plans were made for the night and for the next morning, the young South African woman walked down the dormitory hallway, saying goodbye and saying thank you to us.

I walked behind them, the young Irish woman and the young South African, and I saw them walking as sisters walk, hugging each other, and whispering and sure of each other and I felt how it was not who they were but what they both know and what they were both preparing to do about what they know that was going to make them both free at last.

And I look out the windows of the plane and I see clouds that will not kill me and I know that someday soon other clouds may erupt to kill us all.

And I tell the stewardess No thanks to the cocktails she offers me. But I look about the cabin at the hundred strangers drinking as they fly and I think even here and even now I must make the connection real between me and these strangers everywhere before those other clouds unify this ragged bunch of us, too late.

ANGRY WOMEN ARE BUILDING

4

Issues and Struggles Facing American Indian Women Today

Paula Gunn Allen

The central issue that confronts American Indian women throughout the hemisphere is survival, *literal survival*, both on a cultural and biological level.

From Paula Gunn Allen, *The Sacred Hoop: Recovering the Feminism in American Indian Traditions* (Boston: Beacon Press, 1986), pp. 189–93. Reprinted with permission.

According to the 1980 census, population of American Indians is just over one million. This figure, which is disputed by some American Indians, is probably a fair estimate, and it carries certain implications.*

Some researchers put our pre-contact population at more than 45 million, while others put it at around 20 million. The U.S. government long put it at 450,000—a comforting if imaginary figure, though at one point it was put at around 270,000. If our current population is around one million; if, as some researchers estimate, around 25 percent of Indian women and 10 percent of Indian men in the United States have been sterilized without informed consent; if our average life expectancy is, as the best-informed research presently says, 55 years; if our infant mortality rate continues at well above national standards; if our average unemployment for all segments of our population—male, female, young, adult, and middle-aged—is between 60 and 90 percent; if the U.S. government continues its policy of termination, relocation, removal, and assimilation along with the destruction of wilderness, reservation land, and its resources, and severe curtailment of hunting, fishing, timber harvesting and water-use rights—then existing tribes are facing the threat of extinction which for several hundred tribal groups has already become fact in the past five hundred years.

In this nation of more than 200 million, the Indian people constitute less than one-half of one percent of the population.† In a nation that offers refuge, sympathy, and billions of dollars in aid from federal and private sources in the form of food to the hungry, medicine to the sick, and comfort to the dying, the indigenous subject population goes hungry, homeless, impoverished, cut out of the American deal, new, old, and in between. Americans are daily made aware of the worldwide slaughter of native peoples such as the Cambodians, the Palestinians, the Armenians, the Jews—who constitute only a few groups faced with genocide in this century. . . . The American Indian people are in a situation comparable to the imminent genocide in many parts of the world today. The plight of our people north and south of us is no better; to the south it is considerably worse. Consciously or unconsciously, deliberately, as a matter of national policy, or accidentally as a matter of "fate," *every single government*, right, left, or centrist in the western hemisphere is consciously or subconsciously dedicated to the extinction of those tribal people who live within its borders.

Within this geopolitical charnel house, American Indian women struggle on every front for the survival of our children, our people, our self-respect, our value systems, and our way of life. The past five hundred years testify to our skill at waging this struggle: for all the varied weapons of extinction pointed at our heads, we endure.

---

*Editors' note: The 2000 population is 2,434,000.

†Editors' note: In 2000, this figure was 1 percent.

We survive war and conquest; we survive colonization, acculturation, assimilation; we survive beating, rape, starvation, mutilation, sterilization, abandonment, neglect, death of our children, our loved ones, destruction of our land, our homes, our past, and our future. We survive, and we do more than just survive. We bond, we care, we fight, we teach, we nurse, we bear, we feed, we earn, we laugh, we love, we hang in there, no matter what.

Of course, some, many of us, just give up. Many are alcoholics, many are addicts. Many abandon the children, the old ones. Many commit suicide. Many become violent, go insane. Many go "white" and are never seen or heard from again. But enough hold on to their traditions and their ways so that even after almost five hundred brutal years, we endure. And we even write songs and poems, make paintings and drawings that say "We walk in beauty. Let us continue."

Currently our struggles are on two fronts: physical survival and cultural survival. For women this means fighting alcoholism and drug abuse (our own and that of our husbands, lovers, parents, children); poverty; affluence—a destroyer of people who are not traditionally socialized to deal with large sums of money; rape, incest, battering by Indian men; assaults on fertility and other health matters by the Indian Health Service and the Public Health Service; high infant mortality due to substandard medical care, nutrition, and health information; poor educational opportunities or education that takes us away from our traditions, language, and communities; suicide, homicide, or similar expressions of self-hatred; lack of economic opportunities; substandard housing; sometimes violent and always virulent racist attitudes and behaviors directed against us by an entertainment and educational system that wants only one thing from Indians: our silence, our invisibility, and our collective death.

A headline in the *Navajo Times* . . . reported that rape was the number one crime on the Navajo reservation. In a professional mental health journal of the Indian Health Services, Phyllis Old Dog Cross reported that incest and rape are common among Indian women seeking services and that their incidence is increasing. "It is believed that at least 80 percent of the Native Women seen at the regional psychiatric service center (five state area) have experienced some sort of sexual assault."[1] Among the forms of abuse being suffered by Native American women, Old Dog Cross cites a recent phenomenon, something called "training." This form of gang rape is "a punitive act of a group of males who band together and get even or take revenge on a selected woman."[2]

These and other cases of violence against women are powerful evidence that the status of women within the tribes has suffered grievous decline since contact, and the decline has increased in intensity in recent years. The amount of violence against women, alcoholism, and violence, abuse, and neglect by women against their children and their aged relatives have all increased.

These social ills were virtually unheard of among most tribes fifty years ago, popular American opinion to the contrary. As Old Dog Cross remarks:

> Rapid, unstable and irrational change was required of the Indian people if they were to survive. Incredible loss of all that had meaning was the norm. In-human treatment, murder, death, and punishment was a typical experience for all the tribal groups and some didn't survive.
>
> The dominant society devoted its efforts to the attempt to change the In-dian into a white-Indian. No inhuman pressure to effect this change was over-looked. These pressures included starvation, incarceration and enforced edu-cation. Religious and healing customs were banished.
>
> In spite of the years of oppression, the Indian and the Indian spirit sur-vived. Not, however, without adverse effect. One of the major effects was the loss of cultured values and the concomitant loss of personal identity . . . The Indian was taught to be ashamed of being Indian and to emulate the non-Indian. In short, "white was right." For the Indian male, the only route to be successful, to be good, to be right, and to have an identity was to be as much like the white man as he could.[3]

Often it is said that the increase of violence against women is a result of various sociological factors such as oppression, racism, poverty, hopelessness, emasculation of men, and loss of male self-esteem as their own place within traditional society has been systematically destroyed by increasing urban-ization, industrialization, and institutionalization, but seldom do we notice that for the past forty to fifty years, American popular media have depicted American Indian men as bloodthirsty savages devoted to treating women cru-elly. While traditional Indian men seldom did any such thing—and in fact among most tribes abuse of women was simply unthinkable, as was abuse of children or the aged—the lie about "usual" male Indian behavior seems to have taken root and now bears its brutal and bitter fruit.

Image casting and image control constitute the central process that American Indian women must come to terms with, for on that control rests our sense of self, our claim to a past and to a future that we define and that we build. Images of Indians in media and educational materials profoundly influence how we act, how we relate to the world and to each other, and how we value ourselves. They also determine to a large extent how our men act to-ward us, toward our children, and toward each other. The popular American media image of Indian people as savages with no conscience, no compassion, and no sense of the value of human life and human dignity was hardly true of the tribes—however true it was of the invaders. But as Adolf Hitler noted a little over fifty years ago, if you tell a lie big enough and often enough, it will be believed. Evidently, while Americans and people all over the world have been led into a deep and unquestioned belief that American Indians are cruel savages, a number of American Indian men have been equally deluded into internalizing that image and acting on it. Media images, literary images, and artistic images, particularly those embedded in popular culture, must be changed before Indian women will see much relief from the violence that destroys so many lives.

To survive culturally, American Indian women must often fight the United States government, the tribal governments, women and men of their tribe or their urban community who are virulently misogynist or who are threatened by attempts to change the images foisted on us over the centuries by whites. The colonizers' revisions of our lives, values, and histories have devastated us at the most critical level of all—that of our own minds, our own sense of who we are.

Many women express strong opposition to those who would alter our life supports, steal our tribal lands, colonize our cultures and cultural expressions, and revise our very identities. We must strive to maintain tribal status; we must make certain that the tribes continue to be legally recognized entities, sovereign nations within the larger United States, and we must wage this struggle in many ways—political, educational, literary, artistic, individual, and communal. We are doing all we can: as mothers and grandmothers; as family members and tribal members; as professionals, workers, artists, shamans, leaders, chiefs, speakers, writers, and organizers, we daily demonstrate that we have no intention of disappearing, of being silent, or of quietly acquiescing in our extinction.

NOTES

1. Phyllis Old Dog Cross, "Sexual Abuse, a New Threat to the Native American Woman: An Overview," *Listening Post: A Periodical of the Mental Health Programs of Indian Health Services*, vol. 6, no. 2 (April 1982), p. 18.
2. Old Dog Cross, p. 18.
3. Old Dog Cross, p. 20.

OPPRESSION   5

Marilyn Frye

It is a fundamental claim of feminism that women are oppressed. The word "oppression" is a strong word. It repels and attracts. It is dangerous and dangerously fashionable and endangered. It is much misused, and sometimes not innocently.

From Marilyn Frye, *The Politics of Reality* (Trumansburg, NY: Crossing Press, 1983), pp. 1–16. Reprinted by permission.

The statement that women are oppressed is frequently met with the claim that men are oppressed too. We hear that oppressing is oppressive to those who oppress as well as to those they oppress. Some men cite as evidence of their oppression their much-advertised inability to cry. It is tough, we are told, to be masculine. When the stresses and frustrations of being a man are cited as evidence that oppressors are oppressed by their oppressing, the word "oppression" is being stretched to meaninglessness; it is treated as though its scope includes any and all human experience of limitation or suffering, no matter the cause, degree, or consequence. Once such usage has been put over on us, then if ever we deny that any person or group is oppressed, we seem to imply that we think they never suffer and have no feelings. We are accused of insensitivity; even of bigotry. For women, such accusation is particularly intimidating, since sensitivity is one of the few virtues that has been assigned to us. If we are found insensitive, we may fear we have no redeeming traits at all and perhaps are not real women. Thus are we silenced before we begin: the name of our situation drained of meaning and our guilt mechanisms tripped.

But this is nonsense. Human beings can be miserable without being oppressed, and it is perfectly consistent to deny that a person or group is oppressed without denying that they have feelings or that they suffer. . . .

The root of the word "oppression" is the element "press." *The press of the crowd; pressed into military service; to press a pair of pants; printing press; press the button.* Presses are used to mold things or flatten them or reduce them in bulk, sometimes to reduce them by squeezing out the gases or liquids in them. Something pressed is something caught between or among forces and barriers which are so related to each other that jointly they restrain, restrict or prevent the thing's motion or mobility. Mold. Immobilize. Reduce.

The mundane experience of the oppressed provides another clue. One of the most characteristic and ubiquitous features of the world as experienced by oppressed people is the double bind—situations in which options are reduced to a very few and all of them expose one to penalty, censure, or deprivation. For example, it is often a requirement upon oppressed people that we smile and be cheerful. If we comply, we signal our docility and our acquiescence in our situation. We need not, then, be taken note of. We acquiesce in being made invisible, in our occupying no space. We participate in our own erasure. On the other hand, anything but the sunniest countenance exposes us to being perceived as mean, bitter, angry, or dangerous. This means, at the least, that we may be found "difficult" or unpleasant to work with, which is enough to cost one one's livelihood; at worst, being seen as mean, bitter, angry, or dangerous has been known to result in rape, arrest, beating, and murder. One can only choose to risk one's preferred form and rate of annihilation.

Another example: It is common in the United States that women, especially younger women, are in a bind where neither sexual activity nor sexual inactivity is all right. If she is heterosexually active, a woman is open to censure and punishment for being loose, unprincipled, or a whore. The "punishment" comes in the form of criticism, snide and embarrassing remarks, being treated as an easy lay by men, scorn from her more restrained female friends. She may have to lie and hide her behavior from her parents. She must juggle the risks of unwanted pregnancy and dangerous contraceptives. On the other hand, if she refrains from heterosexual activity, she is fairly constantly harassed by men who try to persuade her into it and pressure her to "relax" and "let her hair down"; she is threatened with labels like "frigid," "uptight," "man-hater," "bitch" and "cocktease." The same parents who would be disapproving of her sexual activity may be worried by her inactivity because it suggests she is not or will not be popular, or is not sexually normal. She may be charged with lesbianism. If a woman is raped, then if she has been heterosexually active she is subject to the presumption that she liked it (since her activity is presumed to show that she likes sex), and if she has not been heterosexually active, she is subject to the presumption that she liked it (since she is supposedly "repressed and frustrated"). Both heterosexual activity and heterosexual nonactivity are likely to be taken as proof that you wanted to be raped, and hence, of course, weren't *really* raped at all. You can't win. You are caught in a bind, caught between systematically related pressures.

Women are caught like this, too, by networks of forces and barriers that expose one to penalty, loss, or contempt whether one works outside the home or not, is on welfare or not, bears children or not, raises children or not, marries or not, stays married or not, is heterosexual, lesbian, both, or neither. Economic necessity; confinement to racial and/or sexual job ghettos; sexual harassment; sex discrimination; pressures of competing expectations and judgments about *women, wives,* and *mothers* (in the society at large, in racial and ethnic subcultures, and in one's own mind); dependence (full or partial) on husbands, parents, or the state; commitment to political ideas; loyalties to racial or ethnic or other "minority" groups; the demands of self-respect and responsibilities to others. Each of these factors exists in complex tension with every other, penalizing or prohibiting all of the apparently available options. And nipping at one's heels, always, is the endless pack of little things. If one dresses one way, one is subject to the assumption that one is advertising one's sexual availability; if one dresses another way, one appears to "not care about oneself" or to be "unfeminine." If one uses "strong language," one invites categorization as a whore or slut; if one does not, one invites categorization as a "lady"—one too delicately constituted to cope with robust speech or the realities to which it presumably refers.

The experience of oppressed people is that the living of one's life is confined and shaped by forces and barriers which are not accidental or occasional and hence avoidable, but are systematically related to each other in such a way as to catch one between and among them and restrict or penalize motion in any direction. It is the experience of being caged in: all avenues, in every direction, are blocked or booby trapped.

Cages. Consider a birdcage. If you look very closely at just one wire in the cage, you cannot see the other wires. If your conception of what is before you is determined by this myopic focus, you could look at that one wire, up and down the length of it, and be unable to see why a bird would not just fly around the wire any time it wanted to go somewhere. Furthermore, even if, one day at a time, you myopically inspected each wire, you still could not see why a bird would have trouble going past the wires to get anywhere. There is no physical property of any one wire, *nothing* that the closest scrutiny could discover, that will reveal how a bird could be inhibited or harmed by it except in the most accidental way. It is only when you step back, stop looking at the wires one by one, microscopically, and take a macroscopic view of the whole cage, that you can see why the bird does not go anywhere; and then you will see it in a moment. It will require no great subtlety of mental powers. It is perfectly *obvious* that the bird is surrounded by a network of systematically related barriers, no one of which would be the least hindrance to its flight, but which, by their relations to each other, are as confining as the solid walls of a dungeon.

It is now possible to grasp one of the reasons why oppression can be hard to see and recognize: one can study the elements of an oppressive structure with great care and some good will without seeing the structure as a whole, and hence without seeing or being able to understand that one is looking at a cage and that there are people there who are caged, whose motion and mobility are restricted, whose lives are shaped and reduced. . . .

As the cageness of the birdcage is a macroscopic phenomenon, the oppressiveness of the situations in which women live their various and different lives is a macroscopic phenomenon. Neither can be *seen* from a microscopic perspective. But when you look macroscopically you can see it—a network of forces and barriers which are systematically related and which conspire to the immobilization, reduction, and molding of women and the lives we live. . . .

A DIFFERENT MIRROR 6

Ronald T. Takaki

I had flown from San Francisco to Norfolk and was riding in a taxi to my hotel to attend a conference on multiculturalism. Hundreds of educators from across the country were meeting to discuss the need for greater cultural diversity in the curriculum. My driver and I chatted about the weather and the tourists. The sky was cloudy, and Virginia Beach was twenty minutes away. The rearview mirror reflected a white man in his forties. "How long have you been in this country?" he asked. "All my life," I replied, wincing. "I was born in the United States." With a strong southern drawl, he remarked: "I was wondering because your English is excellent!" Then, as I had many times before, I explained: "My grandfather came here from Japan in the 1880s. My family has been here, in America, for over a hundred years." He glanced at me in the mirror. Somehow I did not look "American" to him; my eyes and complexion looked foreign.

Suddenly, we both became uncomfortably conscious of a racial divide separating us. An awkward silence turned my gaze from the mirror to the passing landscape, the shore where the English and the Powhatan Indians first encountered each other. Our highway was on land that Sir Walter Raleigh had renamed "Virginia" in honor of Elizabeth I, the Virgin Queen. In the English cultural appropriation of America, the indigenous peoples themselves would become outsiders in their native land. Here, at the eastern edge of the continent, I mused, was the site of the beginning of multicultural America. Jamestown, the English settlement founded in 1607, was nearby: the first twenty Africans were brought here a year before the Pilgrims arrived at Plymouth Rock. Several hundred miles offshore was Bermuda, the "Bermoothes" where William Shakespeare's Prospero had landed and met the native Caliban in *The Tempest*. Earlier, another voyager had made an Atlantic crossing and unexpectedly bumped into some islands to the south. Thinking he had reached Asia, Christopher Columbus mistakenly identified one of the islands as "Cipango" (Japan). In the wake of the admiral, many peoples would come to America from different shores, not only from Europe but also Africa and Asia. One of them would be my grandfather. My mental wandering across terrain and time ended abruptly as we arrived at my destination. I said good-bye to my driver and went into the hotel, carrying a vivid reminder of why I was attending this conference.

Questions like the one my taxi driver asked me are always jarring, but I can understand why he could not see me as American. He had a narrow but widely shared sense of the past—a history that has viewed American as European in ancestry. "Race," Toni Morrison explained, has functioned as a "metaphor" necessary to the "construction of Americanness": in the creation of our national identity, "American" has been defined as "white."[1]

But America has been racially diverse since our very beginning on the Virginia shore, and this reality is increasingly becoming visible and ubiquitous. Currently, one-third of the American people do not trace their origins to Europe; in California, minorities are fast becoming a majority. They already predominate in major cities across the country—New York, Chicago, Atlanta, Detroit, Philadelphia, San Francisco, and Los Angeles.

This emerging demographic diversity has raised fundamental questions about America's identity and culture. In 1990, *Time* published a cover story on "America's Changing Colors." "Someday soon," the magazine announced, "white Americans will become a minority group." How soon? By 2056, most Americans will trace their descent to "Africa, Asia, the Hispanic world, the Pacific Islands, Arabia—almost anywhere but white Europe." This dramatic change in our nation's ethnic composition is altering the way we think about ourselves. "The deeper significance of America's becoming a majority non-white society is what it means to the national psyche, to individuals' sense of themselves and their nation—their idea of what it is to be American." . . . [2]

What is fueling the debate over our national identity and the content of our curriculum is America's intensifying racial crisis. The alarming signs and symptoms seem to be everywhere—the killing of Vincent Chin in Detroit, the black boycott of a Korean grocery store in Flatbush, the hysteria in Boston over the Carol Stuart murder, the battle between white sportsmen and Indians over tribal fishing rights in Wisconsin, the Jewish-black clashes in Brooklyn's Crown Heights, the black-Hispanic competition for jobs and educational resources in Dallas, which *Newsweek* described as "a conflict of the have-nots," and the Willie Horton campaign commercials, which widened the divide between the suburbs and the inner cities.[3]

This reality of racial tension rudely woke America like a fire bell in the night on April 29, 1992. Immediately after four Los Angeles police officers were found not guilty of brutality against Rodney King, rage exploded in Los Angeles. Race relations reached a new nadir. During the nightmarish rampage, scores of people were killed, over two thousand injured, twelve thousand arrested, and almost a billion dollars' worth of property destroyed. The live televised images mesmerized America. The rioting and the murderous melee on the streets resembled the fighting in Beirut and the West Bank. The thousands of fires burning out of control and the dark smoke filling the skies brought back images of the burning oil fields of Kuwait during Desert Storm. Entire sections of Los Angeles looked like a bombed city. "Is this America?" many shocked viewers asked. "Please, can we get along here," pleaded Rodney King, calling for calm. "We all can get along. I mean, we're all stuck here for a while. Let's try to work it out."[4]

But how should "we" be defined? Who are the people "stuck here" in America? One of the lessons of the Los Angeles explosion is the recognition of the fact that we are a multiracial society and that race can no longer be defined in the binary terms of white and black. "We" will have to include Hispanics and Asians. While blacks currently constitute 13 percent of the Los Angeles population, Hispanics represent 40 percent. The 1990 census revealed that South Central Los Angeles, which was predominantly black in 1965 when the Watts rebellion occurred, is now 45 percent Hispanic. A majority of the first 5,438 people arrested were Hispanic, while 37 percent were black. Of the fifty-eight people who died in the riot, more than a third were Hispanic, and about 40 percent of the businesses destroyed were Hispanic-owned. Most of the other shops and stores were Korean-owned. The dreams of many Korean immigrants went up in smoke during the riot: two thousand Korean-owned businesses were damaged or demolished, totaling about $400 million in losses. There is evidence indicating they were targeted. "After all," explained a black gang member, "we didn't burn our community, just *their* stores."[5]

"I don't feel like I'm in America anymore," said Denisse Bustamente as she watched the police protecting the firefighters. "I feel like I am far away." Indeed, Americans have been witnessing ethnic strife erupting around the world—the rise of neo-Nazism and the murder of Turks in Germany, the ugly "ethnic cleansing" in Bosnia, the terrible and bloody clashes between Muslims and Hindus in India. Is the situation here different, we have been nervously wondering, or do ethnic conflicts elsewhere represent a prologue for America? What is the nature of malevolence? Is there a deep, perhaps primordial, need for group identity rooted in hatred for the other? Is ethnic pluralism possible for America? But answers have been limited. Television reports have been little more than thirty-second sound bites. Newspaper articles have been mostly superficial descriptions of racial antagonisms and the current urban malaise. What is lacking is historical context; consequently, we are left feeling bewildered.[6]

How did we get to this point, Americans everywhere are anxiously asking. What does our diversity mean, and where is it leading us? *How* do we work it out in the post–Rodney King era?

Certainly one crucial way is for our society's various ethnic groups to develop a greater understanding of each other. For example, how can African Americans and Korean Americans work it out unless they learn about each other's cultures, histories, and also economic situations? This need to share knowledge about our ethnic diversity has acquired new importance and has given new urgency to the pursuit for a more accurate history. . . .

While all of America's many groups cannot be covered [here], the English immigrants and their descendants require attention, for they possessed inordinate power to define American culture and make public policy. What men like John Winthrop, Thomas Jefferson, and Andrew Jackson thought as well as did mattered greatly to all of us and was consequential for everyone. A broad range of groups [is important]: African Americans, Asian Americans, Chicanos, Irish, Jews, and Indians. While together they help to explain general patterns in our society, each has contributed to the making of the United States.

African Americans have been the central minority throughout our country's history. They were initially brought here on a slave ship in 1619. Actually, these first twenty Africans might not have been slaves; rather, like most of the white laborers, they were probably indentured servants. The transformation of Africans into slaves is the story of the "hidden" origins of slavery. How and when was it decided to institute a system of bonded black labor? What happened, while freighted with racial significance, was actually conditioned by class conflicts within white society. Once established, the "peculiar institution" would have consequences for centuries to come. During the nineteenth century, the political storm over slavery almost destroyed the nation. Since the Civil War and emancipation, race has continued to be largely defined in relation to African Americans—segregation, civil rights, the underclass, and affirmative action. Constituting the largest minority group in our society, they have been at the cutting edge of the Civil Rights Movement. Indeed, their struggle has been a constant reminder of America's moral vision as a country committed to the principle of liberty. Martin Luther King clearly understood this truth when he wrote from a jail cell: "We will reach the goal of freedom in Birmingham and all over the nation, because the goal of America is freedom. Abused and scorned though we may be, our destiny is tied up with America's destiny."[7]

Asian Americans have been here for over one hundred and fifty years, before many European immigrant groups. But as "strangers" coming from a "different shore," they have been stereotyped as "heathen," exotic, and unassimilable. Seeking "Gold Mountain," the Chinese arrived first, and what happened to them influenced the reception of the Japanese, Koreans, Filipinos, and Asian Indians as well as the Southeast Asian refugees like the Vietnamese and the Hmong. The 1882 Chinese Exclusion Act was the first law that prohibited the entry of immigrants on the basis of nationality. The Chinese condemned this restriction as racist and tyrannical. "They call us 'Chink,'" complained a Chinese immigrant, cursing the "white demons." "They think we no good! America cuts us off. No more come now, too bad!" This precedent later provided a basis for the restriction of European immigrant groups such as Italians, Russians, Poles, and Greeks. The Japanese painfully discovered that their accomplishments in America did not lead to acceptance, for during World War II, unlike Italian Americans and German Americans, they were placed in internment camps. Two-thirds of them were citizens by birth. "How could I as a 6-month-old child born in this country," asked Congressman Robert Matsui years later, "be declared by my own Government to be an enemy alien?" Today, Asian Americans represent the fastest-growing ethnic group. They have also become the focus of much mass media attention as "the Model Minority" not only for blacks and Chicanos, but also for whites on welfare and even middle-class whites experiencing economic difficulties.[8]

Chicanos represent the largest group among the Hispanic population, which is projected to outnumber African Americans. They have been in the United States for a long time, initially incorporated by the war against

Mexico. The treaty had moved the border between the two countries, and the people of "occupied" Mexico suddenly found themselves "foreigners" in their "native land." As historian Albert Camarillo pointed out, the Chicano past is an integral part of America's westward expansion, also known as "manifest destiny." But while the early Chicanos were a colonized people, most of them today have immigrant roots. Many began the trek to El Norte in the early twentieth century. "As I had heard a lot about the United States," Jesus Garza recalled, "it was my dream to come here." "We came to know families from Chihuahua, Sonora, Jalisco, and Durango," stated Ernesto Galarza. "Like ourselves, our Mexican neighbors had come this far moving step by step, working and waiting, as if they were feeling their way up a ladder." Nevertheless, the Chicano experience has been unique, for most of them have lived close to their homeland—a proximity that has helped reinforce their language, identity, and culture. This migration to El Norte has continued to the present. Los Angeles has more people of Mexican origin than any other city in the world, except Mexico City. A mostly mestizo people of Indian as well as African and Spanish ancestries, Chicanos currently represent the largest minority group in the Southwest, where they have been visibly transforming culture and society.[9]

The Irish came here in greater numbers than most immigrant groups. Their history has been tied to America's past from the very beginning. Ireland represented the earliest English frontier: the conquest of Ireland occurred before the colonization of America, and the Irish were the first group that the English called "savages." In this context, the Irish past foreshadowed the Indian future. During the nineteenth century, the Irish, like the Chinese, were victims of British colonialism. While the Chinese fled from the ravages of the Opium Wars, the Irish were pushed from their homeland by "English tyranny." Here they became construction workers and factory operatives as well as the "maids" of America. Representing a Catholic group seeking to settle in a fiercely Protestant society, the Irish immigrants were targets of American nativist hostility. They were also what historian Lawrence J. McCaffrey called "the pioneers of the American urban ghetto," "previewing" experiences that would later be shared by the Italians, Poles, and other groups from southern and eastern Europe. Furthermore, they offer contrast to the immigrants from Asia. The Irish came about the same time as the Chinese, but they had a distinct advantage: the Naturalization Law of 1790 had reserved citizenship for "whites" only. Their compatible complexion allowed them to assimilate by blending into American society. In making their journey successfully into the mainstream, however, these immigrants from Erin pursued an Irish "ethnic" strategy: they promoted "Irish" solidarity in order to gain political power and also to dominate the skilled blue-collar occupations, often at the expense of the Chinese and blacks.[10]

Fleeing pogroms and religious persecution in Russia, the Jews were driven from what John Cuddihy described as the "Middle Ages into the

Anglo-American world of the *goyim* 'beyond the pale.'" To them, America represented the Promised Land. This vision led Jews to struggle not only for themselves but also for other oppressed groups, especially blacks. After the 1917 East St. Louis race riot, the Yiddish *Forward* of New York compared this antiblack violence to a 1903 pogrom in Russia: "Kishinev and St. Louis—the same soil, the same people." Jews cheered when Jackie Robinson broke into the Brooklyn Dodgers in 1947. "He was adopted as the surrogate hero by many of us growing up at the time," recalled Jack Greenberg of the NAACP Legal Defense Fund. "He was the way we saw ourselves triumphing against the forces of bigotry and ignorance." Jews stood shoulder to shoulder with blacks in the Civil Rights Movement: two-thirds of the white volunteers who went south during the 1964 Freedom Summer were Jewish. Today Jews are considered a highly successful "ethnic" group. How did they make such great socioeconomic strides? This question is often reframed by neoconservative intellectuals like Irving Kristol and Nathan Glazer to read: if Jewish immigrants were able to lift themselves from poverty into the mainstream through self-help and education without welfare and affirmative action, why can't blacks? But what this thinking overlooks is the unique history of Jewish immigrants, especially the initial advantages of many of them as literate and skilled. Moreover, it minimizes the virulence of racial prejudice rooted in American slavery.[11]

Indians represent a critical contrast, for theirs was not an immigrant experience. The Wampanoags were on the shore as the first English strangers arrived in what would be called "New England." The encounters between Indians and whites not only shaped the course of race relations, but also influenced the very culture and identity of the general society. The architect of Indian removal, President Andrew Jackson told Congress: "Our conduct toward these people is deeply interesting to the national character." Frederick Jackson Turner understood the meaning of this observation when he identified the frontier as our transforming crucible. At first, the European newcomers had to wear Indian moccasins and shout the war cry. "Little by little," as they subdued the wilderness, the pioneers became "a new product" that was "American." But Indians have had a different view of this entire process. "The white man," Luther Standing Bear of the Sioux explained, "does not understand the Indian for the reason that he does not understand America." Continuing to be "troubled with primitive fears," he has "in his consciousness the perils of this frontier continent. . . . The man from Europe is still a foreigner and an alien. And he still hates the man who questioned his path across the continent." Indians questioned what Jackson and Turner trumpeted as "progress." For them, the frontier had a different "significance": their history was how the West was lost. But their story has also been one of resistance. As Vine Deloria declared, "Custer died for your sins."[12]

By looking at these groups from a multicultural perspective, we can comparatively analyze their experiences in order to develop an understanding of their differences and similarities. Race, we will see, has been a social construction that has historically set apart racial minorities from European immigrant groups. Contrary to the notions of scholars like Nathan Glazer and Thomas Sowell, race in America has not been the same as ethnicity. A broad comparative focus also allows us to see how the varied experiences of different racial and ethnic groups occurred within shared contexts.

During the nineteenth century, for example, the Market Revolution employed Irish immigrant laborers in New England factories as it expanded cotton fields worked by enslaved blacks across Indian lands toward Mexico. Like blacks, the Irish newcomers were stereotyped as "savages," ruled by passions rather than "civilized" virtues such as self-control and hard work. The Irish saw themselves as the "slaves" of British oppressors, and during a visit to Ireland in the 1840s, Frederick Douglass found that the "wailing notes" of the Irish ballads reminded him of the "wild notes" of slave songs. The United States annexation of California, while incorporating Mexicans, led to trade with Asia and the migration of "strangers" from Pacific shores. In 1870, Chinese immigrant laborers were transported to Massachusetts as scabs to break an Irish immigrant strike; in response, the Irish recognized the need for interethnic working-class solidarity and tried to organize a Chinese lodge of the Knights of St. Crispin. After the Civil War, Mississippi planters recruited Chinese immigrants to discipline the newly freed blacks. During the debate over an immigration exclusion bill in 1882, a senator asked: If Indians could be located on reservations, why not the Chinese? [13]

Other instances of our connectedness abound. In 1903, Mexican and Japanese farm laborers went on strike together in California: their union officers had names like Yamaguchi and Lizarras, and strike meetings were conducted in Japanese and Spanish. The Mexican strikers declared that they were standing in solidarity with their "Japanese brothers" because the two groups had toiled together in the fields and were now fighting together for a fair wage. Speaking in impassioned Yiddish during the 1909 "uprising of twenty thousand" strikers in New York, the charismatic Clara Lemlich compared the abuse of Jewish female garment workers to the experience of blacks: "[The bosses] yell at the girls and 'call them down' even worse than I imagine the Negro slaves were in the South." During the 1920s, elite universities like Harvard worried about the increasing numbers of Jewish students, and new admissions criteria were instituted to curb their enrollment. Jewish students were scorned for their studiousness and criticized for their "clannishness." Recently, Asian-American students have been the targets of similar complaints: they have been called "nerds" and told there are "too many" of them on campus.[14]

Indians were already here, while blacks were forcibly transported to America, and Mexicans were initially enclosed by America's expanding border. The other groups came here as immigrants: for them, America represented liminality—a new world where they could pursue extravagant urges and do things they had thought beyond their capabilities. Like the land itself, they found themselves "betwixt and between all fixed points of classification." No longer fastened as fiercely to their old countries, they felt a stirring to become new people in a society still being defined and formed.[15]

These immigrants made bold and dangerous crossings, pushed by political events and economic hardships in their homelands and pulled by America's demand for labor as well as by their own dreams for a better life. "By all means let me go to America," a young man in Japan begged his parents. He had calculated that in one year as a laborer here he could save almost a thousand yen—an amount equal to the income of a governor in Japan. "My dear Father," wrote an immigrant Irish girl living in New York, "Any man or woman without a family are fools that would not venture and come to this plentyful Country where no man or woman ever hungered." In the shtetls of Russia, the cry "To America!" roared like "wildfire." "America was in everybody's mouth," a Jewish immigrant recalled. "Businessmen talked [about] it over their accounts; the market women made up their quarrels that they might discuss it from stall to stall; people who had relatives in the famous land went around reading their letters." Similarly, for Mexican immigrants crossing the border in the early twentieth century, El Norte became the stuff of overblown hopes. "If only you could see how nice the United States is," they said, "that is why the Mexicans are crazy about it."[16]

The signs of America's ethnic diversity can be discerned across the continent—Ellis Island, Angel Island, Chinatown, Harlem, South Boston, the Lower East Side, places with Spanish names like Los Angeles and San Antonio or Indian names like Massachusetts and Iowa. Much of what is familiar in America's cultural landscape actually has ethnic origins. The Bing cherry was developed by an early Chinese immigrant named Ah Bing. American Indians were cultivating corn, tomatoes, and tobacco long before the arrival of Columbus. The term *okay* was derived from the Choctaw word *oke*, meaning "it is so." There is evidence indicating that the name *Yankee* came from Indian terms for the English—from *eankke* in Cherokee and *Yankwis* in Delaware. Jazz and blues as well as rock and roll have African-American origins. The "Forty-Niners" of the Gold Rush learned mining techniques from the Mexicans; American cowboys acquired herding skills from Mexican *vaqueros* and adopted their range terms—such as *lariat* from *la reata*, *lasso* from *lazo*, and *stampede* from *estampida*. Songs like "God Bless America," "Easter Parade," and "White Christmas" were written by a Russian-Jewish immigrant named Israel Baline, more popularly known as Irving Berlin.[17]

Furthermore, many diverse ethnic groups have contributed to the building of the American economy, forming what Walt Whitman saluted as "a vast, surging, hopeful army of workers." They worked in the South's cotton fields, New England's textile mills, Hawaii's canefields, New York's garment factories, California's orchards, Washington's salmon canneries, and Arizona's copper mines. They built the railroad, the great symbol of America's industrial triumph. . . .

Moreover, our diversity was tied to America's most serious crisis: the Civil War was fought over a racial issue—slavery. . . .

. . . The people in our study have been actors in history, not merely victims of discrimination and exploitation. They are entitled to be viewed as subjects—as men and women with minds, wills, and voices.

> *In the telling and retelling*
> *of their stories,*
> *They create communities*
> *of memory.*

They also re-vision history. "It is very natural that the history written by the victim," said a Mexican in 1874, "does not altogether chime with the story of the victor." Sometimes they are hesitant to speak, thinking they are only "little people." "I don't know why anybody wants to hear my history," an Irish maid said apologetically in 1900. "Nothing ever happened to me worth the tellin'." [18]

But their stories are worthy. Through their stories, the people who have lived America's history can help all of us, including my taxi driver, understand that Americans originated from many shores, and that all of us are entitled to dignity. "I hope this survey do a lot of good for Chinese people," an immigrant told an interviewer from Stanford University in the 1920s. "Make American people realize that Chinese people are humans. I think very few American people really know anything about Chinese." But the remembering is also for the sake of the children. "This story is dedicated to the descendants of Lazar and Goldie Glauberman," Jewish immigrant Minnie Miller wrote in her autobiography. "My history is bound up in their history and the generations that follow should know where they came from to know better who they are." Similarly, Tomo Shoji, an elderly Nisei woman, urged Asian Americans to learn more about their roots: "We got such good, fantastic stories to tell. All our stories are different." Seeking to know how they fit into America, many young people have become listeners; they are eager to learn about the hardships and humiliations experienced by their parents and grandparents. They want to hear their stories, unwilling to remain ignorant or ashamed of their identity and past. [19]

The telling of stories liberates. By writing about the people on Mango Street, Sandra Cisneros explained, "the ghost does not ache so much." The place no longer holds her with "both arms. She sets me free." Indeed, stories may not be as innocent or simple as they seem to be. Native-American novelist Leslie Marmon Silko cautioned:

> *I will tell you something about stories . . .*
> *They aren't just entertainment.*
> *Don't be fooled.*

Indeed, the accounts given by the people in this study vibrantly re-create moments, capturing the complexities of human emotions and thoughts. They also provide the authenticity of experience. After she escaped from slavery, Harriet Jacobs wrote in her autobiography: "[My purpose] is not to tell you what I have heard but what I have seen—and what I have suffered." In their sharing of memory, the people in this study offer us an opportunity to see ourselves reflected in a mirror called history.[20]

In his recent study of Spain and the New World, *The Buried Mirror*, Carlos Fuentes points out that mirrors have been found in the tombs of ancient Mexico, placed there to guide the dead through the underworld. He also tells us about the legend of Quetzalcoatl, the Plumed Serpent: when this god was given a mirror by the Toltec deity Tezcatlipoca, he saw a man's face in the mirror and realized his own humanity. For us, the "mirror" of history can guide the living and also help us recognize who we have been and hence are. In *A Distant Mirror*, Barbara W. Tuchman finds "phenomenal parallels" between the "calamitous fourteenth century" of European society and our own era. We can, she observes, have "greater fellow-feeling for a distraught age" as we painfully recognize the "similar disarray," "collapsing assumptions," and "unusual discomfort."[21]

But what is needed in our own perplexing times is not so much a "distant" mirror, as one that is "different." While the study of the past can provide collective self-knowledge, it often reflects the scholar's particular perspective or view of the world. What happens when historians leave out many of America's peoples? What happens, to borrow the words of Adrienne Rich, "when someone with the authority of a teacher" describes our society, and "you are not in it"? Such an experience can be disorienting—"a moment of psychic disequilibrium, as if you looked into a mirror and saw nothing."[22]

Through their narratives about their lives and circumstances, the people of America's diverse groups are able to see themselves and each other in our common past. They celebrate what Ishmael Reed has described as a society "unique" in the world because "the world is here"—a place "where the cultures of the world crisscross." Much of America's past, they point out, has been riddled with racism. At the same time, these people offer hope, affirming the struggle for equality as a central theme in our country's history. At its conception, our nation was dedicated to the proposition of equality. What has given concreteness to this powerful national principle has been our coming together in the creation of a new society. "Stuck here" together, workers of different backgrounds have attempted to get along with each other.

People harvesting
Work together unaware
Of racial problems,

wrote a Japanese immigrant describing a lesson learned by Mexican and Asian farm laborers in California.[23]

Finally, how do we see our prospects for "working out" America's racial crisis? Do we see it as through a glass darkly? Do the televised images of racial hatred and violence that riveted us in 1992 during the days of rage in Los Angeles frame a future of divisive race relations—what Arthur Schlesinger Jr. has fearfully denounced as the "disuniting of America"? Or will Americans of diverse races and ethnicities be able to connect themselves to a larger narrative? Whatever happens, we can be certain that much of our society's future will be influenced by which "mirror" we choose to see ourselves. America does not belong to one race or one group. . . . Americans have been constantly redefining their national identity from the moment of first contact on the Virginia shore. By sharing their stories, they invite us to see ourselves in a different mirror.[24]

NOTES

1. Toni Morrison, *Playing in the Dark: Whiteness in the Literary Imagination* (Cambridge, Mass., 1992), p. 47.

2. William A. Henry III, "Beyond the Melting Pot," in "America's Changing Colors," *Time*, vol. 135, no. 15 (April 9, 1990), pp. 28–31.

3. "A Conflict of the Have-Nots," *Newsweek*, December 12, 1988, pp. 28–29.

4. Rodney King's statement to the press, *New York Times*, May 2, 1992, p. 6.

5. Tim Rutten, "A New Kind of Riot," *New York Times Review of Books*, June 11, 1992, pp. 52–53; Maria Newman, "Riots Bring Attention to Growing Hispanic Presence in South-Central Area," *New York Times*, May 11, 1992, p. A10; Mike Davis, "In L.A. Burning All Illusions," *The Nation*, June 1, 1992, pp. 744–745; Jack Viets and Peter Fimrite, "S.F. Mayor Visits Riot-Torn Area to Buoy Businesses," *San Francisco Chronicle*, May 6, 1992, p. A6.

6. Rick DelVecchio, Suzanne Espinosa, and Carl Nolte, "Bradley Ready to Lift Curfew," *San Francisco Chronicle*, May 4, 1992, p. A1.

7. Abraham Lincoln, "The Gettysburg Address," in *The Annals of America*, vol. 9, *1863–1865: The Crisis of the Union* (Chicago, 1968), pp. 462–463; Martin Luther King, *Why We Can't Wait* (New York, 1964), pp. 92–93.

8. Interview with old laundryman, in "Interviews with Two Chinese," circa 1924, Box 326, folder 325, Survey of Race Relations, Stanford University, Hoover Institution Archives; Congressman Robert Matsui, speech in the House of Representatives on the 442 bill for redress and reparations, September 17, 1987, *Congressional Record* (Washington, D.C., 1987), p. 7584.

9. Albert Camarillo, *Chicanos in a Changing Society: From Mexican Pueblos to American Barrios in Santa Barbara and Southern California, 1848–1930* (Cambridge, Mass., 1979), p. 2; Juan Nepornuceno Seguín, in David J. Weber (ed.), *Foreigners in Their Native Land: Historical Roots of the Mexican Americans* (Albuquerque, N. M., 1973), p. vi; Jesus Garza, in Manuel Garnio, *The Mexican Immigrant: His Life Story* (Chicago, 1931),

p. 15; Ernesto Galarza, *Barrio Boy: The Story of a Boy's Acculturation* (Notre Dame, Ind., 1986), p. 200.

10. Lawrence J. McCaffrey, *The Irish Diaspora in America* (Washington, D.C., 1984), pp. 6, 62.

11. John Murray Cuddihy, *The Ordeal of Civility: Freud, Marx, Levi Strauss, and the Jewish Struggle with Modernity* (Boston, 1987), p. 165; Jonathan Kaufman, *Broken Alliance: The Turbulent Times between Blacks and Jews in America* (New York, 1989), pp. 28, 82, 83–84, 91, 93, 106.

12. Andrew Jackson, First Annual Message to Congress, December 8, 1829, in James D. Richardson (ed.), *A Compilation of the Messages and Papers of the Presidents, 1789–1897* (Washington, D.C., 1897), vol. 2, p. 457; Frederick Jackson Turner, "The Significance of the Frontier in American History," in *The Early Writings of Frederick Jackson Turner* (Madison, Wis., 1938), pp. 185ff.; Luther Standing Bear, "What the Indian Means to America," in Wayne Moquin (ed.), *Great Documents in American Indian History* (New York, 1973), p. 307; Vine Deloria, Jr., *Custer Died for Your Sins: An Indian Manifesto* (New York, 1969).

13. Nathan Glazer, *Affirmative Discrimination: Ethnic Inequality and Public Policy* (New York, 1978); Thomas Sowell, *Ethnic America: A History* (New York, 1981); David R. Roediger, *The Wages of Whiteness: Race and the Making of the American Working Class* (London, 1991), pp. 134–136; Dan Caldwell, "The Negroization of the Chinese Stereotype in California," *Southern California Quarterly*, vol. 33 (June 1971), pp. 123–131.

14. Thomas Almaguer, "Racial Domination and Class Conflict in Capitalist Agriculture: The Oxnard Sugar Beet Workers' Strike of 1903," *Labor History*, vol. 25, no. 3 (summer 1984), p. 347; Howard M. Sachar, *A History of the Jews in America* (New York, 1992), p. 183.

15. For the concept of liminality, see Victor Turner, *Dramas, Fields, and Metaphors: Symbolic Action in Human Society* (Ithaca, N.Y., 1974), pp. 232, 237; and Arnold Van Gennep, *The Rites of Passage* (Chicago, 1960). What I try to do is to apply liminality to the land called America.

16. Kazuo Ito, *Issei: A History of Japanese Immigrants in North America* (Seattle, 1973), p. 33; Arnold Schrier, *Ireland and the American Emigration, 1850–1900* (New York, 1970), p. 24; Abraham Cahan, *The Rise of David Levinsky* (New York, 1960; originally published in 1917), pp. 59–61; Mary Antin, quoted in Howe, *World of Our Fathers* (New York, 1983), p. 27; Lawrence A. Cardoso, *Mexican Emigration to the United States, 1897–1931* (Tucson, Ariz., 1981), p. 80.

17. Ronald Takaki, *Strangers from a Different Shore: A History of Asian Americans* (Boston, 1989), pp. 88–89; Jack Weatherford, *Native Roots: How the Indians Enriched America* (New York, 1991), pp. 210, 212; Carey McWilliams, *North from Mexico: The Spanish-Speaking People of the United States* (New York, 1968), p. 154; Stephan Themstrom (ed.), *Harvard Encyclopedia of American Ethnic Groups* (Cambridge, Mass., 1980), p. 22; Sachar, *A History of the Jews in America*, p. 367.

18. Weber (ed.), *Foreigners in Their Native Land*, p. vi; Hamilton Holt (ed.), *The Life Stories of Undistinguished Americans as Told by Themselves* (New York, 1906), p. 143.

19. "Social Document of Pany Lowe, interviewed by C. H. Burnett, Seattle, July 5, 1924," p. 6, Survey of Race Relations, Stanford University, Hoover Institution Archives; Minnie Miller, "Autobiography," private manuscript, copy from Richard Balkin; Tomo Shoji, presentation, Obana Cultural Center, Oakland, California, March 4, 1988.

20. Sandra Cisneros, *The House on Mango Street* (New York, 1991), pp. 109–110; Leslie Marmon Silko, *Ceremony* (New York, 1978), p. 2; Harriet A. Jacobs, *Incidents in the Life of a Slave Girl, written by herself* (Cambridge, Mass., 1987; originally published in 1857), p. xiii.

21. Carlos Fuentes, *The Buried Mirror: Reflections on Spain and the New World* (Boston, 1992), pp. 10, 11, 109; Barbara W. Tuchman, *A Distant Mirror: The Calamitous 14th Century* (New York, 1978), pp. xiii, xiv.

22. Adrienne Rich, *Blood, Bread, and Poetry: Selected Prose, 1979–1985* (New York, 1986), p. 199.

23. Ishmael Reed, "America: The Multinational Society," in Rick Simonson and Scott Walker (eds.), *Multi-cultural Literacy* (St. Paul, 1988), p. 160; Ito, *Issei*, p. 497.

24. Arthur M. Schlesinger, Jr., *The Disuniting of America: Reflections on a Multicultural Society* (Knoxville, Tenn., 1991); Carlos Bulosan, *America Is in the Heart: A Personal History* (Seattle, 1981), pp. 188–189.

AGE, RACE, CLASS, AND SEX

7

Women Redefining Difference

Audre Lorde

Much of Western European history conditions us to see human differences in simplistic opposition to each other: dominant/subordinate, good/bad, up/down, superior/inferior. In a society where the good is defined in terms of profit rather than in terms of human need, there must always be some group of people who, through systematized oppression, can be made to feel surplus, to occupy the place of the dehumanized inferior. Within this society, that group is made up of Black and Third World people, working-class people, older people, and women.

As a forty-nine-year-old Black lesbian feminist socialist mother of two, including one boy, and a member of an interracial couple, I usually find myself a part of some group defined as other, deviant, inferior, or just plain wrong. Traditionally, in american society, it is the members of oppressed, objectified

Paper delivered at the Copeland Colloquium, Amherst College, April 1980.

From Audre Lorde, *Sister Outsider* (Freedom, CA: Crossing Press, 1984), pp. 114–23. Reprinted by permission.

groups who are expected to stretch out and bridge the gap between the actu-
alities of our lives and the consciousness of our oppressor. For in order to survive, those of us for whom oppression is as american as apple pie have always
had to be watchers, to become familiar with the language and manners of the
oppressor, even sometimes adopting them for some illusion of protection.
Whenever the need for some pretense of communication arises, those who
profit from our oppression call upon us to share our knowledge with them. In
other words, it is the responsibility of the oppressed to teach the oppressors
their mistakes. I am responsible for educating teachers who dismiss my children's culture in school. Black and Third World people are expected to educate white people as to our humanity. Women are expected to educate men.
Lesbians and gay men are expected to educate the heterosexual world. The
oppressors maintain their position and evade responsibility for their own actions. There is a constant drain of energy which might be better used in
redefining ourselves and devising realistic scenarios for altering the present
and constructing the future.

Institutionalized rejection of difference is an absolute necessity in a profit
economy which needs outsiders as surplus people. As members of such an
economy, we have *all* been programmed to respond to the human differences
between us with fear and loathing and to handle that difference in one of three
ways: ignore it, and if that is not possible, copy it if we think it is dominant, or
destroy it if we think it is subordinate. But we have no patterns for relating
across our human differences as equals. As a result, those differences have
been misnamed and misused in the service of separation and confusion.

Certainly there are very real differences between us of race, age, and sex.
But it is not those differences between us that are separating us. It is rather our
refusal to recognize those differences, and to examine the distortions which
result from our misnaming them and their effects upon human behavior and
expectation.

*Racism, the belief in the inherent superiority of one race over all others and
thereby the right to dominance. Sexism, the belief in the inherent superiority of one
sex over the other and thereby the right to dominance. Ageism. Heterosexism. Elitism.
Classism.*

It is a lifetime pursuit for each one of us to extract these distortions from
our living at the same time as we recognize, reclaim, and define those differences upon which they are imposed. For we have all been raised in a society
where those distortions were endemic within our living. Too often, we pour
the energy needed for recognizing and exploring difference into pretending
those differences are insurmountable barriers, or that they do not exist at all.
This results in a voluntary isolation, or false and treacherous connections.
Either way, we do not develop tools for using human difference as a spring-
board for creative change within our lives. We speak not of human difference,
but of human deviance.

Somewhere, on the edge of consciousness, there is what I call a *mythical norm*, which each one of us within our hearts knows "that is not me." In america, this norm is usually defined as white, thin, male, young, heterosexual, christian, and financially secure. It is with this mythical norm that the trappings of power reside within this society. Those of us who stand outside that power often identify one way in which we are different, and we assume that to be the primary cause of all oppression, forgetting other distortions around difference, some of which we ourselves may be practicing. By and large within the women's movement today, white women focus upon their oppression as women and ignore differences of race, sexual preference, class, and age. There is a pretense to a homogeneity of experience covered by the word *sisterhood* that does not in fact exist.

Unacknowledged class differences rob women of each others' energy and creative insight. Recently a women's magazine collective made the decision for one issue to print only prose, saying poetry was a less "rigorous" or "serious" art form. Yet even the form our creativity takes is often a class issue. Of all the art forms, poetry is the most economical. It is the one which is the most secret, which requires the least physical labor, the least material, and the one which can be done between shifts, in the hospital pantry, on the subway, and on scraps of surplus paper. Over the last few years, writing a novel on tight finances, I came to appreciate the enormous differences in the material demands between poetry and prose. As we reclaim our literature, poetry has been the major voice of poor, working class, and Colored women. A room of one's own may be a necessity for writing prose, but so are reams of paper, a typewriter, and plenty of time. The actual requirements to produce the visual arts also help determine, along class lines, whose art is whose. In this day of inflated prices for material, who are our sculptors, our painters, our photographers? When we speak of a broadly based women's culture, we need to be aware of the effect of class and economic differences on the supplies available for producing art.

As we move toward creating a society within which we can each flourish, ageism is another distortion of relationship which interferes without vision. By ignoring the past, we are encouraged to repeat its mistakes. The "generation gap" is an important social tool for any repressive society. If the younger members of a community view the older members as contemptible or suspect or excess, they will never be able to join hands and examine the living memories of the community, nor ask the all important question, "Why?" This gives rise to a historical amnesia that keeps us working to invent the wheel every time we have to go to the store for bread.

We find ourselves having to repeat and relearn the same old lessons over and over that our mothers did because we do not pass on what we have learned, or because we are unable to listen. For instance, how many times has this all been said before? For another, who would have believed that once again our daughters are allowing their bodies to be hampered and purgatoried by girdles and high heels and hobble skirts?

Ignoring the differences of race between women and the implications of those differences presents the most serious threat to the mobilization of women's joint power.

As white women ignore their built-in privilege of whiteness and define *woman* in terms of their own experience alone, then women of Color become "other," the outsider whose experience and tradition is too "alien" to comprehend. An example of this is the signal absence of the experience of women of Color as a resource for women's studies courses. The literature of women of Color is seldom included in women's literature courses and almost never in other literature courses, nor in women's studies as a whole. All too often, the excuse given is that the literatures of women of Color can only be taught by Colored women, or that they are too difficult to understand, or that classes cannot "get into" them because they come out of experiences that are "too different." I have heard this argument presented by white women of otherwise quite clear intelligence, women who seem to have no trouble at all teaching and reviewing work that comes out of the vastly different experiences of Shakespeare, Molière, Dostoyevsky, and Aristophanes. Surely there must be some other explanation.

This is a very complex question, but I believe one of the reasons white women have such difficulty reading Black women's work is because of their reluctance to see Black women as women and different from themselves. To examine Black women's literature effectively requires that we be seen as whole people in our actual complexities—as individuals, as women, as human—rather than as one of those problematic but familiar stereotypes provided in this society in place of genuine images of Black women. And I believe this holds true for the literatures of other women of Color who are not Black.

The literatures of all women of Color re-create the textures of our lives, and many white women are heavily invested in ignoring the real differences. For as long as any difference between us means one of us must be inferior, then the recognition of any difference must be fraught with guilt. To allow women of Color to step out of stereotypes is too guilt provoking, for it threatens the complacency of those women who view oppression only in terms of sex.

Refusing to recognize difference makes it impossible to see the different problems and pitfalls facing us as women.

Thus, in a patriarchal power system where whiteskin privilege is a major prop, the entrapments used to neutralize Black women and white women are not the same. For example, it is easy for Black women to be used by the power structure against Black men, not because they are men, but because they are Black. Therefore, for Black women, it is necessary at all times to separate the needs of the oppressor from our own legitimate conflicts within our communities. This same problem does not exist for white women. Black women and men have shared racist oppression and still share it, although in different ways. Out of that shared oppression we have developed joint defenses and joint vulnerabilities to each other that are not duplicated in the white community, with the exception of the relationship between Jewish women and Jewish men.

On the other hand, white women face the pitfall of being seduced into joining the oppressor under the pretense of sharing power. This possibility does not exist in the same way for women of Color. The tokenism that is sometimes extended to us is not an invitation to join power; our racial "otherness" is a visible reality that makes that quite clear. For white women there is a wider range of pretended choices and rewards for identifying with patriarchal power and its tools.

Today, with the defeat of ERA, the tightening economy, and increased conservatism, it is easier once again for white women to believe the dangerous fantasy that if you are good enough, pretty enough, sweet enough, quiet enough, teach the children to behave, hate the right people, and marry the right men, then you will be allowed to co-exist with patriarchy in relative peace, at least until a man needs your job or the neighborhood rapist happens along. And true, unless one lives and loves in the trenches it is difficult to remember that the war against dehumanization is ceaseless.

But Black women and our children know the fabric of our lives is stitched with violence and with hatred, that there is no rest. We do not deal with it only on the picket lines, or in dark midnight alleys, or in the places where we dare to verbalize our resistance. For us, increasingly, violence weaves through the daily tissues of our living—in the supermarket, in the classroom, in the elevator, in the clinic and the schoolyard, from the plumber, the baker, the saleswoman, the bus driver, the bank teller, the waitress who does not serve us.

Some problems we share as women, some we do not. You fear your children will grow up to join the patriarchy and testify against you, we fear our children will be dragged from a car and shot down in the street, and you will turn your backs upon the reasons they are dying.

The threat of difference has been no less blinding to people of Color. Those of us who are Black must see that the reality of our lives and our struggle does not make us immune to the errors of ignoring and misnaming difference. Within Black communities where racism is a living reality, differences among us often seem dangerous and suspect. The need for unity is often misnamed as a need for homogeneity, and a Black feminist vision mistaken for betrayal of our common interests as a people. Because of the continuous battle against racial erasure that Black women and Black men share, some Black women still refuse to recognize that we are also oppressed as women, and that sexual hostility against Black women is practiced not only by the white racist society, but implemented within our Black communities as well. It is a disease striking the heart of Black nationhood, and silence will not make it disappear. Exacerbated by racism and the pressures of powerlessness, violence against Black women and children often becomes a standard within our communities, one by which manliness can be measured. But these woman-hating acts are rarely discussed as crimes against Black women.

As a group, women of Color are the lowest paid wage earners in america. We are the primary targets of abortion and sterilization abuse, here and abroad. In certain parts of Africa, small girls are still being sewed shut between their legs to keep them docile and for men's pleasure. This is known as female circumcision, and it is not a cultural affair as the late Jomo Kenyatta insisted, it is a crime against Black women.

Black women's literature is full of the pain of frequent assault, not only by a racist patriarchy, but also by Black men. Yet the necessity for and history of shared battle have made us, Black women, particularly vulnerable to the false accusation that anti-sexist is anti-Black. Meanwhile, womanhating as a recourse of the powerless is sapping strength from Black communities, and our very lives. Rape is on the increase, reported and unreported, and rape is not aggressive sexuality, it is sexualized aggression. As Kalamu ya Salaam, a Black male writer points out, "As long as male domination exists, rape will exist. Only women revolting and men made conscious of their responsibility to fight sexism can collectively stop rape."*

Differences between ourselves as Black women are also being misnamed and used to separate us from one another. As a Black lesbian feminist comfortable with the many different ingredients of my identity, and a woman committed to racial and sexual freedom from oppression, I find I am constantly being encouraged to pluck out some one aspect of myself and present this as the meaningful whole, eclipsing or denying the other parts of self. But this is a destructive and fragmenting way to live. My fullest concentration of energy is available to me only when I integrate all the parts of who I am, openly, allowing power from particular sources of my living to flow back and forth freely through all my different selves, without the restrictions of externally imposed definition. Only then can I bring myself and my energies as a whole to the service of those struggles which I embrace as part of my living.

A fear of lesbians, or of being accused of being a lesbian, has led many Black women into testifying against themselves. It has led some of us into destructive alliances, and others into despair and isolation. In the white women's communities, heterosexism is sometimes a result of identifying with the white patriarchy, a rejection of that interdependence between women-identified women which allows the self to be, rather than to be used in the service of men. Sometimes it reflects a die-hard belief in the protective coloration of heterosexual relationships, sometimes a self-hate which all women have to fight against, taught us from birth.

---

*From "Rape: A Radical Analysis, An African-American Perspective" by Kalamu ya Salaam in *Black Books Bulletin*, vol. 6, no. 4 (1980).

Although elements of these attitudes exist for all women, there are particular resonances of heterosexism and homophobia among Black women. Despite the fact that woman-bonding has a long and honorable history in the African and African-american communities, and despite the knowledge and accomplishments of many strong and creative women-identified Black women in the political, social and cultural fields, heterosexual Black women often tend to ignore or discount the existence and work of Black lesbians. Part of this attitude has come from an understandable terror of Black male attack within the close confines of Black society, where the punishment for any female self-assertion is still to be accused of being a lesbian and therefore unworthy of the attention or support of the scarce Black male. But part of this need to misname and ignore Black lesbians comes from a very real fear that openly women-identified Black women who are no longer dependent upon men for their self-definition may well reorder our whole concept of social relationships.

Black women who once insisted that lesbianism was a white woman's problem now insist that Black lesbians are a threat to Black nationhood, are consorting with the enemy, are basically un-Black. These accusations, coming from the very women to whom we look for deep and real understanding, have served to keep many Black lesbians in hiding, caught between the racism of white women and the homophobia of their sisters. Often, their work has been ignored, trivialized, or misnamed, as with the work of Angelina Grimke, Alice Dunbar-Nelson, Lorraine Hansberry. Yet women-bonded women have always been some part of the power of Black communities, from our unmarried aunts to the amazons of Dahomey.

And it is certainly not Black lesbians who are assaulting women and raping children and grandmothers on the streets of our communities.

Across this country, as in Boston during the spring of 1979 following the unsolved murders of twelve Black women, Black lesbians are spearheading movements against violence against Black women.

What are the particular details within each of our lives that can be scrutinized and altered to help bring about change? How do we redefine difference for all women? It is not our differences which separate women, but our reluctance to recognize those differences and to deal effectively with the distortions which have resulted from the ignoring and misnaming of those differences.

As a tool of social control, women have been encouraged to recognize only one area of human difference as legitimate, those differences which exist between women and men. And we have learned to deal across those differences with the urgency of all oppressed subordinates. All of us have had to learn to live or work or coexist with men, from our fathers on. We have recognized and negotiated these differences, even when this recognition only continued the old dominant/subordinate mode of human relationship, where the oppressed must recognize the masters' difference in order to survive.

But our future survival is predicated upon our ability to relate within equality. As women, we must root out internalized patterns of oppression within ourselves if we are to move beyond the most superficial aspects of so-cial change. Now we must recognize differences among women who are our equals, neither inferior nor superior, and devise ways to use each others' dif-ference to enrich our visions and our joint struggles.

The future of our earth may depend upon the ability of all women to iden-tify and develop new definitions of power and new patterns of relating across difference. The old definitions have not served us, nor the earth that supports us. The old patterns, no matter how cleverly rearranged to imitate progress, still condemn us to cosmetically altered repetitions of the same old exchanges, the same old guilt, hatred, recrimination, lamentation, and suspicion.

For we have, built into all of us, old blueprints of expectation and re-sponse, old structures of oppression, and these must be altered at the same time as we alter the living conditions which are a result of those structures. For the master's tools will never dismantle the master's house.

As Paulo Freire shows so well in *The Pedagogy of the Oppressed*,* the true focus of revolutionary change is never merely the oppressive situations which we seek to escape, but that piece of the oppressor which is planted deep within each of us, and which knows only the oppressors' tactics, the oppressors' relationships.

Change means growth, and growth can be painful. But we sharpen self-definition by exposing the self in work and struggle together with those whom we define as different from ourselves, although sharing the same goals. For Black and white, old and young, lesbian and heterosexual women alike, this can mean new paths to our survival.

> *We have chosen each other*
> *and the edge of each others battles*
> *the war is the same*
> *if we lose*
> *someday women's blood will congeal*
> *upon a dead planet*
> *if we win*
> *there is no telling*
> *we seek beyond history*
> *for a new and more possible meeting.†*

---

*Seabury Press, New York, 1970.
†From "Outlines," unpublished poem.

Thinking Further

After reading the articles in this section, you will find it helpful to think about the following:

As you begin to think about race, class, and gender, it is helpful to center your thinking in the lives of people different from yourself. Suppose that you were writing a narrative like those in this section. Where do you see connections between your experiences and those of any of these authors? How have those experiences been shaped by the social and historical context of your life and how has that context been shaped by race, class, gender, or any of the other factors revealed in these narratives?

Suggested Readings

Collins, Patricia Hill. 2000. *Black Feminist Thought: Knowledge, Consciousness, and the Politics of Empowerment*, 2d ed. New York: Routledge.

Romero, Mary, Pierrette Hodagneu-Sotelo, and Vilma Ortiz, eds. 1997. *Challenging Fronteras: Structuring Latina and Latino Lives in the U.S.* New York: Routledge.

Takaki, Ronald. 1993. *A Different Mirror: A History of Multicultural America*. Boston: Little, Brown.

 ## *InfoTrac College Edition: Search Terms*

Use your access to the online library, InfoTrac College Edition, to learn more about subjects covered in this section. The following are some suggested terms for searches:

African American studies
ethnic studies
feminist studies
multicultural studies/multiculturalism

 ## *InfoTrac College Edition: Bonus Reading*

You can use the InfoTrac College Edition feature to find an additional reading pertinent to this section and can also search on the author's name or keywords to locate the following reading:

Guy-Sheftall, Beverly. 1997. "Whither Black Women's Studies: Interview." By Evelynn M. Hammonds. *differences: A Journal of Feminist Cultural Studies* 9 (Fall 1997): 31–45.

Beverly Guy-Sheftall is a faculty member at Spelman College—a historically Black women's college. She discusses the specific challenges of developing Black women's studies and also articulates the importance of these studies for feminism in general. How does developing such an inclusive curriculum shift the assumptions of more exclusionary thinking?

Wadsworth Sociology Resource Center: Virtual Society

For additional links, quizzes, and learning tools related to this section, see the Wadsworth Sociology Resource Center, Virtual Society, at www.wadsworth .com/sociology_d/

II

Conceptualizing Race, Class, and Gender

Understanding the intersections between race, class, and gender requires knowing how to conceptualize each. Although we would rather not treat them separately, it is necessary to do so to learn what each means and how each is manifested in different group experiences. In this part, we analyze race, class, and gender, though the readings in this section also examine the connections among them. As we review each in turn, you will also notice several common themes.

First, each is a socially constructed category. That is, their significance stems not from some "natural" state, but from the significance they have taken on as the result of social and historical processes. Second, notice how each tends to construct groups in binary (or polar opposite) terms: "man/woman" or "Black/White" or "rich/poor," thereby creating the "otherness" that we examined in "Shifting the Center." Third, each is a category of individual and group identity, but note—and this is important—they are also social structures. That is, they are not just about identity but are about group location in a system of stratification and institutional forms. Thus, in examining race, class, and gender, it is important to study patterns in the labor market, family structures, state institutions (such as the government and the law), mass media, and so forth. This is a key difference, as we have seen, in a model that focuses solely on difference and one that focuses on the matrix of domination. Finally, neither race, class, nor gender is a fixed category. Because they are social constructions, their form—and their interrelationship—changes over time. This also means that social change is possible.

As you learn about race, class, and gender, you should keep the intersectional model in mind. Although we will be focusing on each one in turn, we continue to emphasize how they are interrelated. To picture this, think of a typical college basketball game. This will probably seem familiar: the players on the court, the cheerleaders moving about on the side, the band playing, fans cheering, boosters watching from the best seats, and—if the team is ranked—perhaps a television crew. Everybody seems to have a place in the game. Everybody seems to be following the rules. But what are the "rules" of this game? What explains the patterns that we see and don't see?

Race clearly matters. The predominance of young African American men on many college basketball teams is noticeable. Why do so many young Black men play basketball? Some people argue that African Americans are better in areas requiring physical skills such as sports and are less capable of doing intellectual work in fields such as physics, law, and medicine. Others look to Black culture for explanations, suggesting that African Americans would be perfectly happy just playing ball and partying. But these perspectives fail to take into account the continuing effects of racism. Lack of access to decent jobs, inadequate housing, poor-quality education, and insufficient health care are also manifestations of the systematic disadvantage of race—and class. Thus, rates of poverty for African Americans and Hispanics are higher than for other groups. In 2001, 23 percent of African Americans, 21 percent of Hispanics, and 10 percent of Asian Americans were poor, compared with 8 percent of non-Hispanic whites (Proctor and Dalaker 2002). For young Black men growing up in communities with few opportunities, sports are perceived as an attractive mobility route. Perceived promises of high salaries, endorsements, and merchandise can make young people believe sports are a path to success. One study has found that two-thirds of African American boys between ages thirteen and eighteen believe they can earn a living playing professional sports. But, the odds of actually doing so are extremely slim. Of the 40,000 African American boys playing high school basketball, only thirty-five will make it to the NBA (National Basketball Association) and only seven of those will be starters (Eitzen 1999). This makes the odds of success 0.000175!

But, as important as race is, does a racial analysis fully explain the "rules" of college basketball? Not really. Black men are not the only players. White men also play college basketball, raising questions about the significance of social class in explaining a basketball game. Who benefits from college basketball? Yes, players get scholarships and are offered a chance to earn college degrees, so players reap the rewards, but this misses the point of who really benefits. College athletics is big business, and the players make far less from it than many people believe. As amateur athletes, they are forbidden to take any payment for their skills. They are offered the hope of an NBA contract when they turn pro, or at least a college degree if they graduate. But, few actually turn pro. Indeed, few even graduate from college. Among Division I Black basketball players, only 35 percent graduate from college, compared with 53 percent of White male basketball players. Interestingly, however, graduation rates among Black college athletes (including all sports) are higher than among nonathletes, most likely because of the scholarship support they receive. (Black basketball players have the same graduation rates as all Black students.) Women student-athletes also have higher graduation rates than nonathletes and higher graduation rates than male student-athletes. This is true regardless of race (National Collegiate Athletic Association 2002).

So who actually benefits from college basketball? The colleges that recruit the athletes certainly benefit. For the university, winning teams garner increased admissions applications, alumni giving, corporate support, and television revenues. Athletics is also a big business. Corporate sponsors want their names and products identified with winning teams and athletes; advertisers want their products promoted by members of winning teams. Even though college athletes are forbidden to promote products, corporations create and market products in conjunction with prevailing excitement about basketball, sustained by the players' achievements. Products such as athletic shoes, workout clothing, cars, and beer all target the consumer dollars of those who enjoy watching basketball. Also, consider how many full-time jobs are supported by the revenues generated from the enterprise of college basketball. Referees, sports reporters—both at the games and on local media outlets—athletic trainers, coaches, and health personnel all benefit. Unlike the players, these people all get paid for their contributions to college basketball. Thus, class matters. The companies and organizations that profit the most—whether schools, product manufacturers, advertisers—are part of a class system where there are differential benefits depending on your "rank" within that system.

So, do race and class fully explain the "rules" of basketball? Sometimes what we don't see can be just as revealing as what we do see. One other feature of the game on the court is so familiar that it may go unquestioned— or even unseen. Where are the women in college basketball (or pro basketball, for that matter)? Only in a few schools does women's basketball draw as large an audience as men's. And certainly in the media, men's basketball is generally the public's focal point, even though women's sports are increasingly popular. In college basketball, like the pros, most of the coaches and support personnel are men, as are the camera crew and announcers.

Where are the women? A few are coaches, rarely paid what the men receive—even on the most winning teams. Those closest to the action on the court may be cheerleaders—tumbling, dancing, and being thrown into the air in support of the exploits of the athletes. Others may be in the band. Some women are in the stands, cheering the team—many of them accompanied by their husbands, partners, boyfriends, parents, and children. Many work in the concession stands, fulfilling women's roles of serving others. Still others are even more invisible, left to clean the restrooms, locker rooms, and stands after the crowd goes home. Women remain on the sidelines in other ways as well. The treatment of women basketball players differs markedly from that of their male counterparts: Women have many fewer opportunities for scholarships and professional careers in athletics. The centrality of men's activities in basketball mirrors the centrality afforded men's activities in society as a whole; thus, women's seeming invisibility in basketball ironically highlights the salience of gender.

Men's behavior reveals a gendered dimension to basketball, as well. Where else are men able to put their arms around each other, slap one another's buttocks, hug each other, or cry in public without having their "masculinity" questioned? Sportscasters, too, bring gender into the play of sports, such as when they talk about men's heroic athletic achievements but talk about women athletes' looks or their connection to children. For that matter, look at the prominence given to men's teams in sports pages of the daily newspaper, compared with sports news about women, who are typically relegated to the back pages—if their athletic accomplishments are reported at all.

This discussion of college basketball demonstrates how race, class, and gender each provide an important, yet partial, perspective on the action on the court. If we use an intersectional model, we not only see each of them in turn but also the connections among them. In fact, race, class, and gender are so inextricably intertwined that gaining a comprehensive understanding of a basketball game requires thinking about all of them and how they work together—in other words, thinking inclusively. Then you will ask why most of those serving the food in concession stands are likely to be women and men of color. How are norms of masculinity played out through sport? What class and racial ideologies are promoted through assuming that sports are a mobility route for those who try hard enough? Building from the example of basketball, you might then ask, "If race, class, and gender relations are embedded in something as familiar and widespread as college basketball, to what extent are other social practices, institutions, relations, and social issues similarly structured?"

Race, gender, and class divisions are deeply embedded in the structure of social institutions such as work, family, education, and the state. They shape human relationships, identities, social institutions, and the social issues that emerge from within institutions. Evelyn Nakano Glenn postulates that you can see the intersections of race, class, and gender in three realms of society: the representational realm, the realm of social interaction, and the social structural realm. The representional realm includes the symbols, language, and images that convey racial meanings in society; social interaction refers to the norms and behaviors observable in human relationships; the social structural realm involves the institutional sites where power and resources are distributed in society (Glenn 2002: 12).

This means that race, class, and gender affect all levels of our experience—our consciousness and ideas, our interaction with others, and the social institutions we live within. And, because they are interrconnected, no one can be subsumed under the other. In this section of the book, although we focus on each one to provide conceptual grounding, keep in mind that they are connected and overlapping—in all three realms of society: the realm of ideas, interaction, and institutions.

You might begin by considering a few facts:

- The United States is in the midst of a sizable redistribution of wealth, with a greater concentration of wealth and income in the hands of a few than at most previous periods of time. At the same time, a declining share of income is going to the middle class—a class that finds its position slipping, relative to years past (Krugman 2002; DeNavas-Walt and Cleveland 2002).
- Within class groups, racial group experiences are widely divergent. Thus, although there has been substantial growth of an African American and Latino middle class, they have a more tenuous hold on this class status

than groups with more stable footing in the middle class. Furthermore, there is significant class differentiation within different racial groups (Pattillo-McCoy 1999; Massey 1993).

- Women in the top 25 percent of income groups have seen the highest wage growth of any group over the last twenty years; the lowest earning groups of women, like men, have seen wages fall while the middle has remained flat (Mishel, Bernstein, and Schmitt 2001). Class differences within gender are hidden by thinking of women as a monolithic group.

- Women of color, including Latinas, African American women, Native American women, and Asian American women are concentrated in the bottom rungs of the labor market along with recent immigrant women (U.S. Department of Labor 2002).

- Although poverty in the United States had been on the decline since 1993, it is now rising. Poverty is particularly severe among women, especially among women of color and their children (Proctor and Dalaker 2002).

- While the mass media extol the virtues of recent reforms in welfare legislation and herald a "decline in the welfare rolls," studies show that increases in family income among former welfare recipients are meager, and there has been an increase in the number of such families evicted from housing because of falling behind on rent. Families also report an increase in other material hardships—phones and utilities being cut off, for example (Lewis, Stephens, and Slack 2002; Acker, Morgen, and Gonzales 2002).

- Welfare reform is only one dimension of the shrinkage of social support systems from federal and state assistance. The shrinkage of social support is not only affecting the very poor, however. Job benefits in the form of health insurance, pensions, and so forth for all workers have declined. Following job loss, less than half of U.S. workers are currently eligible for unemployment insurance (Emsellem et al. 2002).

- At both ends of the economic spectrum there is a growth of gated communities: well-guarded, locked neighborhoods for the rich and prisons for the poor—particularly Latinos and African American men. At the same time, growth in the rate of imprisonment is highest among women (Harrison and Beck 2002; Collins and Veskel 2000).

None of these facts can be explained through an analysis that focuses only on class or race or gender. Clearly, class matters. Race matters. Gender matters. And they matter together.

RACE AND RACISM

In this volume we examine race, class, and gender relations from an institutional or structural perspective. Locating racial oppression in the structure of social institutions provides a different frame of analysis from what would be obtained by analyzing individuals only. *Individual racism* is one person's belief in the superiority of one race over another. Individual racism is related to prejudice, a hostile attitude toward a person who is presumed to have negative characteristics associated with a group to which he or she belongs. Racism is more systematic than this, however.

Racism is not the same thing as prejudice. *Prejudice* refers to people's attitudes. *Racism* is a system of power and privilege; it can be manifested in people's attitudes but is rooted in society's structure and is reflected in the different advantages and disadvantages that groups experience, based on their location in this societal system. Racism is structured into society, not just in people's minds. As such, it is built into the very fabric of dominant institutions in the United States and has been since the founding of the nation. Joe Feagin refers to this as *systemic racism*, meaning the "complex array of antiblack practices, the unjustly gained political-economic power of whites, the continuing economic and other resource inequalities along racial lines, and the white racial ideologies and attitudes created to maintain and rationalize white privilege and power" (2000: 6).

In this definition of institutional racism, notice first that racism is part of society's structure, not just present in individual bigots. Seen in this light, people may not be individually racist but can still benefit from a system that is organized to benefit some at the expense of others. As Gloria Yamato discusses in "Something about the Subject Makes It Hard to Name," racism can be intentional or unintentional. In a racist system, well-meaning White people benefit from racism even if they have no intention of acting or thinking like a "racist." Thus, institutional racism creates a built-in system of privilege. As Yamato suggests, different groups internalize it in different forms of consciousness. Peggy McIntosh's essay, "White Privilege," describes how the system of racial privilege becomes invisible to those who benefit from it, even though it structures the everyday life of both White people and people of color.

Second, racism shapes everyday social relations. In other words, despite its systematic nature, institutional racism depends on the presence of individual racists acting daily in order to continue. If you are a person of color, even being middle class may not protect you from the everyday realities of racism

(Pattillo-McCoy 1999). Patricia J. Williams, a noted African American legal scholar, illustrates this in her discussion of persistent discrimination in housing ("Of Race and Risk"). Despite her middle-class status, systemic racism confronts her—and other African Americans—in daily encounters. Practices of everyday racism are part of the edifice of institutional racism; yet we often misread their meaning.

Many people believe that being nonracist means being color-blind—that is, refusing to recognize or treat as significant a person's racial background and identity. But to ignore the significance of race in a society where racial groups have distinct historical and contemporary experiences is to deny the reality of their group experience. Being color-blind in a society structured on racial privilege means assuming that everybody is "White," which is why people of color might be offended by friends who say, for example, "But I never think of you as Black." Such practices of everyday racism are powerful because, instead of seeing them as components of patterns of institutional racism, we experience these interactions as ordinary occurrences.

Discrimination is one of the driving forces of racism. Though perhaps not as overt as, for example, during Jim Crow segregation in the South, discrimination can still be seen in various patterns and practices. Indeed, segregation, though not mandated by law as it was during Jim Crow, is as stark as ever and, in many cities, has actually increased over recent years with huge inequities in schooling and housing as a result. Research studies known as audit studies also show that people continue to discriminate based on race, even though they will not overtly say so. In audit studies, researchers, one White and one Black, are matched in credentials and appearance and they pose—in person—as job or housing applicants. These studies find significant discrimination is an ongoing fact. White job applicants in such studies are offered the job almost half of the time; Black applicants, only 11 percent of the time. White applicants are often told things such as "You are just what we are looking for"—Black applicants, on the other hand, as not having the right attributes for the job. Studies of employers have also found that employers hold considerable stereotypes about Black workers that prevent them from hiring them (see the article by Philip Moss and Chris Tilly in Part III as one example; also, Moss and Tilly 2001).

Another dimension of racism is that its forms change over time. Racial discrimination is no longer legal, but racism nonetheless continues to structure relations between groups and to differentiate the power that different groups have. The changing character of racism is also evident in the fact that specific racial group histories differ, but different racial groups share common experiences of racial oppression. Thus, Chinese Americans were never enslaved, but they experienced forced residential segregation and economic exploitation based on their presumed racial characteristics. Mexican Americans were never placed on federal reservations as Native Americans have been, but in some regards both groups share the experience of colonization by White settlers. Both have experienced having their lands appropriated by White settlers—Native Americans as they were removed from their lands and forced into reservations, if not killed. Chicanos originally held land in what is now the American Southwest, but it was taken following the Mexican American War; in 1848 Mexico ceded huge parts of what are now California, New Mexico, Nevada, Colorado, Arizona, and Utah to the United States for $15 million. Mexicans living there were one day Mexicans, the next living in the United States, though without all the rights of citizens.

In a racist society, the very meaning of race reflects institutionalized racist practices and beliefs. Most people assume that race is biologically fixed, an assumption that is fueled by arguments about the presumed biological basis for different forms of inequality. But the concept of race is more social than biological; scientists working on the human genome project have even found that there is no "race" gene. But, you should not conclude from this that race is not "real." It is just that its reality stems from its social significance. That is, the meaning and significance of race stems from specific social, historical, and political contexts. It is these contexts that make race meaningful, not just whatever physical differences may exist between groups.

To understand this, think about how racial categories are created, by whom, and for what purposes. Racial classification systems reflect prevailing views of race, thereby establishing groups that are presumed to be "natural." These constructed racial categories then serve as the basis for allocating resources; furthermore, once defined, the categories frame political issues and conflicts (Omi and Winant 1994). Omi and Winant define *racial formation* as "the sociohistorical process by which racial categories are created, inhabited, transformed, and destroyed" (1994: 55). In Nazi Germany, Jews were considered to be a race—a social construction that became the basis for the Holocaust. Abby L. Ferber's essay ("What White Supremacists Taught a Jewish Scholar about Identity") shows the complexities that evolve in the social construction of race. As someone who studies White supremacist groups, she sees how White racism defines her as Jewish, even while she lives in society as White. Her reflections reveal, too, the interconnections between racism and anti-Semitism (the hatred of Jewish people), reminding us of the interplay between different systems of oppression.

In understanding racial formation, we see that societies construct rules and practices that define groups in racial terms. Moreover, racial meanings constantly change as institutions evolve and as different groups contest prevailing racial definitions. Some groups are "racialized"; others, are not. Where, for example, did the term *Caucasian* come from? Although many take it to be "real" and don't think about its racist connotations, the term has quite racist origins. It was developed in the late eighteenth century by a German anthropologist, Johann Blumenbach. He developed a racial classification scheme that put people from the Russian Caucasus at the top of the racial hierarchy because he thought "Caucasians" were the most beautiful and sophisticated people; darker people were put on the bottom of the list: Asians, Africans, Polynesians, and Native Americans (Hannaford 1996). It is amazing when you think about it that this term remains with us, with few questioning its racist connotations.

Consider also the changing definitions of race in the U.S. census. Given the large number of multiracial groups and the increasing diversity brought about by immigration, we can no longer think of race in mutually exclusive terms. In 1860, only three "races" were presumed to exist—Whites, Blacks, and mulattoes. By 1890, however, these original three "races" had been joined by five others—quadroon, octoroon, Chinese, Japanese, and Indian. Ten short years later, this list shrank to five "races"—White, Black, Chinese, Japanese, and Indian—a situation reflecting the growth of strict segregation in the South (O'Hare 1996). Now people of mixed racial heritage present a challenge to census classifications. In the 2000 census, the U.S. government for the first time allowed people to check multiple boxes to identify themselves as more than one race. In addition, you could check "Hispanic" as a separate category. This change in the census reflects the growing number of multiracial people in the United States. The census categories are not just a matter of accurate statistics; they have significant consequences in the apportionment of societal resources. Thus, while some might argue that we should not "count" race at all, doing so is important because data on racial groups are used to enforce voting rights, to regulate equal employment opportunities, and to determine various governmental supports, among other things. Although it may be easy for some people to say, "Race doesn't matter; we should be a color-blind society," Cornel West argues that race clearly matters and it matters a lot.

Shifting definitions of race are grounded in shifting relations of power. Recent decades have seen additional revisions to definitions of race. In particular, the experiences of Latino groups in the United States challenge long-standing racial categories of "Black" and "White." Elizabeth Martinez (in "Seeing More than Black and White") notes that White-Black relations have defined racism in the United States for centuries but that a rapidly changing population that includes diverse Latino groups is forcing Americans to reconsider the nature of racism. Color, she argues, has been the marker of race, but she challenges the dualistic thinking that has promoted this racist thinking.

The overarching structure of racial power relations means that placement in this structure leads to differences in outlook regarding the very presence of racism and what can be done about it. The reappearance of racial hostilities on college campuses is certainly evidence of the continuation of racist practices and beliefs; yet despite this and other evidence, Whites continue to be optimistic in their assessment of racial progress. They say that they are tired of hearing about racism and that they have done all they can to eliminate racial discrimination. People of color are less sanguine about racial progress and are more aware of the nuances of racism. Marked differences by race are still evident in employment, political representation, schooling, and other basic measures of group well-being.

Racism does not exist in a vacuum. As we have said, race, gender, and class are intersecting systems—experienced simultaneously, not separately. It is a mistake to think of any one category in the absence of the others. People's experiences with race and racism are framed by their location in this overarching system of race, class, and gender privileges and penalties. Race possesses not only objective dimensions that result from institutional racism; it also has subjective dimensions that relate to how people experience it. For example, some people of color have class privilege; yet this does not eliminate racism, as Williams's experience in getting a mortgage shows us. Although class differentiation has increased within racial groups and such class difference is significant, this does not mean that people are immune from the effects of racism. Class differences within racial groups show how race and class together configure group experiences differently. All people of color encounter institutional racism, but their actual experiences with racism vary, depending on social class, gender, age, sexuality, and other markers of social position.

CLASS

Like race, the social class system is grounded in social institutions and practices. Rather than thinking of social class as a rank held by an individual, think of social class as a series of relations that pervade the entire society and shape our social institutions and relationships with one another. Although class shapes identity and individual well-being, class is a system that differentially structures group access to economic, political, cultural, and social resources. Within the United States, the class system evolves from patterns of capitalist development, and those patterns intersect with race and gender.

To begin with, the class system in the United States is marked by striking differences in income. *Income* is the amount of money brought into a household in one year. Measures of income in the United States are based on annually reported census data drawn from a sample of the population. These data show quite dramatic differences in class standing when taking gender and race into account. *Median income* is the income level above and below which half of the population lies. It is the best measure of group income standing. Thus in 2001, median income for non-Hispanic White households was $46,305 (meaning half of such households earned more than this and half below); this is the "middle." Black households had a median income of $29,470; Hispanic households, $33,565; Asian and Pacific Islander households, $53,635 (DeNavas-Walt and Cleveland 2002; see also Figure 1).

But this tells only part of the story. Household income is the income of a total household. What about individual earners? This is where you can see the confounding influence of gender. Among workers who were employed full-time and year-round in 2001, White men earned $43,194; Asian/Pacific Islander men, $42,695; Black men, $31,921; White non-Hispanic women, $31,794; Asian/Pacific Islander women, $31,284; Black women, $27,297; Hispanic men, $25,271; Hispanic women, $21,973 (see also Figure 2). Note that this array of income levels does not fall solely along lines of race or gender, because Black men, in the aggregate, earn slightly more than White, non-Hispanic women, but Black women earn more than Hispanic men.

Something to keep in mind is that because household income results from the income of individual workers, some households need more workers than others to reach median levels of income. Also, current data show that the most important source of income growth for all households is the increased number of hours that people are working. Black and Hispanic

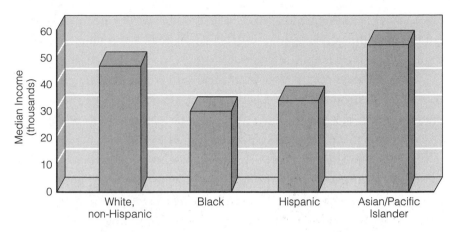

Source: DeNavas-Walt, Carmen, and Robert W. Cleveland. 2002. *Money Income in the United States: 2001.* Washington, DC: U.S. Census Bureau, pp. 16–18.

FIGURE 1 Median Household Income 2001 (by race)

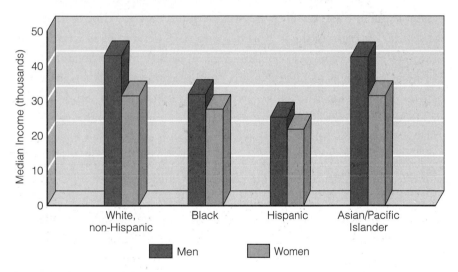

Source: U.S. Census Bureau. 2002. *Income: Detailed Historical Tables, Tables P36B-P36E.* Washington, DC: U.S. Census Bureau. Web site: www.census.gov

FIGURE 2 Median Income for Year-Round, Full-Time Workers 2001 (by race and gender)

families work more hours than White families; the greatest increase in working hours is among women of all races (Mishel, Bernstein, and Schmitt 2001).

Even more significant than income differences in revealing class are differences in patterns of wealth. *Wealth* is determined by adding all of one's financial assets and subtracting all debt. Income and wealth are related but are not the same thing. As important as income can be in determining one's class status, wealth is even more significant.

Consider this: Imagine two recent college graduates. They graduate in the same year, from the same college, with the same major and the identical grade point average. Both get jobs with the same salary in the same company. But, one student's parents paid all college expenses and gave her a car upon graduation. The other student worked while in school and has graduated with substantial debt from student loans. This student's family has no money to help support the new worker. Who is better off? Same salary, same credentials, but one person has a clear advantage—one that will be played out many times over as the young worker buys a home, finances her own children's education, and possibly inherits additional assets. This shows you the significance of wealth—not just income—in structuring social class.

Thus, income data indicate quite dramatic differences in class, race, and gender standing. But, furthermore, wealth differences are also startling. The wealthiest 1 percent of the population control 38 percent of all wealth—the bottom 80 percent, only 17 percent (Mishel, Bernstein, and Schmitt 2001). For most Americans, debt, not wealth, is more common. Furthermore, one-quarter of White households, 61 percent of Black households, and 54 percent of Hispanic households have no financial assets at all (Oliver and Shapiro 1995)—indicative of the vast differences in wealth holdings among different racial groups. In fact, the median net worth of White households is more than ten times that of African American and Latino households.

Wealth is especially significant because it provides a *cumulative* advantage to those who have it. Wealth helps pay for college costs for children and down payments on houses; it can cushion the impact of emergencies, such as unexpected unemployment or sudden health problems, as Dalton Conley explains in "Wealth Matters." He shows how even small amounts of wealth can provide the cushion that averts economic disaster for families. Buying a home, investing, being free of debt, sending one's children to college, and transferring economic assets to the next generation are all instances of class advantage that add up over time and produce advantage even beyond one's current income level. Sociologists Melvin Oliver and Thomas Shapiro (1995) have found, for example, that even Black and White Americans at the same income level, with the same educational and occupational assets, still have a substantial difference in their financial assets—an average difference of $43,143 per year! This means that, even when earning the same income, the two groups are in quite different class situations—although both may be considered "middle class." Furthermore, wealth produces more wealth, because inheritance allows people to transmit economic status from one generation to the next. This results in "the sedimentation of racial inequality" (Oliver and Shapiro 1995: 5).

Social class is a complex system. There are wide differences in the class status of Whites and people of color, but we should be careful not to see all Whites and Asian Americans as well-off and all African Americans, Native Americans, and Latinos as poor. Consider the range of social class experiences just among Whites. Although on average White households possess higher accumulated wealth and have higher incomes than Black, Hispanic, and Native American households, large numbers of White households do not. White people also account for 46 percent of the nation's poor (Proctor and Dalaker 2002). In addition, class experiences across racial groups can vary widely, as shown by Mary Pattillo-McCoy's research on the Black middle class in "Black Picket Fences." She shows that the Black middle class continues to experience racial segregation and, as one result, is more exposed to the risks that the Black poor experience than would be true of the White middle class.

These facts should caution us about conclusions based on *aggregate data* (that is, data that represent whole groups). Such data give you a broad picture of group differences, but they are not attentive to the more nuanced picture you see when taking into account class, race, and gender (along with other factors, such as age, level of education, occupation, and so forth). Aggregate data on Asian Americans, for example, show them as a group to be relatively well off. But this portrayal, like the stereotypes of the "model minority," obscures significant differences both when comparing Asian Americans with other groups and among Asian American groups. So, for example, although Asian American median income is—in the aggregate—higher than for White Americans, this does not mean all Asian American families are better off than White families. If you look at poverty rates, you get a different picture. Ten percent of Asian American/Pacific Islanders are poor, compared with a little less than 8 percent of White, non-Hispanic families (Proctor and Dalaker 2002). The proportion of Asian Americans living in poverty has also increased substantially since the 1980s, particularly among the most recent immigrant groups, including Laotians, Cambodians, Vietnamese, Chinese, and Korean immigrants. Filipino and Asian Indian families had lower rates of poverty (Lee 1995).

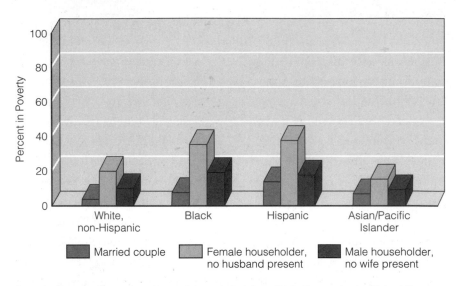

Source: Proctor, Bernadette D., and Joseph Dalaker, 2002. *Poverty in the United States: 2001.* Washington, DC: U.S. Census Bureau, p. 3.

FIGURE 3 Median Household Income 2001 (by race)

Poverty is a problem for many groups (see Figure 3). Women and their children are especially hard hit by poverty. Thirty-seven percent of Hispanic families headed by women are poor, as are 35 percent of Black families, 15 percent of Asian/Pacific Islander families, and 22 percent of White families headed by women. Poverty rates among children are especially disturbing: 16 percent of all children in the United States live below the *poverty line* ($17,960 in 2001 for a family of four, including two children). When adding race, the figures are even more disturbing: 30 percent of African American children, 28 percent of Hispanic children, 11.5 percent of Asian/Pacific Islander children, and 9.5 percent of White (non-Hispanic) children (those under 18 years of age) are poor—astonishing figures for one of the most affluent nations in the world (Proctor and Dalaker 2002). Simplistic solutions suggested by current welfare policy imply that women would not be poor if they would just get married or get a job. But, as James Jennings and Louis Kushnick argue in "Poverty as Race, Power, and Wealth," the root causes of poverty lie in the distribution of wealth and capital, coupled with low wages and high unemployment among certain groups—groups whose social location is the result of race, gender, and class stratification. Keep in mind that among the poor, 38 percent were working (11.5 percent worked full-time, year-round). And these figures count only those whose earnings fell below the official poverty line. If you calculate the income received by someone working full-time (40 hours a week) and year-round (52 weeks, no vacation) at the federal minimum wage ($5.50 per hour), you will see that the dollars earned ($11,440) do not even come close to the federal poverty line for a family of four ($17,960 in 2001).

In the United States, the social class system is also marked by differences in power. Social class is not just a matter of material difference; it is a pattern of domination in which some groups have more power than others. *Power* is the ability to influence and dominate others. This means not just interpersonal power but refers to the structural power that some groups have because of their position in the class system. Groups with vast amounts of wealth, for example, have the ability to influence systems like the media and the political process in ways that less powerful groups cannot. Privilege in social class thus encompasses both a position of material advantage and the ability to control and influence others.

The class system is currently undergoing some profound changes, as detailed by Collins and Veskel ("Economic Apartheid in America"). These changes are intimately linked to patterns of economic transformation in the political economy—changes that are both global and domestic. Jobs are being exported overseas as vast multinational corporations seek to enhance their profits by promoting new markets and cutting the cost of labor. Within the United States, there is a shift from a manufacturing-based economy to a service economy, with corresponding changes in the types of jobs available and the wages attached to these jobs. Fewer skilled, decent-paying manufacturing jobs exist today than in the past. Fewer workers are covered by job benefits and unemployment insurance; only 40 percent of workers are now eligible for unemployment following job loss (Mishel, Bernstein, and Schmitt 2001; Emsellem et al. 2002). Wages are flat for most workers, except those at the very top. Millions of people are left with jobs that do not pay enough, in part-time or temporary work, or without any work at all.

All told, class divisions in the United States are becoming more marked. There is a growing gap between the "haves" and "have-nots." Income growth has been greatest for those at the top end of the population—the upper 20 percent and the upper 5 percent of all income groups, regardless of race. For everyone else, income growth has remained flat. Although in every racial group, the top earners have seen the most growth in income, Black

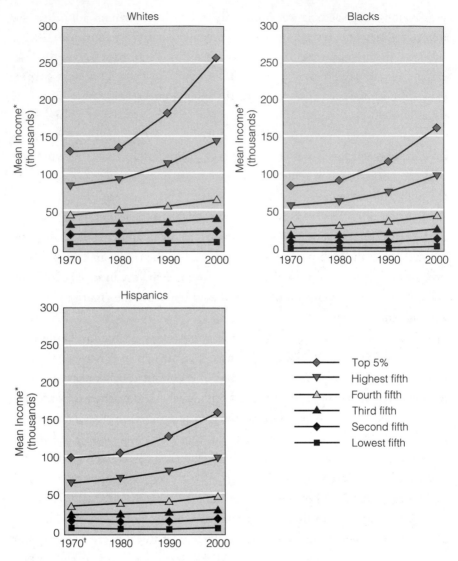

Source: U.S. Census Bureau. 2002. *Historical Income Tables—Households.* Web site: www.census.gov/hhes/income/histinc/

FIGURE 4 Mean Income* (thousands)

and Hispanic high-earners still earn less overall than Whites (see Figure 4.) At the same time, in the nation's cities and towns, homelessness has become increasingly apparent even to casual observers. The number of home-

---

*In 2000 CPI-adjusted dollars.

†Data not available for Hispanics in 1970; 1972 data used.

less in a given year is estimated to be about two million, with families being the largest segment of the homeless population. Half of the homeless are African American; about 20 percent are children (National Coalition for the Homeless 2002).

If social class is so important in shaping life chances, why don't more people realize its significance? The answer lies in how dominant groups use ideology to explain the class system and other systems of inequality. Ideology created by dominant groups refers to a system of beliefs that simultaneously distort reality and justify the status quo. As we learn in "Tired of Playing Monopoly?" by Donna Langston, the class system in the United States has been supported through the myth that we live in a classless society. This myth serves the dominant class, making class privilege seem like something that one earns, not something that is deeply embedded in the institutions of society. Langston also suggests that the system of privilege and inequality (by race, class, and gender) is least visible to those who are most privileged and who, in turn, control the resources to define the dominant cultural belief systems. Perhaps this is why the privileged, not the poor, are more likely to believe that one gets ahead through hard work. It may also explain why men more than women deny that patriarchy exists, and why Whites more than Blacks believe racism is disappearing.

Overall, the effects of race, class, and gender manifest themselves in patterns of advantage and disadvantage. For example, in some ways—such as in rates of poverty and the racial gap in earnings in the labor market—women of color share a class position with men of color. In other ways, women of different races share a common class position: They all make less on average than men of their racial group. These trends point to the need for analyzing class, race, and gender together in thinking about inequality.

GENDER

Gender, like race, is a social construction, not a biological imperative. *Gender* is rooted in social institutions and results in patterns within society that structure the relationships between women and men and that give them differing positions of advantage and disadvantage within institutions. As an identity, gender is learned; that is, through gender socialization, people construct definitions of themselves and others that are marked by gender. Like race, however, gender cannot be understood at the individual level alone. Gender is structured in social institutions, including work, families, mass media, and education.

You can see this if you think about the concept of a gendered institution. *Gendered institution* is now used to define the total patterns of gender relations that are "present in the processes, practices, images, and ideologies, and distribution of power in the various sectors of social life" (Acker 1992: 567). This brings a much more structural analysis of gender to the forefront. Rather than seeing gender only as a matter of interpersonal relationships and learned identities, this framework focuses the analysis of gender on relations of power—just as thinking about institutional racism focuses on power relations and economic and political subordination—not just interpersonal relations. Changing gender relations is not just a matter of changing individuals. As with race and class, change requires transformation of institutional structures.

Gender, however, is not a monolithic category. Maxine Baca Zinn, Pierrette Hondagneu-Sotelo, and Michael Messner argue ("Gender through the Prism of Difference") that, although gender is grounded in specific power relations, it is important to understand gender as constructed differently depending on the specific social locations of diverse groups. Thus race, class, nationality, sexual orientation, and other factors produce varying social and economic consequences that cannot be understood by looking at gender differences alone. They ask us to move beyond studying differences and instead to use multiple "prisms" to see and comprehend the complexities of multiple systems of domination—each of which shapes and is shaped by gender. Seen in this way, men appear as a less monolithic and unidimensional group as well.

Gender patterns in society are also supported through the ideology of sexism, just as racial oppression is supported through racism. Sexism is the belief that men are somehow superior to women; though few overtly believe this, it shows in how women are perceived and treated. But, sexism does not exist in a vacuum. It intersects with racism and class—and sexuality—as Yen Le Espiritu argues in "Ideological Racism and Cultural Resistance." Controlling images of Asian American men and women are both racialized and sexualized, perpetuating ideas of "otherness." She locates the construction of these images in the institutions that produce popular culture, showing how such institutions have a stake in perpetuating race, gender, and sexual domination. Espiritu's work shows that gender stereotypes take unique forms for particular groups. Jewish American women, for example, are stereotyped as "JAPs" (Jewish American princesses), as if all were rich and privileged. This stereotype simultaneously promotes anti-Semitism and misogyny (defined as the hatred of women), at the same time that it uses an anti-Asian stereotype ("Jap") to denigrate others.

Gender, race, class, and sexuality *together* construct stereotypes. Each gains meaning in relationship to the others (Glenn 2002). Thus, for Julia Alvarez ("A White Woman of Color") gender identity is intricately part of her status as Dominican, and, to complicate things further, ideals of beauty in her narrative intertwine with ideas about color. You cannot understand her experience as a woman without also locating her in the ethnic, racial, national, and migration experiences that are also part of her life.

Thus, gender oppression is maintained through multiple systems, systems that are reflected in group stereotypes. And, these stereotypes also sexualize groups in different, but particular, ways. African American men are stereotyped as hypermasculine and oversexed; African American women as promiscuous, bad mothers, and nurturing "mammies" who care for everyone else, but not their own children. Latinos are stereotyped as "macho" and, like African American men, sexually passionate, but out of control. Latinas are stereotyped as either "hot" or virgin-like. Similarly, White women are sexually stereotyped in dichotomous terms, as "madonnas" or "whores." Class and sexuality intermingle with race and gender in these stereotypes. Working-class women are more likely to be seen as "sluts" and upper-class women as frigid and cold (Andersen and Taylor 2002: 360). Here we can see that controlling images of sexuality are part of the architecture of race, class, and gender oppression (Collins 2000). These stereotypes reveal the interlocking systems of race, class, gender, and sexuality.

Michael Messner's essay "Masculinities and Athletic Careers" explores how sport reproduces gender ideology. Messner's study examines how boys from different racial and social class backgrounds learn about masculinity through their involvement in sport. Messner demonstrates that there is no one set of beliefs about masculinity to which all men subscribe; instead, sport shapes understandings of masculinity and men's experiences in sports in race- and class-specific ways.

We also see in the articles here how gender oppression works with another major system of oppression, that of sexuality. *Homophobia*—the fear and hatred of homosexuality—is part of the system of social control that legitimates and enforces gender oppression. It supports the institutionalized power and privilege accorded to heterosexual behavior and identification. If only heterosexual forms of gender identity are labeled "normal," then gays, lesbians, and bisexuals become ostracized, oppressed, and defined as "socially deviant." Homophobia affects heterosexuals as well because it is part of the gender ideology used to distinguish "normal" men and women from those deemed deviant. Thus, young boys learn a rigid view of masculinity—one often associated with violence, bullying, and degrading others—to avoid being perceived as a "fag." The oppression of lesbians and gay men is then linked to the structure of gender in everyone's lives.

Throughout Part II, you should keep the concept of social structure in mind. Remembering Marilyn Frye's analogy of the birdcage, be aware that race, class, and gender form a structure of social relations. This structure is supported by ideological beliefs that make it appear "normal" and "acceptable" and that often cloud our awareness of how the structure operates. Thus, many believe that women now have it made, but the facts tell us otherwise. True, the gap between women's and men's income has closed, although most analysts agree that the narrowing of the gap reflects a drop in men's wages more than an increase in women's wages. And, women are more present in professional jobs—those that have become defined as the stereotypical "working woman." Despite this new image, though, most women remain concentrated in gender-segregated occupations with low wages, little opportunity for mobility, and stressful conditions. This is particularly true for women of color, who are more likely to be in occupations that are both race- and gender-segregated. And, as we have seen, among women heading their own households, poverty persists at alarmingly high rates. Income and occupational data, however, do not tell the full story for women. High rates of violence against women—whether in the home, on campus, in the workplace, or on the streets—indicate the continuing devaluation of and danger for women in this society. And, as Loretta Ross, Sarah Brownlee, Dazon Dixon Diallo, and Luz Rodriquez show through their work with the SisterSong Women of Color Reproductive Health Network ("Just Choices"), justice for women is incomplete without also addressing the race, gender, and class discrimination that takes place in women's health care—including reproductive health care.

As you read the articles in this section (and others), continue to ask yourself how different groups of women are experiencing the changes in gender relations that have characterized recent history. And, remember that although race, class, and gender are often discussed in terms of cultural difference, they are part of the institutional framework of society. A structural analysis studies the intersections of race, class, and gender within institutions and within individual's experiences in those institutions. Although we have divided the articles into sections on race, class, and gender, many theoretically fit in multiple categories. Thinking inclusively means recognizing that people and practices rarely belong to only one category. A more accurate way of viewing the world is to see race, class, and gender as interconnected.

Further Resources

For additional materials relating to this section, see the features on pages 213–14.

References

Acker, Joan, Sandra Morgen, and Lisa Gonzales. 2002. "Welfare Restructuring, Work & Poverty: Policy from Oregon." Working paper, Eugene, OR: Center for the Study of Women in Society.

Acker, Joan. 1992. "Gendered Institutions: From Sex Roles to Gendered Institutions." *Contemporary Sociology* 21 (September): 565–569.

Andersen, Margaret L., and Howard F. Taylor. 2002. *Sociology: Understanding a Diverse Society*. 2d ed. Belmont, CA: Wadsworth.

Collins, Chuck, and Felice Veskel. 2000. *Economic Apartheid in America: A Primer on Economic Inequality and Insecurity*. New York: The New Press.

Collins, Patricia Hill. 2000. *Black Feminist Thought: Knowledge, Consciousness, and the Politics of Empowerment*. New York: Routledge.

DaNavas-Walt, Carmen, and Robert W. Cleveland. 2002. *Money Income in the United States: 2001*. Washington, DC: U.S. Census Bureau.

Eitzen, D. Stanley. 1999. *Fair and Foul: Beyond the Myths and Paradoxes of Sport*. Lanham, MD: Rowman and Littlefield.

Emsellem, Murray, Jessica Goldberg, Rick McHugh, Wendell Primus, Rebecca Smith, and Jeffrey Wenger. 2002. *Failing the Unemployed*. Washington, DC: Economic Policy Institute. Web site: www.epinet.org

Feagin, Joe R. 2000. *Racist America: Roots, Current Realities, and Future Reparations*. New York: Routledge.

Glenn, Evelyn Nakano. 2002. *Unequal Freedom: How Race and Gender Shaped American Citizenship and Labor*. Cambridge, MA: Harvard University Press.

Hannaford, Ivan. 1996. *Race: The History of an Idea in the West*. Baltimore: Johns Hopkins University Press.

Harrison, Paige M., and Allen J. Beck. 2002. *Prisoners in 2001*. Washington, DC: U.S. Bureau of Justice Statistics. Web site: www.ojp.usdoj.gov/bjs

Krugman, Paul. 2002. "For Richer: How the Permissive Capitalism of the Boom Destroyed American Equality." *The New York Times Magazine* (October 20): 62 ff.

Lee, Sharon. 1995. "Poverty and the U.S. Asian Population." *Social Science Quarterly* 75 (September): 541–559.

Lewis, Dan A., Amy Bush Stevens, and Kristen Shook Slack. 2002. *Illinois Families Study, Welfare Reform in Illinois: Is the Moderate Approach Working?* Evanston, IL: Institute for Poverty Research, Northwestern University.

Massey, Douglass S. 1993. "Latino Poverty Research: An Agenda for the 1990s." *Social Science Research Council Newsletter* 47 (March): 7–11.

Mishel, Lawrence, Jared Bernstein, and John Schmitt. 2001. *The State of Working America 2000–2001.* Washington, DC: Economic Policy Institute.

Moss, Philip I., and Chris Tilly. 2001. *Stories Employers Tell: Race, Skill, and Hiring in America (Multi City Study of Urban Inequality).* New York: Russell Sage Foundation.

National Coalition for the Homeless. 2002. *Facts about Homelessness.* Web site: www.nationalhomeless.org

National Collegiate Athletic Association. 2002. *2002 Graduation Rates Report for NCAA-Division I Schools.* Indianapolis, IN: NCAA. Web site: www.ncaa.org

O'Hare, William P. 1996. *A New Look at Poverty in America.* Washington, DC: Population Reference Bureau.

Oliver, Melvin L., and Thomas M. Shapiro. 1995. *Black Wealth/White Wealth: A New Perspective on Racial Inequality.* New York: Routledge.

Omi, Michael, and Howard Winant. 1994. *Racial Formation in the United States: From the 1960s to the 1990s.* 2d ed. New York: Routledge.

Pattillo-McCoy, Mary. 1999. *Black Picket Fences: Privilege and Peril among the Black Middle Class.* Chicago: University of Chicago Press.

Proctor, Bernadette D., and Joseph Dalaker. 2002. *Poverty in the United States: 2001.* Washington, DC: U.S. Census Bureau.

U.S. Census Bureau. 2002. *Income: Detailed Historical Tables, Tables P36B–P36E.* Washington, DC: U.S. Census Bureau. Web site: www.census.gov

U.S. Department of Labor. 2002. *Employment and Earnings.* Washington, DC: U.S. Department of Labor.

Race and Racism

SOMETHING ABOUT THE SUBJECT MAKES IT HARD TO NAME

8

Gloria Yamato

Racism—simple enough in structure, yet difficult to eliminate. Racism—pervasive in the U.S. culture to the point that it deeply affects all the local town folk and spills over, negatively influencing the fortunes of folk around the world. Racism is pervasive to the point that we take many of its manifestations for granted, believing "that's life." Many believe that racism can be dealt with effectively in one hellifying workshop, or one hour-long heated discussion. Many actually believe this monster, racism, that has had at least a few hundred years to take root, grow, invade our space and develop subtle variations . . . this mind-funk that distorts thought and action, can be merely wished away. I've run into folks who really think that we can beat this devil, kick this habit, be healed of this disease in a snap. In a sincere blink of a well-intentioned eye, presto—poof—racism disappears. "I've dealt with my racism . . . (envision a laying on of hands) . . . Hallelujah! Now I can go to the beach." Well, fine. Go to the beach. In fact, why don't we all go to the beach and continue to work on the sucker over there? Cuz you can't even shave a little piece off this thing called racism in a day, or a weekend, or a workshop.

When I speak of *oppression*, I'm talking about the systematic, institutionalized mistreatment of one group of people by another for whatever reason. The oppressors are purported to have an innate ability to access economic resources, information, respect, etc., while the oppressed are believed to have a

From Jo Whitehorse Cochran, Donna Langston, and Carolyn Woodward, eds., *Changing Our Power: An Introduction to Women's Studies* (Dubuque, IA: Kendall-Hunt, 1988), pp. 3–6. Reprinted by permission.

corresponding negative innate ability. The flip side of oppression is *internalized oppression*. Members of the target group are emotionally, physically, and spiritually battered to the point that they begin to actually believe that their oppression is deserved, is their lot in life, is natural and right, and that it doesn't even exist. The oppression begins to feel comfortable, familiar enough that when mean ol' Massa lay down de whip, we got's to pick up and whack ourselves and each other. Like a virus, it's hard to beat racism, because by the time you come up with a cure, it's mutated to a "new cure-resistant" form. One shot just won't get it. Racism must be attacked from many angles.

The forms of racism that I pick up on these days are 1) aware/blatant racism, 2) aware/covert racism, 3) unaware/unintentional racism, and 4) unaware/self-righteous racism. I can't say that I prefer any one form of racism over the others, because they all look like an itch needing a scratch. I've heard it said (and understandably so) that the aware/blatant form of racism is preferable if one must suffer it. Outright racists will, without apology or confusion, tell us that because of our color we don't appeal to them. If we so choose, we can attempt to get the hell out of their way before we get the sweat knocked out of us. Growing up, aware/covert racism is what I heard many of my elders bemoaning "up north," after having escaped the overt racism "down south." Apartments were suddenly no longer vacant or rents were outrageously high, when black, brown, red, or yellow persons went to inquire about them. Job vacancies were suddenly filled, or we were fired for very vague reasons. It still happens, though the perpetrators really take care to cover their tracks these days. They don't want to get gummed to death or slobbered on by the toothless laws that supposedly protect us from such inequities.

Unaware/unintentional racism drives usually tranquil white liberals wild when they get called on it, and confirms the suspicions of many people of color who feel that white folks are just plain crazy. It has led white people to believe that it's just fine to ask if they can touch my hair (while reaching). They then exclaim over how soft it is, how it does not scratch their hand. It has led whites to assume that bending over backwards and speaking to me in high-pitched (terrified), condescending tones would make up for all the racist wrongs that distort our lives. This type of racism has led whites right to my doorstep, talking 'bout, "We're sorry/we love you and want to make things right," which is fine, and further, "We're gonna give you the opportunity to fix it while we sleep. Just tell us what you need. 'Bye!!"—which *ain't* fine. With the best of intentions, the best of educations, and the greatest generosity of heart, whites, operating on the misinformation fed to them from day one, will behave in ways that are racist, will perpetuate racism by being "nice" the way we're taught to be nice. You can just "nice" somebody to death with naïveté and lack of awareness of privilege. Then there's guilt and the desire to end racism and how the two get all tangled up to the point that people, morbidly fascinated with their guilt, are immobilized. Rather than deal with ending racism, they sit and ponder their guilt and hope nobody notices how awful they are. Meanwhile, racism picks up momentum and keeps on keepin' on.

Now, the newest form of racism that I'm hip to is unaware/self-righteous racism. The "good white" racist attempts to shame Blacks into being blacker, scorns Japanese-Americans who don't speak Japanese, and knows more about the Chicano/a community than the folks who make up the community. They assign themselves as the "good whites," as opposed to the "bad whites," and are often so busy telling people of color what the issues in the Black, Asian, Indian, Latino/a communities should be that they don't have time to deal with their errant sisters and brothers in the white community. Which means that people of color are still left to deal with what the "good whites" don't want to . . . racism.

Internalized racism is what really gets in my way as a Black woman. It influences the way I see or don't see myself, limits what I expect of myself or others like me. It results in my acceptance of mistreatment, leads me to believe that being treated with less than absolute respect, at least this once, is to be expected because I am Black, because I am not white. "Because I am (*you fill in the color*), you think, "Life is going to be hard." The fact is life may be hard, but the color of your skin is not the cause of the hardship. The color of your skin may be used as an excuse to mistreat you, but there is no reason or logic involved in the mistreatment. If it seems that your color is the reason, if it seems that your ethnic heritage is the cause of the woe, it's because you've been deliberately beaten down by agents of a greedy system until you swallowed the garbage. That is the internalization of racism.

Racism is the systematic, institutionalized mistreatment of one group of people by another based on racial heritage. Like every other oppression, racism can be internalized. People of color come to believe misinformation about their particular ethnic group and thus believe that their mistreatment is justified. With that basic vocabulary, let's take a look at how the whole thing works together. Meet "the Ism Family," racism, classism, ageism, adultism, elitism, sexism, heterosexism, physicalism, etc. All these ism's are systematic, that is, not only are these parasites feeding off our lives, they are also dependent on one another for foundation. Racism is supported and reinforced by classism, which is given a foothold and a boost by adultism, which also feeds sexism, which is validated by heterosexism, and so it goes on. You cannot have the "ism" functioning without first effectively installing its flip-side, the internalized version of the ism. Like twins, as one particular form of the ism grows in potency, there is a corresponding increase in its internalized form within the population. Before oppression becomes a specific ism like racism, usually all hell breaks loose. War. People fight attempts to enslave them, or to subvert their will, or to take what they consider theirs, whether that is territory or dignity. It's true that the various elements of racism, while repugnant, would not be able to do very much damage, but for one generally overlooked key piece: power/privilege.

While in one sense we all have power we have to look at the fact that, in our society, people are stratified into various classes and some of these classes have more privilege than others. The owning class has enough power and privilege to not have to give a good whinney what the rest of the folks have on their minds. The power and privilege of the owning class provides the ability to pay off enough of the working class and offer that paid-off group, the middle class, just enough privilege to make it agreeable to do various and sundry oppressive things to other working-class and outright disenfranchised folk, keeping the lid on explosive inequities, at least for a minute. If you're at the bottom of this heap, and you believe the line that says you're there because that's all you're worth, it is at least some small solace to believe that there are others more worthless than you, because of their gender, race, sexual preference . . . whatever. The specific form of power that runs the show here is the power to intimidate. The power to take away the most lives the quickest, and back it up with legal and "divine" sanction, is the very bottom line. It makes the difference between who's holding the racism end of the stick and who's getting beat with it (or beating others as vulnerable as they are) on the internalized racism end of the stick. What I am saying is, while people of color are welcome to tear up their own neighborhoods and each other, everybody knows that you cannot do that to white folks without hell to pay. People of color can be prejudiced against one another and whites, but do not have an ice-cube's chance in hell of passing laws that will get whites sent to relocation camps "for their own protection and the security of the nation." People who have not thought about or refuse to acknowledge this imbalance of power/privilege often want to talk about the racism of people of color. But then that is one of the ways racism is able to continue to function. You look for someone to blame and you blame the victim, who will nine times out of ten accept the blame out of habit.

So, what can we do? Acknowledge racism for a start, even though and especially when we've struggled to be kind and fair, or struggled to rise above it all. It is hard to acknowledge the fact that racism circumscribes and pervades our lives. Racism must be dealt with on two levels, personal and societal, emotional and institutional. It is possible—and most effective—to do both at the same time. We must reclaim whatever delight we have lost in our own ethnic heritage or heritages. This so-called melting pot has only succeeded in turning us into fast-food gobbling "generics" (as in generic "white folks" who were once Irish, Polish, Russian, English, etc. and "black folks," who were once Ashanti, Bambara, Baule, Yoruba, etc.). Find or create safe places to actually *feel* what we've been forced to repress each time we were a victim of, witness to or perpetrator of racism, so that we do not continue, like puppets, to act out the past in the present and future. Challenge oppression. Take a stand against it. When you are aware of something oppressive going down, stop the show. At least call it. We become so numbed to racism that we don't even think twice about it, unless it is immediately life-threatening.

Whites who want to be allies to people of color: You can educate yourselves via research and observation rather than rigidly, arrogantly relying solely on interrogating people of color. Do not expect that people of color should teach you how to behave non-oppressively. Do not give into the pull to be lazy. Think, hard. Do not blame people of color for your frustration about racism, but do appreciate the fact that people of color will often help you get in touch with that frustration. Assume that your effort to be a good friend is appreciated, but don't expect or accept gratitude from people of color. Work on racism for your sake, not "their" sake. Assume that you are needed and capable of being a good ally. Know that you'll make mistakes and commit yourself to correcting them and continuing on as an ally, no matter what. Don't give up.

People of color, working through internalized racism: Remember always that you and others like you are completely worthy of respect, completely capable of achieving whatever you take a notion to do. Remember that the term "people of color" refers to a variety of ethnic and cultural backgrounds. These various groups have been oppressed in a variety of ways. Educate yourself about the ways different peoples have been oppressed and how they've resisted that oppression. Expect and insist that whites are capable of being good allies against racism. Don't give up. Resist the pull to give out the "people of color seal of approval" to aspiring white allies. A moment of appreciation is fine, but more than that tends to be less than helpful. Celebrate yourself. Celebrate yourself. Celebrate the inevitable end of racism.

WHITE PRIVILEGE   9

Unpacking the Invisible Knapsack

Peggy McIntosh

Through work to bring materials from Women's Studies into the rest of the curriculum, I have often noticed men's unwillingness to grant that they are

---

overprivileged, even though they may grant that women are disadvantaged. They may say they will work to improve women's status, in the society, the university, or the curriculum, but they can't or won't support the idea of lessening men's. Denials which amount to taboos surround the subject of advantages which men gain from women's disadvantages. These denials protect male privilege from being fully acknowledged, lessened, or ended.

Thinking through unacknowledged male privilege as a phenomenon, I realized that since hierarchies in our society are interlocking, there was most likely a phenomenon of white privilege which was similarly denied and protected. As a white person, I realized I had been taught about racism as something which puts others at a disadvantage, but had been taught not to see one of its corollary aspects, white privilege, which puts me at an advantage.

I think whites are carefully taught not to recognize white privilege, as males are taught not to recognize male privilege. So I have begun in an untutored way to ask what it is like to have white privilege. I have come to see white privilege as an invisible package of unearned assets which I can count on cashing in each day, but about which I was "meant" to remain oblivious. White privilege is like an invisible weightless knapsack of special provisions, maps, passports, codebooks visas, clothes, tools, and blank checks.

Describing white privilege makes one newly accountable. As we in Women's Studies work to reveal male privilege and ask men to give up some of their power, so one who writes about having white privilege must ask, "Having described it, what will I do to lessen or end it?"

After I realized the extent to which men work from a base of unacknowledged privilege, I understood that much of their oppressiveness was unconscious. Then I remembered the frequent charges from women of color that white women whom they encounter are oppressive. I began to understand why we are justly seen as oppressive, even when we don't see ourselves that way. I began to count the ways in which I enjoy unearned skin privilege and have been conditioned into oblivion about its existence.

My schooling gave me no training in seeing myself as an oppressor, as an unfairly advantaged person, or as a participant in a damaged culture. I was taught to see myself as an individual whose moral state depended on her individual moral will. My schooling followed the pattern my colleague Elizabeth Minnich has pointed out: whites are taught to think of their lives as morally neutral, normative, and average, and also ideal, so that when we work to benefit others, this is seen as work which will allow "them" to be more like "us."

I decided to try to work on myself at least by identifying some of the daily effects of white privilege in my life. I have chosen those conditions which I think in my case *attach somewhat more to skin-color privilege* than to class, religion, ethnic status, or geographical location, though of course all these other factors are intricately intertwined. As far as I can see, my African American co-workers, friends and acquaintances with whom I come into daily or frequent contact in this particular time, place, and line of work cannot count on most of these conditions.

1. I can if I wish arrange to be in the company of people of my race most of the time.
2. If I should need to move, I can be pretty sure of renting or purchasing housing in an area which I can afford and in which I would want to live.
3. I can be pretty sure that my neighbors in such a location will be neutral or pleasant to me.
4. I can go shopping alone most of the time, pretty well assured that I will not be followed or harassed.
5. I can turn on the television or open to the front page of the paper and see people of my race widely represented.
6. When I am told about our national heritage or about "civilization," I am shown that people of my color made it what it is.
7. I can be sure that my children will be given curricular materials that testify to the existence of their race.
8. If I want to, I can be pretty sure of finding a publisher for this piece on white privilege.
9. I can go into a music shop and count on finding the music of my race represented, into a supermarket and find the staple foods which fit with my cultural traditions, into a hairdresser's shop and find someone who can cut my hair.
10. Whether I use checks, credit cards, or cash, I can count on my skin color not to work against the appearance of financial reliability.
11. I can arrange to protect my children most of the time from people who might not like them.
12. I can swear, or dress in secondhand clothes, or not answer letters, without having people attribute these choices to the bad morals, the poverty, or the illiteracy of my race.
13. I can speak in public to a powerful male group without putting my race on trial.
14. I can do well in a challenging situation without being called a credit to my race.
15. I am never asked to speak for all the people of my racial group.
16. I can remain oblivious of the language and customs of persons of color who constitute the world's majority without feeling in my culture any penalty for such oblivion.
17. I can criticize our government and talk about how much I fear its policies and behavior without being seen as a cultural outsider.
18. I can be pretty sure that if I ask to talk to "the person in charge," I will be facing a person of my race.

19. If a traffic cop pulls me over or if the IRS audits my tax return, I can be sure I haven't been singled out because of my race.
20. I can easily buy posters, postcards, picture books, greeting cards, dolls, toys, and children's magazines featuring people of my race.
21. I can go home from most meetings of organizations I belong to feeling somewhat tied in, rather than isolated, out-of-place, outnumbered, unheard, held at a distance, or feared.
22. I can take a job with an affirmative action employer without having co-workers on the job suspect that I got it because of my race.
23. I can choose public accommodation without fearing that people of my race cannot get in or will be mistreated in the places I have chosen.
24. I can be sure that if I need legal or medical help, my race will not work against me.
25. If my day, week, or year is going badly, I need not ask of each negative episode or situation whether it has racial overtones.
26. I can choose blemish cover or bandages in "flesh" color and have them more or less match my skin.

I repeatedly forgot each of the realizations on this list until I wrote it down. For me white privilege has turned out to be an elusive and fugitive subject. The pressure to avoid it is great, for in facing it I must give up the myth of meritocracy. If these things are true, this is not such a free country; one's life is not what one makes it; many doors open for certain people through no virtues of their own.

In unpacking this invisible knapsack of white privilege, I have listed conditions of daily experience which I once took for granted. Nor did I think of any of these perquisites as bad for the holder. I now think that we need a more finely differentiated taxonomy of privilege, for some of these varieties are only what one would want for everyone in a just society, and others give license to be ignorant, oblivious, arrogant and destructive.

I see a pattern running through the matrix of white privilege, a pattern of assumptions which were passed on to me as a white person. There was one main piece of cultural turf; it was my own turf, and I was among those who could control the turf. *My skin color was an asset for any move I was educated to want to make.* I could think of myself as belonging in major ways, and of making social systems work for me. I could freely disparage, fear, neglect, or be oblivious to anything outside of the dominant cultural forms. Being of the main culture, I could also criticize it fairly freely.

In proportion as my racial group was being made confident, comfortable, and oblivious, other groups were likely being made inconfident, uncomfortable, and alienated. Whiteness protected me from many kinds of hostility, distress, and violence, which I was being subtly trained to visit in turn upon people of color.

For this reason, the word "privilege" now seems to me misleading. We usually think of privilege as being a favored state, whether earned or conferred by birth or luck. Yet some of the conditions I have described here work to systematically overempower certain groups. Such privilege simply *confers dominance* because of one's race or sex.

I want, then, to distinguish between earned strength and unearned power conferred systemically. Power from unearned privilege can look like strength when it is in fact permission to escape or to dominate. But not all of the privileges on my list are inevitably damaging. Some, like the expectation that neighbors will be decent to you, or that your race will not count against you in court, should be the norm in a just society. Others, like the privilege to ignore less powerful people, distort the humanity of the holders as well as the ignored groups.

We might at least start by distinguishing between positive advantages which we can work to spread, and negative types of advantages which unless rejected will always reinforce our present hierarchies. For example, the feeling that one belongs within the human circle, as Native Americans say, should not be seen as privilege for a few. Ideally it is an *unearned entitlement*. At present, since only a few have it, it is an *unearned advantage* for them. This paper results from a process of coming to see that some of the power which I originally saw as attendant on being a human being in the U.S. consisted in *unearned advantage* and *conferred dominance*.

I have met very few men who are truly distressed about systemic, unearned male advantage and conferred dominance. And so one question for me and others like me is whether we will be like them, or whether we will get truly distressed, even outraged, about unearned race advantage and conferred dominance and if so, what we will do to lessen them. In any case, we need to do more work in identifying how they actually affect our daily lives. Many, perhaps most, of our white students in the U.S. think that racism doesn't affect them because they are not people of color; they do not see "whiteness" as a racial identity. In addition, since race and sex are not the only advantaging systems at work, we need similarly to examine the daily experience of having age advantage, or ethnic advantage, or physical ability, or advantage related to nationality, religion, or sexual orientation.

Difficulties and dangers surrounding the task of finding parallels are many. Since racism, sexism, and heterosexism are not the same, the advantaging associated with them should not be seen as the same. In addition, it is hard to disentangle aspects of unearned advantage which rest more on social class, economic class, race, religion, sex and ethnic identity than on other factors. Still, all of the oppressions are interlocking, as the Combahee River Collective Statement of 1977 continues to remind us eloquently.

One factor seems clear about all of the interlocking oppressions. They take both active forms which we can see and embedded forms which as a member of the dominant group one is taught not to see. In my class and place, I did not see myself as a racist because I was taught to recognize racism only in individual acts of meanness by members of my group, never in invisible systems conferring unsought racial dominance on my group from birth.

Disapproving of the systems won't be enough to change them. I was taught to think that racism could end if white individuals changed their attitudes. [But] a "white" skin in the United States opens many doors for whites whether or not we approve of the way dominance has been conferred on us. Individual acts can palliate, but cannot end, these problems.

To redesign social systems we need first to acknowledge their colossal unseen dimensions. The silences and denials surrounding privilege are the key political tool here. They keep the thinking about equality or equity incomplete, protecting unearned advantage and conferred dominance by making these taboo subjects. Most talk by whites about equal opportunity seems to me now to be about equal opportunity to try to get into a position of dominance while denying that *systems* of dominance exist.

It seems to me that obliviousness about white advantage, like obliviousness about male advantage, is kept strongly inculturated in the United States so as to maintain the myth of meritocracy, the myth that democratic choice is equally available to all. Keeping most people unaware that freedom of confident action is there for just a small number of people props up those in power, and serves to keep power in the hands of the same groups that have most of it already.

Though systemic change takes many decades, there are pressing questions for me and I imagine for some others like me if we raise our daily consciousness on the perquisites of being light-skinned. What will we do with such knowledge? As we know from watching men, it is an open question whether we will choose to use unearned advantage to weaken hidden systems of advantage, and whether we will use any of our arbitrarily-awarded power to try to reconstruct power systems on a broader base.

OF RACE AND RISK   **10**

Patricia J. Williams

Several years ago, at a moment when I was particularly tired of the unstable lifestyle that academic careers sometimes require, I surprised myself and

From *The Nation*, December 29, 1997. Reprinted with permission.

bought a real house. Because the house was in a state other than the one where I was living at the time, I obtained my mortgage by telephone. I am a prudent little squirrel when it comes to things financial, always tucking away stores of nuts for the winter, and so I meet the criteria of a quite good credit risk. My loan was approved almost immediately.

A little while later, the contract came in the mail. Among the papers the bank forwarded were forms documenting compliance with the Fair Housing Act, which outlaws racial discrimination in the housing market. The act monitors lending practices to prevent banks from redlining—redlining being the phenomenon whereby banks circle certain neighborhoods on the map and refuse to lend in those areas. It is a practice for which the bank with which I was dealing, unbeknownst to me, had been cited previously—as well as since. In any event, the act tracks the race of all banking customers to prevent discrimination. Unfortunately, and with the creative variability of all illegality, some banks also use the racial information disclosed on the fair housing forms to engage in precisely the discrimination the law seeks to prevent.

I should repeat that to this point my entire mortgage transaction had been conducted by telephone. I should also note that I speak a Received Standard English, regionally marked as Northeastern perhaps, but not easily identifiable as black. With my credit history, my job as a law professor and, no doubt, with my accent, I am not only middle class but apparently match the cultural stereotype of a good white person. It is thus, perhaps, that the loan officer of the bank, whom I had never met, had checked off the box on the fair housing form indicating that I *was* white.

Race shouldn't matter, I suppose, but it seemed to in this case, so I took a deep breath, crossed out "white" and sent the contract back. That will teach them to presume too much, I thought. A done deal, I assumed. But suddenly the transaction came to a screeching halt. The bank wanted more money, more points, a higher rate of interest. Suddenly I found myself facing great resistance and much more debt. To make a long story short, I threatened to sue under the act in question, the bank quickly backed down and I procured the loan on the original terms.

What was interesting about all this was that the reason the bank gave for its new-found recalcitrance was not race, heaven forbid. No, it was all about economics and increased risk: The reason they gave was that property values in that neighborhood were suddenly falling. They wanted more money to buffer themselves against the snappy winds of projected misfortune.

Initially, I was surprised, confused. The house was in a neighborhood that was extremely stable. I am an extremely careful shopper; I had uncovered absolutely nothing to indicate that prices were falling. It took my realtor to make me see the light. "Don't you get it," he sighed. "This is what always happens."

And even though I suppose it was a little thick of me, I really hadn't gotten it: For of course, I was the reason the prices were in peril.

The bank's response was driven by demographic data that show that any time black people move into a neighborhood, whites are overwhelmingly likely to move out. In droves. In panic. In concert. Pulling every imaginable resource with them, from school funding to garbage collection to social workers who don't want to work in black neighborhoods. The imagery is awfully catchy, you had to admit: the neighborhood just tipping on over like a terrible accident, whoops! Like a pitcher, I suppose. All that nice fresh wholesome milk spilling out, running away . . . leaving the dark, echoing, upended urn of the inner city.

In retrospect, what has remained so fascinating to me about this experience was the way it so exemplified the problems of the new rhetoric of racism. For starts, the new rhetoric of race never mentions race. It wasn't race but risk with which the bank was so considered.

Second, since financial risk is all about economics, my exclusion got reclassified as just a consideration of class. There's no law against class discrimination, goes the argument, because that would represent a restraint on that basic American freedom, the ability to contract or not. If schools, trains, buses, swimming pools and neighborhoods remain segregated, it's no longer a racial problem if someone who just happens to be white keeps hiking up the price for someone who accidentally and purely by the way happens to be black. Black people end up paying higher prices for the attempt to integrate, even as the integration of oneself threatens to lower the value of one's investment.

By this measure of mortgage-worthiness, the ingredient of blackness is cast not just as a social toll but as an actual tax. A fee, an extra contribution at the door, an admission charge for the high costs of handling my dangerous propensities, my inherently unsavory properties. I was not judged based on my independent attributes or financial worth; not even was I judged by statistical profiles of what my group actually does. (For in fact, anxiety-stricken, middle-class people make grovelingly good cake-baking neighbors when not made to feel defensive by the unfortunate historical strategies of bombs, burnings or abandonment.) Rather, I was being evaluated based on what an abstraction of White Society writ large thinks we—or I—do, and that imagined "doing" was treated and thus established as a self-fulfilling prophecy. It is a dispiriting message: that some in society apparently not only devalue black people but devalue *themselves* and their homes just for having us as part of their landscape.

"I bet you'll keep your mouth shut the next time they plug you into the computer as white," laughed a friend when he heard my story. It took me aback, this postmodern pressure to "pass," even as it highlighted the intolerable logic of it all. For by these "rational" economic measures, an investment in my property suggests the selling of myself.

SEEING MORE THAN BLACK AND WHITE 11

Elizabeth Martinez

The racial and ethnic landscape has changed too much in recent years to view it with the same eyes as before. We are looking at a multi-dimensional reality in which race, ethnicity, nationality, culture and immigrant status come together with breathtakingly new results. We are also seeing global changes that have a massive impact on our domestic situation, especially the economy and labor force. For a group of Korean restaurant entrepreneurs to hire Mexican cooks to prepare Chinese dishes for mainly African-American customers, as happened in Houston, Texas, has ceased to be unusual.

The ever-changing demographic landscape compels those struggling against racism and for a transformed, non-capitalist society to resolve several strategic questions. Among them: doesn't the exclusively Black-white framework discourage the perception of common interests among people of color and thus sustain White Supremacy? Doesn't the view that only African Americans face serious institutionalized racism isolate them from potential allies? Doesn't the Black-white model encourage people of color to spend too much energy understanding our lives in relation to whiteness, obsessing about what white society will think and do?

That tendency is inevitable in some ways: the locus of power over our lives has long been white (although big shifts have recently taken place in the color of capital, as we see in Japan, Singapore and elsewhere). The oppressed have always survived by becoming experts on the oppressor's ways. But that can become a prison of sorts, a trap of compulsive vigilance. Let us liberate ourselves, then, from the tunnel vision of whiteness and behold the many colors around us! Let us summon the courage to reject outdated ideas and stretch our imaginations into the next century.

For a Latina to urge recognizing a variety of racist models is not, and should not be, yet another round in the Oppression Olympics. We don't need more competition among different social groups for the gold medal of "Most Oppressed." We don't need more comparisons of suffering between women and Blacks, the disabled and the gay, Latino teenagers and white seniors, or whatever. Pursuing some hierarchy of oppression leads us down dead-end streets where we will never find the linkage between different oppressions and how

---

to overcome them. To criticize the exclusively Black-white framework, then, is not some resentful demand by other people of color for equal sympathy, equal funding, equal clout, equal patronage or other questionable crumbs. Above all, it is not a devious way of minimizing the centrality of the African-American experience in any analysis of racism.

The goal in re-examining the Black-white framework is to find an effective strategy for vanquishing an evil that has expanded rather than diminished. Racism has expanded partly as a result of the worldwide economic recession that followed the end of the post-war boom in the early 1970s, with the resulting capitalist restructuring and changes in the international division of labor. Those developments generated feelings of insecurity and a search for scapegoats. In the United States racism has also escalated as whites increasingly fear becoming a weakened, minority population in the next century. The stage is set for decades of ever more vicious divide-and-conquer tactics.

What has been the response from people of color to this ugly White Supremacist agenda? Instead of uniting, based on common experience and needs, we have often closed our doors in a defensive, isolationist mode, each community on its own. A fire of fear and distrust begins to crackle, threatening to consume us all. Building solidarity among people of color is more necessary than ever—but the exclusively Black-white definition of racism makes such solidarity more difficult than ever.

We urgently need twenty-first-century thinking that will move us beyond the Black-white framework without negating its historical role in the construction of U.S. racism. We need a better understanding of how racism developed both similarly and differently for various peoples, according to whether they experienced genocide, enslavement, colonization or some other structure of oppression. At stake is the building of a united anti-racist force strong enough to resist White Supremacist strategies of divide-and-conquer and move forward toward social justice for all. . . .

. . . African Americans have reason to be uneasy about where they, as a people, will find themselves politically, economically and socially with the rapid numerical growth of other folk of color. The issue is not just possible job loss, a real question that does need to be faced honestly. There is also a feeling that after centuries of fighting for simple recognition as human beings, Blacks will be shoved to the back of history again (like the back of the bus). Whether these fears are real or not, uneasiness exists and can lead to resentment when there's talk about a new model of race relations. So let me repeat: in speaking here of the need to move beyond the bipolar concept, the goal is to clear the way for stronger unity against White Supremacy. The goal is to identify our commonalities of experience and needs so we can build alliances.

The commonalities begin with history, which reveals that again and again peoples of color have had one experience in common: European colonization and/or neo-colonialism with its accompanying exploitation. This is true for all indigenous peoples, including Hawaiians. It is true for all Latino peoples, who were invaded and ruled by Spain or Portugal. It is true for people in Africa, Asia and the Pacific Islands, where European powers became the colonizers. People of color were victimized by colonialism not only externally but also through internalized racism—the "colonized mentality."

Flowing from this shared history are our contemporary commonalities. On the poverty scale, African Americans and Native Americans have always been at the bottom, with Latinos nearby. In 1995 the U.S. Census found that Latinos have the highest poverty rate, 24 percent. Segregation may have been legally abolished in the 1960s, but now the United States is rapidly moving toward resegregation as a result of whites moving to the suburbs. This leaves people of color—especially Blacks and Latinos—with inner cities that lack an adequate tax base and thus have inadequate schools. Not surprisingly, Blacks and Latinos finish college at a far lower rate than whites. In other words, the victims of U.S. social ills come in more than one color. Doesn't that indicate the need for new, inclusive models for fighting racism? Doesn't that speak to the absolutely urgent need for alliances among peoples of color?

With greater solidarity, justice for people of color could be won. And an even bigger prize would be possible: a U.S. society that advances beyond "equality," beyond granting people of color a respect equal to that given to Euro-Americans. Too often "equality" leaves whites still at the center, still embodying the Americanness by which others are judged, still defining the national character. . . .

. . . Innumerable statistics, reports and daily incidents should make it impossible to exclude Latinos and other non-Black populations of color when racism is discussed, but they don't. Police killings, hate crimes by racist individuals and murders with impunity by border officials should make it impossible, but they don't. With chilling regularity, ranch owners compel migrant workers, usually Mexican, to repay the cost of smuggling them into the United States by laboring the rest of their lives for free. The 45 Latino and Thai garment workers locked up in an El Monte, California, factory, working 18 hours a day seven days a week for $299 a month, can also be considered slaves (and one must ask why it took three years for the Immigration and Naturalization Service to act on its own reports about this horror) (*San Francisco Examiner*, August 8, 1995). Abusive treatment of migrant workers can be found all over the United States. In Jackson Hole, Wyoming, for example, police and federal agents rounded up 150 Latino workers in 1997, inked numbers on their arms and hauled them off to jail in patrol cars and a horse trailer full of manure (*Los Angeles Times*, September 6, 1997).

These experiences cannot be attributed to xenophobia, cultural prejudice or some other, less repellent term than racism. Take the case of two small Latino children in San Francisco who were found in 1997 covered from head to toe with flour. They explained they had hoped to make their skin white enough for school. There is no way to understand their action except as the result of fear in the racist climate that accompanied passage of Proposition 187, which denies schooling to the children of undocumented immigrants. Another example: Mexican and Chicana women working at a Nabisco plant in Oxnard, California, were not allowed to take bathroom breaks from the assembly line and were told to wear diapers instead. Can we really imagine white workers being treated that way? (The Nabisco women did file a suit and won, in 1997.)

No "model minority" myth protects Asians and Asian Americans from hate crimes, police brutality, immigrant-bashing, stereotyping and everyday racist prejudice. Scapegoating can even take their lives, as happened with the murder of Vincent Chin in Detroit some years ago. . . .

WHY THE BLACK-WHITE MODEL?

A bipolar model of racism has never been really accurate for the United States. Early in this nation's history. Benjamin Franklin perceived a tri-racial society based on skin color—"the lovely white" (Franklin's words), the Black, and the "tawny," as Ron Takaki tells us in *Iron Cages*. But this concept changed as capital's need for labor intensified in the new nation and came to focus on African slave labor. The "tawny" were decimated or forcibly exiled to distant areas; Mexicans were not yet available to be the main labor force. As enslaved Africans became the crucial labor force for the primitive accumulation of capital, they also served as the foundation for the very idea of whiteness—based on the concept of blackness as inferior.

Three other reasons for the Black-white framework seem obvious: numbers, geography and history. African Americans have long been the largest population of color in the United States; only recently has this begun to change. Also, African Americans have long been found in sizable numbers in most parts of the United States, including major cities, which has not been true of Latinos until recent times. Historically, the Black-white relationship has been entrenched in the nation's collective memory for some 300 years—whereas it is only 150 years since the United States seized half of Mexico and incorporated those lands and their peoples. Slavery and the struggle to end it formed a central theme in this country's only civil war—a prolonged, momentous conflict. Above all, enslaved Africans in the United States and African Americans have created an unmatched heritage of massive, persistent, dramatic and infinitely courageous resistance, with individual leaders of worldwide note.

We also find sociological and psychological explanations of the Black-white model's persistence. From the days of Jefferson onward, Native Americans, Mexicans and later the Asian/Pacific Islanders did not seem as much a threat to racial purity or as capable of arousing white sexual anxieties as did Blacks. A major reason for this must have been Anglo ambiguity about who could be called white. Most of the Mexican *ranchero* elite in California had welcomed the U.S. takeover, and Mexicans were partly European—therefore "semi-civilized"; this allowed Anglos to see them as white, unlike lower-class Mexicans. For years Mexicans were legally white, and even today we hear the ambiguous U.S. Census term "Non-Hispanic Whites."

Like Latinos, Asian Americans have also been officially counted as white in some historical periods. They have been defined as "colored" in others, with "Chinese" being yet another category. Like Mexicans, they were often seen as not really white but not quite Black either. Such ambiguity tended to put Asian Americans along with Latinos outside the prevailing framework of racism.

Blacks, on the other hand, were not defined as white, could rarely become upper-class and maintained an almost constant rebelliousness. Contemporary Black rebellion has been urban: right in the Man's face, scary. Mexicans, by contrast, have lived primarily in rural areas until a few decades ago and "have no Mau-Mau image," as one Black friend said, even when protesting injustice energetically. Only the nineteenth-century resistance heroes labeled "bandits" stirred white fear, and that was along the border, a limited area. Latino stereotypes are mostly silly: snoozing next to a cactus, eating greasy food, always being late and disorganized, rolling big Carmen Miranda eyes, shrugging with self-deprecation "me no speek good eengleesh." In other words, *not serious.* This view may be altered today by stereotypes of the gangbanger, criminal or dirty immigrant, but the prevailing image of Latinos remains that of a debased white, at best. . . .

Among other important reasons for the exclusively Black-white model, sheer ignorance leaps to mind. The oppression and exploitation of Latinos (like Asians) have historical roots unknown to most Americans. People who learn at least a little about Black slavery remain totally ignorant about how the United States seized half of Mexico or how it has colonized Puerto Rico. . . .

One other important reason for the bipolar model of racism is the stubborn self-centeredness of U.S. political culture. It has meant that the nation lacks any global vision other than relations of domination. In particular, the United States refuses to see itself as one among some 20 countries in a hemisphere whose dominant languages are Spanish and Portuguese, not English. It has only a big yawn of contempt or at best indifference for the people, languages and issues of Latin America. It arrogantly took for itself alone the name of half the western hemisphere, America, as was its "Manifest Destiny," of course.

So Mexico may be nice for a vacation and lots of Yankees like tacos, but the political image of Latin America combines incompetence with absurdity, fat corrupt dictators with endless siestas. Similar attitudes extend to Latinos within the United States. My parents, both Spanish teachers, endured decades of being told that students were better off learning French or German. The mass media complain that "people can't relate to Hispanics (or Asians)." It takes mysterious masked rebels, a beautiful young murdered singer or salsa outselling ketchup for the Anglo world to take notice of Latinos. If there weren't a mushrooming, billion-dollar "Hispanic" market to be wooed, the Anglo world might still not know we exist. No wonder that racial paradigm sees only two poles.

The exclusively Black-white framework is also sustained by the "model minority" myth, because it distances Asian Americans from other victims of racism. Portraying Asian Americans as people who work hard, study hard, obey the established order and therefore prosper, the myth in effect admonishes Blacks and Latinos: "See, anyone can make it in this society if you try hard enough. The poverty and prejudice you face are all *your* fault."

The "model" label has been a wedge separating Asian Americans from others of color by denying their commonalities. It creates a sort of racial bourgeoisie, which White Supremacy uses to keep Asian Americans from joining forces with the poor, the homeless and criminalized youth. People then see Asian Americans as a special class of yuppie: young, single, college-educated, on the white-collar track—and they like to shop for fun. Here is a dandy minority group, ready to be used against others.

The stereotype of Asian Americans as whiz kids is also enraging because it hides so many harsh truths about the impoverishment, oppression and racist treatment they experience. Some do come from middle- or upper-class families in Asia, some do attain middle-class or higher status in the U.S., and their community must deal with the reality of class privilege where it exists. But the hidden truths include the poverty of many Asian/Pacific Islander groups, especially women, who often work under intolerable conditions, as in the sweatshops. . . .

THE DEVILS OF DUALISM

Yet another cause of the persistent Black-white conception of racism is dualism, the philosophy that sees all life as consisting of two irreducible elements. Those elements are usually oppositional, like good and evil, mind and body, civilized and savage. Dualism allowed the invaders, colonizers and enslavers of today's United States to rationalize their actions by stratifying supposed opposites along race, color or gender lines. So mind is European, male and rational; body is colored, female and emotional. Dozens of other such pairs can be found, with their clear implications of superior-inferior. In the arena of race, this society's dualism has long maintained that if a person is not totally white (whatever that can mean biologically), he or she must be considered Black. . . .

Racism evolves; our models must also evolve. Today's challenge is to move beyond the Black-white dualism that has served as the foundation of White Supremacy. In taking up this challenge, we have to proceed with both boldness and infinite care. Talking race in these United States is an intellectual minefield; for every observation, one can find three contradictions and four necessary qualifications from five different racial groups. Making your way through that complexity, you have to think: keep your eyes on the prize.

WHAT WHITE SUPREMACISTS TAUGHT
A JEWISH SCHOLAR ABOUT IDENTITY

12

Abby L. Ferber

A few years ago, my work on white supremacy led me to the neo-Nazi tract *The New Order*, which proclaims: "The single serious enemy facing the white man is the Jew." I must have read that statement a dozen times. Until then, I hadn't thought of myself as the enemy.

When I began my research for a book on race, gender, and white supremacy, I could not understand why white supremacists so feared and hated Jews. But after being immersed in newsletters and periodicals for months, I learned that white supremacists imagine Jews as the masterminds behind a great plot to mix races and, thereby, to wipe the white race out of existence.

The identity of white supremacists, and the white racial purity they espouse, requires the maintenance of secure boundaries. For that reason, the literature I read described interracial sex as "the ultimate abomination." White supremacists see Jews as threats to racial purity, the villains responsible for desegregation, integration, the civil-rights movement, the women's movement, and affirmative action—each depicted as eventually leading white women into the beds of black men. Jews are believed to be in control everywhere, staging a multipronged attack against the white race. For *WAR*, the newsletter

---

From *The Chronicle of Higher Education*, May 7, 1999, pp. B6–B7. Reprinted with permission of the author.

of White Aryan Resistance, the Jew "promotes a thousand social ills . . . [f]or which you'll have to foot the bills."

Reading white-supremacist literature is a profoundly disturbing experience, and even more difficult if you are one of those targeted for elimination. Yet, as a Jewish woman, I found my research to be unsettling in unexpected ways. I had not imagined that it would involve so much self-reflection. I knew white supremacists were vehemently anti-Semitic, but I was ambivalent about my Jewish identity and did not see it as essential to who I was. Having grown up in a large Jewish community, and then having attended a college with a large Jewish enrollment, my Jewishness was invisible to me—something I mostly ignored. As I soon learned, to white supremacists, that is irrelevant.

Contemporary white supremacists define Jews as non-white: "not a religion, they are an Asiatic *race*, locked in a mortal conflict with Aryan man," according to *The New Order*. In fact, throughout white-supremacist tracts, Jews are described not merely as a separate race, but as an impure race, the product of mongrelization. Jews, who pose the ultimate threat to racial boundaries, are themselves imagined as the product of mixed-race unions.

Although self-examination was not my goal when I began, my research pushed me to explore the contradictions in my own racial identity. Intellectually, I knew that the meaning of race was not rooted in biology or genetics, but it was only through researching the white-supremacist movement that I gained a more personal understanding of the social construction of race. Reading white-supremacist literature, I moved between two worlds: one where I was white, another where I was the non-white seed of Satan; one where I was privileged, another where I was despised; one where I was safe and secure, the other where I was feared and thus marked for death.

According to white-supremacist ideology, I am so dangerous that I must be eliminated. Yet, when I put down the racist, anti-Semitic newsletters, leave my office, and walk outdoors, I am white.

Growing up white has meant growing up privileged. Sure, I learned about the historical persecutions of Jews, overheard the hushed references to distant relatives lost in the Holocaust. I knew of my grandmother's experiences with anti-Semitism as a child of the only Jewish family in a Catholic neighborhood. But those were just stories to me. Reading white supremacists finally made the history real.

While conducting my research, I was reminded of the first time I felt like an "other." Arriving in the late 1980s for the first day of graduate school in the Pacific Northwest, I was greeted by a senior graduate student with the welcome: "Oh, you're the Jewish one." It was a jarring remark, for it immediately set me apart. This must have been how my mother felt, I thought, when, a generation earlier, a college classmate had asked to see her horns. Having lived in predominantly Jewish communities, I had never experienced my

Jewishness as "otherness." In fact, I did not even *feel* Jewish. Since moving out of my parents' home, I had not celebrated a Jewish holiday or set foot in a synagogue. So it felt particularly odd to be identified by this stranger as a Jew. At the time, I did not feel that the designation described who I was in any meaningful sense.

But whether or not I define myself as Jewish, I am constantly defined by others that way. Jewishness is not simply a religious designation that one may choose, as I once naïvely assumed. Whether or not I see myself as Jewish does not matter to white supremacists.

I've come to realize that my own experience with race reflects the larger historical picture for Jews. As whites, Jews today are certainly a privileged group in the United States. Yet the history of the Jewish experience demonstrates precisely what scholars mean when they say that race is a social construction.

At certain points in time, Jews have been defined as a non-white minority. Around the turn of the last century, they were considered a separate, inferior race, with a distinguishable biological identity justifying discrimination and even genocide. Today, Jews are generally considered white, and Jewishness is largely considered merely a religious or ethnic designation. Jews, along with other European ethnic groups, were welcomed into the category of "white" as beneficiaries of one of the largest affirmative-action programs in history—the 1944 GI Bill of Rights. Yet, when I read white-supremacist discourse, I am reminded that my ancestors were expelled from the dominant race, persecuted, and even killed.

Since conducting my research, having heard dozens of descriptions of the murders and mutilations of "race traitors" by white supremacists, I now carry with me the knowledge that there are many people out there who would still wish to see me dead. For a brief moment, I think that I can imagine what it must feel like to be a person of color in our society . . . but then I realize that, as a white person, I cannot begin to imagine that.

Jewishness has become both clearer and more ambiguous for me. And the questions I have encountered in thinking about Jewish identity highlight the central issues involved in studying race today. I teach a class on race and ethnicity, and usually, about midway through the course, students complain of confusion. They enter my course seeking answers to the most troubling and divisive questions of our time, and are disappointed when they discover only more questions. If race is not biological or genetic, what is it? Why, in some states, does it take just one black ancestor out of 32 to make a person legally black, yet those 31 white ancestors are not enough to make that person white? And, always, are Jews a race?

I have no simple answers. As Jewish history demonstrates, what is and is not a racial designation, and who is included within it, is unstable and changes over time—and that designation is always tied to power. We do not have to look far to find other examples: The Irish were also once considered non-white in the United States, and U.S. racial categories change with almost every census.

My prolonged encounter with the white-supremacist movement forced me to question not only my assumptions about Jewish identity, but also my assumptions about whiteness. Growing up "white," I felt raceless. As it is for most white people, my race was invisible to me. Reflecting the assumption of most research on race at the time, I saw race as something that shaped the lives of people of color—the victims of racism. We are not used to thinking about whiteness when we think about race. Consequently, white people like myself have failed to recognize the ways in which our own lives are shaped by race. It was not until others began identifying me as the Jew, the "other," that I began to explore race in my own life.

Ironically, that is the same phenomenon shaping the consciousness of white supremacists: They embrace their racial identity at the precise moment when they feel their privilege and power under attack. Whiteness historically has equaled power, and when that equation is threatened, their own whiteness becomes visible to many white people for the first time. Hence, white supremacists seek to make racial identity, racial hierarchies, and white power part of the natural order again. The notion that race is a social construct threatens that order. While it has become an academic commonplace to assert that race is socially constructed, the revelation is profoundly unsettling to many, especially those who benefit most from the constructs.

My research on hate groups not only opened the way for me to explore my own racial identity, but also provided insight into the question with which I began this essay: Why do white supremacists express such hatred and fear of Jews? This ambiguity in Jewish racial identity is precisely what white supremacists find so threatening. Jewish history reveals race as a social designation, rather than a God-given or genetic endowment. Jews blur the boundaries between whites and people of color, failing to fall securely on either side of the divide. And it is ambiguity that white supremacists fear most of all.

I find it especially ironic that, today, some strict Orthodox Jewish leaders also find that ambiguity threatening. Speaking out against the high rates of intermarriage among Jews and non-Jews, they issue dire warnings. Like white supremacists, they fear assaults on the integrity of the community and fight to secure its racial boundaries, defining Jewishness as biological and restricting it only to those with Jewish mothers. For both white supremacists and such Orthodox Jews, intermarriage is tantamount to genocide.

For me, the task is no longer to resolve the ambiguity, but to embrace it. My exploration of white-supremacist ideology has revealed just how subversive doing so can be: Reading white-supremacist discourse through the lens of Jewish experience has helped me toward new interpretations. White supremacy is not a movement just about hatred, but even more about fear: fear of the vulnerability and instability of white identity and privilege. For white supremacists, the central goal is to naturalize racial identity and hierarchy, to establish boundaries.

Both my own experience and Jewish history reveal that to be an impossible task. Embracing Jewish identity and history, with all their contradictions, has given me an empowering alternative to white-supremacist conceptions of race. I have found that eliminating ambivalence does not require eliminating ambiguity.

RACE MATTERS

13

Cornel West

Since the beginning of the nation, white Americans have suffered from a deeper inner uncertainty as to who they really are. One of the ways that has been used to simplify the answer has been to seize upon the presence of black Americans and use them as a marker, a symbol of limits, a metaphor for the "outsider." Many whites could look at the social position of blacks and feel that color formed an easy and reliable gauge for determining to what extent one was or was not American. Perhaps that is why one of the first epithets that many European immigrants learned when they got off the boat was the term "nigger"—it made them feel instantly American. But this is tricky magic. Despite his racial difference and social status, something indisputably American about Negroes not only raised doubts about the white man's value system but aroused the troubling suspicion that whatever else the true American is, he is also somehow black.

Ralph Ellison, "What America Would Be Like Without Blacks" (1970)

What happened in Los Angeles in April of 1992 was neither a race riot nor a class rebellion. Rather, this monumental upheaval was a multiracial, trans-class, and largely male display of justified social rage. For all its ugly, xenophobic resentment, its air of adolescent carnival, and its downright barbaric behavior, it signified the sense of powerlessness in American society. Glib attempts to reduce its meaning to the pathologies of the black underclass, the criminal actions of hoodlums, or the political revolt of the oppressed urban masses miss the mark. Of those arrested, only 36 percent were black, more than a third had full-time jobs, and most claimed to shun political affiliation. What we witnessed in Los Angeles was the consequence of a lethal linkage of economic decline, cultural decay, and political lethargy in American life. Race was the visible catalyst, not the underlying cause.

The meaning of the earthshaking events in Los Angeles is difficult to grasp because most of us remain trapped in the narrow framework of the dominant liberal and conservative views of race in America, which with its worn-out vocabulary leaves us intellectually debilitated, morally disempowered, and personally depressed. The astonishing disappearance of the event from public dialogue is testimony to just how painful and distressing a serious engagement with race is. Our truncated public discussions of race suppress the best of who and what we are as a people because they fail to confront the complexity of the issue in a candid and critical manner. The predictable pitting of liberals against conservatives, Great Society Democrats against self-help Republicans, reinforces intellectual parochialism and political paralysis.

The liberal notion that more government programs can solve racial problems is simplistic—precisely because it focuses *solely* on the economic dimension. And the conservative idea that what is needed is a change in the moral behavior of poor black urban dwellers (especially poor black men, who, they say, should stay married, support their children, and stop committing so much crime) highlights immoral actions while ignoring public responsibility for the immoral circumstances that haunt our fellow citizens.

The common denominator of these views of race is that each still sees black people as a "problem people," in the words of Dorothy I. Height, president of the National Council of Negro Women, rather than as fellow American citizens with problems. Her words echo the poignant "unasked question" of W. E. B. Du Bois, who, in *The Souls of Black Folk* (1903), wrote:

> They approach me in a half-hesitant sort of way, eye me curiously or compassionately, and then instead of saying directly, How does it feel to be a problem? they say, I know an excellent colored man in my town. . . . Do not these Southern outrages make your blood boil? At these I smile, or am interested, or reduce the boiling to a simmer, as the occasion may require. To the real question, How does it feel to be a problem? I answer seldom a word.

Nearly a century later, we confine discussions about race in America to the "problems" black people pose for whites rather than consider what this way of viewing black people reveals about us as a nation.

This paralyzing framework encourages liberals to relieve their guilty consciences by supporting public funds directed at "the problems"; but at the same time, reluctant to exercise principled criticism of black people, liberals deny them the freedom to err. Similarly, conservatives blame the "problems" on black people themselves—and thereby render black social misery invisible or unworthy of public attention.

Hence, for liberals, black people are to be "included" and "integrated" into "our" society and culture, while for conservatives they are to be "well behaved" and "worthy of acceptance" by "our" way of life. Both fail to see that the presence and predicaments of black people are neither additions to nor defections from American life, but rather *constitutive elements of that life.*

To engage in a serious discussion of race in America, we must begin not with the problems of black people but with the flaws of American society—flaws rooted in historic inequalities and longstanding cultural stereotypes. How we set up the terms for discussing racial issues shapes our perception and response to these issues. As long as black people are viewed as a "them," the burden falls on blacks to do all the "cultural" and "moral" work necessary for healthy race relations. The implication is that only certain Americans can define what it means to be American—and the rest must simply "fit in."

The emergence of strong black-nationalist sentiments among blacks, especially among young people, is a revolt against this sense of having to "fit in." The variety of black-nationalist ideologies, from the moderate views of Supreme Court Justice Clarence Thomas in his youth to those of Louis Farrakhan today, rest upon a fundamental truth: white America has been historically weak-willed in ensuring racial justice and has continued to resist fully accepting the humanity of blacks. As long as double standards and differential treatment abound—as long as the rap performer Ice-T is harshly condemned while former Los Angeles Police Chief Daryl F. Gates's antiblack comments are received in polite silence, as long as Dr. Leonard Jeffries's anti-Semitic statements are met with vitriolic outrage while presidential candidate Patrick J. Buchanan's anti-Semitism receives a genteel response—black nationalisms will thrive.

Afrocentrism, a contemporary species of black nationalism, is a gallant yet misguided attempt to define an African identity in a white society perceived to be hostile. It is gallant because it puts black doings and sufferings, not white anxieties and fears, at the center of discussion. It is misguided because—out of fear of cultural hybridization and through silence on the issue of class, retrograde views on black women, gay men, and lesbians, and a reluctance to link race to the common good—it reinforces the narrow discussions about race.

To establish a new framework, we need to begin with a frank acknowledgement of the basic humanness and Americanness of each of us. And we must acknowledge that as a people—*E Pluribus Unum*—we are on a slippery slope toward economic strife, social turmoil, and cultural chaos. If we go down, we go down together. The Los Angeles upheaval forced us to see not only that we are not connected in ways we would like to be but also, in a more profound sense, that this failure to connect binds us even more tightly together. The paradox of race in America is that our common destiny is more pronounced and imperiled precisely when our divisions are deeper. The Civil War and its legacy speak loudly here. And our divisions are growing deeper. Today, 86 percent of white suburban Americans live in neighborhoods that are less than 1 percent black, meaning that the prospects for the country depend largely on how its cities fare in the hands of a suburban electorate. There is no escape from our interracial interdependence, yet enforced racial hierarchy dooms us as a nation to collective paranoia and hysteria—the unmaking of any democratic order.

The verdict in the Rodney King case which sparked the incidents in Los Angeles was perceived to be wrong by the vast majority of Americans. But whites have often failed to acknowledge the widespread mistreatment of black people, especially black men, by law enforcement agencies, which helped ignite the spark. The verdict was merely the occasion for deep-seated rage to come to the surface. This rage is fed by the "silent" depression ravaging the country—in which real weekly wages of all American workers since 1973 have declined nearly 20 percent, while at the same time wealth has been upwardly distributed.

The exodus of stable industrial jobs from urban centers to cheaper labor markets here and abroad, housing policies that have created "chocolate cities and vanilla suburbs" (to use the popular musical artist George Clinton's memorable phrase), white fear of black crime, and the urban influx of poor Spanish-speaking and Asian immigrants—all have helped erode the tax base of American cities just as the federal government has cut its supports and programs. The result is unemployment, hunger, homelessness, and sickness for millions.

And a pervasive spiritual impoverishment grows. The collapse of meaning in life—the eclipse of hope and absence of love of self and others, the breakdown of family and neighborhood bonds—leads to the social deracination and cultural denudement of urban dwellers, especially children. We have created rootless, dangling people with little link to the supportive networks—family, friends, school—that sustain some sense of purpose in life. We have witnessed the collapse of the spiritual communities that in the past helped Americans face despair, disease, and death and that transmit through the generations dignity and decency, excellence and elegance.

The result is lives of what we might call "random nows," of fortuitous and fleeting moments preoccupied with "getting over"—with acquiring pleasure, property, and power by any means necessary. (This is not what Malcolm X meant by this famous phrase.) Post-modern culture is more and more a market culture dominated by gangster mentalities and self-destructive wantonness. This culture engulfs all of us—yet its impact on the disadvantaged is devastating, resulting in extreme violence in everyday life. Sexual violence against women and homicidal assaults by young black men on one another are only the most obvious signs of this empty quest for pleasure, property, and power.

Last, this rage is fueled by a political atmosphere in which images, not ideas, dominate, where politicians spend more time raising money than debating issues. The functions of parties have been displaced by public polls, and politicians behave less as thermostats that determine the climate of opinion than as thermometers registering the public mood. American politics has been rocked by an unleashing of greed among opportunistic public officials— who have followed the lead of their counterparts in the private sphere, where, as of 1989, 1 percent of the population owned 37 percent of the wealth and 10 percent of the population owned 86 percent of the wealth—leading to a profound cynicism and pessimism among the citizenry.

And given the way in which the Republican Party since 1968 has appealed to popular xenophobic images—playing the black, female, and homophobic cards to realign the electorate along race, sex, and sexual-orientation lines— it is no surprise that the notion that we are all part of one garment of destiny is discredited. Appeals to special interests rather than to public interests reinforce this polarization. The Los Angeles upheaval was an expression of utter fragmentation by a powerless citizenry that includes not just the poor but all of us.

What is to be done? How do we capture a new spirit and vision to meet the challenges of the post-industrial city, post-modern culture, and post-party politics?

First, we must admit that the most valuable sources for help, hope, and power consist of ourselves and our common history. As in the ages of Lincoln, Roosevelt, and King, we must look to new frameworks and languages to understand our multilayered crisis and overcome our deep malaise.

Second, we must focus our attention on the public square—the common good that undergirds our national and global destinies. The vitality of any public square ultimately depends on how much we *care* about the quality of our lives together. The neglect of our public infrastructure, for example—our water and sewage systems, bridges, tunnels, highways, subways, and streets— reflects not only our myopic economic policies, which impede productivity, but also the low priority we place on our common life.

The tragic plight of our children clearly reveals our deep disregard for public well-being. About one out of every five children in this country lives in poverty, including one out of every two black children and two out of every five Hispanic children. Most of our children—neglected by overburdened parents and bombarded by the market values of profit-hungry corporations— are ill-equipped to live lives of spiritual and cultural quality. Faced with these facts, how do we expect ever to constitute a vibrant society?

One essential step is some form of large-scale public intervention to ensure access to basic social goods—housing, food, health care, education, child care, and jobs. We must invigorate the common good with a mixture of government, business, and labor that does not follow any existing blueprint. After a period in which the private sphere has been sacralized and the public square gutted, the temptation is to make a fetish of the public square. We need to resist such dogmatic swings.

Last, the major challenge is to meet the need to generate new leadership. The paucity of courageous leaders—so apparent in the response to the events in Los Angeles—requires that we look beyond the same elites and voices that recycle the older frameworks. We need leaders—neither saints nor sparkling television personalities—who can situate themselves within a larger historical narrative of this country and our world, who can grasp the complex dynamics of our peoplehood and imagine a future grounded in the best of our past, yet who are attuned to the frightening obstacles that now perplex us. Our ideals of freedom, democracy, and equality must be invoked to invigorate all of us, especially the landless, propertyless, and luckless. Only a visionary leadership that can motivate "the better angels of our nature," as Lincoln said, and activate possibilities for a freer, more efficient, and stable America—only that leadership deserves cultivation and support.

This new leadership must be grounded in grass-roots organizing that highlights democratic accountability. Whoever *our* leaders will be as we approach the twenty-first century, their challenge will be to help Americans determine whether a genuine multiracial democracy can be created and sustained in an era of global economy and a moment of xenophobic frenzy.

Let us hope and pray that the vast intelligence, imagination, humor, and courage of Americans will not fail us. Either we learn a new language of empathy and compassion, or the fire this time will consume us all.

Class and Inequality

ECONOMIC APARTHEID IN AMERICA

14

Chuck Collins and Felice Veskel

Most of us experience our daily lives in terms of family, work, friends, and our local community. We do not always see the connections between the struggles we may be facing and changes in the larger economy such as longer working hours, having multiple jobs, and the inability to save money. Even when we understand the link between the larger political and economic forces and our daily lives, we do not necessarily believe that we can do anything to improve our circumstances. Often we simply hear: "That's just the way things are" or "There is no alternative." However, this is not true.

There are alternatives—and there were times in our country's history when we were not so economically divided. The U.S. economy was significantly fairer in the thirty years following World War II. Prosperity was better shared among almost everyone in society than it is today. There is no reason why this should no longer hold true in the present. During the last few decades, the economic rules of the game were changed, by wealthy individuals and corporations, and they can be changed back by people like us. In the 1930s, the New Deal ushered in an era of greater economic security and a rising standard of living for seniors and working families. In the 1960s, we launched a War on Poverty that improved health care for seniors and the poor and provided a "Head Start" to low-income children. Today, at the beginning of a new century, we need a new set of policies and priorities to ensure a more broadly shared prosperity.

From Chuck Collins and Felice Veskel, *Economic Apartheid in America: a Primer of Inequality and Insecurity* (New York: The New Press, 2000). Reprinted by permission of the New Press.

There have also been cycles of great inequality in our country's past. In the 1880s, as the United States went through the industrial revolution, there was grotesque inequality as the richest 1 percent owned an estimated 50 percent of all private wealth. The ostentatious fortunes of the robber barons were built on the backs of impoverished workers and by extracting wealth from the natural environment. The industrial workforce was subject to enormous dangers and sweatshop conditions. Children worked in factories. Cities were crowded and disease thrived in unsanitary conditions.

As a result, a coalition of workers, farmers, and urban reformers created a social movement pushing for fundamental reforms to make the economy fairer and the distribution of income and wealth more equitable. The Populist movement united farmers and urban workers, across racial lines, to build a powerful political movement that became a countervailing force to the agenda of big corporations and wealth-holders. The Populists pushed for democratic reforms such as the direct election of U.S. senators. They fought for reforms like antitrust legislation and a constitutional amendment for the first income tax to break up overconcentrations of wealth. They established alternative economic institutions including a national network of agricultural producer cooperatives. They pressed for reforms in tariffs and cuts in special subsidies for the big trusts, a nineteenth-century version of cutting corporate welfare. The Populists and their reforms changed the political and economic face of the United States.

Today, we find ourselves in a world similar to that of the farmers and workers in the 1880s before the peak of the Populist movement. Inequality is on the rise. Power has shifted into the hands of fewer wealthy individuals and corporations, and economic insecurity is growing for most. It is time to build an economic fairness movement to act as countervailing force to the power of concentrated wealth and large corporations. . . .

Some people argue that inequality is a relative problem. We should not be so concerned with the gap as with the overall improvement in our standard of living. Someone living in the bottom half of the income spectrum today has a much higher standard of living than someone in the bottom half of the income spectrum forty years ago. Apologists for inequality point to mobility and the notion that people move up the economic ladder over their lifetimes. Others might say that poverty is the problem that we should address. We should not concern ourselves with the top end of the income and wealth ladder because it is irrelevant. . . .

. . . Does the increased availability of consumer goods, twenty-four-hour communications and entertainment networks, cheap air travel, and more college graduates add up to a great economy for all? Or have we paid too great a price for this "prosperity" by tolerating a surge in economic inequality, a hollowing out of the economic and social middle? . . .

Let's start by looking close to home at the impact of the changing economy. There are new pressures and some alarming trends facing a growing number of households in this country. These include

- less free time and more working hours,
- fewer households with health insurance,
- rising personal debt,
- declining personal savings,
- diminishing retirement security,
- growing number of temporary jobs,
- rising college costs.

THE DECLINE OF LEISURE TIME
AND THE BREAKDOWN OF CIVIL SOCIETY

One powerful consequence of growing inequality is an erosion in the amount of free time that families have. As individuals and families struggle to stay afloat and remain secure in the changing economy, they are spending more hours at work.

Falling wages in the 1970s and early 1980s were masked by the entry of a second wage earner in many households—usually a woman—into the workforce. Families now have to work longer hours to make up for falling wages. At the same time, temporary and part-time workers generally do not have paid vacations, and their numbers in the workforce are growing. The number of overall hours worked per year, per household has increased since 1972. The number of hours worked per person each year has increased from 1,905 in 1980 to 1,960 in 1998.[1] According to economist Julie Schor, since 1972, U.S. workers had an increase of eighteen and one-half work days or more than three and a half weeks of work! The average two-earner middle-income family has seen its annual paid workload grow from 3,206 hours in 1989 to 3,335 in 1997.[2] According to the Council of Economic Advisors, this has left us with a dramatic 14 percent decrease in people's free time and leisure time in the last several decades.[3] . . .

Technological advances in the workplace, instead of increasing the amount of free time, are having the opposite effect. Productivity has been steadily increasing, but these gains have not translated into increased wages or increased free time. Blue-collar and service workers have experienced a veritable "speed up" in the workplace as employers squeeze productivity gains into increased profits. Not only are people working longer hours, but in many cases the tempo and scope of their jobs have become more harried.

For middle-class service and white-collar workers, technological advances like computers, pagers, faxes, and electronic mail can contribute to a breakdown of leisure time and an erosion of vacation time. Today's modern white-collar worker goes on summer vacation for a much needed break, but brings along a cell phone, a computer with modem, and a portable fax machine. Employers may encourage the ways in which technology breaks down the division between work and leisure.

U.S. vacations are among the shortest in the industrialized world, averaging only two weeks. They are also not legally mandated, as they are in many Western European countries where workers get between four and six weeks of vacation. . . .

This loss of leisure time has a direct impact on the quality of people's lives. As people have less free time, they have less time to care for children and elders, less time to be involved in schools and education and less time to volunteer to help others. This dramatically increases stress, particularly on women whose unpaid labor is largely performed in this nonmonetary "caring economy." Stress leads to increased illness and the unraveling of communities, the breakdown in voluntary mutual aid systems, and the fragmentation and isolation of people. It leads to "latchkey" children and a lack of support for the next generation, all of which we have seen may have very serious consequences.

Overwork feeds the breakdown of civil society and civic institutions. A free democratic society depends on strong voluntary institutions such as religious congregations, political parties, neighborhood block clubs, fraternal or women's organizations, nonprofit service organizations, and other nongovernmental, nonbusiness organizations. A number of sociologists have pointed to the decline of "social capital" that comes from strong voluntary institutions as contributing to the erosion of democracy.[4] As these institutions weaken, so does the power of ordinary citizens to defend their interests as consumers, workers, and communities. As we will discuss later, this contributes to the power imbalance that underlies growing inequality.

FEWER HOUSEHOLDS WITH HEALTH INSURANCE

. . . Having a job does not guarantee health insurance. The percentage of Americans covered through employer-sponsored health insurance declined from 67.9 percent in 1990 to 64.6 percent in 1995. Twenty-four million of those without insurance are employed, 17 percent of whom are in the temporary or contingent workforce. Even as the unemployment rate declines, the rate of workers without health insurance increases. Even with a sustained decrease in unemployment, current trends indicate that the number of uninsured non-elderly Americans is growing by 1 million a year and will approach 47 million by the year 2005, equal to one in five Americans under the age of sixty-five.[5] Thirty-four percent of the nation's 31 million Hispanic people have no health insurance, compared to 22 percent of blacks and 12 percent of non-Hispanic whites. Over 44 million people in the nation lack health insurance, up from 31 million a decade ago.[6]

RISING PERSONAL DEBT

Another way families continue to make up for falling wages in order to maintain (or in some cases attain) a certain standard of living is by going deeper into debt. Sixty percent of all American households carry credit card balances. The balances average over $7,000, costing these households more than $1,000 per year in interest and fees.[7] Total credit card debt increased from $243 billion in 1990 to $560 billion in 1997.[8] Some of this is rooted in consumerism, but a central reason for growing debt lies in declining or stagnant wages. People are now using credit cards for things like food and medicine—items previously paid for with cash. One downside of growing personal indebtedness is the increasing number of personal bankruptcies, which are now at an all-time high. . . .

DECLINING PERSONAL SAVINGS

The flip side of greater debt is less savings. The savings rate is the percentage of annual income that is actually saved each year. Since 1980, the savings rate has generally fallen. The United States has the lowest savings rates of any industrialized country. In 1999, the U.S. savings rate was 2.3 percent, down from 4.5 percent in 1997 and 10.6 percent in 1984.[9] In contrast, the savings rate in 1998 in Japan was 13 percent and in Germany it was 11.5 percent.

Why does the United States have such a low savings rate? It is *not* because the average person has no personal restraint or will to save. According to researchers from the Boston Federal Reserve, the main reason for the lower savings rate is linked to rising cost of health care and the increase in other involuntary costs such as day care and bank service fees.[10]

DIMINISHING RETIREMENT SECURITY

Those employees lucky enough to have pensions have ones that are less secure. The percentage of workers in the private labor force in monthly pension plans (where companies bear the risk of falling markets) has declined from 38 percent in 1980 to 24 percent in 1995. Meanwhile, the growth in private 401(k) pension plans (where the employee bears all the risk) has failed to broaden overall pension coverage. In fact, there is more inequality in 401(k) plans than in defined benefit plans, as they tend to be held by higher income workers. The decline in both savings and guaranteed pensions are two concrete indicators of growing insecurity. . . .

GROWING NUMBER OF TEMPORARY JOBS

The greatest percentage of new jobs in the workforce are filled by temporary and part-time workers. Currently about 30 percent of the workforce is what is called "contingent," including temporary, part-time, contract, and day laborers.[11] Some of these workers are voluntarily part-time. But a growing number are involuntarily part-time, often holding two to three jobs to pay their bills. According to the Bureau of Labor Statistics, two-thirds of contingent workers would like traditional permanent jobs. . . .

HIGHER EDUCATION, HIGHER REACH

The cost of going to college has risen, while government support to ensure access to higher education has failed to keep pace. Going to college is more of a privilege than it was twenty years ago. College loans have now replaced college grants, rising from 41.4 percent of student financing in 1982 to 58.9 percent in 1996. Student debt has dramatically increased from an average of $8,200 per student in 1991 to $18,800 in 1997.[12] Student debt between 1991 and 1997 totaled $140 billion—more than the total combined student borrowing for the 1960s, 1970s, and 1980s.[13] Many students will still be paying off college loans in their mid-thirties.

The rising cost of college affects who gets to go to college. In 1995, 83.4 percent of high school students in the top fifth of income-earning households went to college compared to 56.1 percent of students in the middle three quintiles and 41.3 percent in the bottom quintile.[14] As of 1998, 25 percent of whites have completed 4 years of college compared with 14.7 percent of blacks and 11 percent of Latinos.[15]

Some of these stresses on families are invisible and not widely discussed or shared. Nor do we often think of them as the result of a more polarized society. Many of us experience them as a result of our individual failing. . . . The effects of inequality on other aspects of our lives . . . undermine our basic sense of well being.

INEQUALITY AND THE THREAT TO PROSPERITY

Economic inequality is bad for the economy and for economic growth. Inequality means that some people have too much money burning holes in their pockets while some people don't have enough. As the stable middle class in this country loses savings and security, inequality has a downside for everyone, particularly smaller businesses. Small business owners should be alarmed about the economic divide because it directly affects their consumers. Unless you're a Lexus car dealer or a diamond jeweler, inequality is a threat to the wider prosperity of our economy. . . .

Since 1980, there has been a visible stratification in the retail market. Stores that used to serve the middle class are going out of business. Retailers must either remake themselves as bargain outlets to compete in the low-end market with stores like Walmart or K-Mart, or they must become upscale, like Bloomingdale's, Neiman-Marcus, or Brookstone, to chase the plentiful but concentrated cash of the richest 10 percent of the population. . . .

Products are now designed with this two-tier market in mind. Take the refrigerator market, for example. Manufacturers of refrigerators create a stripped-down basic model that they sell to working families. This no-frills model has a couple of shelves and plastic vegetable drawers. However, they also design an upscale refrigerator that dispenses ice, mineral water, and includes an erasable message board that retails for 100 to 200 percent higher than the basic model. . . .

INEQUALITY AND DEMOCRACY

. . . Americans spent a good part of a century throwing off the rule of the British monarchy and establishing our own self-governing democracy. Yet we are now at risk of becoming subjects again. Too much economic inequality is a fundamental threat to our democracy.

Once a household accumulates wealth in excess of $10 million, placing it in the richest one half of one percent, they have moved beyond meeting the needs and wildest desires of themselves and their heirs. As the character Bud Fox asked speculator Gordon Gekko in the 1987 film *Wall Street*, "How many yachts can you water-ski behind?" After a certain point, the accumulation of wealth is about amassing power—the power to influence society and the rules of our economy.

This influence, however, has not always been all bad. Carnegie's libraries and Rockefeller's national parks stand as testaments to how personal wealth has contributed to the commonwealth. Nonetheless, concentrated wealth can have great corrupting power when it extends to the ownership of the media and the power to shape public opinion, through the shaping of philanthropic priorities and through political contributions. Such concentrated power mocks democracy and undermines the power and participation of the broader society.

Large corporations use their concentrated wealth and power to lobby for rule changes in legislatures across the land. Corporate lobbyists, with millions in campaign contributions at their disposal, roam the halls of Congress, influencing the writing or unraveling of environmental regulations, tax laws, and health, safety, and consumer protections. . . .

Imagine that we are lining up all of the people in the United States, from the lowest income to the highest income. Visualize all these people standing in a line as they are divided into five equal size groups. [Each group takes one step for every 10 percent increase or decrease in income over the last twenty years. The top fifth takes four steps ahead; the bottom, a half step backwards.]

The biggest leap in real growth in income over the last twenty years goes to households with incomes in the top one percent of households. The rich are getting richer, but it's not just the poor who are getting poorer. Half the population is standing still if not moving backwards.

The inequality of the last two decades becomes even more dramatic when we contrast the most recent twenty years to the thirty years after World War II. . . . From 1947 to 1979, all of the quintiles saw their incomes increase 90 to 100 percent. The bottom quintile actually grew most rapidly, its income growing over 116 percent. Economic prosperity was shared—we grew together. However, in the last twenty years we have grown dramatically apart.

RACE, GENDER, AND INCOME

The postwar years were not a panacea. Racism and sexism were powerful social forces. . . . However, in the post–World War II years, people of color did share in the income gains of the society, though by no means equally. Incomes of blacks, Latinos, and other nonwhite groups remain significantly lower than that of white households. But while the disparity of wealth and financial assets is great between people of color and whites, the trends in income disparity within black and Latino populations are very similar to the patterns for the population as a whole.

The lowest fifth of black income earners saw their incomes fall 9.5 percent between 1979 and 1997. Meanwhile, the wealthiest fifth of black income earners saw their incomes go up 21.4 percent, with the wealthiest five percent going up 30.8 percent.[16] As sociologist William Julius Wilson points out, "In 1992, the highest fifth of black families. . . . secured a record 48.8 percent of the total black family income compared to 43.8 percent of the total white income received by the highest fifth of white families, also a record. So, while income inequality has widened generally in America since 1975, the divide is even more dramatic among African Americans."[17]

In the thirty years following World War II, race was a principal barrier to the economic advancement for people of color. Research on the economic mobility and advancement of blacks, according to Wilson, "could uncover no evidence of class effects on occupation or income achievements that could rival the effect of race." Beginning in the mid-1960s, "class began to affect career and generational mobility for Blacks as it had regularly done for whites."[18] In other words, prosperous blacks began to move ahead and low-income blacks began to move backwards. Race, while still a major factor, was diminishing as a factor in determining economic security, while class was becoming more significant.

This explains why the Reverend Jesse Jackson, when asked to comment on President Clinton's proposal to have a national dialogue on race, suggested that we should also have a national dialogue on class.

We've looked at income and race; what about income and gender? The gap between male and female incomes has closed in the last two decades, as efforts to enforce pay equity have succeeded. At the beginning of the 1980s, women earned 59 cents for every dollar of men's earnings. Today, the pay gap has narrowed to 78 cents on the dollar. Unfortunately, only half of this gain comes from higher women's wages. The other half is because men's wages have dropped.[19]

The fact that at least two incomes are required to maintain a financially stable household in the last decade means that many more women have entered the workforce, albeit at a lower wage than their male counterparts. The persistent wage gap between men and women means that households headed by single women wage earners make up an enormous percentage of the families in poverty.

THE GAP BETWEEN HIGHEST AND AVERAGE WAGE EARNERS

One measurable facet of inequality is the widening gap between highest and average wage earners, both within individual firms and within the entire economy. In 1975, the gap between the highest and average paid worker in a firm was 41 to 1. In 1975, the median wage was $13.14 in 1997 dollars, about $27,300 a year.[20] This means that the average top manager pay was about $1,120,579 in 1997 dollars. By 1997, the ratio between highest and average worker pay had risen to 212 to 1. In 1998, the gap catapulted up to 419 to 1.[21]

In its 1997 survey, *Business Week* called executive compensation "out of control." In 1998, the average salary and bonus for top executives at the 365 largest U.S. corporations dropped slightly from $2.3 million to $2.1 million. But CEOs weren't concerned. That's because if you add in the value of long-term stock options, executive pay climbed 36 percent in 1998 to an average of $10.6 million, increasing, in *Business Week*'s words, an astounding 442 percent over the 1990 average of $2 million.[22]

Executive or CEO compensation is extremely visible because CEOs sit at the top of publicly traded corporations that report their top salaries in the proxy statements that accompany annual reports to shareholders. But top professional salaries have also risen, especially for bond traders, lawyers, medical specialists, high tech wizards, sports and entertainment stars, and others who primarily work in the area of corporate finance. Many of these salaries are not subject to public scrutiny and attention. . . .

In this era of corporate restructuring and downsizing, many chief executives have a perverse incentive to cut jobs because they get a pay bounce. Examining the thirty biggest downsizing firms in 1996, one study found CEOs who presided over a significant downsizing received a 67 percent pay increase, compared to 54 percent for executives at the top 365 U.S. firms. . . .

THE INEQUALITY OF WEALTH

What Is Wealth and Why Is It Important?

Income and wages are one index of how prosperity is distributed; the ownership of wealth is another. If income is the stream of money that comes into our lives each year, wealth is our reservoir. From our income stream, we take our bucket and pull out our housing costs, food expenses, and health care costs. At the end of the year, any leftover goes into our reservoir, becoming accumulated wealth.

Wealth is comprised of assets (all the stuff you own) minus liabilities (what you owe). This remainder is called net worth. If you own a car worth $4,000, but you owe $5,000 in car loans, you are in the red. You might feel wealthy driving that car down the road, but you actually have $1,000 in negative wealth. (Warning: Do not think about this when you are driving 65 mph.)

Wealth for working-class people is usually in the form of savings, consumer goods like cars and, if they are fortunate, equity in a home. As people move up the economic ladder, wealth takes the form of greater savings and home value—but also includes second homes, recreational equipment (like boats), shares in corporations (called securities or stocks), bonds and commercial real estate. It may also include luxury items like high-priced artwork, racehorses, jewelry, and antiques.

Wealth is important because it is what people have to fall back on and pass on to their children. Yet today, about one out of five households in the United States has zero or negative wealth. This means they literally have no financial reserves to fall back on in times of trouble or, in fact, they owe more than they own. The percentage of households with zero or negative net worth doubled in the last thirty years, increasing from 9.2 percent in 1962 to 18.5 percent in 1995.[23]

Financial planners advise people that they should strive to have at least six months of financial reserves to help them weather a job transition, major illness, or other unexpected change. In reality, 45 percent of the population has only three months or less of financial reserves, even if they were to drop their spending to that of the poverty level. Thirty-eight percent of white households have three months or less of reserves. Even more dramatically, 81 percent of African American households and 79 percent of Latino households have less than three months of financial reserves.[24]

It is in the area of wealth and asset accumulation that the legacy of racial discrimination has left its most profound mark. While black and Latino *incomes* have begun to catch up to white earnings, black and Latino *wealth* continues to lag dramatically behind that of white households. Since wealth accumulates over generations, discrimination has taken its toll on wealth accumulation for people of color. Blacks were prohibited from building wealth by slavery and prohibitions against owning property after the Civil War. Bank lending practices have also discriminated against black and Latino households and kept them from home ownership, business development, or other asset-building measures. . . .

The Concentration of Wealth

In the last twenty years, the overall wealth pie has grown, but virtually all the new growth in wealth has gone to the richest one percent of the population.[25]

In 1976, the wealthiest 1 percent of the population owned just under 20 percent of all the private wealth. The top 10 percent of the population owned about 50 percent of all private wealth.[26] By 1999, the richest 1 percent's share had increased to over 40 percent of all wealth, increasing the top 10 percent's share to 70 percent. The top 1 percent of households now has more wealth than the entire bottom 95 percent. As of 1999, to join the top 1 percent club you need at least $2.45 million in net worth.[27]

Financial wealth is even more concentrated. The top 1 percent of households has nearly half of all financial wealth (net worth minus net equity in owner-occupied housing). Wealth is further concentrated at the top of the top 1 percent. The richest 0.5 percent of households have 42 percent of the financial wealth in the United States.[28]

In the last two decades, what has happened to the wealth of different sectors of the United States? Between 1983 and 1995, the inflation-adjusted net worth of the top 1 percent grew by 17 percent. The bottom 40 percent of households lost an astounding 80 percent. Their net worth shrank from $4,400 to an even more meager $900. The middle fifth of Americans lost over 11 percent. Only the top 5 percent gained any net worth during this period. The top 5 percent now have more than 60 percent of all household wealth.[29] . . .

During the same time, the relative share of wealth owned by the bottom 90 percent of the population declined from 51 percent to 28 percent. What does this loss of wealth actually mean? It means less savings, growing personal debt, less retirement security, and fewer households having access to home ownership. It means fewer financial reserves to fall back on in the event of a setback or job loss. . . .

As the twenty-first century unfolds, the pace of change may seem slow. And conditions of economic inequality and polarization may worsen before they get better. We trust, however, that the seeds of a social movement that is being sown today will blossom. We are at an early stage of movement formation. We each must find our role and devote part of our waking hours to agitating for change. It took a long time for things to become this bad; it will take some time for them to improve.

NOTES

1. U.S. Bureau of the Census, Statistical Abstract of the United States: 1999 (119th Edition), Washington, D.C. (U.S. Government Printing Office), Table 662, 418. For non-agricultural wage and salary employees. Yearly hours arrived at by multiplying weekly hours by 50 weeks.

2. Mishel, Bernstein, and Schmitt, *The State of Working America*, 1996–97 (Armonk, N.Y.: M. E. Sharpe, 1997).

3. Juliet B. Schor, *The Overworked American* (New York: Basic Books, 1993).

4. Robert D. Putnam, "The Strange Disappearance of Civic America," *American Prospect*, Winter 1996. epn.org/prospect/24/24putn.html. Also see Putnam, "Bowling Alone: America's Declining Social Capital," *Journal of Democracy*, January 1995. Much of Putnam's research is rooted in a major study on social capital and democracy in Italy: Robert Putnam, Roberto Leonardi, Raffaella Y. Nanetti, *Making Democracy Work: Civic Traditions in Modern Italy* (Princeton, N.J.: Princeton University Press, 1994).

5. Kenneth E. Thorpe, "The Rising Number of Uninsured Workers: An Approaching Crisis in Health Care Financing" (Tulane University Medical Center, Institute for Health Services Research, October 1997).

6. Peter Kilborn, "Denver's Hispanic Residents Point to Ills of the Uninsured," *New York Times*, 9 April 1999.

7. Stephen Brobeck, The Consumer Federation of America, "Recent Trends in Bank Credit Card Marketing and Indebtedness" (Washington D.C., Consumer Federation of America, 1998), 3.

8. U.S. Bureau of the Census, *Statistical Abstract of the United States: 1998* (118th edition), Washington D.C. (U.S. Government Printing Office), Table 822, 523.

9. U.S. Department of Commerce, Bureau of Economic Analysis. In February 2000, the Commerce Department began using a narrower definition of income that slightly magnified the savings rate decrease.

10. Gordon Matthews, "Drop in Savings Rate Isn't for Lack of Trying," *American Banker*, 12 November 1996. This article summarizes the report by Federal Reserve analysts Lynn Elaine Browne and Joshua Gleason: "The Saving Mystery, or Where Did the Money Go?" *New England Economic Review*, September–October 1996.

11. Economic Policy Institute, "Non-Standard Work, Substandard Jobs: Flexible Work Arrangements in the U.S." (Economic Policy Institute, Washington, D.C., August 1997). Also see Susan N. Houseman, "Temporary, Part-Time, and Contract Employment in the United States: A Report on the W. E. Upjohn Institute's Employer Survey on Flexible Staffing Policies" (Kalamazoo, Michigan: W. E. Upjohn Institute for Employment Research, November 1996; revised June 1997).

12. Nellie Mae, "The College Board," as reported in *Boston Globe,* 23 October 1997.

13. *U.S. News and World Report,* 9 June 1997.

14. U.S. Department of Commerce, Bureau of the Census, Current Population Survey, October 1998. From "Percentage of high school completes ages 16–24 who were enrolled in college after completing high school, by type of institution, family income, and race/ethnicity: October 1972–1996."

15. Bureau of Census, 1998 Current Population Survey.

16. U.S. Census, "Mean Income Received by Each Fifth and Top 5 Percent of Black Families," from March 1998 Current Population Survey.

17. William Julius Wilson, *When Work Disappears* (New York: Vintage Books, 1996), 195.

18. Ibid., 195.

19. Mishel, Bernstein, and Schmitt, *The State of Working America, 1998–99* (Ithaca, N.Y.: Cornell University Press, 1999), 134.

20. Economic Policy Institute, epinet.org/datazone, 1999.

21. "Executive Pay: Special Report," *Business Week,* 20 April 1998, 58.

22. "Executive Pay: Special Report," *Business Week,* 19 April 1999, 72–74.

23. 1995 data from Ed Wolff, "Recent Trends in Wealth Ownership," a paper prepared for the Conference of Benefits and Mechanisms for Spreading Asset Ownership in the United States, New York University, December 10–12, 1998; Table 1 "Mean and Median Wealth and Income, 1983–1997. Data for 1962 from Ferdinand Lundbert, *The Rich and the Superrich* (New York: Lyle Stuart, 1968), citing Federal Reserve, "Survey of Financial Characteristics of Consumers," 1962.

24. Mel Oliver and Tom Shapiro, *Black Wealth/White Wealth* (New York: Routledge, 1995). See also: Dalton Conley, *Being Black, Living in the Red: Race, Wealth, and Social Policy* (Los Angeles and Berkeley: University of California Press, 1999).

25. For an in-depth discussion of wealth inequality, see Chuck Collins, Betsy Leondar-Wright, Holly Sklar, *Shifting Fortunes: The Perils of the Growing American Wealth Gap* (Boston: United for a Fair Economy, 1999). See also: Edward N. Wolff, *Top Heavy: The Increasing Inequality of Wealth in America and What Can Be Done About It* (New York: The New Press, 1996).

26. Edward N. Wolff, *Top Heavy: The Increasing Inequality of Wealth in America and What Can Be Done About It* (New York: The New Press, 1996), 78–79.

27. Edward N. Wolff, "Recent Trends in Wealth Ownership." Computations are based on the Federal Reserve Surveys of Consumer Finances, 1983, 1989, 1992, and 1995.

28. Ibid.

29. Ibid.

TIRED OF PLAYING MONOPOLY?   **15**

Donna Langston

I. Magnin, Nordstrom, The Bon, Sears, Penneys, K mart, Goodwill, Salvation Army. If the order of this list of stores makes any sense to you, then we've begun to deal with the first question which inevitably arises in any discussion of class here in the U.S.—huh? Unlike our European allies, we in the U.S. are reluctant to recognize class differences. This denial of class divisions functions to reinforce ruling class control and domination. America is, after all, the supposed land of equal opportunity where, if you just work hard enough, you can get ahead, pull yourself up by your bootstraps. What the old bootstraps theory overlooks is that some were born with silver shoe horns. Female-headed households, communities of color, the elderly, disabled and children find themselves, disproportionately, living in poverty. If hard work were the sole determinant of your ability to support yourself and your family, surely we'd have a different outcome for many in our society. We also, however, believe in luck and, on closer examination, it certainly is quite a coincidence that the "unlucky" come from certain race, gender and class backgrounds. In order to perpetuate racist, sexist and classist outcomes, we also have to believe that the current economic distribution is unchangeable, has always existed, and probably exists in this form throughout the known universe; i.e., it's "natural." Some people explain or try to account for poverty or class position by focusing on the personal and moral merits of an individual. If people are poor, then it's something they did or didn't do; they were lazy, unlucky, didn't try hard enough, etc. This has the familiar ring of blaming the victims. Alternative explanations focus on the ways in which poverty and class position are due to structural, systematic, institutionalized economic and political power relations. These power relations are based firmly on dynamics such as race, gender, and class.

In the myth of the classless society, ambition and intelligence alone are responsible for success. The myth conceals the existence of a class society, which serves many functions. One of the main ways it keeps the working class and poor locked into a class-based system in a position of servitude is by cruelly creating false hope. It perpetuates the false hope among the working class and poor that they can have different opportunities in life. The hope that they can escape the fate that awaits them due to the class position they were born into. Another way the rags-to-riches myth is perpetuated is by creating enough visible tokens so that oppressed persons believe they, too, can get ahead. The creation of hope through tokenism keeps a hierarchical structure in place and lays the blame for not succeeding on those who don't. This keeps us from resisting and changing the class-based system. Instead, we accept it as inevitable, something we just have to live with. If oppressed people believe in equality of opportunity, then they won't develop class consciousness and will internalize the blame for their economic position. If the working class and poor do not recognize the way false hope is used to control them, they won't get a chance to control their lives by acknowledging their class position, by claiming that identity and taking action as a group.

The myth also keeps the middle class and upper class entrenched in the privileges awarded in a class-based system. It reinforces middle- and upper-class beliefs in their own superiority. If we believe that anyone in society really can get ahead, then middle- and upper-class status and privileges must be deserved, due to personal merits, and enjoyed—and defended at all costs. According to this viewpoint, poverty is regrettable but acceptable, just the outcome of a fair game: "There have always been poor people, and there always will be."

Class is more than just the amount of money you have; it's also the presence of economic security. For the working class and poor, working and eating are matters of survival, not taste. However, while one's class status can be defined in important ways in terms of monetary income, class is also a whole lot more—specifically, class is also culture. As a result of the class you are born into and raised in, class is your understanding of the world and where you fit in; it's composed of ideas, behavior, attitudes, values, and language; class is how you think, feel, act, look, dress, talk, move, walk; class is what stores you shop at, restaurants you eat in; class is the schools you attend, the education you attain; class is the very jobs you will work at throughout your adult life. Class even determines when we marry and become mothers. Working-class women become mothers long before middle-class women receive their bachelor's degrees. We experience class at every level of our lives; class is who our friends are, where we live and work, even what kind of car we drive, if we own one, and what kind of health care we receive, if any. Have I left anything out? In other words, class is socially constructed and all-encompassing. When we experience classism, it will be because of our lack of money (i.e., choices and power in this society) and because of the way we talk, think, act, move—because of our culture.

Class affects what we perceive as and what we have available to us as choices. Upon graduation from high school, I was awarded a scholarship to attend any college, private or public, in the state of California. Yet it never occurred to me or my family that it made any difference which college you went to.

I ended up just going to a small college in my town. It never would have occurred to me to move away from my family for school, because no one ever
had and no one would. I was the first person in my family to go to college. I
had to figure out from reading college catalogs how to apply—no one in my
family could have sat down and said, "Well, you take this test and then you really should think about. . . ." Although tests and high school performance had
shown I had the ability to pick up white middle-class lingo, I still had quite an
adjustment to make—it was lonely and isolating in college. I lost my friends
from high school—they were at the community college, vo-tech school,
working, or married. I lasted a year and a half in this foreign environment before I quit college, married a factory worker, had a baby and resumed living
in a community I knew. One middle-class friend in college had asked if I'd like
to travel to Europe with her. Her father was a college professor and people in
her family had actually travelled there. My family had seldom been able to
take a vacation at all. A couple of times my parents were able—by saving all
year—to take the family over to the coast on their annual two-week vacation.
I'd seen the time and energy my parents invested in trying to take a family vacation to some place a few hours away; the idea of how anybody ever got to
Europe was beyond me.

If class is more than simple economic status but one's cultural background
as well, what happens if you're born and raised middle class, but spend some
of your adult life with earnings below a middle-class income bracket—are you
then working-class? Probably not. If your economic position changes, you
still have the language, behavior, educational background, etc., of the middle
class, which you can bank on. You will always have choices. Men who consciously try to refuse male privilege are still male; whites who want to challenge white privilege are still white. I think those who come from middle-class
backgrounds need to recognize that their class privilege does not float out
with the rinse water. Middle-class people can exert incredible power just by
being nice and polite. The middle-class way of doing things is the standard—
they're always right, just by being themselves. Beware of middle-class people
who deny their privilege. Many people have times when they struggle to get
shoes for the kids, when budgets are tight, etc. This isn't the same as long-
term economic conditions without choices. Being working class is also generational. Examine your family's history of education, work, and standard of living. It may not be a coincidence that you share the same class status as your
parents and grandparents. If your grandparents were professionals, or your
parents were professionals, it's much more likely you'll be able to grow up to
become a yuppie, if your heart so desires, or even if you don't think about it.

How about if you're born and raised poor or working class, yet through
struggle, usually through education, you manage to achieve a different economic level: do you become middle class? Can you pass? I think some working-
class people may successfully assimilate into the middle class by learning to

dress, talk, and act middle class—to accept and adopt the middle-class way of doing things. It all depends on how far they're able to go. To succeed in the middle-class world means facing great pressures to abandon working-class friends and ways.

Contrary to our stereotype of the working class—white guys in overalls—the working class is not homogeneous in terms of race or gender. If you are a person of color, if you live in a female-headed household, you are much more likely to be working class or poor. The experience of Black, Latino, American Indian or Asian American working classes will differ significantly from the white working classes, which have traditionally been able to rely on white privilege to provide a more elite position within the working class. Working-class people are often grouped together and stereotyped, but distinctions can be made among the working class, working poor, and poor. Many working-class families are supported by unionized workers who possess marketable skills. Most working-poor families are supported by nonunionized, unskilled men and women. Many poor families are dependent on welfare for their income.

Attacks on the welfare system and those who live on welfare are a good example of classism in action. We have a "dual welfare" system in this country whereby welfare for the rich in the form of tax-free capital gain, guaranteed loans, oil depletion allowances, etc., is not recognized as welfare. Almost everyone in America is on some type of welfare; but, if you're rich, it's in the form of tax deductions for "business" meals and entertainment, and if you're poor, it's in the form of food stamps. The difference is the stigma and humiliation connected to welfare for the poor, as compared to welfare for the rich, which is called "incentives." . . . A common focal point for complaints about "welfare" is the belief that most welfare recipients are cheaters—goodness knows there are no middle-class income tax cheaters out there. Imagine focusing the same anger and energy on the way corporations and big business cheat on their tax revenues. Now, there would be some dollars worth quibbling about. The "dual welfare" system also assigns a different degree of stigma to programs that benefit women and children . . . and programs whose recipients are primarily male, such as veterans' benefits. The implicit assumption is that mothers who raise children do not work and therefore are not deserving of their daily bread crumbs.

Anti-union attitudes are another prime example of classism in action. At best, unions have been a very progressive force for workers, women and people of color. At worst, unions have reflected the same regressive attitudes which are out there in other social structures: classism, racism, and sexism. Classism exists within the working class. The aristocracy of the working class—unionized, skilled workers—have mainly been white and male and have viewed themselves as being better than unskilled workers, the unemployed, and the poor, who are mostly women and people of color. The white working class must commit itself to a cultural and ideological transformation of racist attitudes. The history of working people, and the ways we've resisted many types of oppressions, are not something we're taught in school. Missing from our education is information about workers and their resistance.

Working-class women's critiques have focused on the following issues:

Education: White middle-class professionals have used academic jargon to rationalize and justify classism. The whole structure of education is a classist system. Schools in every town reflect class divisions: like the store list at the beginning of this article, you can list schools in your town by what classes of kids attend, and in most cities you can also list by race. The classist system is perpetuated in schools with the tracking system, whereby the "dumbs" are tracked into homemaking, shop courses and vocational school futures, while the "smarts" end up in advanced math, science, literature, and college-prep courses. If we examine these groups carefully, the coincidence of poor and working-class backgrounds with "dumbs" is rather alarming. The standard measurement of supposed intelligence is white middle-class English. If you're other than white middle class, you have to become bilingual to succeed in the educational system. If you're white middle class, you only need the language and writing skills you were raised with, since they're the standard. To do well in society presupposes middle-class background, experiences and learning for everyone. The tracking system separates those from the working class who can potentially assimilate to the middle class from all our friends, and labels us "college bound."

After high school, you go on to vocational school, community college, or college—public or private—according to your class position. Apart from the few who break into middle-class schools, the classist stereotyping of the working class as being dumb and inarticulate tracks most into vocational and low-skilled jobs. A few of us are allowed to slip through to reinforce the idea that equal opportunity exists. But for most, class position is destiny—determining our educational attainment and employment. Since we must overall abide by middle-class rules to succeed, the assumption is that we go to college in order to "better ourselves"—i.e., become more like them. I suppose it's assumed we have "yuppie envy" and desire nothing more than to be upwardly mobile individuals. It's assumed that we want to fit into their world. But many of us remain connected to our communities and families. Becoming college educated doesn't mean we have to, or want to, erase our first and natural language and value system. It's important for many of us to remain in and return to our communities to work, live, and stay sane.

Jobs: Middle-class people have the privilege of choosing careers. They can decide which jobs they want to work, according to their moral or political commitments, needs for challenge or creativity. This is a privilege denied the working class and poor, whose work is a means of survival, not choice (see Hartsock). Working-class women have seldom had the luxury of choosing between work in the home or market. We've generally done both, with little ability to purchase services to help with this double burden. Middle- and upper-class women can often hire other women to clean their houses, take care of their children, and cook their meals. Guess what class and race those "other" women are? Working a double or triple day is common for working-class women. Only middle-class women have an array of choices such as: parents put you through school, then you choose a career, then you choose when and if to have babies, then you choose a support system of working-class women to take care of your kids and house if you choose to resume your career. After the birth of my second child, I was working two part-time jobs— one loading trucks at night—and going to school during the days. While I was quite privileged because I could take my colicky infant with me to classes and the day-time job, I was in a state of continuous semi-consciousness. I had to work to support my family; the only choice I had was between school or sleep: Sleep became a privilege. A white middle-class feminist instructor at the university suggested to me, all sympathetically, that I ought to hire someone to clean my house and watch the baby. Her suggestion was totally out of my reality, both economically and socially. I'd worked for years cleaning other peoples' houses. Hiring a working-class woman to do the shit work is a middle-class woman's solution to any dilemma which her privileges, such as a career, may present her.

Mothering: The feminist critique of families and the oppressive role of mothering has focused on white middle-class nuclear families. This may not be an appropriate model for communities of class and color. Mothering and families may hold a different importance for working-class women. Within this context, the issue of coming out can be a very painful process for working-class lesbians. Due to the homophobia of working-class communities, to be a lesbian is most often to be excommunicated from your family, neighborhood, friends and the people you work with. If you're working class, you don't have such clearly demarcated concepts of yourself as an individual, but instead see yourself as part of a family and community that forms your survival structure. It is not easy to be faced with the risk of giving up ties which are so central to your identity and survival.

Individualism: Preoccupation with one's self—one's body, looks, relationships—is a luxury working-class women can't afford. Making an occupation out of taking care of yourself through therapy, aerobics, jogging, dressing for success, gourmet meals and proper nutrition, etc., may be responses that are directly rooted in privilege. The middle class has the leisure time to be preoccupied with their own problems, such as their waistlines, planning their vacations, coordinating their wardrobes, or dealing with what their mother said to them when they were five—my!

The white middle-class women's movement has been patronizing to working-class women. Its supporters think we don't understand sexism. What we don't understand is white middle-class feminism. They act as though they invented the truth, the light, and the way, which they merely need to pass along to us lower-class drudges. What they invented is a distorted form of what working-class women already know—if you're female, life sucks. Only at least we were smart enough to know that it's not just being female, but also being a person of color or class, which makes life a quicksand trap. The class system weakens all women. It censors and eliminates images of female strength. The idea of women as passive, weak creatures totally discounts the strength, self-dependence and inter-dependence necessary to survive as working-class and poor women. My mother and her friends always had a less-than-passive, less-than-enamoured attitude toward their spouses, male bosses, and men in general. I know from listening to their conversations, jokes and what they passed on to us, their daughters, as folklore. When I was five years old, my mother told me about how Aunt Betty had hit Uncle Ernie over the head with a skillet and knocked him out because he was raising his hand to hit her, and how he's never even thought about doing it since. This story was told to me with a good amount of glee and laughter. All the men in the neighborhood were told of the event as an example of what was a very acceptable response in the women's community for that type of male behavior. We kids in the neighborhood grew up with these stories of women giving husbands, bosses, the welfare system, schools, unions and men in general hell, whenever they deserved it. For me there were many role models of women taking action, control and resisting what was supposed to be their lot. Yet many white middle-class feminists continue to view feminism like math homework, where there's only supposed to be one answer. Never occurs to them that they might be talking algebra while working-class women might be talking metaphysics.

Women with backgrounds other than white middle-class experience compounded, simultaneous oppressions. We can't so easily separate our experiences by categories of gender, or race, or class, i.e., "I remember it well: on Saturday, June 3, I was experiencing class oppression, but by Tuesday, June 6, I was caught up in race oppression, then all day Friday, June 9, I was in the middle of gender oppression. What a week!" Sometimes, for example, gender and class reinforce each other. When I returned to college as a single parent after a few years of having kids and working crummy jobs—I went in for vocational testing. Even before I was tested, the white middle-class male vocational counselor looked at me, a welfare mother in my best selection from the Salvation Army racks, and suggested I quit college, go to vo-tech school and become a grocery clerk. This was probably the highest paying female working-class occupation he could think of. The vocational test results suggested I become an attorney. I did end up quitting college once again, not because of his suggestion, but because I was tired of supporting my children in ungenteel poverty. I entered vo-tech school for training as an electrician and, as one of the first women in a non-traditional field, was able to earn a living wage at a job which had traditionally been reserved for white working-class males. But this is a story for another day. Let's return to our little vocational counselor example. Was he suggesting the occupational choice of grocery clerk to me because of my gender or my class? Probably both. Let's imagine for a moment what this same vocational counselor might have advised, on sight only, to the following people:

1. A white middle-class male: doctor, lawyer, engineer, business executive.
2. A white middle-class female: close to the same suggestion as #1 if the counselor was not sexist, or, if sexist, then: librarian, teacher, nurse, social worker.
3. A middle-class man of color: close to the same suggestions as #1 if the counselor was not racist, or, if racist, then: school principal, sales, management, technician.
4. A middle-class woman of color: close to the same suggestions as #3 if counselor was not sexist; #2 if not racist; if not racist or sexist, then potentially #1.
5. A white working-class male: carpenter, electrician, plumber, welder.
6. A white working-class female—well, we already know what he told me, although he could have also suggested secretary, waitress and dental hygienist (except I'd already told him I hated these jobs).
7. A working-class man of color: garbage collector, janitor, fieldhand.
8. A working-class woman of color: maid, laundress, garment worker.

Notice anything about this list? As you move down it, a narrowing of choices, status, pay, working conditions, benefits and chances for promotions occurs. To be connected to any one factor, such as gender or class or race, can make life difficult. To be connected to multiple factors can guarantee limited economic status and poverty.

WAYS TO AVOID FACING CLASSISM

Deny Deny Deny: Deny your class position and the privileges connected to it. Deny the existence or experience of the working class and poor. You can even set yourself up (in your own mind) as judge and jury in deciding who qualifies as working class by your white middle-class standards. So if someone went to college, or seems intelligent to you, not at all like your stereotypes, they must be middle class.

Guilt Guilt Guilt: "I feel so bad, I just didn't realize!" is not helpful, but is a way to avoid changing attitudes and behaviors. Passivity—"Well, what can I do about it, anyway?"—and anger—"Well, what do they want!"—aren't too helpful either. Again, with these responses, the focus is on you and absolving the white middle class from responsibility. A more helpful remedy is to take action. Donate your time and money to local foodbanks. Don't cross picket lines. Better yet, go join a picket line.

HOW TO CHALLENGE CLASSISM

If you're middle class, you can begin to challenge classism with the following:

1. Confront classist behavior in yourself, others and society. Use and share the privileges, like time or money, which you do have.
2. Make demands on working-class and poor communities' issues—anti-racism, poverty, unions, public housing, public transportation, literacy and day care.
3. Learn from the skills and strength of working people—study working and poor people's history; take some Labor Studies, Ethnic Studies, Women Studies classes. Challenge elitism. There are many different types of intelligence: white middle-class, academic, professional intellectualism being one of them (reportedly). Finally, educate yourself, take responsibility and take action.

If you're working class, just some general suggestions (it's cheaper than therapy—free, less time-consuming and I won't ask you about what your mother said to you when you were five):

1. Face your racism! Educate yourself and others, your family, community, any organizations you belong to; take responsibility and take action. Face your classism, sexism, heterosexism, ageism, able-bodiness, adultism. . . .
2. Claim your identity. Learn all you can about your history and the history and experience of all working and poor peoples. Raise your children to be anti-racist, anti-sexist and anti-classist. Teach them the language and culture of working peoples. Learn to survive with a fair amount of anger and lots of humor, which can be tough when this stuff isn't even funny.
3. Work on issues which will benefit your community. Consider remaining in or returning to your communities. If you live and work in white middle-class environments, look for working-class allies to help you survive with your humor and wits intact. How do working-class people spot each other? We have antennae.

We need not deny or erase the differences of working-class cultures but can embrace their richness, their variety, their moral and intellectual heritage. We're not at the point yet where we can celebrate differences—not having money for a prescription for your child is nothing to celebrate. It's not time yet to party with the white middle class, because we'd be the entertainment ("Aren't they quaint? Just love their workboots and uniforms and the way they cuss!"). We need to overcome divisions among working people, not by ignoring the multiple oppressions many of us encounter, or by oppressing each other, but by becoming committed allies on all issues which affect working people: racism, sexism, classism, etc. An injury to one is an injury to all. Don't play by ruling-class rules, hoping that maybe you can live on Connecticut Avenue instead of Baltic, or that you as an individual can make it to Park Place and Boardwalk. Tired of Monopoly? Always ending up on Mediterranean Avenue? How about changing the game?

WEALTH MATTERS

16

Dalton Conley

Property is theft.

Pierre Joseph Proudhon, 1809–65

. . . Contrast the situations of two hypothetical families. Let's say that both households consist of married parents, in their thirties, with two young children.[1] Both families are low-income—that is, the total household income of each family is approximately the amount that the federal government has "declared" to be the poverty line for a family of four (with two children). In 1996, this figure was $15,911.*

Brett and Samantha Jones (family 1) earned about $12,000 that year. Brett earned this income from his job at a local fast-food franchise (approximately two thousand hours at a rate of $6 per hour). He found himself employed at this low-wage job after being laid off from his relatively well-paid position as a sheet metal worker at a local manufacturing plant, which closed because of fierce competition from companies in Asia and Latin America.

From Dalton Conley, *Being Black, Living in the Red: Race, Wealth, and Social Policy in America* (Berkeley, CA: University of California Press, 1999). Reprinted with permission of the author and the publisher.

*Editors' note: In 2001, the poverty line for a family of four with two children was $17,960.

After six months of unemployment, the only work Brett could find was flipping burgers alongside teenagers from the local high school.

Fortunately for the Jones family, however, they owned their own home. Fifteen years earlier, when Brett graduated from high school, married Samantha, and landed his original job as a sheet metal worker, his parents had lent the newlyweds money out of their retirement nest egg that enabled Brett and Samantha to make a 10 percent down payment on a house. With Samantha's parents cosigning—backed by the value of their own home—the newlyweds took out a fifteen-year mortgage for the balance of the cost of their $30,000 home. Although money was tight in the beginning, they were nonetheless thrilled to have a place of their own. During those initial, difficult years, an average of $209 of their $290.14 monthly mortgage payment was tax deductible as a home mortgage interest deduction. In addition, their annual property taxes of $800 were completely deductible, lowering their annual taxable income by a total of $3,308 per year. This more than offset the payments they were making to Brett's parents for the $3,000 they had borrowed for the down payment.

After four years, Brett and Samantha had paid back the $3,000 loan from his parents. At that point, the total of their combined mortgage payment ($290.14), monthly insurance premium ($50), and monthly property tax payment ($67), minus the tax savings from the deductions for mortgage interest and local property taxes, was less than the $350 that the Smiths (family 2) were paying to rent a unit the same size as the Joneses' house on the other side of town.

That other neighborhood, on the "bad" side of town, where David and Janet Smith lived, had worse schools and a higher crime rate and had just been chosen as a site for a waste disposal center. Most of the residents rented their housing units from absentee landlords who had no personal stake in the community other than profit. A few blocks from the Smiths' apartment was a row of public housing projects. Although they earned the same salaries and paid more or less the same monthly costs for housing as the Joneses did, the Smiths and their children experienced living conditions that were far inferior on every dimension, ranging from the aesthetic to the functional (buses ran less frequently, large supermarkets were nowhere to be found, and class size at the local school was well over thirty).

Like Brett Jones, David Smith had been employed as a sheet metal worker at the now-closed manufacturing plant. Unfortunately, the Smiths had not been able to buy a home when David was first hired at the plant. With little in the way of a down payment, they had looked for an affordable unit at the time, but the real estate agents they saw routinely claimed that there was nothing available at the moment, although they promised to "be sure to call as soon as something comes up. . . ." The Smiths never heard back from the agents and eventually settled into a rental apartment.

David spent the first three months after the layoffs searching for work, drawing down the family's savings to supplement unemployment insurance—savings that were not significantly greater than those of the Joneses, since both families had more or less the same monthly expenses. After several months of searching, David managed to land a job. Unfortunately, it was of the same variety as the job Brett Jones found: working as a security guard at the local mall, for about $12,000 a year. Meanwhile, Janet Smith went to work part time, as a nurse's aide for a home health care agency, grossing about $4,000 annually.

After the layoffs, the Joneses experienced a couple of rough months, when they were forced to dip into their small cash savings. But they were able to pay off the last two installments of their mortgage, thus eliminating their single biggest living expense. So, although they had some trouble adjusting to their lower standard of living, they managed to get by, always hoping that another manufacturing job would become available or that another company would buy out the plant and reopen it. If worst came to worst, they felt that they could always sell their home and relocate in a less expensive locale or an area with a more promising labor market.

The Smiths were a different case entirely. As renters, they had no latitude in reducing their expenses to meet their new economic reality, and they could not afford their rent on David's reduced salary. The financial strain eventually proved too much for the Smiths, who fought over how to structure the family budget. After a particularly bad row when the last of their savings had been spent, they decided to take a break; both thought life would be easier and better for the children if Janet moved back in with her mother for a while, just until things turned around economically—that is, until David found a better-paying job. With no house to anchor them, this seemed to be the best course of action.

Several years later, David and Janet Smith divorced, and the children began to see less and less of their father, who stayed with a friend on a "temporary" basis. Even though together they had earned more than the Jones family (with total incomes of $16,000 and $12,000, respectively), the Smiths had a rougher financial, emotional, and family situation, which, we may infer, resulted from a lack of property ownership.

What this comparison of the two families illustrates is the inadequacy of relying on income alone to describe the economic and social circumstances of families at the lower end of the economic scale. With a $16,000 annual income, the Smiths were just above the poverty threshold. In other words, they were not defined as "poor," in contrast to the Joneses, who were.[2] Yet the Smiths were worse off than the Joneses, despite the fact that the U.S. government and most researchers would have classified the Jones family as the one who met the threshold of neediness, based on that family's lower income.

These income-based poverty thresholds differ by family size and are adjusted annually for changes in the average cost of living in the United States.

In 1998, more than two dozen government programs—including food stamps, Head Start, and Medicaid—based their eligibility standards on the official poverty threshold. Additionally, more than a dozen states currently link their needs standard in some way to this poverty threshold. The example of the Joneses and the Smiths should tell us that something is gravely wrong with the way we are measuring economic hardship—poverty—in the United States. By ignoring assets, we not only give a distorted picture of life at the bottom of the income distribution but may even create perverse incentives.

Of course, we must be cautious and remember that the Smiths and the Joneses are hypothetically embellished examples that may exaggerate differences. Perhaps the Smiths would have divorced regardless of their economic circumstances. The hard evidence linking modest financial differences to a propensity toward marital dissolution is thin; however, a substantial body of research shows that financial issues are a major source of marital discord and relationship strain.[3] It is also possible that the Smiths, with nothing to lose in the form of assets, might have easily slid into the world of welfare dependency. A wide range of other factors, not included in our examples, affect a family's well-being and its trajectory. For example, the members of one family might have been healthier than those of the other, which would have had important economic consequences and could have affected family stability. Perhaps one family might have been especially savvy about using available resources and would have been able to take in boarders, do under-the-table work, or employ another strategy to better its standard of living. Nor do our examples address educational differences between the two households.

But I have chosen not to address all these confounding factors for the purpose of illustrating the importance of asset ownership *per se*. Of course, homeownership, savings behavior, and employment status all interact with a variety of other measurable and unmeasurable factors. This interaction, however, does not take away from the importance of property ownership itself.

. . . In order to understand a family's well-being and the life chances of its children—in short, to understand its class position—we not only must consider income, education, and occupation but also must take into account accumulated wealth (that is, property, assets, or net worth). . . . As you might have guessed, an important detail is missing from the preceding descriptions of the two families: the Smiths are black and have fewer assets than the Joneses, who are white.

At all income, occupational, and education levels, black families on average have drastically lower levels of wealth than similar white families. The situation of the Smiths may help us to understand the reason for this disparity of wealth between blacks and whites. For the Smiths, it was not discrimination in hiring or education that led to a family outcome vastly different from that of the Joneses; rather, it was a relative lack of assets from which they could draw. In contemporary America, race and property are intimately linked and form the nexus for the persistence of black-white inequality.

Let us look again at the Smith family, this time through the lens of race. Why did real estate agents tell the Smiths that nothing was available, thereby hindering their chances of finding a home to buy? This well-documented practice is called "steering," in which agents do not disclose properties on the market to qualified African American home seekers, in order to preserve the racial makeup of white communities—with an eye to maintaining the property values in those neighborhoods. Even if the Smiths had managed to locate a home in a predominantly African American neighborhood, they might well have encountered difficulty in obtaining a home mortgage because of "redlining," the procedure by which banks code such neighborhoods "red"—the lowest rating—on their loan evaluations, thereby making it next to impossible to get a mortgage for a home in these districts. Finally, and perhaps most important, the Smiths' parents were more likely to have been poor and without assets themselves (being black and having been born early in the century), meaning that it would have been harder for them to amass enough money to loan their children a down payment or to cosign a loan for them. The result is that while poor whites manage to have, on average, net worths of over $10,000, impoverished blacks have essentially no assets whatsoever.[4]

Since wealth accumulation depends heavily on intergenerational support issues such as gifts, informal loans, and inheritances, net worth has the ability to pick up both the current dynamics of race and the legacy of past inequalities that may be obscured in simple measures of income, occupation, or education. This thesis has been suggested by the work of sociologists Melvin Oliver and Thomas Shapiro in their recent book *Black Wealth/White Wealth*.[5] They claim that wealth is central to the nature of black-white inequality and that wealth—as opposed to income, occupation, or education—represents the "sedimentation" of both a legacy of racial inequality as well as contemporary, continuing inequalities. . . .

NOTES

1. These family descriptions were extrapolated from profiles of specific families who were interviewed for this study. The age, racial, income, family size, wealth, housing tenure, and divorce descriptions of these families come directly from cases 4348 and 1586 of the PSID 1984 wave (inflation-adjusted to 1996 dollars). The names and other details are fictitious but are in line with previous research that would suggest such profiles.

2. Neither family received health insurance from an employer. Since the Smiths' income was under 185 percent of the poverty line, their children were eligible for Medicaid. (In most states, the Joneses' children would also have been eligible for

Medicaid since that family's wealth was in the form of a home, which is excluded from the asset limits of many states.)

3. See, e.g., G. Levinger and O. Moles, eds., *Divorce and Separation: Contexts, Causes, and Consequences* (New York: Basic Books, 1979); and R. Conger, G. H. Elder, et al., "Linking Economic Hardship to Marital Quality and Instability," *Journal of Marriage and the Family* 52 (1990): 643–56.

4. The terms "black" and "African American" are used interchangeably, as are the terms "Hispanic" and "Latino." Black people of Caribbean origin make up a negligible portion of the data sample.

5. M. Oliver and T. Shapiro, *Black Wealth/White Wealth: A New Perspective on Racial Inequality* (New York: Routledge, 1995).

POVERTY AS RACE, POWER, AND WEALTH   17

James Jennings and Louis Kushnick

. . . Racial and gender hierarchies are important features for the maintenance of wealth and political power in this country. In addition, . . . poor people are not politically passive. . . . The persistence of poverty cannot be understood completely, nor can it be reduced significantly, without also examining the question of who has political influence in the United States. . . . The unchanging distribution of wealth and power is the greatest determinant in the persistence of massive impoverization of some sectors in this society. Furthermore, the racializing of poverty serves to strengthen institutional arrangements and social relations that generally keep high proportions of women and people of color in economically vulnerable and impoverished positions.

. . . The problem of poverty cannot be analyzed or responded to effectively if it is perceived simply as a behavioral issue for poor people or, in what William Ryan described several decades ago as a "blaming-the-victim" approach. We believe that persistent poverty, and poverty that is highly concentrated in some communities, reflects the interplay of an array of national and international factors, as well as specific economic arrangements that support social relations at least partially defined by race and gender. At a broad level,

From Louis Kushnick and James Jennings, eds., *A New Introduction to Poverty: The Role of Race, Power, and Politics* (New York: New York University Press, 1999). Reprinted with permission of the publisher.

three factors contribute to continuing poverty and high rates of impoveriza-tion among certain groups: the increasing imbalance in the distribution of wealth, with the rich continually becoming richer; the unbridled mobility of capital, both in finance and in production; and the prevalence of low wages coupled with levels of relatively high unemployment among certain groups. These are the root causes of poverty in many societies.

These causes of poverty are exacerbated by national policies that allow corporate leaders to pursue profits without consideration of the social costs incurred by their strategies. Militarization also aggravates the causes of pov-erty by appropriating resources that could be utilized to expand socially beneficial productivity and using them instead to produce tools for human de-struction. The major causes of poverty also contribute to the fiscal incapacity of national and local governments, at times raising the level of political and social tensions and popular dissatisfaction as a result. . . .

In the United States, at least three mechanisms work to mute class ten-sions that arise from the subsidization of the rich by the working- and middle-class sectors of the population. One ideological mechanism is the belief that diligent individuals can make it economically and achieve the good life if they work hard and put God and country above everything; this is the popular Horatio Alger myth. Second is the supply-side proposition, which maintains that economic development and progress will result from allowing wealthy in-terests to accumulate more wealth; policies that result in lower taxes and less regulation of income mean, according to this thinking, that the wealthy will have more resources to save and invest, raising the standard of living for everyone by producing a healthier economy.

Race is the third mechanism by which class tensions and popular dissatis-faction with economic policies have been managed by wealthy and powerful interests. Certain racial mechanisms have been manipulated by some interests to divide poor and working-class people who are black from poor and working-class people who are not black. Tools used for this purpose throughout U.S. history include segregation and discrimination; the belief and practices asso-ciated with the presumption of white cultural superiority; political and racial scapegoating; and even the public presentation and dialogue of welfare reform in the contemporary period. The media also contribute to this process by ag-gressively racializing poverty. Journalists and commentators frequently pre-sent the faces of poor people as black, Puerto Rican, or of other minority ra-cial heritage, and focus on problems like teen births, family instability, crime and drug use, low educational achievement, and poor health attributes—as if poor whites did not exhibit similar characteristics. The racialization of pov-erty is strengthened and facilitated by private and public policies that continue to result in residential and educational segregation.

Indeed, there are significant historical and contemporary differences among black, white, Latino, Asian, and Native American poverty in this country. Black poverty tends to be more highly concentrated, and black people tend to be impoverished for longer periods of times than whites or some Latino groups. Black poor people also tend to be poorer than white poor people. Black children and youth, as well as some groups of Latino children, remain the most impoverished in this society. But the root causes of long-lasting and persistent poverty are linked directly to the political and economic factors suggested here. If poor people were politically capable of raising and sustaining alliances on the basis of their poverty, then these issues would be highlighted. In a few episodic cases, such alliances have actually been forged. . . . It is precisely this political possibility, however, that is discouraged and resisted by political and economic leadership in response to the interests of wealth and corporate power.

To a large extent, the current literature and the public dialogue on poverty in the United States ignore the possibility that the poor might forge political alliances. There is generally an absence of a perspective that considers power or the relationship between racial hierarchy and divisions, poverty, and the distribution and management of wealth. The connection between wealth and power, on the one hand, and poverty and homelessness, on the other, has been ignored or overlooked by politicians, the media, and academe. . . .

The work of some poverty researchers is myopic regarding political factors that have led to and that sustain persistent impoverishment for many Americans. But politics . . . is a major factor requiring examination by those seeking to determine the causes of poverty, as well as the means for its resolution, especially among communities of color in the United States. As James Jennings writes, . . . politics is not "explored for its explanatory possibilities; it is merely assumed by some researchers that poverty has more to do with pathology among groups like Puerto Ricans and blacks, or with politically irrelevant structural cataclysmic changes in the economy, than with the lack of political power that characterizes impoverished groups in this nation."[1] . . . The lack of political analysis regarding the problem of poverty on the part of the research community and the media encourages the use of the notion of dependency as a major, yet undefined, variable in explaining the nature of poverty. And yet his unexamined concept . . . is consistently used in devising policy responses to poverty.

The myth of a dependency culture as created by income-transfer programs that benefit poor people became the new orthodoxy in mainstream media and intellectual circles, as reflected in the work of Thomas Sowell, George Gilder, Mickey Kaus, Charles Murray, and others. Yet, the policy initiatives of the current Democratic administration reflect the ideas and suggestions of these very same conservative writers. In fact, there has been little fundamental difference between Republicans and Democrats in how they approach the notions of work and personal responsibility. Both electoral camps ignore the basic causes of poverty and instead propose that this problem reflects the lack of a work ethic or a sense of personal responsibility among poor people. Although the problem of poverty has been presented and discussed in many sectors in terms of dependency or the absence of a work ethic, domestic poverty in the United States is in fact much more complex, reflecting the interplay of political and economic factors, rather than the particular behavior of poor people. . . .

. . . Race has been utilized to detract attention from wealth inequality and from the consequence of poverty and economic dislocation for significant numbers of people. Thus, although poverty is not solely a "black" problem, it is politically effective (and perhaps even methodologically neater) to represent it as such because it serves to divert dialogue from issues related to the accumulation, management, and distribution of wealth. . . . Historically, race and racial divisions have been fomented in order to keep people from realizing the political, or class, nature of persistent poverty in the United States.

The use of race and ethnicity as political dividers of working-class and poor people is also a common tactic in the current period. Some evidence for the usefulness of highlighting race and ethnicity to keep people divided is found in a *CBS News/New York Times Poll* reported on December 6–9, 1994:

> attitudes about the poor and about welfare recipients have a racial element: Americans are more likely to think that people on welfare (and all those who are poor in this country) are more likely to be black than they are to be white. And the characterizations of welfare recipients differ dramatically based on the racial images one has of them. Those who say welfare recipients are mostly white (18% say this) are more likely to think people on welfare want to work, that they are on welfare because of circumstances beyond their control, and that they really need help. On the other hand, people who think most welfare recipients are black (44%) say they're on welfare because of lack of effort, that they don't want to work, and that they could get along without welfare benefits.

. . . Distorted media images have been central to the attack on poor people and the few programs that even slightly benefit them, images of black and Puerto Rican women with loads of "illegitimate" children and of black and Latino men, lazy and dangerous, as suggested by Edward C. Banfield in *The Unheavenly City;*[2] frequent photos of poor black women with babies bolstering these myths in the *New York Times* and other major news outlets; and similar images that will continue to pervade the literature. . . . This catalog of distorted images was crucial in Ronald Reagan's presidential campaigns, as it was in the Ku Klux Klan spokesperson David Duke's gubernatorial and senatorial campaigns in Louisiana and in the right-wing politician Oval Fordyce's gubernatorial campaign in Mississippi. And such images continue to be useful for both Republicans and Democrats. The portrait of the *underclass*, a pejorative term used by conservatives and liberals alike, has been created as another myth to prove the deleterious consequences of programs that benefit the poor, thereby justifying policies and containment and reminding good, normal, hard-working middle-class citizens that they have nothing in common with "them." . . .

NOTES

1. James Jennings, "Missing Links in the Study of Puerto Rican Poverty in the United States," in *A New Introduction to Poverty: The Role of Race, Power, and Politics*, Louis Kushnick and James Jennings, eds., pp. 89–101 (New York: New York University Press, 1999).

2. Edward C. Banfield, *The Unheavenly City* (Boston: Little, Brown, 1968).

BLACK PICKET FENCES   18
Privilege and Peril among the Black Middle Class

Mary Pattillo-McCoy

Much of the research and media attention on African Americans is on the black poor. Welfare debates, discussions of crime and safety, urban policy initiatives, and even the cultural uproar over things like rap music are focused on the situation of poor African Americans. With more than one in four African Americans living below the official poverty line (versus approximately one in nine whites), this is a reasonable and warranted bias. But rarely do we hear the stories of the other three-fourths, or the majority of African Americans, who may be the office secretary, the company's computer technician, a project manager down the hall, or the person who teaches our children. The growth of the black middle class has been hailed as one of the major triumphs of the civil rights movement, but if we have so little information on who makes up this group and what their lives are like, how can we be so sure that triumphant progress is the full story? The optimistic assumption of the 1970s and 1980s was that upwardly mobile African Americans were quietly integrating

From Mary Pattillo-McCoy, *Black Picket Fences: Privilege and Peril among the Black Middle Class* (Chicago: University of Chicago Press, 1999). Reprinted by permission of the University of Chicago Press and the author.

formerly all-white occupations, businesses, neighborhoods, and social clubs. Black middle- and working-class families were moving out of all-black urban neighborhoods and into the suburbs. With these suppositions, the black middle class dropped from under the scientific lens and off the policy agenda, even though basic evidence suggests that the public celebration of black middle-class ascendance has perhaps been too hasty.

We know, for example, that a more appropriate socioeconomic label for members of the black middle class is "lower middle class." The one black doctor who lives in an exclusive white suburb and the few African American lawyers who work at a large firm are not representative of the black middle class overall (but neither are their experiences identical to those of their white colleagues). And although most white Americans are also not doctors or lawyers, the lopsided distribution of occupations for whites does favor such professional and managerial jobs, whereas the black middle class is clustered in the sales and clerical fields. Because one's occupation affects one's income, African Americans have lower earnings. Yet the inequalities run even deeper than just income. Compound and exponentiate the current differences over a history of slavery and Jim Crow, and the nearly fourteenfold wealth advantage that whites enjoy over African Americans—regardless of income, education, or occupation—needs little explanation.

We also know that the black middle class faces housing segregation to the same extent as the black poor. African Americans are more segregated from whites than any other racial or ethnic group. In fact, the black middle class likely faces the most blatant racial discrimination, in that many in its ranks can actually afford to pay for housing in predominantly white areas. Real estate agents and apartment managers can easily turn away poor African Americans by simply quoting prohibitive home costs or high rents. It takes more purposive creativity, however, to consistently steer middle-class blacks into already established African American neighborhoods by such tactics as disingenuously asserting that an apartment has just been rented when the prospective renters who show up at the property manager's door are, to his or her surprise, black. Racial segregation means that racial inequalities in employment, education, income, and wealth are inscribed in space. Predominantly white neighborhoods benefit from the historically determined and contemporarily sustained edge that whites enjoy.

Finally, we know that middle-class African Americans do not perform as well as whites on standardized tests (in school or in employment); are more likely to be incarcerated for drug offenses; are less likely to marry, and more likely to have a child without being married; and are less likely to be working. Liberals bumble when addressing these realities because, unlike housing segregation or job discrimination, of which middle-class African Americans are the clear victims, earning low grades in school or getting pregnant without a husband can easily be attributed to the bad behaviors of blacks themselves. For middle-class blacks, who ostensibly do not face the daily disadvantages of poverty, it is even more difficult to explain why they do not measure up to whites. To resolve this quandary it is essential to continuously refer back to the ways in which the black middle class *is not equal* to the white middle class. . . .

The lives of the families in Groveland* provide some answers to these questions. Groveland's approximately ninety square blocks contain a population of just under twelve thousand residents, over 95 percent of whom are African American. The 1990 median annual family income in the neighborhood is nearly $40,000, while the comparable figure for Chicago as a whole is just over $30,000. More than 70 percent of Groveland families own their own homes. By income and occupational criteria, as well as the American dream of homeownership, Groveland qualifies as a "middle-class neighborhood."

Yet this sterile description does not at all capture the neighborhood's diversity, which is critical to correctly portraying the neighborhood context of the black middle class. Groveland's unemployment rate is 12 percent, which is higher than the citywide rate, but *lower* than the percentage of unemployed residents in the neighborhoods that border Groveland. Twelve percent of Groveland's families are poor, which again makes it a bit *more* advantaged than the surrounding areas, but worse off than most of Chicago's predominantly white neighborhoods. The geography of Groveland is typical of black middle-class areas, which often sit as a kind of buffer between core black poverty areas and whites. Contrary to popular discussion, the black middle class has not out-migrated to unnamed neighborhoods outside of the black community. Instead, they are an overlooked population still rooted in the contemporary "Black Belts" of cities across the country. Some of the questions about why middle-class blacks are not at parity with middle-class whites can be answered once this fact is recognized. . . .

By the end of my research tenure in Groveland, I had seen three groups of eighth-graders graduate to high school, high school kids go on to college, and college graduates start their careers. I also heard too many stories and read too many obituaries of the teenagers who were jailed or killed along the way. The son of a police detective in jail for murder. The grandson of a teacher shot while visiting his girlfriend's house. The daughter of a park supervisor living with a drug dealer who would later be killed at a fast-food restaurant. These events were jarring, and all-too-frequent, discontinuities in the daily routine of Groveland residents. Why were some Groveland youth following a path to success, while others had concocted a recipe for certain failure? After all, these are not the stories of poor youth caught in a trap of absent opportunities, low aspirations, and harsh environments. Instead, Groveland is a neighborhood of single-family homes, old stately churches,

---

*Editors' note: Groveland is the name given to the Black middle-class neighborhood where Pattillo-McCoy conducted her research for three years.

tree-lined streets, active political and civic organizations, and concerned parents trying to maintain a middle-class way of life. These black middle-class families are a hidden population in this country's urban fabric.

The evening news hour in every major American city is filled with reports of urban crime and violence. Newspapers fill in the gaps of the more sensational tragedies about which the television could provide only a few sound bites. Rounding out the flow of urban Armageddon stories are the gossip and hearsay passed informally between neighbors, church friends, and drinking buddies. For many middle-class white Americans, the incidents they hear about in distant and troubled inner cities provide a constant symbolic threat, but an infrequent reality. For the families who live on the corner of the crime scene—overwhelmingly black or Latino, and poor—daily life is organized to avoid victimization. In the middle of these two geographically and socially distant groups lives the black middle class.

African American social workers and teachers, secretaries and nurses, entrepreneurs and government bureaucrats are in many ways the buffer between the black poor and the white middle class. When neighborhoods are changing, white middle-class families may find themselves living near low-income black families, but one group is inevitably displaced. The neighborhood becomes, once again, racially homogeneous. More than thirty years after the civil rights movement, racial segregation remains a reality in most American cities. Middle-income black families fill the residential gap between the neighborhoods that house middle-class whites and the neighborhoods where poor African Americans live. Unlike most whites, middle-class black families must contend with the crime, dilapidated housing, and social disorder in the deteriorating poor neighborhoods that continue to grow in their direction. Residents attempt to fortify their neighborhoods against this encroachment, and limit their travel and associations to other middle-class neighborhoods in the city and suburbs. Yet even with these efforts, residents of black middle-class neighborhoods share schools, grocery stores, hospitals, nightclubs, and parks with their poorer neighbors, ensuring frequent interaction within and outside the neighborhood.

The in-between position of the black middle class sets up certain crossroads for its youth. This peculiar limbo begins to explain the disparate outcomes of otherwise similar young people in Groveland. The right and wrong paths are in easy reach of neighborhood youth. Working adults are models of success. Some parents even work two jobs; while still others combine work and school to increase their chances of on-the-job promotions. All of the positive knowledge, networking, and role-model benefits that accrue to working parents are operative for many families in Groveland. But at the same time the rebellious nature of adolescence inevitably makes the wrong path a strong temptation, and there is no shortage of showy drug dealers and cocky gang members who make dabbling in deviance look fun. Youth walk a fine line between preparing for success and youthful delinquent experimentation, the consequences of which can be especially serious for black youth. . . .

. . . The black middle class is connected to the black poor through friendship and kinship ties, as well as geographically. Policies that hurt the black poor will ultimately negatively affect the black middle class. At the same time, the black middle class sits at the doorstep of middle-class privilege. Continued affirmative action, access to higher education, a plan to create real family-wage jobs, and the alleviation of residential segregation should be at the forefront of policy initiatives to support the gains already made by the black middle class. . . .

"Middle class" is a notoriously elusive category based on a combination of socioeconomic factors (mostly income, occupation, and education) and normative judgments (ranging from where people live, to what churches or clubs they belong to, to whether they plant flowers in their gardens). Among African Americans, where there has historically been less income and occupational diversity, the question of middle-class position becomes even more murky. . . .

Conversations with Groveland residents . . . underscore the fluid and complex nature of class categories among African Americans. Although most Groveland residents settle on a label somewhere between "lower middle class" and "middle class" to describe their own class position, the intermediate descriptors are plentiful. Some classification schemes focus on inequality. One resident resolved that there are the "rich," and everyone else falls into the categories of "poor, poorer, and poorest." Other words, like *ghetto, bourgie* (the shortened version of *bourgeois*), and *uppity* are normative terms that Grovelandites use to describe the intersection of standard socioeconomic measures and normative judgments of lifestyles and attitudes. Still other people talk about class in geographic terms, delineating a hierarchy of places rather than of incomes or occupations. . . .

. . . Despite continuing social and political ties, the reality of class schisms cannot be ignored. In *The Declining Significance of Race* (1978), William Julius Wilson argued that the African American community was splitting in two, with middle-class blacks improving their position relative to whites, and poor blacks becoming ever more marginalized. Civil rights legislation, especially affirmative action, worked well for African Americans poised to take advantage of educational and employment opportunities. The unsolved problem was what to do about African Americans in poverty. They were doing poorly not primarily because they were black, Wilson argued, but because they were unskilled and because the structure of the labor market had changed around them. Grounded in the conviction that social structure influences the nature of race relations, Wilson saw the growth in high-wage employment and the rise of political liberalism as fueling the diminution of race as a factor in the stratification process. The life chances of blacks were becoming more dependent on their class position. African Americans with a college education were positioned to take advantage of jobs in a service-producing economy—jobs in trade and finance, public management, and social services. And because of affirmative action legislation, firms were motivated to hire these qualified blacks.

At the same time, the situation for the black poor was stagnating, if not deteriorating. Black unemployment began to rise in the 1950s. There was not much difference in the unemployment rate for blacks and whites in 1930, but by the mid-1950s the ratio of black to white unemployment reached 2 to 1 (Farley 1985). These changes, Wilson and others argued, were the result of shifts in the mode of production. The number of well-paying manufacturing jobs in the central city had declined as a result of both technological changes and relocation. These changes permanently relegated unskilled blacks to low-wage, menial, and dead-end jobs, or pushed them out of the workforce altogether. Wilson's contribution was to direct attention to changes in the nature of production that disadvantaged unskilled blacks. His prognosis for the black middle class was relatively optimistic, a position for which he was criticized by other African American scholars. Wilson's critics rushed to prove him wrong and show that members of the new black middle class continued to face obstacles because of their race (Pinkney 1984; Willie 1979; Washington 1979, 1980; see Morris 1996 for a review).

. . . To be sure, the obstacles faced by poor blacks in a changing economy and the persistence of black poverty more generally are intolerable facts that merit considerable research and government resources. However, the research pendulum swung to the extreme, virtually ignoring the majority of African Americans who are not poor. . . .

. . . The declining interest in the status of nonpoor blacks was premature. The African American community was in a short time transformed from a population almost uniform in its poverty to one with a nascent middle class—this as recently as the 1950s. But racial disparities in occupation, income, and intergenerational mobility were not eradicated by the few years of progress. The brief period of growth spawned a kind of dismissive optimism, but the economic and social purse strings were once again pulled tight, stalling the advances made by some African Americans. The continuing inequalities between middle-class whites and African Americans attest to the persistence of racism and discrimination, albeit in quite different forms than in the Jim Crow era (Bobo, Kluegel, and Smith 1997). . . . The same stages that characterize the socioeconomic past and present of African Americans—overwhelming disadvantage, followed by progress and optimism, followed by stagnation and retrenchment—are mirrored in the spatial history (the *where*) of the black middle class. . . .

While some black families *have* integrated white neighborhoods as many commentators had predicted, the black middle class overall remains as segregated from whites as the black poor (Farley 1991). This means that the search for better neighborhoods has taken place *within* a segregated housing market. As a result, black middle-class neighborhoods are often located next to predominantly black areas with much higher poverty rates. Blacks of all socioeconomic statuses tend to be confined to a limited geographic space, which is formally designated by the discriminatory practices of banks, insurance companies, and urban planners, and symbolically identified by the formation of cultural and social institutions. . . .

. . . Historic and contemporary black residential patterns suggest the following. African Americans have long attempted to translate socioeconomic success into residential mobility, making them similar to other ethnic groups (Massey and Denton 1985). They desire to purchase better homes, safer neighborhoods, higher quality schools, and more amenities with their increased earnings. Out-migration has been a constant process. The black middle class has *always* attempted to leave poor neighborhoods, but has never been able to get very far. However, when the relative *size* of the black middle class grew, the size of its residential enclaves grew as well. *The increase in the number of black middle-class persons has led to growth in the size of black middle-class enclaves, which in turn increases the spatial distance between poor and middle-class African Americans. This greater physical separation within a segregated black community accounts for the popular belief that black middle-class out-migration is a recent and alarming trend.* . . .

The problems confronting middle-class African Americans are not solved by simply moving away from a low-income black family and next door to a middle-class white family. The fact that a neighborhood's racial makeup is frequently a proxy for the things that really count—quality of schools, security, appreciation of property values, political clout, and availability of desirable amenities—attests to the ways in which larger processes of discrimination penalize blacks at the neighborhood level. Racial inequalities perpetuate the higher poverty rate among blacks and ensure that segregated black communities will bear nearly the full burden of such inequality. The argument for residential integration is not to allow the black middle class to easily abandon black neighborhoods. Instead, more strict desegregation laws would also open the door for low-income blacks to move to predominantly white neighborhoods, where jobs and resources are unfairly clustered. Yet we need not wait for whites to accept blacks into their neighborhoods, and think of integration as the panacea for current problems. Aggressive measures must be taken to improve the socioeconomic conditions of African Americans *where they are*, by highlighting *where* the black middle class lives, it becomes apparent that concentrated urban poverty has repercussions not only for poor African Americans, but for middle-class blacks as well, while a majority of middle-class whites move farther into the hinterlands. A comprehensive antipoverty agenda would have positive benefits for African Americans as a group, and therefore for the residential environs of the black middle class—although it leaves unchallenged the desire of many blacks and even more whites to live with others of the same race.

References

Bobo, Lawrence, James R. Kluegel, and Ryan A. Smith. 1997. "Laissez-Faire Racism: The Crystallization of a Kinder, Gentler, Antiblack Ideology." In *Racial Attitudes in the 1980s*, edited by Steven A. Tuch and Jack K. Martin, 15–42. Westport, CT: Praeger.

Farley, Reynolds. 1991. "Residential Segregation of Social and Economic Groups among Blacks, 1970–1980." In *The Urban Underclass*, edited by Christopher Jencks and Paul E. Peterson, 274–298. Washington, DC: The Brookings Institution.

Farley, Reynolds. 1985. "Three Steps Forward and Two Back? Recent Changes in the Social and Economic Status of Blacks." *Ethnic and Racial Studies* 8: 4–28.

Massey, Douglas, and Nancy Denton. 1985. "Spatial Assimilation as a Socioeconomic Outcome." *American Sociological Review* 50: 94–106.

Morris, Aldon. 1996. "What's Race Got to Do With it?" *Contemporary Sociology* 25: 309–313.

Pinkney, Alfonso. 1984. *The Myth of Black Progress*. New York: Cambridge University Press.

Washington, Joseph R. 1980. *Dilemmas of the New Black Middle Class*. Philadelphia: University of Pennsylvania Afro-American Studies Program.

Washington, Joseph R., ed. 1979. *The Declining Significance of Race: A Dialogue among Black and White Social Scientists*. Philadelphia: University of Pennsylvania Afro-American Studies Program.

Willie, Charles Vert. 1979. *The Caste and Class Controversy*. Bayside, NY: General Hall.

Wilson, William Julius. 1978. *The Declining Significance of Race: Blacks and Changing American Institutions*. Chicago: University of Chicago Press.

Gender and Sexism

GENDER THROUGH THE PRISM OF DIFFERENCE

19

Maxine Baca Zinn, Pierrette Hondagneu-Sotelo, and Michael A. Messner

"Men can't cry." "Women are victims of patriarchal oppression." "After divorce, single mothers are downwardly mobile, often moving into poverty." "Men don't do their share of housework and childcare." "Professional women face barriers such as sexual harassment and a 'glass ceiling' that prevent them from competing equally with men for high-status positions and high salaries." "Heterosexual intercourse is an expression of men's power over women." Sometimes, the students in our sociology and gender studies courses balk at these kinds of generalizations. And they are right to do so. After all, some men are more emotionally expressive than some women, some women have more power and success than some men, some men do their share—or more—of housework and childcare, and some women experience sex with men as both pleasurable and empowering. Indeed, contemporary gender relations are complex, changing in various directions, and, as such, we need to be wary of simplistic, if handy, slogans that seem to sum up the essence of relations between women and men.

On the other hand, we think it is a tremendous mistake to conclude that "all individuals are totally unique and different" and that therefore all generalizations about social groups are impossible or inherently oppressive. In fact, we are convinced that it is this very complexity, this multifaceted nature of contemporary gender relations that fairly begs for a sociological analysis of gender. . . . We use the image of "the prism of difference" to illustrate our approach to developing this sociological perspective on contemporary gender relations. *The American Heritage Dictionary* defines *prism*, in part, as "a homogeneous transparent solid, usually with triangular bases and rectangular sides, used to produce or analyze a continuous spectrum." Imagine a ray of light—which to the naked eye, appears to be only one color—refracted through a prism onto a white wall. To the eye, the result is not an infinite, disorganized scatter of individual colors. Rather, the refracted light displays an order, a structure of relationships among the different colors—a rainbow. Similarly, we propose to use the "prism of difference" . . . to analyze a continuous spectrum of people, in order to show how gender is organized and experienced differently when refracted through the prism of sexual, racial/ethnic, social class, physical abilities, age, and national citizenship differences.

EARLY WOMEN'S STUDIES: CATEGORICAL VIEWS OF "WOMEN" AND "MEN"

. . . It is possible to make good generalizations about women and men. But these generalizations should be drawn carefully, by always asking the questions "*which* women?" and "*which* men?" Scholars of sex and gender have not always done this. In the 1960s and 1970s, women's studies focused on the differences *between* women and men rather than *among* women and men. The very concept of gender, women's studies scholars demonstrated, is based on socially defined difference between women and men. From the macro level of social institutions like the economy, politics, and religion, to the micro level of interpersonal relations, distinctions between women and men structure social relations. Making males and females *different* from one another is the essence of gender. It is also the basis of men's power and domination. Understanding this was profoundly illuminating. Knowing that difference produced domination enabled women to name, analyze, and set about changing their victimization.

In the 1970s, riding the wave of a resurgent feminist movement, colleges and universities began to develop women's studies courses that aimed first and foremost to make women's lives visible. The texts that were developed for these courses tended to stress the things that women shared under patriarchy—having the responsibility for housework and childcare, the experience or fear of men's sexual violence, a lack of formal or informal access to education, exclusion from high-status professional and managerial jobs, political office, and religious leadership positions (Brownmiller 1975; Kanter 1977).

The study of women in society offered new ways of seeing the world. But the 1970s approach was limited in several ways. Thinking of gender primarily in terms of differences between women and men led scholars to overgeneralize about both. The concept of patriarchy led to a dualistic perspective of male privilege and female subordination. Women and men were cast as opposites. Each was treated as a homogeneous category with common characteristics and experiences. This approach *essentialized* women and men. Essentialism, simply put, is the notion that women's and men's attributes are categorically different. From this perspective, male control and coercion of women produced conflict between the sexes. The feminist insight originally introduced by Simone de Beauvoir in 1953—that women, as a group, had been socially defined as the "other" and that men had constructed themselves as the subjects of history, while constructing women as their objects—fueled an energizing sense of togetherness among many women. As college students read books like *Sisterhood Is Powerful* (Morgan 1970), many of them joined organizations that fought, with some success, for equality and justice for women.

THE VOICES OF "OTHER" WOMEN

Although this view of women as an oppressed "other" was empowering for certain groups of women, some women began to claim that the feminist view of universal sisterhood ignored and marginalized their major concerns. It soon became apparent that treating women as a group united in its victimization by patriarchy was biased by too narrow a focus on the experiences and perspectives of women from more privileged social groups. "Gender" was treated as a generic category, uncritically applied to women. Ironically, this analysis, which was meant to unify women, instead produced divisions between and among them. The concerns projected as "universal" were removed from the realities of many women's lives. For example, it became a matter of faith in second-wave feminism that women's liberation would be accomplished by breaking down the "gendered public–domestic split." Indeed, the feminist call for women to move out of the kitchens and into the workplaces resonated in the experiences of many of the college-educated white women who were inspired by Betty Friedan's 1963 book, *The Feminine Mystique*. But the idea that women's movement into workplaces was itself empowering or liberating seemed absurd or irrelevant to many working-class women and women of color. They were already working for wages, as had many of their mothers and grandmothers, and did not consider access to jobs and public life as "liberating." For many of these women, liberation had more to do with organizing in communities and workplaces—often alongside men—for better schools, better pay, decent benefits, and other policies to benefit their neighborhoods, jobs, and families. The feminism of the 1970s did not seem to address these issues.

As more and more women analyzed their own experiences, they began to address the power relations creating differences among women and the part that privileged women played in the oppression of others. For many women of color, working-class women, lesbians, and women in contexts outside the United States (especially women in non-Western societies), the focus on male domination was a distraction from other oppressions. Their lived experiences could support neither a unitary theory of gender nor an ideology of universal sisterhood. As a result, finding common ground in a universal female victimization was never a priority for many groups of women.

Challenges to gender stereotypes soon emerged. Women of varied races, classes, national origins, and sexualities insisted that the concept of gender be broadened to take their differences into account (Baca Zinn et al. 1986; Hartmann 1976; Rich 1980; Smith 1977). Many women began to argue that their lives are affected by their location in a number of different hierarchies: as African Americans, Latinas, Native Americans, or Asian Americans in the race hierarchy; as young or old in the age hierarchy; as heterosexual, gay, lesbian, or bisexual in the sexual orientation hierarchy; and as women outside of the Western, industrialized nations, in subordinated geopolitical contexts. These arguments make it clear that women were not victimized by gender alone but by the historical and systematic denial of rights and privileges based on other differences as well.

MEN AS GENDERED BEINGS

As the voices of "other" women in the mid- to late 1970s began to challenge and expand the parameters of women's studies, a new area of scholarly inquiry was beginning to stir—a critical examination of men and masculinity. To be sure, in those early years of gender studies, the major task was to conduct studies and develop courses about the lives of women in order to begin to correct centuries of scholarship that rendered women's lives, problems, and accomplishments invisible. But the core idea of feminism—that "femininity" and women's subordination is a social construction—logically led to an examination of the social construction of "masculinity" and men's power. Many of the first scholars to take on this task were psychologists, who were concerned with looking at the social construction of "the male sex role" (e.g., Pleck 1981). By the late 1980s, there was a growing interdisciplinary collection of studies of men and masculinity, much of it by social scientists (Brod 1987; Kaufman 1987; Kimmel 1987; Kimmel & Messner 1989).

Reflecting developments in women's studies, the scholarship on men's lives tended to develop three themes: First, what we think of as "masculinity" is not a fixed, biological essence of men but, rather, is a social construction that shifts and changes over time, as well as between and among various national and cultural contexts. Second, power is central to understanding gender as a relational construct, and the dominant definition of masculinity is largely about expressing difference from—and superiority over—anything considered "feminine." Third, there is no singular "male sex role." Rather, at any given time there are various masculinities. R. W. Connell (1987, 1995) has been among the most articulate advocates of this perspective. Connell argues that hegemonic masculinity (the dominant form of masculinity at any given moment) is constructed in relation to femininities *as well as* in relation to various subordinated or marginalized masculinities. For example, in the United States, various racialized masculinities (e.g., as represented by African American men, Latino immigrant men, etc.) have been central to the construction of hegemonic (White, middle-class) masculinity. . . . This "othering" of racialized masculinities helps to shore up the material privileges that have been historically connected to hegemonic masculinity. When viewed this way, we can better understand hegemonic masculinity as part of a system that includes gender, as well as racial, class, sexual, and other relations of power.

The new literature on men and masculinities also begins to move us beyond the simplistic, falsely categorical, and pessimistic view of men simply as a privileged sex class. When race, social class, sexual orientation, physical abilities, and immigrant or national status are taken into account, we can see that in some circumstances, "male privilege" is partly—sometimes substantially— muted (Kimmel & Messner 1998). Although it is unlikely that we will soon see a "men's movement" that aims to undermine the power and privileges that are connected with hegemonic masculinity, when we begin to look at "masculinities" through the prism of difference, we can begin to see similarities and possible points of coalition between and among certain groups of women and men (Messner 1998). Certain kinds of changes in gender relations—for instance, a national family leave policy for working parents—might serve as a means of uniting particular groups of women and men.

GENDER IN INTERNATIONAL CONTEXTS

It is an increasingly accepted truism that late-twentieth-century increases in transnational trade, international migration, and global systems of production and communication have diminished both the power of nation-states and the significance of national borders. A much more ignored issue is the extent to which gender relations—in the United States and elsewhere in the world— are increasingly linked to patterns of global economic restructuring. Decisions made in corporate headquarters located in Los Angeles, Tokyo, or

London may have immediate repercussions in how women and men thousands of miles away organize their work, community, and family lives (Sassen 1991). It is no longer possible to study gender relations without attention to global processes and inequalities. . . .

Around the world, women's paid and unpaid labor is key to global development strategies. Yet it would be a mistake to conclude that gender is molded from the "top down." What happens on a daily basis in families and workplaces simultaneously constitutes and is constrained by structural transnational institutions. For instance, in the second half of the twentieth century young, single women, many of them from poor rural areas, have been recruited for work in export assembly plants along the U.S.–Mexico border, in East and Southeast Asia, in Silicon Valley, in the Caribbean, and in Central America. While the profitability of these multinational factories depends, in part, on management's ability to manipulate the young women's ideologies of gender, the women . . . do not respond passively or uniformly, but they actively resist, challenge, and accommodate. At the same time, the global diversion of the assembly line has concentrated corporate facilities in many U.S. cities, making available myriad managerial, administrative, and clerical jobs for college educated women. Women's paid labor is used at various points along this international system of production. Not only employment, but also consumption, embodies global interdependencies. There is a high probability that the clothes you wear and the computer you use originated in multinational corporate headquarters and in assembly plants scattered around third-world nations. And if these items were actually manufactured in the United States, they were probably assembled by Latin American and Asian-born women.

Worldwide, international labor migration and refugee movements are creating new types of multiracial societies. Although these developments are often discussed and analyzed with respect to racial differences, gender typically remains absent. As several commentators have noted, the White feminist movement in the United States has not addressed issues of immigration and nationality. Gender, however, has been fundamental in shaping immigration policies (Chang 1994; Hondagneu-Sotelo 1994). Direct labor recruitment programs generally solicit either male or female labor (e.g., Filipina nurses, Mexican male farm workers); national disenfranchisement has particular repercussions for women and men; and current immigrant laws are based on very gendered notions of what constitutes "family unification." As Chandra Mohanty suggests, "analytically these issues are the contemporary metropolitan counterpart of women's struggles against colonial occupation in the geographical third world" (1991: 23). Moreover, immigrant and refugee women's daily lives often challenge familiar feminist paradigms. The occupations in which immigrant and refugee women concentrate—paid domestic work, informal sector street vending, assembly or industrial piece work performed in the home—often blur the ideological distinction between work and family and between public and private spheres.

FROM PATCHWORK QUILT TO PRISM

All of these developments—the voices of "other" women, the study of men and masculinities, and the examination of gender in transnational contexts—have helped redefine the study of gender. By working to develop knowledge that is inclusive of the experiences of all groups, new insights about gender have begun to emerge. Examining gender in the context of other differences makes it clear that nobody experiences themselves as solely gendered. Instead, gender is configured through cross-cutting forms of difference that carry deep social and economic consequences.

By the mid-1980s, thinking about gender had entered a new stage that was more carefully grounded in the experiences of diverse groups of women and men. This perspective is a general way of looking at women and men and understanding their relationships to the structure of society. Gender is no longer viewed simply as a matter of two opposite categories of people—males and females—but as a range of social relations among differently situated people. Because centering on difference is a radical challenge to the conventional gender framework, it raises several concerns. Does the recognition that gender can only be understood contextually (meaning that there is no singular "gender" per se) make women's studies and gender studies newly vulnerable to critics in the academy? Does the immersion in difference throw us into a whirlwind of "spiraling diversity" (Hewitt 1992: 316) where multiple identities and locations shatter the categories *women* and *men?*

. . . We take a position directly opposed to an empty pluralism. Although the categories *woman* and *man* have multiple meanings, this does not reduce gender to a "postmodern kaleidoscope of lifestyles. Rather, it points to the *relational* character of gender" (Connell 1992: 736). Not only are masculinity and femininity relational, but different masculinities and femininities are interconnected also through other social structures such as race, class, and nation. Groups are created by their relationships with each other. The meaning of *woman* is defined by the existence of women of different races and classes. Being a White woman in the United States is meaningful only insofar as it is set apart from and in contradistinction to women of color.

Just as masculinity and femininity each depend on the definitions of the other to produce domination, differences *among* women and *among* men are also created in the context of structured relationships. Some women derive benefits from their race and class position, and from their location in the global economy, while they are simultaneously restricted by gender. In other words, such women are subordinated by patriarchy, yet their relatively privileged positions within hierarchies of race, class, and the global political economy intersect to create for them an expanded range of opportunities, choices, and ways of living. They may even use their race and class advantage to minimize some of the consequences of patriarchy and/or to oppose other women. Similarly, one can become a man in opposition to other men. For example, "the relation between heterosexual and homosexual men is central, carrying heavy symbolic freight. To many people, homosexuality is the *negation* of masculinity. . . . Given that assumption, antagonism toward homosexual men may be used to define masculinity" (Connell 1992: 736).

In the past decade, viewing gender through the prism of difference has profoundly reoriented the field (Acker 1999; Glenn 1999; Messner 1996; West & Fenstermaker 1995). Yet analyzing the multiple constructions of gender does not mean studying just groups of women and groups of men as different. It is clearly time to go beyond what we call the "patchwork quilt" phase in the study of women and men—that is, the phase in which we have acknowledged the importance of examining differences within constructions of gender, but do so largely by collecting a study here on African American women, a study there on gay men, and a study on working-class Chicanas. This patchwork quilt approach too often amounts to no more than "adding difference and stirring." The result may be a lovely mosaic, but like a patchwork quilt it still tends to overemphasize boundaries, rather than highlighting bridges of interdependency. In addition, this approach too often does not explore the ways that social constructions of femininities and masculinities are based on, and reproduce relations of, power. In short, we think that the substantial quantity of research that has now been done on various groups and subgroups needs to be analyzed within a framework that emphasizes differences and inequalities not as discrete areas of separation, but as interrelated bands of color that together make up a spectrum.

References

Acker, Joan. 1999. "Rewriting Class, Race and Gender: Problems in Feminist Rethinking." In Myra Marx Ferree, Judith Lorber, and Beth B. Hess (eds.), *Revisioning Gender*, 44–69. Thousand Oaks, CA: Sage.

Baca Zinn, M., Weber Cannon, L., Higginbotham, E., Thornton Dill, B. 1986. "The Costs of Exclusionary Practices in Women's Studies," *Signs: Journal of Women in Culture and Society* 11: 290–303.

Brod, Harry (ed.). 1987. *The Making of Masculinities: The New Men's Studies.* Boston: Allen & Unwin.

Brownmiller, Susan. 1975. *Against Our Will: Men, Women, and Rape.* New York: Simon and Schuster.

Chang, Grace. 1994. "Undocumented Latinas: The New 'Employable Mothers.'" In Evelyn Nakano Glenn, Grace Chang, and Linda Rennie Forcey (eds.), *Mothering, Ideology, Experience, and Agency*, 259–285. New York and London: Routledge.

Connell, R. W. 1987. *Gender and Power*. Stanford, CA: Stanford University Press.

Connell, R. W. 1992. "A Very Straight Gay: Masculinity, Homosexual Experience, and the Dynamics of Gender," *American Sociological Review* 57: 735–751.

Connell, R. W. 1995. *Masculinities*. Berkeley: University of California Press.

De Beauvoir, Simone. 1953. *The Second Sex*. New York: Knopf.

Glenn, Evelyn Nakano. 1999. "The Social Construction and Institutionalization of Gender and Race: An Integrative Framework." In Myra Marx Ferree, Judith Lorber, and Beth B. Hess (eds.), *Revisioning Gender*, 3–43. Thousand Oaks, CA: Sage.

Hartmann, Heidi. 1976. "Capitalism, Patriarchy, and Job Segregation by Sex," *Signs: Journal of Women in Culture and Society* 1: (3), part 2, Spring: 137–167.

Hewitt, Nancy A. 1992 "Compounding Differences," *Feminist Studies* 18: 313–326.

Hondagneu-Sotelo, Pierrette. 1994. *Gendered Transitions: Mexican Experiences of Immigration*. Berkeley: University of California Press.

Kanter, Rosabeth Moss. 1977. *Men and Women of the Corporation*. New York: Basic Books.

Kaufman, Michael. 1987. *Beyond Patriarchy: Essays by Men on Pleasure, Power, and Change*. Toronto and New York: Oxford University Press.

Kimmel, Michael S. (ed.). 1987. *Changing Men: New Directions in Research on Men and Masculinity*. Newbury Park, CA: Sage.

Kimmel, Michael S., Michael A. Messner (eds.). 1989. *Men's Lives*. New York: Macmillan.

Kimmel, Michael S., Michael A. Messner (eds.). 1998. *Men's Lives*, 4th edition. Allyn & Bacon.

Messner, Michael A. 1996. "Studying Up on Sex," *Sociology of Sport Journal* 13: 221–237.

Messner, Michael A. 1998. *Politics of Masculinities: Men in Movements*. Thousand Oaks, CA: Sage.

Mohanty, Chandra Talpade. 1991. "Cartographies of Struggle: Third World Women and the Politics of Feminism." In Chandra Talpade Mohanty, Ann Russo, and Lourdes Torres (eds.), *Third World Women and the Politics of Feminism*, 51–80. Bloomington: Indiana University Press.

Morgan, Robin. 1970. *Sisterhood Is Powerful: An Anthology of Writing from the Women's Liberation Movement*. New York: Vintage Books.

Pleck, Joseph. 1981. *Thy Myth of Masculinity*. Cambridge, MA: M.I.T. Press.

Rich, Adrienne. 1980. "Compulsory Heterosexuality and the Lesbian Experience," *Signs: Journal of Women in Culture and Society* 5: 631–660.

Sassen, Saskia. 1991. *The Global City: New York, London, Tokyo*. Princeton, NJ: Princeton University Press.

Smith, Barbara. 1977. *Toward a Black Feminist Criticism*. Freedom, CA: Crossing Press.

West, Candace, Sarah Fenstermaker. 1995. "Doing Difference," *Gender & Society* 9: 8–37.

IDEOLOGICAL RACISM
AND CULTURAL RESISTANCE

20

Constructing Our Own Images

Yen Le Espiritu

. . . A central aspect of racial exploitation centers on defining people of color as "the other" (Said 1979). The social construction of Asian American "otherness"—through such controlling images as the Yellow Peril, the model minority, the Dragon Lady, and the China Doll—is "the precondition for their cultural marginalization, political impotence, and psychic alienation from mainstream American society" (Hamamoto 1994, p. 5). As indicated by these stereotypes, representations of gender and sexuality figure strongly in the articulation of racism. These racist stereotypes collapse gender and sexuality: Asian men have been constructed as hypermasculine, in the image of the "Yellow Peril," but also as effeminate, in the image of the "model minority," and Asian women have been depicted as superfeminine, in the image of the "China Doll," but also as castrating, in the image of the "Dragon Lady" (Mullings 1994, pp. 279–280; Okihiro 1995). As Mary Ann Doane (1991) suggested, sexuality is "indissociable from the effects of polarization and differentiation, often linking them to structures of power and domination" (p. 217). In the Asian American case, the gendering of ethnicity—the process whereby white ideology assigns selected gender characteristics to various ethnic "others"—cast Asian American men and women as simultaneously masculine and feminine but also as neither masculine nor feminine. On the one hand, as part of the Yellow Peril, Asian American men and women have been depicted as a *masculine* threat that needs to be contained. On the other hand, both sexes have been skewed toward the female side: an indication of the group's marginalization in U.S. society and its role as the compliant "model minority" in contemporary U.S. cultural ideology. Although an apparent disjunction, both the feminization and masculinization of Asian men and women exist to define and confirm the white man's superiority (Kim 1990).

THE YELLOW PERIL

In the United States, Asia and America—East and West—are viewed as mutually exclusive binaries (Kim 1993, p. viii). Within this exclusive binary system,

---

From Yen Le Espiritu, *Asian American Women and Men* (Thousand Oaks, CA: Sage, 1997). Reprinted by permission of Rowman & Littlefield Publishers, Inc.

Asian Americans, even as citizens, are designated Asians, not Americans. Characterizing Asian Americans as "permanent houseguests in the house of America," Sau-Ling Cynthia Wong (1993) stated that "Asian Americans are put in the niche of the 'unassimilable alien': . . . they are alleged to be self-disqualified from full American membership by materialistic motives, questionable political allegiance, and, above all, outlandish, overripe, 'Oriental' cultures" (p. 6). Sonia Shah (1994) defined this form of "cultural discrimination" as a "peculiar blend of cultural and sexist oppression based on our accents, our clothes, our foods, our values and our commitments" (p. 182). This cultural discrimination brands Asians as perpetual foreigners and thus perpetuates the notion of their alleged racial unassimilability. For example, although Japanese Americans have lived in the United States since the turn of the century, many television programs, such as *Happy Days* (1974–1984) and *Gung Ho* (1986–1987), have continued to portray them as newly arrived foreigners (Hamamoto 1994, p. 13).

As the unassimilable alien, Asian Americans embody for many other Americans the "Yellow Peril"—the threat that Asians will one day unite and conquer the world. This threat includes military invasion and foreign trade from Asia, competition to white labor from Asian labor, the alleged moral degeneracy of Asian people, and potential miscegenation between whites and Asians (Wu 1982, p. 1). Between 1850 and 1940, U.S. popular media consistently portrayed Asian men as a military threat to the security and welfare of the United States *and* as a sexual danger to innocent white women (Wu 1982). In numerous dime novels, movies, and comic strips, Asians appeared as feral, rat-faced men lusting after virginal white women. Arguing for racial purity, these popular media depicted Asian-white sexual union as "at best, a form of beastly sodomy, and, at worst, a Satanic marriage" (Hoppenstand 1983, p. 174). In these popular depictions, the white man was the desirable sexual partner and the hero who rescued the white woman from "a fate worse than death" (Hoppenstand 1983, pp. 174–175). By the mid-1880s, hundreds of garishly illustrated and garishly written dime novels were being disseminated among a wide audience, sporting such sensational titles as *The Bradys and the Yellow Crooks*, *The Chase for the Chinese Diamonds*, *The Opium Den Detective*, and *The Stranglers of New York*. As portrayed in these dime novels, the Yellow Peril was the Chinatown district of a big city "in which decent, honest white folk never ventured" (Hoppenstand 1983, p. 177).

In 20th-century U.S. popular media, the Japanese joined the Chinese as a perceived threat to Europe and the United States (Wu 1982, p. 2). In 1916, William Randolph Hearst produced and distributed *Petria*, a movie about a group of fanatical Japanese who invade the United States and attempt to rape a white woman (Quinsaat 1976, p. 265). After the Japanese bombing of Pearl Harbor on December 7, 1941, the entire Yellow Peril stereotype became incorporated in the nation's war propaganda, quickly whipping white Americans into a war fever. Along with the print media, Hollywood cranked up its anti-Japanese propaganda and produced dozens of war films that centered on the Japanese menace. The fiction of the Yellow Peril stereotype became intertwined with the fact of the United States' war with Japan, and the two became one in the mind-set of the American public (Hoppenstand 1983, pp. 182–183). It was fear of the Yellow Peril—fear of the rise of nonwhite people and their contestation of white supremacy—that led to the declaration of martial law in Hawaii on December 7, 1941, and to the internment of over 110,000 Japanese on the mainland in concentration camps (Okihiro 1994, p. 137). In subsequent decades, reflecting changing geopolitical concerns, U.S. popular media featured a host of new Yellow Peril stereotypes. During the 1950s Cold War years, in television programs as well as in movies, the Communist Chinese evildoers replaced the Japanese monster; during the Vietnam war of the 1970s, the Vietnamese Communists emerged as the new Oriental villains.

Today, Yellow Perilism takes the forms of the greedy, calculating, and clever Japanese businessman aggressively buying up U.S. real estate and cultural institutions *and* the superachieving but nonassimilating Asian Americans (Hagedorn 1993, p. xxii). In a time of rising economic powers in Asia, declining economic opportunities in the United States, and growing diversity among America's people, this new Yellow Perilism—the depiction of Asian and Asian Americans as economic and cultural threats to mainstream United States—supplies white Americans with a united identity and provides ideological justification for U.S. isolationist policy toward Asia, increasing restrictions against Asian (and Latino) immigration, and the invisible institutional racism and visible violence against Asians in the United States (Okihiro 1994, pp. 138–139).

THE RACIAL CONSTRUCTION OF ASIAN AMERICAN MANHOOD

Like other men of color, Asian American men have been excluded from white-based cultural notions of the masculine. Whereas white men are depicted both as virile and as protectors of women, Asian men have been characterized both as asexual *and* as threats to white women. It is important to note the historical contexts of these seemingly divergent representations of Asian American manhood. The racist depictions of Asian men as "lascivious and predatory" were especially pronounced during the nativist movement against Asians at the turn of the [20th] century (Frankenberg 1993, pp. 75–76). The exclusion of Asian women from the United States and the subsequent establishment of bachelor societies eventually reversed the construction of Asian masculinity from "hypersexual" to "asexual" and even "homosexual." The contemporary model-minority stereotype further emasculates Asian American men as passive and malleable. Disseminated and perpetuated through the popular media, these stereotypes of the emasculated Asian male construct a reality in which social and economic discrimination against these men appears defensible. As an example, the desexualization of Asian men naturalized their inability to establish conjugal families in pre-World War II United States. Gliding over race-based exclusion laws that banned the immigration of most Asian women and antimiscegenation laws that prohibited men of color from marrying white women, these dual images of the eunuch and the rapist attributed the "womanless households" characteristics of pre-war Asian America to Asian men's lack of sexual prowess and desirability.

A popular controlling image applied to Asian American men is that of the sinister Oriental—a brilliant, powerful villain who plots the destruction of Western civilization. Personified by the movie character of Dr. Fu Manchu, this Oriental mastermind combines Western science with Eastern magic and commands an army of devoted assassins (Hoppenstand 1983, p. 178). Though ruthless, Fu Manchu lacks masculine heterosexual prowess (Wang 1988, p. 19), thus privileging heterosexuality. Frank Chin and Jeffrey Chan (1972), in a critique of the desexualization of Asian men in Western culture, described how the Fu Manchu character undermines Chinese American virility:

> Dr. Fu, a man wearing a long dress, batting his eyelashes, surround by muscular black servants in loin cloths, and with his habit of caressingly touching white men on the leg, wrist, and face with his long fingernails is not so much a threat as he is a frivolous offense to white manhood. (p. 60)

In another critique that glorifies male aggression, Frank Chin (1972) contrasted the neuterlike characteristics assigned to Asian men to the sexually aggressive images associated with other men of color. "Unlike the white stereotype of the evil black stud, Indian rapist, Mexican macho, the evil of the evil Dr. Fu Manchu was not sexual, but homosexual" (p. 66). However, Chin failed to note that as a homosexual, Dr. Fu (and by extension, Asian men) threatens and offends white masculinity—and therefore needs to be contained ideologically and destroyed physically.

Whereas the evil Oriental stereotype marks Asian American men as the white man's enemy, the stereotype of the sexless Asian sidekick—Charlie Chan, the Chinese laundryman, the Filipino houseboy—depicts Asian men as devoted and impotent, eager to please. William Wu (1982) reported that the Chinese servant "is the most important single image of the Chinese immigrants" in American fiction about Chinese Americans between 1850 and 1940 (p. 60). More recently, such diverse television programs as *Bachelor Father* (1957–1962), *Bonanza* (1959–1973), *Star Trek* (1966–1969), and *Falcon Crest* (1981–1990) all featured the stock Chinese bachelor domestic who dispenses sage advice to his superiors in addition to performing traditional female functions within the household (Hamamoto 1994, p. 7). By trapping

Chinese men (and by extension, Asian men) in the stereotypical "feminine" tasks of serving white men, American society erases the figure of the Asian "masculine" plantation worker in Hawaii or railroad construction worker in the western United States, thus perpetuating the myth of the androgynous and effeminate Asian man (Goellnicht 1992, p. 198). This feminization, in turn, confines Asian immigrant men to the segment of the labor force that performs women's work.

The motion picture industry has been key in the construction of Asian men as sexual deviants. In a study of Asians in the U.S. motion pictures, Eugene Franklin Wong (1978) maintained that the movie industry filmically castrates Asian males to magnify the superior sexual status of white males (p. 27). As on-screen sexual rivals of whites, Asian males are neutralized, unable to sexually engage Asian women and prohibited from sexually engaging white women. By saving the white women from sexual contact with the racial "other," the motion picture industry protects the Anglo-American, bourgeois male establishment from any challenges to its hegemony (Marchetti 1993, p. 218). At the other extreme, the industry has exploited on the most potent aspects of the Yellow Peril discourses—the sexual danger of contact between the races—by concocting a sexually threatening portrayal of the licentious and aggressive Yellow Man lusting after the White Woman (Marchetti 1993, p. 3). Heedful of the larger society's taboos against Asian male–white female sexual union, white male actors donning "yellowface"—instead of Asian male actors—are used in these "love scenes." Nevertheless, the message of the perverse and animalistic Asian male attacking helpless white women is clear (Wong 1978). Though depicting sexual aggression, this image of the rapist, like that of the eunuch, casts Asian men as sexually undesirable. As Wong (1978) succinctly stated, in Asian male–white female relations, "There can be rape, but there cannot be romance" (p. 25). Thus, Asian males yield to the sexual superiority of the white males who are permitted filmically to maintain their sexual dominance over both white women and women of color. A young Vietnamese American man describes the damaging effect of these stereotypes on his self-image:

> Every day I was forced to look into a mirror created by white society and its media. As a young Asian man, I shrank before white eyes. I wasn't tall, I wasn't fair, I wasn't muscular, and so on. Combine that with the enormous insecurities any pubescent teenager feels, and I have no difficulty in knowing now why I felt naked before a mass of white people. (Nguyen 1990, p. 23)

White cultural and institutional racism against Asian males is also reflected in the motion picture industry's preoccupation with the death of Asians—a filmic solution to the threats of the Yellow Peril. In a perceptive analysis of Hollywood's view of Asians in films made from the 1930s to the 1960s, Tom Engelhardt (1976) described how Asians, like Native Americans, are seen by the movie industry as inhuman invaders, ripe for extermination. He argued that the theme of the nonhumanness of Asians prepares the audience to accept, without flinching, "the levelling and near-obliteration of three Asian areas in the course of three decades" (Engelhardt 1976, p. 273). The industry's death theme, though applying to all Asians, is mainly focused on Asian males, with Asian females reserved for sexual purposes (Wong 1978, p. 35). Especially in war films, Asian males, however advantageous their initial position, inevitably perish at the hands of the superior white males (Wong 1978, p. 34).

THE RACIAL CONSTRUCTION OF ASIAN AMERICAN WOMANHOOD

Like Asian men, Asian women have been reduced to one-dimensional caricatures in Western representation. The condensation of Asian women's multiple differences into gross character types—mysterious, feminine, and nonwhite—obscures the social injustice of racial, class, and gender oppression (Marchetti 1993, p. 71). Both Western film and literature promote dichotomous stereotypes of the Asian woman: Either she is the cunning Dragon Lady or the servile Lotus Blossom Baby (Tong 1994, p. 197). Though connoting two extremes, these stereotypes are interrelated: Both eroticize Asian women as exotic "others"—sensuous, promiscuous, but untrustworthy. Whereas American popular culture denies "manhood" to Asian men, it endows Asian women with an excess of "womanhood," sexualizing them but also impugning their sexuality. In this process, both sexism and racism have been blended together to produce the sexualization of white racism (Wong 1978, p. 260). Linking the controlling images of Asian men and women, Elaine Kim (1990) suggested that Asian women are portrayed as sexual for the same reasons that men are asexual: "Both exist to define the white man's virility and the white man's superiority" (p. 70).

As the racialized exotic "others," Asian American women do not fit the white-constructed notions of the feminine. Whereas white women have been depicted as chaste and dependable, Asian women have been represented as promiscuous and untrustworthy. In a mirror image of the evil Fu Manchu, the Asian woman was portrayed as the castrating Dragon Lady who, while puffing on her foot-long cigarette holder, could poison a man as easily as she could seduce him. "With her talon-like six-inch fingernails, her skin-tight satin dress slit to the thigh," the Dragon Lady is desirable, deceitful, and dangerous (Ling 1990, p. 11). In the 1924 film *The Thief of Baghdad*, Anna May Wong, a pioneer Chinese American actress, played a handmaid who employed treachery to help an evil Mongol prince attempt to win the hand of the Princess of Baghdad (Tajima 1989, p. 309). In so doing, Wong unwittingly popularized a common Dragon Lady social type: treacherous women who are partners in crime with men of their own kind. The publication of *Daughter of Fu Manchu*

(1931) firmly entrenched the Dragon Lady image in white consciousness. Carrying on her father's work as the champion of Asian hegemony over the white race, Fah Lo Sue exhibited, in the words of American studies scholar, William F. Wu, "exotic sensuality, sexual availability to a white man, and a treacherous nature" (cited in Tong 1994, p. 197). A few years later, in 1934, Milton Caniff inserted into his adventure comic strip *Terry and the Pirates* another version of the Dragon Lady who "combines all the best features of past moustache twirlers with the lure of the handsome wench" (Hoppenstand 1983, p. 178). As such, Caniff's Dragon Lady fuses the image of the evil male Oriental mastermind with that of the Oriental prostitute first introduced some 50 years earlier in the dime novels.

At the opposite end of the spectrum is the Lotus Blossom stereotype, reincarnated throughout the years as the China Doll, the Geisha Girl, the War Bride, or the Vietnamese prostitute—many of whom are the spoils of the last three wars fought in Asia (Tajima 1989, p. 309). Demure, diminutive, and deferential, the Lotus Blossom Baby is "modest, tittering behind her delicate ivory hand, eyes downcast, always walking ten steps behind her man, and, best of all, devot[ing] body and soul to serving him" (Ling 1990, p. 11). Interchangeable in appearance and name, these women have no voice; their "nonlanguage" includes uninterpretable chattering, pidgin English, giggling, or silence (Tajima 1989). These stereotypes of Asian women as submissive and dainty sex objects not only have impeded women's economic mobility but also have fostered an enormous demand for X-rated films and pornographic materials featuring Asian women in bondage, for "Oriental" bathhouse workers in U.S. cities, and for Asian mail-order brides (Kim 1984, p. 64).

SEXISM, RACISM, AND LOVE

The racialization of Asian manhood and womanhood upholds white masculine hegemony. Cast as sexually available, Asian women become yet another possession of the white man. In motion pictures and network programs, interracial sexuality, though rare, occurs principally between a white male and an Asian female. A combination of sexism and racism makes this form of miscegenation more acceptable: Race mixing between an Asian male and a white female would upset not only racial taboos but those that attend patriarchal authority as well (Hamamoto 1994, p. 39). Whereas Asian men are depicted as either the threatening rapist or the impotent eunuch, white men are endowed with the masculine attributes with which to sexually attract the Asian woman. Such popular television shows as *Gunsmoke* (1955–1975) and *How the West Was Won* (1978–1979) clearly articulate the theme of Asian female sexual possession by the white male. In these shows, only white males have the prerogative to cross racial boundaries and to choose freely from among women of color as sex partners. Within a system of racial and gender oppression, the sexual possession of women and men of color by white men becomes yet another means of enforcing unequal power relations (Hamamoto 1994, p. 46).

The preference for white male–Asian female is also prevalent in contemporary television news broadcasting, most recently in the 1993–1995 pairing of Dan Rather and Connie Chung as coanchors of the *CBS Evening News*. Today, virtually every major metropolitan market across the United States today has at least one Asian American female newscaster (Hamamoto 1994, p. 245). While female Asian American anchorpersons—Connie Chung, Tritia Toyota, Wendy Tokuda, and Emerald Yeh—are popular television news figures, there is a nearly total absence of Asian American men. Critics argue that this is so because the white male hiring establishment, and presumably the larger American public, feels more comfortable (i.e., less threatened) seeing a white male sitting next to a minority female at the anchor desk than the reverse. Stephen Tschida of WDBJ-TV (Roanoke, Virginia), one of only a handful of male Asian American television news anchors, was informed early in his career that he did not have the proper "look" to qualify for the anchorperson position. Other male broadcast news veterans have reported being passed over for younger, more beauteous, female Asian Americans (Hamamoto 1994, p. 245). This gender imbalance sustains the construction of Asian American women as more successful, assimilated, attractive, and desirable than their male counterparts. . . .

CONCLUSION

Ideological representations of gender and sexuality are central in the exercise and maintenance of racial, patriarchal, and class domination. In the Asian American case, this ideological racism has taken seemingly contrasting forms: Asian men have been cast as both hypersexual and asexual, and Asian women have been rendered both superfeminine and masculine. Although in apparent disjunction, both forms exist to define, maintain, and justify white male supremacy. The racialization of Asian American manhood and womanhood underscores the interconnections of race, gender, and class. As categories of difference, race and gender relations do not parallel but intersect and confirm each other, and it is the complicity among these categories of difference that enables U.S. elites to justify and maintain their cultural, social, and economic power. Responding to the ideological assaults on their gender identities, Asian American cultural workers have engaged in a wide range of oppositional projects to defend Asian American manhood and womanhood. In the process, some have embraced a masculinist cultural nationalism, a stance that marginalizes Asian American women and their needs. Though sensitive to the emasculation of Asian American men, Asian American feminists have pointed out that Asian American nationalism insists on a fixed masculinist identity, thus obscuring gender differences. Though divergent, both the nationalist and feminist positions advance the dichotomous stance of man or woman, gender or race or class, without recognizing the complex relationality of these categories of oppression. It is only when Asian Americans recognize the intersections of race, gender, and class that we can transform the existing hierarchical structure.

References

Chin, F. (1972). Confessions of the Chinatown cowboy. *Bulletin of Concerned Asian Scholars*, 4(3), 66.

Chin. F., & Chan, J. P. (1972). Racist love. In R. Kostelanetz (Ed.), *Seeing through shuck* (pp. 65–79). New York: Ballantine.

Doane, M. A. (1991). *Femme fatales: Feminism, film theory, psychoanalysis.* New York: Routledge.

Engelhardt, T. (1976). Ambush at Kamikaze Pass. In E. Gee (Ed.), *Counterpoint: Perspectives on Asian America* (pp. 270–279). Los Angeles: University of California at Los Angeles, Asian American Studies Center.

Frankenberg, R. (1993). *White women, race matters: The social construction of whiteness.* Minneapolis: University of Minnesota Press.

Goellnicht, D. C. (1992). Tang Ao in America: Male subject positions in *China Men.* In S. G. Linn & A. Ling (Eds.), *Reading the literatures of Asian America* (pp. 191–212). Philadelphia: Temple University Press.

Hagedorn, J. (1993). Introduction: "Role of dead man require very little acting." In J. Hagedorn (Ed.), *Charlie Chan is dead: An anthology of contemporary Asian American fiction* (pp. xxi–xxx). New York: Penguin.

Hamamoto, D. Y. (1994). *Monitored peril: Asian Americans and the politics of representation.* Minneapolis: University of Minnesota Press.

Hoppenstand, G. (1983). Yellow devil doctors and opium dens: A survey of the yellow peril stereotypes in mass media entertainment. In C. D. Geist & J. Nachbar (Eds.), *The popular culture reader* (3rd ed., pp. 171–185). Bowling Green, OH: Bowling Green University Press.

Kim, E. (1984). Asian American writers: A bibliographical review. *American Studies International*, 22, 2.

Kim, E. (1990). "Such opposite creatures": Men and women in Asian American literature. *Michigan Quarterly Review*, 29, 68–93.

Kim, E. (1993). Preface. In J. Hagedorn (Ed.), *Charlie Chan is dead: An anthology of Contemporary Asian American fiction* (pp. vii–xiv). New York: Penguin.

Ling, A. (1990). *Between worlds: Women writers of Chinese ancestry.* New York: Pergamon.

Marchetti, G. (1993). *Romance and the "Yellow Peril": Race, sex, and discursive strategies in Hollywood fiction.* Berkeley: University of California Press.

Mullings, L. (1994). Images, ideology, and women of color. In M. Baca Zinn & B. T. Dill (Eds.), *Women of color in U.S. society* (pp. 265–289). Philadelphia: Temple University Press.

Nguyen, V. (1990, December 7). Growing up in white America. *Asian Week*, p. 23.

Okihiro, G. Y. (1994). *Margins and mainstreams: Asians in American history and culture.* Seattle: University of Washington Press.

Okihiro, G. Y. (1995, November). *Reading Asian bodies, reading anxieties.* Paper presented at the University of California, San Diego Ethnic Studies Colloquium, La Jolla.

Quinsaat, J. (1976). Asians in the media. The shadows in the spotlight. In E. Gee (Ed.), *Counterpoint: Perspectives on Asian America* (pp. 264–269). Los Angeles: University of California at Los Angeles, Asian American Studies Center.

Said, E. (1979). *Orientalism.* New York: Random House.

Shah, S. (1994). Presenting the Blue Goddess: Toward a national, Pan-Asian Feminist agenda. In K. Aguilar-San Juan (Ed.), *The state of Asian America: Activism and resistance in the 1990s* (pp. 147–158). Boston: South End.

Tajima, R. (1989). Lotus blossoms don't bleed: Images of Asian women. In Asian Women United of California (Ed.), *Making waves: An anthology of writings by and about Asian American women* (pp. 308–317). Boston: Beacon.

Tong, B. (1994). *Unsubmissive women: Chinese prostitutes in nineteenth-century San Francisco.* Norman: University of Oklahoma Press.

Wang, A. (1988). Maxine Hong Kingston's reclaiming of America: The birthright of the Chinese American male. *South Dakota Review, 26,* 18–29.

Wong, E. F. (1978). *On visual media racism: Asians in the American motion pictures.* New York: Arno.

Wong, S.-L. C. (1993). *Reading Asian American literature: From necessity to extravagance.* Princeton, NJ: Princeton University Press.

Wu, W. F. (1982). *The Yellow Peril: Chinese Americans in American fiction 1850–1940.* Hamden, CT: Archon.

A WHITE WOMAN OF COLOR

21

Julia Álvarez

Growing up in the Dominican Republic, I experienced racism within my own family—though I didn't think of it as racism. But there was definitely a hierarchy of beauty, which was the main currency in our daughters-only family. It was not until years later, from the vantage point of this country and this education, that I realized that this hierarchy of beauty was dictated by our coloring. We were a progression of whitening, as if my mother were slowly bleaching the color out of her children.

The oldest sister had the darkest coloring, with very curly hair and "coarse" features. She looked the most like Papi's side of the family and was considered the least pretty. I came next, with "good hair," and skin that back then was a deep olive, for I was a tomboy—another dark mark against me— who would not stay out of the sun. The sister right after me had my skin color, but she was a good girl who stayed indoors, so she was much paler, her hair a golden brown. But the pride and joy of the family was the baby. She was the one who made heads turn and strangers approach asking to feel her silken hair. She was white white, an adjective that was repeated in describing her color as if to deepen the shade of white. Her eyes were brown, but her hair was an unaccountable towheaded blond. Because of her coloring, my father was teased that there must have been a German milkman in our neighborhood. How could *she* be *his* daughter? It was clear that this youngest child resembled Mami's side of the family.

It was Mami's family who were *really* white. They were white in terms of race and white also in terms of class. From them came the fine features, the pale skin, the lank hair. Her brothers and uncles went to schools abroad and had important businesses in the country. They also emulated the manners and habits of North Americans. Growing up, I remember arguments at the supper table on whether or not it was proper to tie one's napkin around one's neck, on how much of one's arm one could properly lay on the table, on whether spaghetti could be eaten with the help of a spoon. My mother, of course, insisted on all the protocol of knives and forks and on eating a little portion of everything served; my father, on the other hand, defended our eating whatever we wanted, with our hands if need be, so we could "have fun" with our food. My mother would snap back that we looked like *jibaritas* who should be living out in the country. Of course, that was precisely where my father's family came from.

Not that Papi's family weren't smart and enterprising, all twenty-five brothers and sisters. (The size of the family in and of itself was considered very country by some members of Mami's family.) Many of Papi's brothers had gone to the university and become professionals. But their education was totally island—no fancy degrees from Andover and Cornell and Yale, no summer camps or school songs in another language. Papi's family still lived in the interior versus the capital, in old-fashioned houses without air conditioning, decorated in ways my mother's family would have considered, well, tasteless. I remember antimacassars on the backs of rocking chairs (which were the living-room set), garish paintings of flamboyant trees, ceramic planters with plastic flowers in bloom. They were *criollos*—creoles—rather than cosmopolitans, expansive, proud, colorful. (Some members had a sixth finger on their right—or was it their left hand?) Their features were less aquiline than Mother's family's, the skin darker, the hair coarse and curly. Their money still had the smell of the earth on it and was kept in a wad in their back pockets, whereas my mother's family had money in the Chase Manhattan Bank, most of it with George Washington's picture on it, not Juan Pablo Duarte's.

It was clear to us growing up then that lighter was better, but there was no question of discriminating against someone because he or she was dark-skinned. Everyone's family, even an elite one like Mami's, had darker-skinned members. All Dominicans, as the saying goes, have a little black behind the ears. So, to separate oneself from those who were darker would have been to divide *una familia*, a sacrosanct entity in our culture. Neither was white blood necessarily a sign of moral or intellectual or political superiority. All one has to do is page through a Dominican history book and look at the number of dark-skinned presidents, dictators, generals, and entrepreneurs to see that power has not resided exclusively or even primarily among the whites on the island. The leadership of our country has been historically "colored."

But being black was something else. A black Dominican was referred to as a "dark Indian" (*indio oscuro*)—unless you wanted to come to blows with him, that is. The real blacks were the Haitians who lived next door and who occupied the Dominican Republic for twenty years, from 1822 to 1844, a fact that can still so inflame the Dominican populace you'd think it had happened last year. The denial of the Afro-Dominican part of our culture reached its climax during the dictatorship of Trujillo, whose own maternal grandmother was Haitian. In 1937, to protect Dominican race purity, Trujillo ordered the overnight genocide of thousands (figures range from 4,000 to 20,000) of Haitians by his military, who committed this atrocity using only machetes and knives in order to make this planned extermination look like a "spontaneous" border skirmish. He also had the Dominican Republic declared a white nation despite of the evidence of the mulatto senators who were forced to pass this ridiculous measure.

So, black was not so good, kinky hair was not so good, thick lips not so good. But even if you were *indio oscuro conpelo malo y una bemba de aquí a Bani*, you could still sit in the front of the bus and order at the lunch counter—or the equivalent thereof. There was no segregation of races in the halls of power. But in the aesthetic arena—the one to which we girls were relegated as females—lighter was better. Lank hair and pale skin and small, fine features were better. All I had to do was stay out of the sun and behave myself and I could pass as a pretty white girl.

Another aspect of my growing up also greatly influenced my thinking on race. Although I was raised in the heart of a large family, my day-to-day caretakers were the maids. Most of these women were dark-skinned, some of Haitian background. One of them, Misiá, had been spared the machetes of the 1937 massacre when she was taken in and hidden from the prowling *guardias* by the family. We children spent most of the day with these women. They tended to us, nursed us when we were sick, cradled us when we fell down and scraped an elbow or knee (as a tomboy, there was a lot of this scraping for me), and most important, they told us stories of *los santos* and *el baroón del cementerio*, of *el cuco* and *las ciguapas*, beautiful dark-skinned creatures who escaped capture because their feet were turned backwards so they left behind a false set of footprints. These women spread the wings of our imaginations and connected us deeply to the land we came from. They were the ones with the stories that had power over us.

We arrived in Nueva York in 1960, before the large waves of Caribbean immigrants created little Habanas, little Santo Domingos, and little San Juans in the boroughs of the city. Here we encountered a whole new kettle of wax—as my malapropping Mami might have said. People of color were treated as if they were inferior, prone to violence, uneducated, untrustworthy, lazy—all the "bad" adjectives we were learning in our new language. Our dark-skinned aunt, Tía Ana, who had lived in New York for several decades and so was the authority in these matters, recounted stories of discrimination on buses and subways. These American were so blind! One drop of black and you were black. Everyone back home would have known that Tía Ana was not black: she had "good hair" and her skin color was a light *indio*. All week, she worked in a *factoria* in the Bronx, and when she came to visit us on Saturdays to sew our school clothes, she had to take three trains to our nice neighborhood where the darkest face on the street was usually her own.

We were lucky we were white Dominicans or we would have had a much harder time of it in this country. We would have encountered a lot more prejudice than we already did, for white as we were, we found that our Latinoness, our accents, our habits and smells, added "color" to our complexion. Had we been darker, we certainly could not have bought our mock Tudor house in Jamaica Estates. In fact, the African American family who moved in across the street several years later needed police protection because of threats. Even so, at the local school, we endured the bullying of classmates. "Go back to where you came from!" they yelled at my sisters and me in the playground. When some of them started throwing stones, my mother made up her mind that we were not safe and began applying to boarding schools where privilege transformed prejudice into patronage.

"So where are you from?" my classmates would ask.

"Jamaica Estates," I'd say, an edge of belligerence to my voice. It was obvious from my accent, if not my looks, that I was not *from* there in the way they meant being from somewhere.

"I mean *originally*."

And then it would come out, the color, the accent, the cousins with six fingers, the smell of garlic.

By the time I went off to college, a great explosion of American culture was taking place on campuses across the country. The civil rights movement, the Vietnam War and subsequent peace movement, the women's movement, were transforming traditional definitions of American identity. Ethnicity was in: my classmates wore long braids like Native Americans and peasant blouses from Mexico and long, diaphanous skirts and dangly earrings from India. Suddenly, my foreignness was being celebrated. This reversal felt affirming but also disturbing. As huipils, serapes, and embroidered dresses proliferated about me, I had the feeling that my ethnicity had become a commodity. I resented it.

When I began looking for a job after college, I discovered that being a white Latina made me a nonthreatening minority in the eyes of these employers. My color was a question *only* of culture, and if I kept my cultural color to myself, I was "no problem." Each time I was hired for one of my countless "visiting appointments"—they were never permanent "invitations," mind you—the inevitable questionnaire would accompany my contract in which I was to check off my RACE: CAUCASIAN, BLACK, NATIVE AMERICAN, ASIAN, HISPANIC, OTHER. How could a Dominican divide herself in this way? Or was I really a Dominican anymore? And what was a Hispanic? A census creation— there is no such culture—how could it define who I was at all? Given this set of options, the truest answer might have been to check off OTHER.

For that was the way I had begun to think of myself. Adrift from any Latino community in this country, my culture had become an internal homeland, periodically replenished by trips "back home." But as a professional woman on my own, I felt less and less at home on the island. My values, the loss of my Catholic faith, my lifestyle, my wardrobe, my hippy ways, and my feminist ideas separated me from my native culture. I did not subscribe to many of the mores and constraints that seemed to be an intrinsic part of that culture. And since my culture had always been my "color," by rejecting these mores I had become not only Americanized but whiter.

If I could have been a part of a Latino community in the United States, the struggle might have been, if not easier, less private and therefore less isolating. These issues of acculturation and ethnicity would have been struggles to share with others like me. But all my North American life I had lived in shifting academic communities—going to boarding schools, then college, and later teaching wherever I could get those yearly appointments—and these communities reflected the dearth of Latinos in the profession. Except for friends in Spanish departments, who tended to have come from their countries of origin to teach rather than being raised in this country as I was, I had very little daily contact with Latinos.

Where I looked for company was where I had always looked for company since coming to this country: in books. At first the texts that I read and taught were the ones prescribed to me, the canonical works which formed the content of the bread-and-butter courses that as a "visiting instructor" I was hired to teach. These texts were mostly written by white male writers from Britain and the United States, with a few women thrown in and no Latinos. Thank goodness for the occasional creative writing workshop where I could bring in the multicultural authors I wanted. But since I had been formed in this very academy, I was clueless where to start. I began to educate myself by reading, and that is when I discovered that there were others out there like me, hybrids who came in a variety of colors and whose ethnicity and race were an evolving process, not a rigid paradigm or a list of boxes, one of which you checked off.

This discovery of my ethnicity on paper was like a rebirth. I had been going through a pretty bad writer's block: the white page seemed impossible to fill with whatever it was I had in me to say. But listening to authors like Maxine Hong Kingston, Toni Morrison, Gwendolyn Brooks, Langston Hughes, Maya Angelou, June Jordan, and to Lorna Dee Cervantes, Piri Thomas, Rudolfo Anaya, Edward Rivera, Ernesto Galarza (that first wave of Latino writers), I began to hear the language "in color." I began to see that literature could reflect the otherness I was feeling, that the choices in fiction and poetry did not have to be bleached out of their color or simplified into either/or. A story could allow for the competing claims of different parts of ourselves and where we came from.

Ironically, it was through my own stories and poems that I finally made contact with Latino communities in this country. As I published more, I was invited to read at community centers and bilingual programs. Latino students, who began attending colleges in larger numbers in the late seventies and eighties, sought me out as a writer and teacher "of color." After the publication of *How the García Girls Lost Their Accents*, I found that I had become a sort of spokesperson for Dominicans in this country, a role I had neither sought nor accepted. Of course, some Dominicans refused to grant me any status as a "real" Dominican because I was "white." With the color word there was also a suggestion of class. My family had not been among the waves of economic immigrants that left the island in the seventies, a generally darker-skinned, working-class group, who might have been the maids and workers in my mother's family house. We had come in 1960, political refugees, with no money but with "prospects": Papi had a friend who was the doctor at the Waldorf Astoria and who helped him get a job; Mami's family had money in the Chase Manhattan Bank they could lend us. We had changed class in America—from Mami's elite family to middle-class spics—but our background and education and most especially our pale skin had made mobility easier for us here. We had not undergone the same kind of race struggles as other Dominicans; therefore, we could not be "real" Dominicans.

What I came to understand and accept and ultimately fight for with my writing is the reality that ethnicity and race are not fixed constructs or measurable quantities. What constitutes our ethnicity and our race—once there is literally no common ground beneath us to define it—evolves as we seek to define and redefine ourselves in new contexts. My Latino-ness is not something someone can take away from me or leave me out of with a definition. It is in my blood: it comes from that mixture of biology, culture, native language, and experience that makes me a different American from one whose family comes from Ireland or Poland or Italy. My Latino-ness is also a political choice. I am choosing to hold on to my ethnicity and native language even if I can "pass." I am choosing to color my Americanness with my Dominican-ness even if it came in a light shade of skin color.

I hope that as Latinos, coming from so many different countries and continents, we can achieve solidarity in this country as the mix that we are. I hope we won't shoot ourselves in the foot in order to maintain some sort of false "purity" as the glue that holds us together. Such an enterprise is bound to fail. We need each other. We can't afford to reject the darker or lighter varieties, and to do so is to have absorbed a definition of ourselves as exclusively one thing or the other. And haven't we learned to fear that word "exclusive"? This reductiveness is absurd when we are talking about a group whose very definition is that of a mestizo race, a mixture of European, indigenous, African, and much more. Within this vast circle, shades will lighten and darken into overlapping categories. If we cut them off, we diminish our richness and we plant a seed of ethnic cleansing that is the root of the bloodshed we have seen in Bosnia and the West Bank and Rwanda and even our own Los Angeles and Dominican Republic.

As we Latinos redefine ourselves in America, making ourselves up and making ourselves over, we have to be careful, in taking up the promises of America, not to adopt its limiting racial paradigms. Many of us have shed customs and prejudices that oppressed our gender, race, or class on our native islands and in our native countries. We should not replace these with modes of thinking that are divisive and oppressive of our rich diversity. Maybe as a group that embraces many races and differences, we Latinos can provide a positive multicultural, multiracial model to a divided America.

MASCULINITIES AND ATHLETIC CAREERS

22

Michael Messner

. . . It is now widely accepted in sport sociology that social institutions such as the media, education, the economy, and (a more recent and controversial

Author's note: Parts of this article were presented as papers at the American Sociological Association Annual Meeting, Chicago, in August 1987, and at the North American Society for the Sociology of Sport Annual Meeting in Edmonton, Alberta, in November 1987. I thank Maxine Baca Zinn, Bob Blauner, Bob Dunn, Pierrette Hondagneu-Sotelo, Carol Jacklin, Michael Kimmel, Judith Lorber, Don Sabo, Barrie Thorne, and Carol Warren for constructive comments on earlier versions of this article.

From *Gender & Society* 3 (March 1989): 71–88. Reprinted by permission of Sage Publications, Inc.

addition to the list) the black family itself all serve to systematically channel disproportionately large numbers of young black men into football, basketball, boxing, and baseball, where they are subsequently "stacked" into low-prestige and high-risk positions, exploited for their skills, and finally, when their bodies are used up, excreted from organized athletics at a young age with no transferable skills with which to compete in the labor market (Edwards 1984; Eitzen and Purdy 1986; Eitzen and Yetman 1977).

While there are racial differences in involvement in sports, class, age, and educational differences seem more significant. Rudman's (1986) initial analysis revealed profound differences between whites' and blacks' orientations to sports. Blacks were found to be more likely than whites to view sports favorably, to incorporate sports into their daily lives, and to be affected by the outcome of sporting events. However, when age, education, and social class were factored into the analysis, Rudman found that race did not explain whites' and blacks' different orientations. Blacks' affinity to sports is best explained by their tendency to be clustered disproportionately in lower-income groups.

The 1980s has ushered in what Wellman (1986, p. 43) calls a "new political linguistics of race," which emphasizes cultural rather than structural causes (and solutions) to the problems faced by black communities. The advocates of the cultural perspective believe that the high value placed on sports by black communities has led to the development of unrealistic hopes in millions of black youths. They appeal to family and community to bolster other choices based upon a more rational assessment of "reality." Visible black role models in many other professions now exist, they say, and there is ample evidence which proves that sports careers are, at best, a bad gamble.

Critics of the cultural perspective have condemned it as conservative and victim blaming. But it can also be seen as a response to the view of black athletes as little more than unreflexive dupes of an all-powerful system, which ignores the importance of agency. Gruneau (1983) has argued that sports must be examined within a theory that views human beings as active subjects who are operating within historically constituted structural constraints. Gruneau's reflexive theory rejects the simplistic views of sports as either a realm of absolute oppression or an arena of absolute freedom and spontaneity. Instead, he argues, it is necessary to construct an understanding of how and why participants themselves actively make choices and construct and define meaning and a sense of identity within the institutions that they find themselves.

None of these perspectives considers the ways that gender shapes men's definitions of meaning and choices. Within the sociology of sport, gender as a process that interacts with race and class is usually ignored or taken for granted—except when it is *women* athletes who are being studied. Sociologists who are attempting to come to grips with the experiences of black men in general, and in organized sports in particular, have almost exclusively focused their analytic attention on the variable "black," while uncritically taking "men" as a given. Hare and Hare (1984), for example, view masculinity as a biologically determined tendency to act as a provider and protector that is thwarted for black men by socioeconomic and racist obstacles. Staples (1982) does view masculinity largely as a socially produced script, but he accepts this script as a given, preferring to focus on black men's blocked access to male role fulfillment. These perspectives on masculinity fail to show how the male role itself, as it interacts with a constricted structure of opportunity, can contribute to locking black men into destructive relationships and life-styles (Franklin 1984; Majors 1986).

This article will examine the relationships among male identity, race, and social class by listening to the voices of former athletes. I will first briefly describe my research. Then I will discuss the similarities and differences in the choices and experiences of men from different racial and social class backgrounds. Together, these choices and experiences help to construct what Connell (1987) calls "the gender order." Organized sports, it will be suggested, is a practice through which men's separation from and power over women is embodied and naturalized at the same time that hegemonic (white, heterosexual, professional-class) masculinity is clearly differentiated from marginalized and subordinated masculinities.

DESCRIPTION OF RESEARCH

Between 1983 and 1985, I conducted 30 open-ended, in-depth interviews with male former athletes. My purpose was to add a critical understanding of male gender identity to Levinson's (1978) conception of the "individual life-course"—specifically, to discover how masculinity develops and changes as a man interacts with the socially constructed world of organized sports. Most of the men I interviewed had played the U.S. "major sports"—football, basketball, baseball, track. At the time of the interview, each had been retired from playing organized sports for at least 5 years. Their ages ranged from 21 to 48, with the median, 33. Fourteen were black, 14 were white, and 2 were Hispanic. Fifteen of the 16 black and Hispanic men had come from poor or working-class families, while the majority (9 of 14) of the white men had come from middle-class or professional families. Twelve had played organized sports through high school, 11 through college, and 7 had been professional athletes. All had at some time in their lives based their identities largely on their roles as athletes and could therefore be said to have had athletic careers.

MALE IDENTITY AND ORGANIZED SPORTS

. . . For the men in my study, the rule-bound structure of organized sports became a context in which they struggled to construct a masculine positional identity.

All of the men in this study described the emotional salience of their earliest experiences in sports in terms of relationships with other males. It was not winning and victories that seemed important at first; it was something "fun" to do with fathers, older brothers or uncles, and eventually with same-aged peers. As a man from a white, middle-class family said, "The most important thing was just being out there with the rest of the guys—being friends." A 32-year-old man from a poor Chicano family, whose mother had died when he was 9 years old, put it more succinctly:

> What I think sports did for me is it brought me into kind of an instant family. By being on a Little League team, or even just playing with kids in the neighborhood, it brought what I really wanted, which was some kind of closeness.

Though sports participation may have initially promised "some kind of closeness," by the ages of 9 or 10, the less skilled boys were already becoming alienated from—or weeded out of—the highly competitive and hierarchical system of organized sports. Those who did experience some early successes received recognition from adult males (especially fathers and older brothers) and held higher status among peers. As a result, they began to pour more and more of their energies into athletic participation. It was only after they learned that they would get recognition from other people for being a good athlete—indeed, that this attention was contingent upon *being a winner*—that performance and winning (the dominant values of organized sports) became extremely important. For some, this created pressures that served to lessen or eliminate the fun of athletic participation (Messner 1987a, 1987b).

While feminist psychoanalytic and developmental theories of masculinity are helpful in explaining boys' early attraction and motivations in organized sports, the imperatives of core gender identity do not fully determine the contours and directions of the life course. As Rubin (1985) and Levinson (1978) have pointed out, an understanding of the lives of men must take into account the processual nature of male identity as it unfolds through interaction between the internal (psychological ambivalences) and the external (social, historical, and institutional) contexts.

To examine the impact of the social contexts, I divided my sample into two comparison groups. In the first group were 10 men from higher-status backgrounds, primarily white, middle-class, and professional families. In the second group were 20 men from lower-status backgrounds, primarily minority, poor, and working-class families. While my data offered evidence for the similarity of experiences and motivations of men from poor backgrounds, independent of race, I also found anecdotal evidence of a racial dynamic that operates independently of social class. However, my sample was not large enough to separate race and class, and so I have combined them to make two status groups.

In discussing these two groups, I will focus mainly on the high school years. During this crucial period, the athletic role may become a master status for a young man, and he is beginning to make assessments and choices about his future. It is here that many young men make a major commitment to—or begin to back away from—athletic careers.

Men from Higher-Status Backgrounds

The boyhood dream of one day becoming a professional athlete—a dream shared by nearly all the men interviewed in this study—is rarely realized. The sports world is extremely hierarchical. The pyramid of sports careers narrows very rapidly as one climbs from high school, to college, to professional levels of competition (Edwards 1984; Harris and Eitzen 1978; Hill and Lowe 1978). In fact, the chances of attaining professional status in sports are approximately 4/100,000 for a white man, 2/100,000 for a black man, and 3/100,000 for a Hispanic man in the United States (Leonard and Reyman 1988). For many young athletes, their dream ends early when coaches inform them that they are not big enough, strong enough, fast enough, or skilled enough to compete at the higher levels. But six of the higher-status men I interviewed did not wait for coaches to weed them out. They made conscious decisions in high school or in college to shift their attentions elsewhere—usually toward educational and career goals. Their decision not to pursue an athletic career appeared to them in retrospect to be a rational decision based on the growing knowledge of how very slim their chances were to be successful in the sports world. For instance, a 28-year-old white graduate student said:

> By junior high I started to realize that I was a good player—maybe even one of the best in my community—but I realized that there were all these people all over the country and how few will get to play pro sports. By high school, I still dreamed of being a pro—I was a serious athlete, I played hard—but I knew it wasn't heading anywhere. I wasn't going to play pro ball.

A 32-year-old white athletic director at a small private college had been a successful college baseball player. Despite considerable attention from professional scouts, he had decided to forgo a shot at a baseball career and to enter graduate school to pursue a teaching credential. As he explained this decision:

> At the time I think I saw baseball as pissing in the wind, really. I was married, I was 22 years old with a kid. I didn't want to spend 4 or 5 years in the minors with a family. And I could see I wasn't a superstar; so it wasn't really worth it. So I went to grad school. I thought that would be better for me.

Perhaps most striking was the story of a high school student body president and top-notch student who was also "Mr. Everything" in sports. He was named captain of his basketball, baseball, and football teams and achieved

All-League honors in each sport. This young white man from a middle-class family received attention from the press and praise from his community and peers for his athletic accomplishments, as well as several offers of athletic scholarships from universities. But by the time he completed high school, he had already decided to quit playing organized sports. As he said:

> I think in my own mind I kind of downgraded the stardom thing. I thought that was small potatoes. And sure, that's nice in high school and all that, but on a broad scale, I didn't think it amounted to all that much. So I decided that my goal's to be a dentist, as soon as I can.

In his sophomore year of college, the basketball coach nearly persuaded him to go out for the team, but eventually he decided against it:

> I thought, so what if I can spend two years playing basketball? I'm not going to be a basketball player forever and I might jeopardize my chances of getting into dental school if I play.

He finished college in three years, completed dental school, and now, in his mid-30s, is again the epitome of the successful American man: a professional with a family, a home, and a membership in the local country club.

How and why do so many successful male athletes from higher-status backgrounds come to view sports careers as "pissing in the wind," or as "small potatoes"? How and why do they make this early assessment and choice to shift from sports and toward educational and professional goals? The white, middle-class institutional context, with its emphasis on education and income, makes it clear to them that choices exist and that the pursuit of an athletic career is not a particularly good choice to make. Where the young male once found sports to be a convenient institution within which to construct masculine status, the postadolescent and young adult man from a higher-status background simply *transfers* these same strivings to other institutional contexts: education and careers.

For the higher-status men who had chosen to shift from athletic careers, sports remained important on two levels. First, having been a successful high school or college athlete enhances one's adult status among other men in the community—but only as a badge of masculinity that is *added* to his professional status. In fact, several men in professions chose to be interviewed in their offices, where they publicly displayed the trophies and plaques that attested to their earlier athletic accomplishments. Their high school and college athletic careers may have appeared to them as "small potatoes," but many successful men speak of their earlier status as athletes as having "opened doors" for them in their present professions and in community affairs. Similarly, Farr's (1988) research on "Good Old Boys Sociability Groups" shows how sports, as part of the glue of masculine culture, continues to facilitate "dominance bonding" among privileged men long after active sports careers end.

The college-educated, career-successful men in Farr's study rarely express overtly sexist, racist, or classist attitudes; in fact, in their relationships with women, they "often engage in expressive intimacies" and "make fun of exaggerated 'machismo'" (p. 276). But though they outwardly conform more to what Pleck (1982) calls "the modern male role," their informal relationships within their sociability groups, in effect, affirm their own gender and class status by constructing and clarifying the boundaries between themselves and women and lower-status men. This dominance bonding is based largely upon ritual forms of sociability (camaraderie, competition), "the superiority of which was first affirmed in the exclusionary play activities of young boys in groups" (Farr 1988, p. 265).

In addition to contributing to dominance bonding among higher-status adult men, sports remains salient in terms of the ideology of gender relations. Most men continued to watch, talk about, and identify with sports long after their own disengagement from athletic careers. Sports as a mediated spectacle provides an important context in which traditional conceptions of masculine superiority—conceptions recently contested by women—are shored up. As a 32-year-old white professional-class man said of one of the most feared professional football players today:

> A woman can do the same job as I can do—maybe even be my boss. But I'll be *damned* if she can go out on the football field and take a hit from Ronnie Lott.

Violent sports as spectacle provide linkages among men in the project of the domination of women, while at the same time helping to construct and clarify differences among various masculinities. The statement above is a clear identification with Ronnie Lott *as a man*, and the basis of the identification is the violent male body. As Connell (1987, p. 85) argues, sports is an important organizing institution for the embodiment of masculinity. Here, men's power over women becomes naturalized and linked to the social distribution of violence. Sports, as a practice, suppresses natural (sex) similarities, constructs differences, and then, largely through the media, weaves a structure of symbol and interpretation around these differences that naturalizes them (Hargreaves 1986, p. 112). It is also significant that the man who made the above statement about Ronnie Lott was quite aware that he (and perhaps 99 percent of the rest of the U.S. male population) was probably as incapable as most women of taking a "hit" from someone like Lott and living to tell of it. For middle-class men, the "tough guys" of the culture industry—the Rambos, the Ronnie Lotts who are fearsome "hitters," who "play hurt"—are the heroes who "prove" that "we men" are superior to women. At the same time, they play the role of the "primitive other," against whom higher-status men define themselves as "modern" and "civilized."

Sports, then, is important from boyhood through adulthood for men from higher-status backgrounds. But it is significant that by adolescence and early adulthood, most of these young men have concluded that sports *careers* are not for them. Their middle-class cultural environment encourages them to decide to shift their masculine strivings in more "rational" directions: education and nonsports careers. Yet their previous sports participation continues to be very important to them in terms of constructing and validating their status within privileged male peer groups and within their chosen professional careers. And organized sports, as a public spectacle, is a crucial locus around which ideologies of male superiority over women, as well as higher-status men's superiority over lower-status men, are constructed and naturalized.

Men from Lower-Status Backgrounds

For the lower-status young men in this study, success in sports was not an added proof of masculinity; it was often their only hope of achieving public masculine status. A 34-year-old black bus driver who had been a star athlete in three sports in high school had neither the grades nor the money to attend college, so he accepted an offer from the U.S. Marine Corps to play on their baseball team. He ended up in Vietnam, where a grenade blew four fingers off his pitching hand. In retrospect, he believed that his youthful focus on sports stardom and his concomitant lack of effort in academics made sense:

> You can go anywhere with athletics—you don't have to have brains. I mean, I didn't feel like I was gonna go out there and be a computer expert, or something that was gonna make a lot of money. The only thing I could do and live comfortably would be to play sports—just to get a contract—doesn't matter if you play second or third team in the pros, you're gonna make big bucks. That's all I wanted, a confirmed livelihood at the end of my ventures, and the only way I could do it would be through sports. So I tried. It failed, but that's what I tried.

Similar, and even more tragic, is the story of a 34-year-old black man who is now serving a life term in prison. After a career-ending knee injury at the age of 20 abruptly ended what had appeared to be a certain road to professional football fame and fortune, he decided that he "could still be rich and famous" by robbing a bank. During his high school and college years, he said, he was nearly illiterate:

> I'd hardly ever go to classes and they'd give me *C*s. My coaches taught some of the classes. And I felt, "So what? They *owe* me that! I'm an *athlete!* I thought that was what I was born to do—to play sports—and everybody understood that.

Are lower-status boys and young men simply duped into putting all their eggs into one basket? My research suggested that there was more than "hope for the future" operating here. There were also immediate psychological reasons that they chose to pursue athletic careers. By the high school years, class and ethnic inequalities had become glaringly obvious, especially for those who attended socioeconomically heterogeneous schools. Cars, nice clothes, and other signs of status were often unavailable to these young men, and this contributed to a situation in which sports took on an expanded importance for them in terms of constructing masculine identities and status. A white, 36-year-old man from a poor, single-parent family who later played professional baseball had been acutely aware of his low-class status in his high school:

> I had one pair of jeans, and I wore them every day. I was always afraid of what people thought of me—that this guy doesn't have anything, that he's wearing the same Levi's all the time, he's having to work in the cafeteria for his lunch. What's going on? I think that's what made me so shy. . . . But boy, when I got into sports, I let it all hang out—[laughs]—and maybe that's why I became so good, because I was frustrated, and when I got into that element, they gave me my uniform in football, basketball, and baseball, and I didn't have to worry about how I looked, because then it was *me* who was coming out, and not my clothes or whatever. And I think that was the drive.

Similarly, a 41-year-old black man who had a 10-year professional football career described his insecurities as one of the few poor blacks in a mostly white, middle-class school and his belief that sports was the one arena in which he could be judged solely on his merit:

> I came from a poor family, and I was very sensitive about that in those days. When people would say things like "Look at him—he has dirty pants on," I'd think about it for a week. [But] I'd put my pants on and I'd go out on the football field with the intention that I'm gonna do a job. And if that calls on me to hurt you, I'm gonna do it. It's as simple as that. I demand respect just like everybody else.

"Respect" was what I heard over and over when talking with the men from lower-status backgrounds, especially black men. I interpret this type of respect to be a crystallization of the masculine quest for recognition through public achievement, unfolding within a system of structured constraints due to class and race inequities. The institutional context of education (sometimes with the collusion of teachers and coaches) and the constricted structure of opportunity in the economy made the pursuit of athletic careers appear to be the most rational choice to these young men.

The same is not true of young lower-status women. Dunkle (1985) points out that from junior high school through adulthood, young black men are far more likely to place high value on sports than are young black women, who are more likely to value academic achievement. There appears to be a gender dynamic operating in adolescent male peer groups that contributes toward their valuing sports more highly than education. Franklin (1986, p. 161) has argued that many of the normative values of the black male peer group (little respect for nonaggressive solutions to disputes, contempt for nonmaterial culture) contribute to the constriction of black men's views of desirable social positions, especially through education. In my study, a 42-year-old black man who did succeed in beating the odds by using his athletic scholarship to get a college degree and eventually becoming a successful professional said:

> By junior high, you either got identified as an athlete, a thug, or a bookworm. It's very important to be seen as somebody who's capable in some area. And you *don't* want to be identified as a bookworm. I was very good with books, but I was kind of covert about it. I was a closet bookworm. But with sports, I was *somebody;* so I worked very hard at it.

For most young men from lower-status backgrounds, the poor quality of their schools, the attitudes of teachers and coaches, as well as the antieducation environment within their own male peer groups, made it extremely unlikely that they would be able to succeed as students. Sports, therefore, became *the* arena in which they attempted to "show their stuff." For these lower-status men, as Baca Zinn (1982) and Majors (1986) argued in their respective studies of Chicano men and black men, when institutional resources that signify masculine status and control are absent, physical presence, personal style, and expressiveness take on increased importance. What Majors (1986, p. 6) calls "cool pose" is black men's expressive, often aggressive, assertion of masculinity. This self-assertion often takes place within a social context in which the young man is quite aware of existing social inequities. As the black bus driver, referred to above, said of his high school years:

> See, the rich people use their money to do what they want to do. I use my ability. If you wanted to be around me, if you wanted to learn something about sports, I'd teach you. But you're gonna take me to lunch. You're gonna let me use your car. See what I'm saying? In high school I'd go where I wanted to go. I didn't have to be educated. I was well-respected. I'd go somewhere, and they'd say, "Hey, that's Mitch Harris,[1] yeah, that's a bad son of a bitch!"

Majors (1986) argues that although "cool pose" represents a creative survival technique within a hostile environment, the most likely long-term effect of this masculine posturing is educational and occupational dead ends. As a result, we can conclude, lower-status men's personal and peer-group responses to a constricted structure of opportunity—responses that are rooted, in part, in the developmental insecurities and ambivalences of masculinity—serve to lock many of these young men into limiting activities such as sports.

SUMMARY AND CONCLUSIONS

This research has suggested that within a social context that is stratified by social class and by race, the choice to pursue—or not to pursue—an athletic career is explicable as an individual's rational assessment of the available means
to achieve a respected masculine identity. For nearly all of the men from
lower-status backgrounds, the status and respect that they received through
sports was temporary—it did not translate into upward mobility. Nonetheless, a strategy of discouraging young black boys and men from involvement
in sports is probably doomed to fail, since it ignores the continued existence
of structural constraints. Despite the increased number of black role models
in nonsports professions, employment opportunities for young black males
have actually deteriorated. . . .

But it would be a mistake to conclude that we simply need to breed socioeconomic conditions that make it possible for poor and minority men to
mimic the "rational choices" of white, middle-class men. If we are to build an
appropriate understanding of the lives of all men, we must critically analyze
white middle-class masculinity, rather than uncritically taking it as a normative standard. To fail to do this would be to ignore the ways in which organized sports serve to construct and legitimate gender differences and inequalities among men and women.

Feminist scholars have demonstrated that organized sports gives men
from all backgrounds a means of status enhancement that is not available to
young women. Sports thus serve the interests of all men in helping to construct and legitimize their control of public life and their domination of
women (Bryson 1987; Hall 1987; Theberge 1987). Yet concrete studies are
suggesting that men's experiences within sports are not all of a piece. Brian
Pronger's (1990) research suggests that gay men approach sports differently
than straight men do, with a sense of "irony." And my research suggests that
although sports are important for men from both higher- and lower-status
backgrounds, there are crucial differences. In fact, it appears that the meaning
that most men give to their athletic strivings has more to do with competing
for status among men than it has to do with proving superiority over women.
How can we explain this seeming contradiction between the feminist claim
that sports links all men in the domination of women and the research findings that different groups of men relate to sports in very different ways?

. . . Connell's (1987) concept of the "gender order" is useful. The gender
order is a dynamic process that is constantly in a state of play. Moving beyond
static gender-role theory and reductionist concepts of patriarchy that view
men as an undifferentiated group which oppresses women, Connell argues
that at any given historical moment, there are competing masculinities—some
hegemonic, some marginalized, some stigmatized. Hegemonic masculinity

(that definition of masculinity which is culturally ascendant) is constructed in relation to various subordinated masculinities as well as in relation to femininities. The project of male domination of women may tie all men together, but men share very unequally in the fruits of this domination.

These are key insights in examining the contemporary meaning of sports. Utilizing the concept of the gender order, we can begin to conceptualize how hierarchies of race, class, age, and sexual preference among men help to construct and legitimize men's overall power and privilege over women. And how, for some black, working-class, or gay men, the false promise of sharing in the fruits of hegemonic masculinity often ties them into their marginalized and subordinate statuses within hierarchies of intermale dominance. For instance, black men's development of what Majors (1986) calls "cool pose" within sports can be interpreted as an example of creative resistance to one form of social domination (racism); yet it also demonstrates the limits of an agency that adopts other forms of social domination (masculinity) as its vehicle. As Majors (1990) points out:

> Cool Pose demonstrates black males' potential to transcend oppressive conditions in order to express themselves *as men*. [Yet] it ultimately does not put black males in a position to live and work in more egalitarian ways with women, nor does it directly challenge male hierarchies.

Indeed, as Connell's (1990) analysis of an Australian "Iron Man" shows, the commercially successful, publicly acclaimed athlete may embody all that is valued in present cultural conceptions of hegemonic masculinity—physical strength, commercial success, supposed heterosexual virility. Yet higher-status men, while they admire the public image of the successful athlete, may also look down on him as a narrow, even atavistic, example of masculinity. For these higher-status men, their earlier sports successes are often status enhancing and serve to link them with other men in ways that continue to exclude women. Their decisions not to pursue athletic careers are equally important signs of their status vis-à-vis other men. Future examinations of the contemporary meaning and importance of sports to men might take as a fruitful point of departure that athletic participation, and sports as public spectacle, serve to provide linkages among men in the project of the domination of women, while at the same time helping to construct and clarify differences and hierarchies among various masculinities.

NOTE

1. "Mitch Harris" is a pseudonym.

References

Baca Zinn, M. 1982. "Chicano Men and Masuclinity." *Journal of Ethnic Studies* 10:29–44.

Bryson, L. 1987. "Sport and the Maintenance of Masculine Hegemony." *Women's Studies International Forum* 10:349–60.

Connell, R. W. 1987. *Gender and Power*. Stanford, CA: Stanford University Press.

——. 1990. "An Iron Man: The Body and Some Contradictions of Hegemonic Masculinity." In *Sport, Men, and the Gender Order: Critical Feminist Perspectives*, edited by M. A. Messner and D. S. Sabo. Champaign, IL: Human Kinetics.

Dunkle, M. 1985. "Minority and Low-Income Girls and Young Women in Athletics." *Equal Play* 5 (Spring-Summer): 12–13.

Edwards, H. 1984. "The Collegiate Athletic Arms Race: Origins and Implications of the 'Rule 48' Controversy." *Journal of Sport and Social Issues* 8:4–22.

Eitzen, D. S. and D. A. Purdy. 1986. "The Academic Preparation and Achievement of Black and White College Athletes." *Journal of Sport and Social Issues* 10:15–29.

Eitzen, D. S. and N. B. Yetman. 1977. "Immune From Racism?" *Civil Rights Digest* 9:3–13.

Farr, K. A. 1988. "Dominance Bonding Through the Good Old Boys Sociability Group." *Sex Roles* 18:259–77.

Franklin, C. W. II. 1984. *The Changing Definition of Masculinity*. New York: Plenum.

——. 1986. "Surviving the Institutional Decimation of Black Males: Causes, Consequences, and Intervention." In *The Making of Masculinities: The New Men's Studies*, edited by H. Brod, pp. 155–70. Winchester, MA: Allen & Unwin.

Gruneau, R. 1983. *Class, Sports, and Social Development*. Amherst: University of Massachusetts Press.

Hall, M. A. (ed.). 1987. "The Gendering of Sport, Leisure, and Physical Education." *Women's Studies International Forum* 10:361–474.

Hare, N. and J. Hare. 1984. *The Endangered Black Family: Coping with the Unisexualization and Coming Extinction of the Black Race*. San Francisco, CA: Black Think Tank.

Hargreaves, J. A. 1986. "Where's the Virtue? Where's the Grace? A Discussion of the Social Production of Gender Through Sport." *Theory, Culture and Society* 3:109–21.

Harris, D. S. and D. S. Eitzen. 1978. "The Consequences of Failure in Sport." *Urban Life* 7:177–88.

Hill, P. and B. Lowe. 1978. "The Inevitable Metathesis of the Retiring Athlete." *International Review of Sport Sociology* 9:5–29.

Leonard, W. M. II and J. M. Reyman. 1988. "The Odds of Attaining Professional Athlete Status: Refining the Computations." *Sociology of Sport Journal* 5:162–69.

Levinson, D. J. 1978. *The Seasons of a Man's Life*. New York: Ballantine.

Majors, R. 1986. "Cool Pose: The Proud Signature of Black Survival." *Changing Men: Issues in Gender, Sex, and Politics* 17:5–6.

——. 1990. "Cool Pose: Black Masculinity in Sports." In *Sport, Men, and the Gender Order: Critical Feminist Perspectives*, edited by M. A. Messner and D. S. Sabo. Champaign, IL: Human Kinetics.

Messner, M. 1987a. "The Meaning of Success: The Athletic Experience and the Development of Male Identity." In *The Making of Masculinities: The New Men's Studies*, edited by H. Brod, pp. 193–209. Winchester, MA: Allen & Unwin.

——. 1987b. "The Life of a Man's Seasons: Male Identity in the Lifecourse of the Athlete." In *Changing Men: New Directions in Research on Men and Masculinity*, edited by M. S. Kimmel, pp. 53–67. Newbury Park, CA: Sage.

Pleck, J. H. 1982. *The Myth of Masculinity*. Cambridge: MIT Press.

Pronger, B. 1990. "Gay Jocks: A Phenomenology of Gay Men in Athletics." In *Sport, Men, and the Gender Order: Critical Feminist Perspectives*, edited by M. A. Messner and D. S. Sabo. Champaign, IL: Human Kinetics.

Rubin, L. B. 1985. *Just Friends: The Role of Friendship in Our Lives*. New York: Harper & Row.

Rudman, W. J. 1986. "The Sport Mystique in Black Culture." *Sociology of Sport Journal* 3:305–19.

Staples, Robert. 1982. *Black Masculinity*. New York: Black Scholar Press.

Theberge, N. 1987. "Sport and Women's Empowerment." *Women's Studies International Forum* 10:387–93.

Wellman, D. 1986. "The New Political Linguistics of Race." *Socialist Review* 87/88:43–62.

JUST CHOICES

23

Women of Color, Reproductive Health, and Human Rights

Loretta J. Ross, Sarah L. Brownlee, Dazon Dixon Diallo, Luz Rodriquez,

and SisterSong Women of Color Reproductive Health Project

Making just choices around reproductive health is difficult for women of color. Just choices are not simply a range of options, but of options that make sense in order to optimize our reproductive health. We don't expect perfect choices, but we want choices that don't violate our sense of dignity, fairness, and justice. Our ability to control what happens to our bodies is constantly challenged by poverty, racism, sexism, homophobia, and injustice in the United States.

From Jael Silliman and Anannya Bhattacharjee, eds., *Policing the National Body: Race, Gender, and Criminalization* (Boston: South End Press, 2002), pp. 147–74. Reprinted by permission of the South End Press.

We face political regimes that seek to restrict our migration, to punish us for poverty, and to incarcerate us for behaviors politicians have criminalized. We are subject to various population control schemes, carriers of our own internalized oppression, and objects of unprincipled medical research. The race, gender, and class discrimination faced by women of color interferes with our ability to acquire services or culturally appropriate reproductive health information, particularly information on reproductive tract infections (RTIs). Mental health issues—such as oppression, depression, substance abuse, physical and sexual violence, lack of education, the lack of availability of services, and poverty—are related to racial, gender, and economic inequalities that specifically limit the potential of women of color to live healthy lives.

Women of color have not been passive victims in the face of this onslaught. In the words of Francesca Miller, we are "repressed but not resigned."[1] Yet organizing around reproductive health issues for women of color has not been easy. As women of color, we have always sought to protect our reproductive freedom, but issues of power and subordination complicate our efforts to bring attention to the reproductive health issues that threaten our lives.

Sometimes we work with predominantly white organizations that marginalize issues of race and class, and privilege abortion rights over other issues of reproductive justice. . . .

Some of us work with people of color organizations that marginalize gender and class issues, and where women's reproductive health issues are tangential to struggles against racism. Occasionally, women's reproductive health issues reach the agenda of these organizations, particularly in the areas of sterilization abuse and population control, but they rarely sustain any long-term momentum.

A number of women of color work with anti-poverty organizations that sometimes neglect race and gender issues altogether, assuming that class issues subsume concerns about reproductive health. Women of color numerically dominate organizations that work on poverty, homelessness, and welfare reform, but rarely do these organizations address the reproductive health needs of women of color. Similarly, programs that address violence against women of color sometimes highlight the connection between violence and poor health, but again, reproductive health issues are not their primary concern.

Few women of color organizations have significant programs that focus on reproductive health issues. Most women of color organizations address gender and racial discrimination, but many of these organizations believe that reproductive rights, particularly abortion rights, are too controversial in their communities. . . .

White women's organizations increasingly attempt to mobilize women of color in order to advance their political agenda and broaden their funding options. Organizations such as Planned Parenthood and NARAL want to present a picture of a diverse, inclusive movement and, at the same time, convince funders that they are capable of organizing women of color. While these efforts may be well intentioned, they are inadequate for including significant numbers of women of color. Moreover, these efforts are prone to the same issues of racism, classism, and homophobia that beset previous attempts to engage women of color in white organizations. . . .

The funding issues are also problematic because of their racial implications. While many women of color organizations suffer from a lack of sufficient funding to achieve organizational stability, many white women's organizations are provided with generous funding for their "women of color" projects, without a visible track record. This trend has several implications for women of color: the funding we desperately need to organize in our own communities is diverted to outside groups and we are forced to compete for funding on the terms set by the agencies with access to more resources. If funders prefer public policy strategies, for example, we have to produce policy papers in order to secure support.

Another consequence of funding white organizations to work on reproductive rights in communities of color is that such efforts often poison the soil, because racism is synonymous with genocide in the minds of many people of color. If issues of white supremacy, racism, or classism are not addressed by the white organizations, these defects contaminate efforts by women of color, making it difficult to avoid the stigmas left in the wake of more visible and better funded organizations. For example, it is difficult to counter charges that abortion is genocide when a white pro-choice organization displays racist or insensitive behavior in a community of color. A clear example was during the debate over immigration reform in 1996. Several pro-choice organizations actually agreed with the white majority in limiting the rights of immigrants, making it harder for women of color to join with them in solidarity in efforts to stop the "partial birth abortion" bills in Congress.

Even more problematic than the efforts of white pro-choice organizations are the inroads made by the religious right in communities of color. Among the religious right organizations that have increased their appeal to women of color are the Christian Coalition, Operation Rescue, Blacks for Life, and other groups opposed to abortion. Many people of color are drawn into supporting the ambitions of the religious right because those groups do not use open racism to promote their agenda. Because religiously cloaked bigotry is well disguised, it is difficult for most people to identify and challenge.

The religious right opposes women's reproductive freedom by cloaking itself in the moral mantle of the civil rights movement, for example by comparing the rights of unborn fetuses to the rights of African-Americans. It is not uncommon for anti-abortionists like Operation Rescue to use imagery of

Martin Luther King, Jr. in their opposition to abortion, even though King was pro-choice. Such appropriation of civil rights language and symbols mocks the integrity of the civil rights movement's struggle for equality and freedom.

Strategists of the religious right use money to appeal to many religious leaders, by providing the funding that progressives fail to supply. In the 1990s, many anti-poverty programs in communities of color were funded by the religious right, often with several caveats such as mandates for organizations to embrace "abstinence only" language for sex education, and calls for the termination of abortion and contraceptive support. Anti-poverty organizations applying for funds from Catholic Charities in the 1990s were forced to sign an agreement that they would not work with pro-choice organizations in their communities, as was reported by anti-poverty groups like the Georgia Citizens' Coalition on Hunger and the Up and Out of Poverty Network. This coercive strategy was lamentably successful as alternative funding from progressive organizations was not available.

This situation is guaranteed to worsen under President George W. Bush, who has already established several faith-based initiatives to provide government funds to religious institutions. Even if one disregards the blurring line between church and state, it is impossible to overlook the fact that such funding will effectively promulgate the president's anti-choice agenda. Not only are women's health issues jeopardized by the president's personal agenda, but increased spending on the military and anti-terrorist campaigns in the wake of the September 11, 2001 attack on the United States will most likely divert significant funds away from addressing all domestic health care concerns.

These attempts to organize women of color demonstrate that much work remains to be done in order to formulate a collaborative, global reproductive health movement that tackles the lack of just choices for women of color. In the twenty-first century, women of color need to examine their previous organizing efforts. For the past decade, women of color working on reproductive health issues have attempted a dual strategy: working within our own communities to advance issues of reproductive health, while also working with the mainstream women's health movement to advance issues for women of color. This dual focus is critical and necessary, but after ten years of making relatively little progress on either front, it is now time to question the splitting of our energies.

Thus, it is critical that women of color again mobilize to defend our reproductive freedom. We are fighting battles on many fronts, and the health of our communities is at stake. We have to use new organizing strategies that do not create an artificial division between public policy advocacy and grassroots mobilizing, and we have to resist the blandishments of the religious right and the mainstream pro-choice movement if they do not truly represent our own agendas. . . .

Native American/Indigenous Community

According to the Indian Health Service (IHS), Native American women account for 15 percent of AIDS cases in the United States, compared with 7 percent for non-minority women. Cervical cancer affects 20 percent of Native American women, a rate more than twice that of the US national average of 8.6 percent. The death rate for Native women from cervical cancer is also higher due to delayed diagnosis. Chlamydia rates are thirteen times higher among Native American women than for other American women.[2]

In part due to a cultural attitude of mistrust that Native people feel toward researchers, many Native Americans do not participate in research studies. Health professionals estimate that the prevalence of RTIs may be as high as 65 percent for Native Americans. The prevalence of myth, mystery, and misinformation about sexuality in Native communities presents another major challenge to the promotion of reproductive health. Alcoholism and drug abuse, as well as physical and emotional abuse, all increase the risk of RTIs. The migrant-worker status of many Native Americans also exacerbates their risk. . . .

Health care is controlled by the federal government for many Native Americans. Just thirty-four Indian Health Service (IHS) clinics serve over 1.3 million Indian people nationwide, leaving 75 percent virtually without any health care. Urban Indians, the majority of all American Indians, receive less than 1 percent of the IHS budget.[3] The government's ban on traditional reproductive health care, such as midwifery, has only exacerbated the RTI-related problems.[4] Native American women with high-risk pregnancies are left with inadequate federal health care and their option for more community-based care is eroded. . . .

Forced sterilization, with the use of Depo-Provera and Norplant, has been utilized as a form of population control. Physicians offer Norplant implants for free but charge more than $300 to have them removed. Women are pressured into a procedure that limits their reproductive capacity. Physicians take advantage of the low-income status of many women by limiting their options for reproductive health.

Oftentimes, women are not aware of the choices they have in health services. Physicians do not take the time to explain health care options, and there is little effort toward public education. Another factor that influences public awareness of health issues is education level. Because many Native American women are not high school graduates, some information that is available is not accessible due to the high reading levels required for the material.

Culturally appropriate health education material that is "tradition-inclusive" is necessary to increase public awareness of health options. Approaches that empower people in the community to take control of their health choices are also paramount. . . .

Asian/Pacific Islander American Community

According to the US Census Bureau, Asian/Pacific Islanders (A/PIs) are the fastest growing minority group in the United States.[5] Consisting of immigrants from over twenty countries who speak more than 120 languages, this population numbered more than 10 million people in 1997.[6] The majority of this population live in urban areas, mostly in California, Hawaii, and Washington. . . .

Because much of the A/PI community is non-English speaking, mainstream health education and disease prevention efforts have had little impact on these communities, and over 11,500 new cases of sexually transmitted diseases were reported for Asian-Americans and Pacific Islanders in 1998.[7] Moreover, health care professionals have limited knowledge of the different cultures and endemic diseases from immigrants' countries of origin, and thus lack the expertise to serve immigrant clients. In addition, A/PI women often cannot afford the care that may be available.

Cultural and family influences also affect the rate of health care access. A/PI women have traditionally placed the health needs of their families at the forefront and put less emphasis on individual needs.[8] Cultural taboos surrounding the body, sexuality, and pregnancy also contribute to the low rates of health service access. Many A/PI women feel uncomfortable discussing health issues.

The misperception of Asian-Americans as a healthy and well-off "model minority" has limited the development of research studies to document any poor status in health of the A/PI community. Contrary to this stereotype, over two-thirds of Asian-American women are sexually active, but less than 40 percent always use protection against STDs. One-quarter have never visited a health care provider for reproductive health services, or received education on reproductive tract infections.[9] . . .

One way to facilitate the provision of adequate care is through more research on nontraditional methods. It is critical for A/PI women to control their bodies, and to have influence over the care provided them. A national effort is necessary to achieve this goal and more voices need to be heard within the policy and medical arenas. Collaborative efforts to make resources known are a crucial first step.

Latina/Hispanic Community

The Latina population has many factors contributing to high risk for RTIs. Approximately 30 percent of Latinos live below the poverty line, and nearly half the poor Hispanic families are female-headed. In 1995, 30 percent of the Hispanic population was not covered by health insurance. Medicaid coverage for Latinos varies by state and in many states poor undocumented Latinos are ineligible for Medicaid and often cannot afford health insurance. Welfare reform has also increased the number of Latinos ineligible for publicly funded health insurance. Women who are left without care then turn to self-medication and share medicine.[10]

Latinas also underutilize available health care services. A lack of United States citizenship often deters undocumented immigrant Latinos from using public clinics and other health facilities for fear of detection and deportation. In addition, Hispanics traditionally seek family members' advice before getting professional health care, which contributes to Latinas' low or delayed utilization of health care services. In many families, women's low status impedes their ability to seek and negotiate for their own health care. Many Latinas also work in the agriculture industry which exposes workers to pesticides that place them at risk for a range of health problems, including RTIs. Hispanics are also more likely to have diabetes, placing them at risk for a range of health problems, including RTIs. Hispanics are also more likely to have diabetes, placing them at greater risk for endogenous RTIs, further complicating and prolonging the treatment regime.[11] . . .

High rates of cesareans and sterilization abuse cause distrust of the medical community among many Latinas. For example, according to a national fertility study conducted in 1970 by Princeton University's Office of Population Control, 20 percent of all Mexican-American women had been sterilized. The disproportionate number of Puerto Rican women who have been sterilized reflects United States government policy that dates back to 1939, when an experimental sterilization campaign was implemented. By the 1970s, more than 35 percent of all Puerto Rican women of child-bearing age had been surgically sterilized.[12]

Unfortunately, structural, cultural, and social influences often prevent Latinas from actively asserting their rights for adequate and appropriate health care services. In many cases, Latinas, particularly immigrants, do not feel they are in a position of power. Religion plays an integral role in inhibiting women from feeling confident about reproductive health needs due to the connection with sexuality. The health beliefs of many Latinas relate to their views about God as the omnipotent creator of the universe, with personal behavior subject to God's judgment. Fatalistic beliefs such as these make it difficult to establish the importance of preventive health behaviors. Cultural factors also influence the spread of HIV infection and AIDS because they often are unwilling to discuss intimate and emotional matters such as illness and sex unless they are able to speak to someone in Spanish.[13]

THE BLACK/AFRICAN-AMERICAN COMMUNITY

The Black population in America consists primarily of African-Americans, although significant numbers of African and African-Caribbean immigrants have become part of this group in the last fifteen years. . . .

Available data on the health status of African-American women indicate that they suffer higher rates of undetected diseases, higher incidences of diseases and illnesses, and more chronic conditions, as well as higher morbidity rates than non-minority American women. According to statistics, African-American women exhibit very high rates of cervical and breast cancer and reproductive track infections (RTIs).[14]

Yet, despite high rates of serious health problems within the community, far too many women have never had a Pap smear or gynecological exam. In 1995, one in four African-American women did not receive a Pap smear, and one-third failed to receive a clinical breast exam. More than half of African-American women age 50 to 64 did not receive mammography screening between 1994 and 1995. One in seven (14 percent) rely on emergency rooms for basic health services.[15] Obesity, a contributing factor for some cancers, is most prevalent among African-American women.[16]

Appropriate mental health services are even worse in many cases. The influence of racial and gender oppression as it relates to depression, substance abuse, physical abuse, access to quality education, and the availability of services is often overlooked. One of the most obvious factors affecting the rate of health care access is the availability of appropriate services. In rural areas of the United States, services geared toward African-American women are frequently non-existent.[17] The lack of health care access in both rural and urban areas has led to a gap in research regarding Black women's health, including in areas of birth control, HIV/AIDS transmission, and RTIs. . . .

The lack of appropriate services, information, and research requires holistic approaches to health care that include advocacy and education. In addition, powerful and positive women motivators from the African-American community need to guide and lead dialogue around reproductive health issues. . . .

RECONCEPTUALIZING
THE HUMAN RIGHTS FRAMEWORK

Global structural adjustment policies imposed by the World Bank on developing countries have resulted in cuts in social services, fees for public services, privatization, and the removal of subsidies for food, medicines, clean water, and transportation. In the United States, welfare reform is, in effect, acting as a structural adjustment program. A high level of poverty forces thousands of girls and women to sell their bodies as a means of survival.[18] Sadly enough, even in monogamous relationships, women's sexuality becomes their only bargaining tool for survival. Violence against women also tends to increase poverty, risks of STDs, unintended pregnancies, drug and alcohol abuse, and mental illness. . . .

Organizations concerned with reproductive health issues in the United States are increasingly drawing inspiration and tools from the international human rights movement, and those working within a traditional human rights framework are gradually including issues related to reproductive health. Women, particularly women of color, have spearheaded this rearticulation of the human rights framework in the United States. . . .

Liberal human rights interpretations often do not recognize that equality of opportunity is an illusion in a society based on competitive individualism in which one ethnic group (whites) has social, political, and economic advantages in relationship to other ethnic groups (people of color). Thus, individualistic rights frameworks often neglect the importance of group rights of ethnicity and culture.[19] For example, while an individual African-American may have the legal freedom and the financial means to purchase a home in any community, that potential buyer must be protected from racial discrimination against African-Americans as a group, in order to enjoy that individual right.

Parental consent laws, for-profit health care, welfare reform policies, and immigration policies impact women's health choices and detrimentally affect the quality of care available. In order to ensure access to health care and treatment, and to address the intersection of class, race, and gender that affects women of color, a comprehensive human rights–based approach is necessary.

This new and comprehensive, human rights–based reproductive health agenda challenges the traditional American liberal human rights framework, by giving economic, social, and cultural rights the same consideration given to civil and political rights. The United States government has an obligation to provide an environment in which policies, laws, and practices enable women to realize their reproductive rights, and to refrain from creating conditions that compromise or restrict such rights.

NOTES

1. Francesca Miller, *Latin American Women and the Search for Social Justice* (Hanover, NH: University Press of New England, 1991), 207.

2. National Institutes of Health, *Women of Color Health Data Book* (1998), 75.

3. Ibid., 2.

4. Ibid., 3.

5. United States Department of Commerce, Bureau of the Census CB97–FS.04, 1997.

6. Ibid.

7. National Asian American Women's Health Organization, "Community Solutions: Meeting the Challenge of STDs in Asian Americans and Pacific Islanders" (2000), 7.

8. Ibid., 13.

9. Ibid.

10. National Institutes of Health, *Women of Color Health Data Book* (1998), 9.

11. Ibid., 11.

12. Angela Davis, *Women, Race and Class* (New York: Knopf, 1981), 219.

13. National Institutes of Health, *Women of Color Health Data Book* (1998), 11.

14. Deborah R. Grayson, "Necessity Was the Midwife to Our Politics: Black Women's Health Activism in the "Post-" Civil Rights Era (1980–1996)," in *Still Lifting, Still Climbing: African American Women's Contemporary Activism*, ed. Kimberly Springer (New York: New York University Press, 1999), 133.

15. Ibid., 70.

16. "Women's Health: Choices and Challenges," *The Commonwealth Fund Quarterly*, Special issue (1996).

17. Susan F. Feiner, *Race and Gender in the American Economy: Views Across the Spectrum* (Englewood Cliffs, NJ: Prentice-Hall, 1994), 315.

18. Kamala Kempadoo and Jo Doezema, *Global Sex Workers: Rights, Resistance and Redefinition* (New York: Routledge Press, 1998), 17.

19. William Felice, *Taking Suffering Seriously: The Importance of Collective Human Rights* (Albany, NY: SUNY Press, 1996), 35.

Thinking Further

After reading the articles in this section, you will find it helpful to think about the following:

This section shows how race, class, and gender are all interrelated axes of social structure. That is, they are not just individual attributes or identities, but part of the fabric of society. Other factors, such as sexuality, age, disability, nationality, and so forth, have been claimed by some to be comparable to race, class, and gender in society. Are they? No doubt some of the processes of discrimination and prejudice work much the same way whatever the characteristic on which they are based. In what ways are race, class, and gender similar to these other sources of differentiation and in what ways are they different?

Suggested Readings

Andersen, Margaret L. 2003. *Thinking about Women: Sociological Perspectives on Sex and Gender*, 6th ed. Boston, MA: Allyn & Bacon.

Collins, Chuck, and Felice Veskel. 2000. *Economic Apartheid in America: A Primer on Economic Inequality and Insecurity*. New York: The New Press.

Ehrenreich, Barbara. 2001. *Nickel and Dimed: On (Not) Getting By in America*. New York: Metropolitan Books.

Espiritu, Yen Le. 1997. *Asian American Women and Men*. Thousand Oaks: Sage.

Feagin, Joe. 2000. *Racist America: Roots, Current Realities, and Future Reparations*. New York: Routledge.

Oliver, Melvin, and Thomas Shapiro. 1995. *Black Wealth / White Wealth: A New Perspective on Racial Inequality*. New York: Routledge.

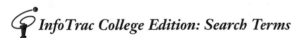

InfoTrac College Edition: Search Terms

Use your access to the online library, InfoTrac College Edition, to learn more about subjects covered in this section. The following are some suggested terms for searches:

gender segregation	wealth
homophobia	White privilege
institutional racism	working poor
model minority	

InfoTrac College Edition: Bonus Reading

You can use the InfoTrac College Edition feature to find an additional reading pertinent to this section and can also search on the author's name or keywords to locate the following reading:

Byng, Michelle D. 1998. "Mediating Discrimination: Resisting Oppression among African-American Muslim Women." *Social Problems* 45 (November): 473–487.

Through detailed interviews with Muslim women in the United States, Michelle Byng uses a race/class/gender framework to show how Muslim women construct their identities even in the face of discrimination. Her research shows not only how discrimination operates, but, just as importantly, how women resist the oppressive consequences of discrimination. What does her research suggest about how people construct their identities and visions of the world even in the context of race, class, and gender oppression? What unique challenges do Muslim women face, and what lessons do other women learn from them?

Wadsworth Sociology Resource Center: Virtual Society

For additional links, quizzes, and learning tools related to this section, see the Wadsworth Sociology Resource Center, Virtual Society, at www.wadsworth .com/sociology_d/

Rethinking Institutions

In the United States and globally, social institutions exert a powerful influence on everyday life. They are also powerful channels for societal penalties and privileges. The type of work you do, the structure of your family, whether your religion will be recognized or suppressed, the kind of education you receive, the images you see of yourself in the media, and how you are treated by the state are all shaped by the institutional structures of the society where you live. People rely on institutions to meet their needs, although social institutions do so better for some groups than others. When a specific institution (such as the economy) fails them, people often appeal to another institution (such as the state) for redress. In this sense, institutions are both sources of support and sources of oppression.

The concept of an institution is abstract, because there is no thing or object that one can point to as an institution. *Social institutions* are the established societal patterns of behavior organized around particular purposes. The economy is an institution, as are the family, the media, education, and the state—important societal institutions examined here. Each is organized around a specific purpose; in the case of the economy, for example, that purpose is the production, distribution, and consumption of goods and services. Within a given institution, there may be various patterns, such as different family structures, but, as a whole, institutions are general patterns of behavior that emerge because of the specific societal conditions in which groups live. Institutions do change over time, both as societal conditions evolve and as groups challenge specific institutional structures, but they are also enduring and persistent, even in the face of active efforts to change them. Institutions confront people from birth and live on after they die.

Across all societies, institutions are patterned by intersections of race, class, gender, sexuality, age, ethnicity, and disability (among others). The effect differs from one society to the next, each society having a distinctive history and institutional configuration of social inequalities that is unique. Within the United States, race, class, and gender constitute fundamental categories that shape social institutions. Thus, social institutions are the fundamental conduits for race, class, and gender oppression, even though they are often presented as entities far removed from these experiences. Moreover, social institutions are interconnected. We think the interrelationships among institutions are especially apparent when studied in the context of race, class, and gender. Race, class, and gender oppression rests on a network of interconnected social institutions. Understanding the interconnections between institutions helps us see that we are all part of one historically created system that finds structural form in interconnected social institutions.

Despite the significance of institutions, dominant ideology in the United States portrays them as neutral in their treatment of different groups; indeed, the liberal framework of the law allegedly makes access to public institutions (like education and work) gender- and race-blind. Still, institutions differentiate on the basis of race, class, and gender. As the articles included here show, institutions are actually structured on the basis of race, class, and gender relations.

As an example, think of the economy. Economic institutions in the United States are founded on capitalism—an economic system based on the pursuit of profit and the principle of private ownership. Such a system creates class inequality because, in simple terms, the profits of some stem from the exploitation of the labor of others. Race and gender further divide the U.S. capitalist economy, resulting in labor and consumer markets that routinely advantage some and disadvantage others. In particular, corporate and government structures create jobs for some while leaving others underemployed or without work.

Those with jobs encounter a dual labor market that includes: (1) a primary labor market characterized by relatively high wages, opportunities for advancement, employee benefits, and rules of due process that protect work-

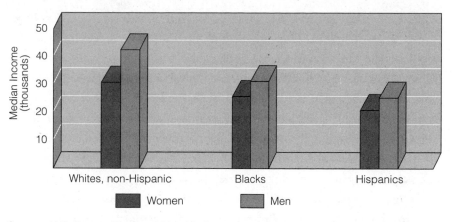

Source: U.S. Census Bureau. 2001. *Median Income of People by Selected Characteristics: 2000, 1999, and 1998*. Washington, DC: U.S. Census Bureau. Web site: www.census.gov/hhes/income/income00/inctab87.html.

FIGURE 1 The Pay Gap: 2000 (year-round, full-time workers)

ers' rights; and, (2) a secondary labor market (where most women and minorities are located) characterized by low wages, little opportunity for advancement, few benefits, and little protection for workers. One result of the dual labor market is the persistent wage gap between men and women (see Figure 1), and between Whites and people of color—even when they have the same level of education. Women tend to be clustered in jobs that employ mostly women workers; such jobs have been economically devalued as a result. Gender segregation and race segregation intersect in the dual labor market, thus women of color are most likely to be working in occupations where most of the other workers are also women of color. At the same time, men of color are segregated into particular segments of the market, too. Indeed, there is a direct connection between gender and race segregation and wages, because wages are lowest in occupations where women of color predominate (U.S. Department of Labor 2002). This is what it means to say that institutions are structured by intersections of race, class, and gender: Institutions are built from and then reflect the historical and contemporary patterns of race, class, and gender relations in society.

As a second example, think of the state. The state refers to the organized system of power and authority in society. This includes the government, the police, the military, and the law (as reflected in both social policy and the civil and criminal justice system). The state is supposed to protect all citizens, regardless of their race, class, and gender (as well as other characteristics, such as disability and age); yet, like other "gendered" institutions, state policies routinely privilege men. This means not only that the majority of powerful people in the state are men (such as elected officials, judges, police, and the military) but, just as important, that the state works to protect men's interests. Policies about reproductive rights provide a good example. They are largely enacted by men, but have a particularly profound effect on women. Similarly, welfare policies designed to encourage people to work are based on the model of men's experiences, because they presume that staying home to care for one's children is not work. In this sense, the state is a "gendered" institution.

Most people do not examine the institutional structure of society when thinking about intersections of race, class, and gender. The individualist framework of the dominant culture sees race, class, and gender as attributes of individuals, instead of seeing them as embedded in institutional structures. People do, of course, have race, class, and gender identities, and these identities have an enormous impact on individual experience. However, seeing race, class, and gender from an individualist viewpoint overlooks how profoundly embedded these identities are in the structure of American institutions. Moving historically marginalized groups to the center of analysis clarifies the importance of social institutions as links between individual experience and larger structures of race, class, and gender.

In this section of the book, we examine how institutions structure race, class, and gender experiences and, in turn, how race, class, and gender intersect to structure institutions. We look at four major institutions: work and the economic system, families, cultural institutions that create and reproduce ideas, and the state and social policy. Each can be shown to have a unique impact on different groups (such as the discriminatory treatment of African American and Latino men by the criminal justice system). At the same time, each institution and its interrelationship with other institutions can be seen as specifically structured through the dynamics of race, class, and gender relations.

WORK AND ECONOMIC TRANSFORMATION

Structural transformations in the global economy have dramatically changed the conditions under which all people work. Three basic transformations currently affect the character of work. First, employers have turned to new technologies and job export to cut the costs of labor and to increase profitability. In some cases, new technologies have made some jobs obsolete. In other cases, where technology could not replace workers, jobs have been exported to countries where employees work for lower wages. Race, class, and gender frame how individual workers encounter these dual processes of deskilling and job export.

Second, the new global economy has spurred the growth of the service sector. Here the dual economy operates by creating high-paid service work for skilled, college-educated workers (accountants, marketing representatives, and so on) and low-paid service work for everyone else (food service workers, nursing home aides, child care workers, and domestic workers, for example). With this growing service sector comes a shift in the kinds of skills required of workers. Here, too, race, gender, and class intersect in privileging and penalizing groups of workers. In "'Soft' Skills and Race," Philip Moss and Chris Tilly show how having the ability to interact with a range of people and demonstrating motivation to employers are increasingly important "soft" skills in the service economy. African American men, in particular, are perceived as lacking these skills and, as a result, are less likely to be hired for certain types of jobs.

Finally, global economic interdependence means that national borders no longer match up with corporations and the products that they make. Instead, products may be manufactured and assembled in a variety of places. This has fostered major population migrations; workers migrate in search of jobs and economic opportunities. The influx of new immigrant workers is nothing new in the United States, which has long incorporated immigrants into its dual labor market structure. Domestic service captures these trends of deskilling, the growth of the service sector, and migration in search of economic opportunity. Pierrette Hondagneu-Sotelo's article, "Doméstica," details how immigrant women from a few non-European nations are the newest generation of domestic workers. Service work has long served as an entry point to the labor market for poor women and women of color.

These changes have brought new opportunities for some groups—those positioned to benefit from the changes—while, for others, they have meant massive economic dislocation. The 30 million people who make up the working poor are one group most affected by these economic trends. Katherine Newman points out in "The Invisible Poor" that powerful media images erroneously equate poverty with Black, undeserving welfare mothers. Instead, the largest group of poor people in the United States is not made up of those on welfare but of the working poor, whose earnings do not allow them to afford decent housing, health care, child care, and nutrition. Many are young workers, mothers with children, or African American or Latino or immigrants. Moreover, Kenneth Brown's experiences with unemployment show that even workers in good jobs can be negatively affected by these economic trends.

FAMILIES

Families are another primary social institution profoundly influenced by intersecting systems of race, class, and gender. Historically, the family has been presumed to be the world of women: The ideology of the family (that is, dominant belief systems about the family) purports that families are places for nurturing, love, and support—characteristics that have been associated with women. This ideal identifies women with the private world of the family and men with the public sphere of work. In this sense, the family ideal identifies the family as a gendered institution. Family ideology, of course, only projects an ideal, because we know that few families actually fit the presumed ideal. Nonetheless, the ideology of the family provides a standard against which all families are judged. Moreover, the ideology of the family is class- and race-specific; that is, it ignores and distorts the family experiences of African Americans, Latinos, and most Whites.

Bonnie Thornton Dill's essay, "Our Mothers' Grief," shows that for African American, Chinese American, and Mexican American women, family structure is deeply affected by the relationships of families to the structures of race, class, and gender. Dill's historical analysis of racial-ethnic women and their families in "Our Mothers' Grief" examines diverse patterns of family organization directly influenced by a group's placement in the larger political economy. Just as the political economy of the nineteenth century affected women's experience in families, the contemporary political economy shapes family relations for women and men of all races.

The articles in this section illustrate several points that have emerged from feminist studies of families (see, for example, Thorne 1992). First, the family is not monolithic. The now widely acknowledged diversity among families refutes the idea that there is a normative family: White, middle class, with children, organized around a heterosexual married couple (preferably with the wife not employed), and needing little support from relatives or neighbors. As the experiences of African Americans, Latinos, lesbians, gays, Asian Americans, and others reveal, this so-called normal family actually represents a minority experience. In "The Diversity of American Families,"

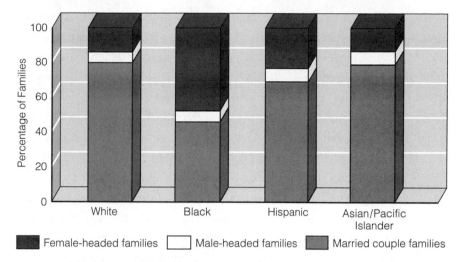

Female-headed families ▢ Male-headed families ▬ Married couple families

Source: Jason Fields, and Lynne M. Casper. 2000. *America's Families and Living Arrangements*. Washington, DC: U.S. Department of Commerce.

FIGURE 2 Diversity in U.S. Families

Eleanor Palo Stoller and Rose Campbell Gibson show how traditional myths about the American family mask the diversity of families in the United States (see also Figure 2). By focusing on older women, they also show us the different contexts of family experience of people in different generations.

Second, in this context of the idealized family, the best way to analyze families is to look at their underlying structures. For example, many presume that families are formed around a heterosexual norm. But as Kath Weston points out in "Straight Is to Gay as Family Is to No Family," gays and lesbians also participate in their families of origin and create families of their own. Claiming sexual identities as gay, lesbian, or bisexual does not mean relinquishing family or, worse yet, being defined as anti-family or as being a threat to family. Gay and lesbian partners often form permanent relationships, often including children. Although such partnerships are not legally recognized, they are part of the diversity among families. Raising children in diverse family environments so that they appreciate not only their own families but also diverse ways of forming families is a challenge in a society that devalues gays and lesbians and people of color.

Third, the dominant ideology of the family that has glorified the family routinely distorts actual behavior of women, people of color, children, and the elderly. Take, for example, the derogatory treatment of African American

men as absent fathers in their children's lives. In contrast, in his essay on being a father to his young daughter, Robin D. G. Kelley demonstrates that
many African American fathers are concerned about their daughters as well as
their sons. His "Countering the Conspiracy to Ignore Black Girls" takes on
several long-standing myths—that African American fathers are uninvolved
in the lives of their daughters and that the concerns of African American boys
are more pressing than those of African American girls—and shows that they
are untrue. His essay demonstrates that, contrary to the myth of the absent
Black father, Kelley is very concerned about his daughter's well-being.

Recentering one's thinking about the family by understanding the interconnections of race, class, and gender reveals new understandings about families and the myths that have pervaded assumptions about family experience.
In particular, ideology that presents families as peaceful, loving, nurturing
spaces masks the underlying conflicts that are embedded in family systems.
Although we do not examine family violence directly here, we know that violence stems from the gender, race, and class conflicts that are encouraged in a
sexist, racist, and class-based society. In addition, we think of family violence
not only as that which occurs between family members but also as what happens to families as they confront the institutional structures of race, class, and
gender oppression.

A fourth theme that has emerged from new studies of families concerns
family boundaries. The boundaries separating family, work, and other institutions are more fluid than was previously believed. No institution is isolated
from another. Changes in the economy, for example, affect family structures;
moreover, this effect is reciprocal, because changes in the family generate
changes in the economy. The situation of working mothers challenges this assumption of the separation of work and family. Instead, working mothers
show how interconnected social institutions really are. In "Racial Safety and
Cultural Maintenance: The Child Care Concerns of Employed Mothers of
Color," Lynet Uttal studies how working mothers of color search for day care
settings where their children will be protected from racially derogatory experiences. Because they work, these mothers cannot spend enough time impart-
ing culture to their children. As a result, they want settings where their children might learn about their own culture and they introduce this requirement
as a standard of care.

The behavior of employed women of color leads to a final theme, that the
presumed dichotomy between the public and private spheres is false. This is

especially evident in light of the experience of people of color. Racial-ethnic families have rarely been provided the protection and privacy that the alleged split between public and private assumes. From welfare policy to reproductive rights, this dichotomy is based on the myth that families are insulated from the society around them.

CULTURAL INSTITUTIONS AND THE PRODUCTION OF IDEAS

Work, family, and other social institutions are patterned by race, class, and gender, yet many people do not see these patterns and even may claim that they do not exist. The effects of economic restructuring on people of color and women and the differences manifested as people try to meet the American family ideal can be difficult to understand. How can race, class, and gender exert such a powerful influence on our everyday lives, yet remain so invisible?

The products of cultural institutions, such as movies, books, television, advertisements in the mass media, the statistics and documents produced by government, and the textbooks, curriculum materials, and teaching methods of schools, explain in part why race, class, and gender relations are obscured. These and other cultural institutions reproduce ideas by identifying which ideas are valuable, which are not, and which should not be heard at all. In this way, the ideas of groups that are privileged within race, class, and gender relations are routinely heard, whereas the ideas of groups who are disadvantaged are silenced.

In part, invisibility and silencing result from deeply entrenched ideas about race, class, and gender in the social fabric of the United States. Because beliefs about race, class, and gender are so ordinary, they often go unnoticed. Robert Moore points this out in his essay on "Racist Stereotyping in the English Language." Moore's analysis of the negative connotations associated with uses of the term "black" and the almost uniformly positive connotations that accompany the term "white" reveal a pattern that, for most people, is hidden from plain sight. In everyday social interaction, people invoke these negative and positive meanings associated with "black" and "white" without stopping to think about them at all. This everyday use obscures the historical origins of these meanings and makes the meanings of "black" and "white" appear to be natural and timeless rather than socially constructed.

Cultural images have the ability to do harm to those stigmatized by them. Placing his analysis in a global human rights context, Ward Churchill discusses how American national identity has been constructed against the interests of Native Americans. He suggests that, while Native American symbolism and imagery is ubiquitous in the United States, the derogatory use of Native American mascots and the like for sports teams stereotypes and stigmatizes contemporary Native Americans. Ironically, while Native Americans continue to experience what Churchill discusses as genocidal policies, distorted ideas about Native American culture continue to be celebrated at college and professional sporting events.

The articles in this section also remind us that the actual ideas produced by these cultural institutions are far from benign for those who are stigmatized by them. Churchill feels so strongly about the damage done to Native Americans by pervasive negative symbolism that he views this practice as criminal. On a more personal note, Judith Ortiz Cofer points out how burdensome it can be to continually encounter the "Myth of the Latin Woman" wherever she goes. Ideas are not simply reproduced by cultural institutions, because there is no thing or object that one can point to as an institution. Because cultural institutions are established societal patterns of behavior, we all participate in reproducing these ideas all of the time.

Certain cultural institutions have gained greater importance in the context of global economic restructuring. With the emergence of a global, mass media that circulates via television, music videos, cable, radio, and the Internet to all parts of the globe, racist and sexist ideas can travel quickly. Thus, those who control mass media are in a position to shape the basic ideas of society that come to be seen as truth. Gregory Mantsios's essay, "Media Magic: Making Class Invisible," explores how mass media have been central in shaping ideas about social class. His work identifies the growing significance of mass media in framing how people think about class, race, and gender.

Sports and leisure constitute another cultural institution that is growing in importance. As a cultural institution, sport has a central place in the social lives of millions. Boys and girls from different racial and social class backgrounds learn about masculinity and femininity through their involvement in sport. Julianne Malveaux's article on basketball captures the contradictions of sports as a major vehicle for American ideas of gender and race. Despite her understanding of the ways in which basketball limits the horizons of its largely African American players and broadens those of predominantly White coaches, corporate sponsors, and television audiences, Malveaux still loves basketball. These are the contradictions of contemporary patterns of race, class, and gender—exploitation can be profitable and entertaining.

Mass media, sports, and other cultural institutions can help reproduce and mask social inequalities of race, class, and gender, but these same cultural institutions can be sites for challenging long-standing oppressions. In this sense, cultural institutions are rarely all good or all bad—more often, they are a little bit of both. Educational institutions illustrate these tendencies. Schools certainly have been central to reproducing ideas about race, class, and gender. At the same time, schools have also formed the focus of much activism by historically excluded groups who pushed for changes to school curricula. Schools are often the site of struggles between those who hold traditional views of race, class, and gender oppression as natural and inevitable and people of color, women, poor people, and other historically excluded groups who challenge these inequalities. Few winners or losers exist in these struggles.

Enid Lee's efforts as one teacher developing an antiracist, multicultural curriculum within her own classroom reflect her choice to work from a position of authority inside an existing social institution. In Barbara Miner's interview with Lee, Lee offers some solid advice to teachers who want to make a difference in their classrooms. While acknowledging the difficulties of working in conditions with inadequate funding and uncooperative colleagues, Lee firmly believes that attitude and commitment still matter. As she points out, "If you don't take multicultural education or antiracist education seriously, you are actually promoting a monocultural or racist education. There is no neutral ground on this issue."

STATE INSTITUTIONS AND SOCIAL POLICY

Race, class, and gender are fundamental in the construction of the state and its policies, just as the state is fundamental in shaping and defining these categories. Because the state is the system of legitimated power and authority in society, government policies determine people's rights. As the history of Jim Crow legislation shows, the U.S. government has often been the basis for extreme exclusionary action. At the same time, as the civil rights movement and subsequent civil rights legislation of the 1960s demonstrate, the American nation-state can be an important albeit reluctant ally in challenging the exclusionary policies resulting from race, class, and gender patterns.

Matthew Snipp's essay, "The First Americans," details the historical role of the state in forcing Native Americans from their homes and subordinating them to a new system of state control. Genocide, forced relocation, and regulation by state law have shaped Native American experience; state processes, whether in the form of war, legislation, or social policy, have contributed to the current place of Native Americans in U.S. society. At the same time, as Snipp notes, groups are resilient as they face adversity and often turn to the state for assistance. The history of civil rights legislation shows that the state can be an invaluable resource for addressing the wrongs of race, class, and gender injustice. In this sense, the state is both a source of oppression and an avenue for seeking justice. For example, state policy has done much to eliminate earlier structures of inequality. At least in law (de jure), citizens' rights are race, class, and gender blind; in practice (de facto), however, these inequities continue.

Social policies can be developed on behalf of disadvantaged groups, and historically the state has had a significant role in developing such policies. The state can negotiate groups' rights and organize people's access to state and societal resources. Government can solve problems that individuals cannot solve on their own, although many people are now cynical about the state's ability to do so. Indeed, in recent years, the state has been more involved in dismantling social supports than in providing assistance for those in need.

In this context, education has emerged as an important part of the state. Education has recently and frequently been described as an institution in crisis. People worry that children are not learning in school. High school dropout rates, especially among students from poor and working-class families, reveal deep problems in the system of education. The sheer cost of post-secondary education makes it inaccessible to many. All of these factors contribute to the continuing disparity in the educational attainment of different groups.

In "Can Education Eliminate Race, Class, and Gender Inequality?" Roslyn Arlin Mickelson and Stephen Samuel Smith show how schools promote inequalities of race, class, and gender. Their analysis reveals the interplay between school policies that perpetuate existing hierarchies (such as credential inflation) and belief systems about race, class, and gender that are embedded in the so-called hidden curriculum. Students encounter these structures and belief systems while young, when their identities assume major importance. Schools thus become key sites where the inequalities of race, class, and gender are fostered and resisted.

State welfare policy constitutes an important site of struggle where intersections of race, class, and gender are highly visible. On the one hand, as

Chuck Collins points out in "Aid to Dependent Corporations," welfare for the rich is rarely identified as such. Because affluent White men run American corporations and the social welfare state itself, social welfare policies that benefit this group disappear from public scrutiny. In contrast, social welfare policies reserved for the least powerful members of U.S. society—people of color, women, children and immigrant groups—become highly visible, publicly debated, and often stigmatized. In "Welfare Reform, Family Hardship, and Women of Color," Linda Burnham details the negative effects of contemporary welfare policy on women. Burnham suggests that the family hardship experienced by women of color in the United States be placed in a human rights context. Collectively, contemporary social welfare policies in the United States have taken on a highly punitive demeanor.

Overall, institutions are powerful mechanisms for perpetuating the race, class, and gender inequities in society. Who controls them, who benefits from institutional resources, and who is best able to negotiate their way through institutional structures all reveal patterns of race, class, and gender inequity. Jael Silliman provides an analysis of state power that demonstrates intersections of race, class, and gender. Silliman links policies such as aggressive policing and the increasing incarceration of African American, Latino, and poor White men, the assault on reproductive rights of women of color, and searching immigrants of color by the Immigration and Naturalization Service under the pretext of national security as part of an overarching system of policing the body politic. She gives a clear view of how power operates in a "top-down" fashion. But it is important to remember that social institutions are always works in progress. They can be changed to become equitable and more humane. Regardless of strategies used, institutional change can take many forms. But change begins with analysis.

Further Resources

For additional materials relating to this section, see the features on pages 393–94 at the conclusion of this section.

References

Thorne, Barrie, with Marilyn Yalom. 1992. *Rethinking the Family: Some Feminist Questions*, rev. ed. Boston: Northeastern University Press.

U.S. Department of Labor. 2002. *Employment and Earnings* (January). Washington, DC: U.S. Government Printing Office.

Work and Economic Transformation

RACE, CLASS, GENDER, AND WOMEN'S WORKS

Teresa Amott and Julie Matthaei

What social and economic factors determine and differentiate women's work lives? Why is it, for instance, that the work experiences of African American women are so different from those of European American women? Why have some women worked outside the home for pay, while others have provided for their families through unpaid work in the home? Why are most of the wealthy women in the United States of European descent, and why are so many women of color poor? . . .

Throughout U.S. history, economic differences among women (and men) have been constructed and organized along a number of social categories. In our analysis, we focus on the three categories which we see as most central— gender, race-ethnicity, and class—with less discussion of others, such as age, sexual preference, and religion. We see these three social categories as interconnected, historical processes of domination and subordination. Thinking about gender, race-ethnicity, and class, then, necessitates thinking historically about power and economic exploitation. . . .

GENDER, RACE-ETHNICITY, AND CLASS PROCESSES: HISTORICAL AND INTERCONNECTED

The concepts of gender, race-ethnicity, and class are neither trans-historical nor independent. Hence, it is artificial to discuss them outside of historical

From Teresa Amott and Julie Matthaei, *Race, Gender, and Work: A Multicultural Economic History of Women in the United States*, 2d ed. (Boston: South End Press, 1996). Reprinted with permission of South End Press.

time and place, and separately from one another. At the same time, without such a set of concepts, it is impossible to make sense of women's disparate economic experiences.

Gender, race-ethnicity, and class are not natural or biological categories which are unchanging over time and across cultures. Rather, these categories are socially constructed: they arise and are transformed in history, and themselves transform history. Although societies rationalize them as natural or god-given, ideas of appropriate feminine and masculine behavior vary widely across history and culture. Concepts and practices of race-ethnicity, usually justified by religion or biology, also vary over time, reflecting the politics, economics, and ideology of a particular time and, in turn, reinforcing or transforming politics, economics, and ideology. For example, nineteenth-century European biologists Louis Agassiz and Count Arthur de Gobineau developed a taxonomy of race which divided humanity into separate and unequal racial species; this taxonomy was used to rationalize European colonization of Africa and Asia, and slavery in the United States. Class is perhaps the most historically specific category of all, clearly dependent upon the particular economic and social constellation of a society at a point in time. Still, notions of class as inherited or genetic continue to haunt us, harkening back to earlier eras in which lowly birth was thought to cause low intelligence and a predisposition to criminal activity.

Central to the historical transformation of gender, race-ethnicity, and class processes have been the struggles of subordinated groups to redefine or transcend them. For example, throughout the development of capitalism, workers' consciousness of themselves as workers and their struggles against class oppression have transformed capitalist-worker relationships, expanding workers' rights and powers. In the nineteenth century, educated white women escaped from the prevailing, domestic view of womanhood by arguing that homemaking included caring for the sick and the needy through volunteer work, social homemaking careers, and political organizing. In the 1960s, the transformation of racial-ethnic identity into a source of solidarity and pride was essential to movements of people of color, such as the Black Power and American Indian movements.

Race-ethnicity, gender, and class are interconnected, interdetermining historical processes, rather than separate systems. This is true in two senses, which we will explore in more detail below. First, it is often difficult to determine whether an economic practice constitutes class, race, or gender oppression: for example, slavery in the U.S. South was at the same time a system of class oppression (of slaves by owners) and of racial-ethnic oppression (of Africans by Europeans). Second, a person does not experience these different processes of domination and subordination independently of one another; in philosopher Elizabeth Spelman's metaphor, gender, race-ethnicity, and class are not separate "pop-beads" on a necklace of identity. Hence, there is no generic gender oppression which is experienced by all women regardless of their race-ethnicity or class. As Spelman puts it:

> . . . in the case of much feminist thought we may get the impression that a woman's identity consists of a sum of parts neatly divisible from one another, parts defined in terms of her race, gender, class, and so on. . . . On this view of personal identity (which might also be called pop-bead metaphysics), my being a woman means the same whether I am white or Black, rich or poor, French or Jamaican, Jewish or Muslim.[1]

The problems of "pop-bead metaphysics" also apply to historical analysis. In our reading of history, there is no common experience of gender across race-ethnicity and class, of race-ethnicity across class and gender lines, or of class across race-ethnicity and gender.

With these caveats in mind, let us examine the processes of gender, class, and race-ethnicity, their importance in the histories of women's works, and some of the ways in which these processes have been intertwined.

Gender

. . . Gender differences in the social lives of men and women are based on, but are not the same thing as, biological differences between the sexes. Gender is rooted in societies' beliefs that the sexes are naturally distinct and opposed social beings. These beliefs are turned into self-fulfilling prophecies through sex-role socialization: the biological sexes are assigned distinct and often unequal work and political positions, and turned into socially distinct genders.

Economists view the sexual division of labor as central to the gender differentiation of the sexes. By assigning the sexes to different and complementary tasks, the sexual division of labor turns them into different and complementary genders. . . . The work of males is at least partially, if not wholly, different from that of females, making "men" and "women" different economic and social beings. Sexual divisions of labor, not sexual difference alone, create difference and complementarity between "opposite" sexes. These differences, in turn, have been the basis for marriage in most societies. . . .

The concept of gender certainly helps us understand women's economic histories. . . . Each racial-ethnic group has had a sexual division of labor which has barred individuals from the activities of the opposite sex. Gender processes do differentiate women's lives in many ways from those of the men in their own racial-ethnic and class group. Further, gender relations in all groups tend to assign women to the intra-familial work of childrearing, as well as to place women in a subordinate position to the men of their class and racial-ethnic group.

But as soon as we have written these generalizations, exceptions pop into mind. Gender roles do not always correspond to sex. Some American Indian tribes allowed individuals to choose among gender roles: a female, for example, could choose a man's role, do men's work, and marry another female who lived out a woman's role. In the nineteenth century, some white females "passed" as men in order to escape the rigid mandates of gender roles. In many of these cases, women lived with and loved other women.

Even though childrearing is women's work in most societies, many women do not have children, and others do not perform their own child care or domestic work. Here, class is an especially important differentiating process. Upper-class women have been able to use their economic power to reassign some of the work of infant care—sometimes even breastfeeding—to lower-class women of their own or different racial-ethnic groups. These women, in turn, may have been forced to leave their infants alone or with relatives or friends. Finally, gender complementarity has not always led to social and economic inequality; for example, many American Indian women had real control over the home and benefited from a more egalitarian sharing of power between men and women.

Since the processes of sex-role socialization are historically distinct in different times and different cultures, they result in different conceptions of appropriate gender behavior. Both African American and Chicana girls, for instance, learn how to be women—but both must learn the specific gender roles which have developed within their racial-ethnic and class group and historical period. For example, for white middle-class homemakers in the 1950s, adherence to the concept of womanhood discouraged paid employment, while for poor Black women it meant employment as domestic servants for white middle-class women. Since racial-ethnic and class domination have differentiated the experiences of women, one cannot assume, as do many feminist theorists and activists, that all women have the same experience of gender oppression—or even that they will be on the same side of a struggle, not even when some women define that struggle as "feminist."

Not only is gender differentiation and oppression not a universal experience which creates a common "women's oppression," the sexual divisions of labor and family systems of people of color have been systematically disrupted by racial-ethnic and class processes. In the process of invasion and conquest, Europeans imposed their notions of male superiority on cultures with more egalitarian forms of gender relations, including many American Indian and African tribes. At the same time, European Americans were quick to abandon their notion of appropriate femininity when it conflicted with profits; for example, slave owners often assigned slave women to backbreaking labor in the fields.

Racial-ethnic and class oppression have also disrupted family life among people of color and the white working class. Europeans interfered with family relations within subordinated racial-ethnic communities through rape and forced cohabitation. Sometimes whites encouraged or forced reproduction, as when slaveowners forced slave women into sexual relations. On the other hand, whites have often used their power to curtail reproduction among peoples of color, and aggressive sterilization programs were practiced against Puerto Ricans and American Indians as late as the 1970s. Beginning in the late nineteenth century, white administrators took American Indian children from their parents to "civilize" them in boarding schools where they were forbidden to speak their own languages or wear their native dress. Slaveowners commonly split up slave families through sale to different and distant new owners. Nevertheless, African Americans were able to maintain strong family ties, and even augmented these with "fictive" or chosen kin. From the mid-nineteenth through the mid-twentieth centuries, many Asians were separated from their spouses or children by hiring policies and restrictions on immigration. Still, they maintained family life within these split households, and eventually succeeded in reuniting, sometimes after generations. Hence, for peoples of color, having children and maintaining families have been an essential part of the struggle against racist oppression. Not surprisingly, many women of color have rejected the white women's movement view of the family as the center of "women's oppression."

These examples reveal the limitations of gender as a single lens through which to view women's economic lives. Indeed, any attempt to understand women's experiences using gender alone cannot only cause misunderstanding, but can also interfere with the construction of broad-based movements against the oppressions experienced by women.

Race-Ethnicity

Like gender, race-ethnicity is based on a perceived physical difference, and rationalized as "natural" or "god-given." But whereas gender creates difference and inequality according to biological sex, race-ethnicity differentiates individuals according to skin color or other physical features.

In all of human history, individuals have lived in societies with distinct languages, cultures, and economic institutions; these ethnic differences have been perpetuated by intermarriage within, but rarely between, societies. However, ethnic differences can exist independently of a conception of race, and without a practice of racial-ethnic domination such as the Europeans practiced over the last three centuries. . . .

Does the concept of race-ethnicity help us understand the economic history of women in the United States? Certainly, for white racial-ethnic domination has been a central force in U.S. history. European colonization of North America entailed the displacement and murder of the continent's indigenous peoples, rationalized by the racist view that American Indians were savage heathens. The economy of the South was based on a racial-ethnic system, in which imported Africans were forced to work for white landowning families as slaves. U.S. military expansion in the nineteenth century brought more lands under U.S. control—the territories of northern Mexico (now the Southwest), the Philippines, and Puerto Rico—incorporating their peoples into the racial-ethnic hierarchy. And from the mid-nineteenth century onward, Asians were brought into Hawaii's plantation system to work for whites as semi-free laborers. In the twentieth century, racial-ethnic difference and inequality have been perpetuated by the segregation of people of color into different and inferior jobs, living conditions, schools, and political positions, and by the prohibition of intermarriage with whites in some states up until 1967.

Race-ethnicity is a key concept in understanding women's economic histories. But it is not without limitations. First, racial-ethnic processes have never operated independently of class and gender. In the previous section on gender, we saw how racial domination distorted gender and family relations among people of color. Racial domination has also been intricately linked to economic or class domination. As social scientists Michael Omi (Asian American) and Howard Winant (European American) explain, the early European arguments that people of color were without souls had direct economic meaning:

> At stake were not only the prospects for conversion, but the types of treatment to be accorded them. The expropriation of property, the denial of political rights, the introduction of slavery and other forms of coercive labor, as well as outright extermination, all presupposed a worldview which distinguished Europeans—children of God, human beings, etc.—from "others." Such a worldview was needed to explain why some should be "free" and others enslaved, why some had rights to land and property while others did not.[2]

Indeed, many have argued that racial theories only developed after the economic process of colonization had started, as a justification for white domination of peoples of color.

The essentially economic nature of early racial-ethnic oppression in the United States makes it difficult to isolate whether peoples of color were subordinated in the emerging U.S. economy because of their race-ethnicity or their economic class. Whites displaced American Indians and Mexicans to obtain their land. Whites imported Africans to work as slaves and Asians to work as contract laborers. Puerto Ricans and Filipinas/os were victims of further U.S. expansionism. Race-ethnicity and class intertwined in the patterns of displacement for land, genocide, forced labor, and recruitment from the seventeenth through the twentieth centuries. While it is impossible, in our minds, to determine which came first in these instances—race-ethnicity or class—it is clear that they were intertwined and inseparable.

Privileging racial-ethnic analysis also leads one to deny the existence of class differences, both among whites and among people of color, which complicate and blur the racial-ethnic hierarchy. A racial-ethnic analysis implies that all whites are placed above all peoples of color. . . .

. . . A minority of the dominated race is allowed some upward mobility and ranks economically above whites. At the same time, however, all whites have some people of color below them. For example, there are upper-class Black, Chicana, and Puerto Rican women who are more economically privileged than poor white women; however, there are always people of color who are less economically privileged than the poorest white woman. Finally, class oppression operates among women of the same racial-ethnic group.

A third problem with the analysis of racial domination is that such domination has not been a homogeneous process. Each subordinated racial-ethnic group has been oppressed and exploited differently by whites: for example, American Indians were killed and displaced, Africans were enslaved, and Filipinas/os and Puerto Ricans were colonized. Whites have also dominated whites; some European immigrant groups, particularly Southern and Eastern Europeans, were subjected to segregation and violence. In some cases, people of color have oppressed and exploited those in another group: Some American Indian tribes had African slaves; some Mexicans and Puerto Ricans displaced and murdered Indians and had African slaves. Because of these differences, racial oppression does not automatically bring unity of peoples of color across their own racial-ethnic differences, and feminists of color are not necessarily in solidarity with one another.

To sum up, we see that, as with gender, the concept of race-ethnicity is essential to our analysis of women's works. However, divorcing this concept from gender and class, again, leads to problems in both theory and practice.

Class

. . . We believe that the concepts of class and exploitation are crucial to understanding the work lives of women in early U.S. history, as well as in the modern, capitalist economy. Up through the nineteenth century, different class relations organized production in different regions of the United States. The South was dominated by slave agriculture; the Northeast by emerging industrial capitalism; the Southwest (then part of Mexico) by the *hacienda* system which carried over many elements of the feudal manor into large-scale production for the market; the rural Midwest by independent family farms that produced on a small scale for the market; and the American Indian West by a variety of tribal forms centered in hunting and gathering or agriculture, many characterized by cooperative, egalitarian economic relations. Living within these different labor systems, and in different class positions within them, women led very different economic lives.

By the late nineteenth century, however, capitalism had become the dominant form of production, displacing artisans and other small producers along with slave plantations and tribal economies. Today, wage labor accounts for over 90 percent of employment; self employment, including family businesses, accounts for the remaining share. With the rise of capitalism, women were brought into the same labor system, and polarized according to the capitalist-wage laborer hierarchy.

At the same time as the wage labor form specific to capitalism became more prevalent, capitalist class relations became more complex and less transparent. Owners of wealth (stocks and bonds) now rarely direct the production process; instead, salaried managers, who may or may not own stock in the company, take on this function. While the capitalist class may be less identifiable, it still remains a small and dominant elite. . . .

Class can be a powerful concept in understanding women's economic lives, but there are limits to class analysis if it is kept separate from race-ethnicity and gender. First, as we saw in the race section above, the class relations which characterized the early U.S. economy were also racial-ethnic and gender formations. Slave owners were white and mostly male, slaves were Black. The displaced tribal economies were the societies of indigenous peoples. Independent family farmers were whites who farmed American Indian lands; they organized production in a patriarchal manner, with women and children's work defined by and subordinated to the male household head and property owner. After establishing their dominance in the pre-capitalist period, white men were able to perpetuate and institutionalize this dominance in the emerging capitalist system, particularly through the monopolization of managerial and other high-level jobs.

Second, the sexual division of labor within the family makes the determination of a woman's class complicated—determined not simply by her relationship to the production process, but also by that of her husband or father. For instance, if a woman is not in the labor force but her husband is a capitalist, then we might wish to categorize her as a member of the capitalist class. But what if that same woman worked as a personnel manager for a large corporation, or as a salesperson in an elegant boutique? Clearly, she derives upper-class status and access to income from her husband, but she is also, in her own right, a worker. Conversely, when women lose their husbands through divorce, widowhood, or desertion, they often change their class position in a downward direction. A second gender-related economic process overlooked by class analysis is the unpaid household labor performed by women for their fathers, husbands, and children—or by other women, for pay.

Third, while all workers are exploited by capitalists, they are not equally exploited, and gender and race-ethnicity play important roles in this differentiation. Men and women of the same racial-ethnic group have rarely performed the same jobs—this sex-typing and segregation is the labor market form of the sexual division of labor we studied above. Further, women of different racial-ethnic groups have rarely been employed at the same job, at least not within the same workplace or region. This racial-ethnic typing and seg-regation has both reflected and reinforced the racist economic practices upon which the U.S. economy was built.

Thus, jobs in the labor force hierarchy tend to be simultaneously race-typed and gender-typed. Picture in your mind a registered nurse. Most likely, you thought of a white woman. Picture a doctor. Again, you imagined a per-son of a particular gender (probably a man), and a race (probably a white per-son). If you think of a railroad porter, it is likely that a Black man comes to mind. Almost all jobs tend to be typed in such a way that stereotypes make it difficult for persons of the "wrong" race and/or gender to train for or obtain the job. Of course, there are regional and historical variations in the typing of jobs. On the West Coast, for example, Asian men performed much of the paid domestic work during the nineteenth century because women were in such short supply. In states where the African American population is very small, such as South Dakota or Vermont, domestic servants and hotel chambermaids are typically white. Nonetheless, the presence of variations in race-gender typing does not contradict the idea that jobs tend to take on racial-ethnic and gender characteristics with profound effects on the labor market opportuni-ties of job-seekers. . . .

. . . The racial-ethnic and gender processes operating in the labor market have transposed white and male domination from pre-capitalist structures into the labor market's class hierarchy. This hierarchy can be described by grouping jobs into different labor market sectors or segments: "primary," "secondary," and "underground." . . . The primary labor market—which has been monopolized by white men—offers high salaries, steady employment, and upward mobility. Its upper tier consists of white-collar salaried or self-employed workers with high status, autonomy, and, often, supervisory capac-ity. Wealth increases access to this sector, since it purchases elite education and provides helpful job connections. . . .

The lower tier of the primary sector, which still yields high earnings but involves less autonomy, contains many unionized blue-collar jobs. White working-class men have used union practices, mob violence, and intimidation to monopolize these jobs. By World War II, however, new ideologies of worker solidarity, embodied in the mass industrial unions, began to overcome the resistance of white male workers to the employment of people of color and white women in these jobs.

In contrast to both these primary tiers, the secondary sector offers low wages, few or no benefits, little opportunity for advancement, and unstable employment. People of color and most white women have been concentrated in these secondary sector jobs, where work is often part-time, temporary, or seasonal, and pay does not rise with increasing education or experience. Jobs in both tiers of the primary labor market have generally yielded family wages, earnings high enough to support a wife and children who are not in the labor force for pay. Men of color in the secondary sector have not been able to earn enough to support their families, and women of color have therefore participated in wage labor to a much higher degree than white women whose husbands held primary sector jobs.

Outside of the formal labor market is the underground sector, where the most marginalized labor force groups, including many people of color, earn their livings from illegal or quasi-legal work. This sector contains a great variety of jobs, including drug trafficking, crime, prostitution, work done by undocumented workers, and sweatshop work which violates labor standards such as minimum wages and job safety regulations.

NOTES

1. Elizabeth V. Spelman, *Inessential Woman: Problems of Exclusion in Feminist Thought* (Boston: Beacon Press, 1988).

2. Michael Omi and Howard Winant, *Racial Formation in the United States* (New York: Routledge & Kegan Paul, 1986), p. 58.

THE INDIGNITIES OF UNEMPLOYMENT   25

Kenneth W. Brown

I am a number, a statistic, a peak on a national bar graph. I am not physically or mentally disabled. But I am financially disabled; I am unemployed.

I am unemployed after nearly 14 years of being a loyal, hardworking employee. I bought into the company philosophy, followed the company credo and worked hard to help accomplish the corporate goal. As I was coddled, supported, and groomed, I grew to believe that I was an important member of the corporate family, only to be unceremoniously dumped after I had served my purpose and was no longer needed or welcomed, like a bastard son at a family reunion.

I am making every effort to find work. I have made countless phone inquiries, answered many classified ads and contacted people I have not spoken

From *Essence* 26 (September 1995): 56. Reprinted by permission.

to in years, looking for job leads. I have attended job fairs in search of another chance to rejoin corporate America, struggling to maintain my sense of dignity as I walked about the crowded rooms—looking along with so many others for the next opportunity to reenter the economic mainstream.

As unemployment refuses to loosen its grasp, I find it increasingly difficult to remain hopeful, positive, and confident. Prolonged unemployment can do a number on your psyche, your soul, and your spirit. Unemployment can make you question your skills and abilities. And it can damage your self-esteem. I don't care how much confidence you have; unemployment can shake that confidence.

As I search for my next position, I do so with growing reluctance and trepidation. When I prepare for the next job interview, I practice my responses to the questions about myself, my strengths, my weaknesses, and my future goals.

I make sure I speak the King's English with perfect enunciation, ever mindful not to let street slang or my ethnicity slip into my conversation.

I often wonder whether my white counterparts are as conscious of their diction as I am of mine. I also wonder whether I am selling out or compromising myself too much. Do white men wonder if they are being too forceful, too eager to please, too assertive? Do they agonize as much as I do over how to appear skilled and capable without intimidating the people who have the power to hire? The thought of always keeping these ideas in mind is exhausting and sad. Is this the cross that African-American men and women must bear in order to be accepted into corporate America?

Unemployment can make you question your faith and spiritual foundation. I pray for my unemployment to end. I pray for direction and guidance. I also pray for the day when I do not have to wonder whether my blackness is a barrier to reentry into the labor force. I am still waiting for a response. Unemployment makes me wonder if anyone is listening.

Being unemployed also makes me question my manhood. I was raised to be a responsible household member who would carry my share of the weight. But now that I am no longer holding up my end of the financial bargain, my wife and I exist on a single income: hers. While I am fortunate that she has been supportive throughout my long period of unemployment, I must fight hard against seeing myself as a failure or loser. I have never before considered myself a loser, and I refuse to be one. Every day I must remind myself that being without a job does not make me less of a man.

Some good things have resulted from my unemployment. I have gained a renewed awareness of life's blessings: the value of true friends and the love, support, encouragement and prayers of family. In many ways I have become a stronger person. I have resolved to survive unemployment. I resolve to become a statistic for working Americans, and not just another African-American male without a job.

"SOFT" SKILLS AND RACE

26

Philip Moss and Chris Tilly

The growth of "soft" or social skills is a factor that has been neglected in research on the racial gap in labor market outcomes, despite the fact that employer surveys have repeatedly identified such skills as the most important hiring criterion for entry-level jobs (Capelli 1995). In this study, therefore, we examined the relationship between employer racial attitudes and their hiring practices, with special attention given to their use of soft skills. We defined soft skills as skills, abilities, and traits that pertain to personality, attitude, and behavior rather than to formal or technical knowledge. Based on 56 face-to-face interviews with employers, we sought to learn how employment gatekeepers conceived of soft skills in relation to hiring Black men for entry-level jobs.

According to our findings, employers reported an increasing need for soft skills—driven, they said, by heightened competitive pressure—and they rated Black men poorly in terms of such skills. Thus, the heightened competition that propels current business restructuring appears to contribute to increased labor market inequality by race. . . .

THE GROWING IMPORTANCE OF SOFT SKILLS

Again, we defined soft skills as skills, abilities, and traits that pertain to personality, attitude, and behavior rather than to formal or technical knowledge. Two clusters of soft skills were important to the employers surveyed. The first, *interaction*, involves ability to interact with customers, coworkers, and supervisors. This cluster includes friendliness, teamwork, ability to fit in, and appropriate affect, grooming, and attire. The interaction category is related to concepts of *emotional labor* (Hochschild 1983; Wharton 1993) and *nurturant social skills* (England 1992; Kilbourne, England, & Beron 1994). A second cluster, *motivation*, takes in characteristics such as enthusiasm, positive work attitude, commitment, dependability, and willingness to learn. We distinguish both from hard skills, including skills in math, reading, and writing; knowledge of particular job procedures; "brightness;" ability to learn; educational attainment; and physical strength.

From *Work and Occupations* 23 (August 1996): 252–76 © 1996 Sage Publications, Inc. Reprinted by permission of Sage Publications, Inc.

Interaction and motivation skills differ from one another, and in much of our analysis, we distinguish between the two. However, we grouped them together under the rubric of soft skills because employers often subsumed both in terms like *attitude* and because many employers viewed both as more immutable than hard skills. Of course, soft skills are in part culturally defined, and therefore employer assessments of soft skills will be confounded by differences in culture and by racial stereotyping. Indeed, the word *skills* is to some extent a misnomer, although employers most definitely conceptualize these attributes as contributing to individual productivity differences.

We asked almost all interviewees to identify the most important qualities they looked for when hiring entry-level workers. As Table 1 shows, interaction skills were by far the most important qualification in retail. Motivation and hard skills received roughly equal emphasis in auto parts and insurance. Only in the public sector were hard skills mentioned most often. Overall, 86% of respondents included soft skills in their list of the most important hiring criteria, and almost half mentioned soft skills first in that list. . . .

Respondents in all four industries stated that competitive pressures have led to growing soft-skill needs (Moss & Tilly 1996). For instance, auto makers are pushing their suppliers to cut costs and increase quality. In response, many of the parts manufacturers are escalating basic and technical skill requirements. Some, however, are demanding more soft skills as well. For instance, a human resource manager at an alloy casting plant told us:

> Hiring used to be based on 90% experience, 10% attitude or work ethic. I find that changing . . . due to the [emphasis on] team work and total quality. Attitudes and people getting along with one another . . . —this is a big part now. I would almost say it's 50% to 60% being experience and the other 40% being attitude, work ethic, teamwork.

In retail clothing, heightened emphasis on customer service—again, spurred by intensified competition—has led companies to screen more carefully for soft skills when hiring sales clerks. One discount clothing chain has

Table 1   Most Important Qualities Looked for in Entry-Level Employees

	Frequency With Which Each Category Was Mentioned (Percent)		
Industry	Hard Skills	Interaction	Motivation
Auto parts manufacturing	58	32	63
Retail clothing	22	78	39
Insurance	67	67	78
Public sector	100	60	60

NOTE: Informants typically mentioned more than one desired quality.

adopted the slogan "fast, fun, and friendly." According to a regional person-nel representative for the chain, this means that now

> I tell my . . . personnel managers, "If they don't smile, don't hire them."
> I don't care how well-educated they are, how well-versed they are in retail,
> if they can't smile, they're not going to make a customer feel welcome.
> And we don't want them in our store.

On the other hand, respondents reported declining hard skill needs among sales clerks, due to optical scanning equipment and computerized cash registers.

Insurance companies increasingly demand computer literacy among their clerical workers. In addition, financial deregulation has heightened competi-tion, leading insurers to adopt several strategies: downsizing, reorganization of work, and greater stress on customer service. As one human resource man-ager put it, "on a scale of 1 to 10 it [customer relations] is a 9.999. . . . There is much more emphasis now being placed on it."

Finally, in the public sector, budget cuts have combined with political de-mands for greater productivity and quality of service. In our public sector in-terviews, two thirds of respondents reported that they were looking for more skilled people. Although some of these respondents noted a greater need for some basic or technical skill, all mentioned a need for greater customer skills—in jobs ranging from clerk-typist to hospital housekeeper.

SOFT SKILLS AND RACE IN THE EYES OF EMPLOYERS

The emphasis employers place on soft skills disadvantages Black male job ap-plicants. This is because many employers see Black men as lacking in precisely the skills they consider increasingly important. Indeed, in our sample, the em-ployers placing the greatest emphasis on soft skills were those most likely to have negative views of Black men as workers. The views employers hold of Black men in this regard were partly stereotype, partly cultural gap, and partly an accurate perception of the skills that many less educated Black men bring to the labor market. . . .

Interaction Skills

Employers voiced two main sets of concerns about the ability of Black men to interact effectively with customers and coworkers. First, a substantial minor-ity of respondents—32%—described Black men as defensive, hostile, or hav-ing a difficult "attitude." The content of these comments ranged widely. A Latino store manager in a Black area of Los Angeles, who hires mostly

Latinos, flatly stated, "You know, a lot of people are afraid, they [Black men] project a certain image that makes you back off. . . . They're really scary." When asked, "How much of that do you think is perception, how much do you think is actually reality?" he responded, "I think 80% is reality. 80% of it's factual."

Other respondents stated that managers see Black men as difficult to control. For example, the Black female personnel manager of a Detroit retail store commented,

> Employers are sometimes intimidated by an uneducated Black male to come in. Their appearance really isn't up to par, their language, how they go about an interview. Whereas females Black or White, most people do feel, "I could control this person." . . . A lot of times people are physically intimidated by Black men. . . . The majority of our employers are not Black. And if you think that person may be a problem, [that] young Black men normally are bad, or [that] the ones in this area [are], you say, "I'm not going to hire that person, because I don't want trouble."

A White female personnel official from a Los Angeles public sector department offered a related perspective, laying part of the blame with White supervisors:

> . . . And I also think that part of the problem is that the supervisors and managers of these people have their own sets of expectations and their own sets of goals that don't address the diversity of these people, and it's kind of like, well, hell, if they're going to come work for me, they're going to damn well do it my way. . . . And my own personal feeling is that a lot of these young Black men who are being tough scare some of their supervisors. And so rather than address their behavior problems and deal with the issues, they will back away until they can find a way to get rid of them. We have a tendency to fear what we're not real familiar with.

Although a few respondents provided this level of detail, most of the negative responses were far briefer: "a lot of Black males have a chip on their shoulder," or "I get a strong sense of, with Black males at times of a hostility, of 'I deserve so and so, this belongs to me.'"

Our questions probed primarily for generalizations, but some respondents noted variation within race and gender categories. Even the store manager who described Black men as "really scary" added, "You know, there's a lot of Black males that are nice; they usually project a different image. They don't want to project the same image as gangsters."

In addition to negative views of the demeanor of Black men, we found that many retail employers saw the racial composition of employment itself as an issue for customers. The Black male personnel manager of a large retail store located in a Detroit-area suburban mall stated that because the labor market area is 90% Black, "we are forced to have an Affirmative Action program for nonminorities in this particular store." In fact, the store has shifted away from walk-in applications to in-store or mail recruiting from the store's customer base. Given that the mall sits in an integrated suburb of Detroit, his statement implies a fear that an all-Black workforce would erode the White suburban customer base.

In subsequent interviews, we asked retail informants explicitly about attempts to keep the racial mix of store employees similar to that of customers. Seven of the 10 retail informants whom we asked, responded that this was indeed a management concern. Not all of them approved of the customer attitudes to which they were responding. For example, a White female personnel manager at a Los Angeles store said,

> At [a store she was posted at previously] we had a lot of customer complaints because it's primarily White, and we were always getting complaints that there were all Black employees and it's because they were Black. That would be the first thing the customer would bring up was "Black." It was because they were Black that they didn't do their job right.

Nonetheless, this informant and others—Black, White, and Latino alike—viewed the goal of race-matching staff with customers as a legitimate management objective.

Motivation

Forty percent of respondents voiced perceptions of Black men as unmotivated employees. Once again, comments varied widely in substance. A Latino female personnel officer of a Los Angeles retail distribution warehouse, whose workforce is 72% Latino and only 6% Black, stated, "Black men are lazy. . . . Who is going to turn over? The uneducated Black." The White male owner of a small Detroit area plastic parts plant (46% Black, 54% White) said that in his experience, Black men "just don't care—like everybody owes them." "Black kids don't want to work," was the opinion of a White male owner of a small auto parts rebuilding shop in Los Angeles, whose workforce was entirely Hispanic female. "Black men are not responsible," added a Latino female personnel supervisor for a Los Angeles auto parts manufacturer located next to a major Black area but with a workforce that was 85% Latino and less than 1% Black.

This is still a minority viewpoint. The majority of respondents stated that they saw no differences in work ethic by race, although surely, in some cases, they were simply proffering the socially approved response. As in the case of interaction, some respondents discussed variation within race/gender categories. A number of employers invoked such variation to dismiss racial differences:

> We have problem employees, who choose not to come to work as often as others, but that cuts across all racial lines, you know, so no, I wouldn't say that's

> the case as a blanket statement. We have problem people, but they're just problem people. Some of them happen to be Black and some of them happen to be White, but I wouldn't say it's any better or worse in any one group or the other.

Other managers attributed apparent racial differences to class or neighborhood effects. Yet others noted distinctions within racial groups but still rated Black employees lower on average. For example, in a Los Angeles area discount store where the main workforce was Latino, the manager opined:

> I think the Hispanic people have a very serious work ethic. I have a lot of respect for them. They take pride in what they do. Some of the Black folks that I've worked with do, but I'd say a majority of them are just there putting in the time and kind of playing around.

In fact, although only a minority of respondents questioned the work ethic of Blacks, a substantial majority agreed with the idea that immigrants have a stronger work ethic than native-born workers—81% of Detroit respondents who ventured an opinion on this agreed, and 88% of Los Angeles respondents. This bodes ill for less-skilled Black workers, particularly in Los Angeles, because they increasingly compete with immigrant workers for jobs. . . .

How and Why Employers Form These Perceptions

Employers indicated that they based their perceptions of Black men on experiences with past and present employees, on impressions of applicants, and on more general impressions from the media and from experiences outside work. About half of respondents referred to their own employees, sometimes arguing that the immediacy of these observations rendered them objective. Stated a store manager,

> I think [Black men] feel things should be given to them and not earned. And because of that, they don't earn the right to keep jobs. Now that may be, you know, someone would say I may have an attitude problem, but that, I just look at pure facts. I mean with the people that I've had work for me. . . .

. . . In a number of cases, informants pleaded ignorance due to a lack of Black applicants: "It's hard for us to say [why Black men do poorly in the labor market], because we don't have a large enough community, so we don't interview very many Blacks, and we don't reject many Blacks."

Other employers, especially those in White or Hispanic areas, referred to contact with Blacks outside the workplace. A White male insurance manager outside Detroit reported:

> I am involved and attend an urban church, and we attend there working with the homeless and the retarded and people of that nature, [and] unless something is done to help these Black males, it is just a sorry situation.

Although no respondent specifically cited the media as a source for their impressions of Black men, the impact of the media was evident, as in these two comments by White Detroit area manufacturing managers:

> Manager 1: We have a lot guys out there with cocaine in their pocket and Uzis in the trunk.

> Manager 2: I think a lot of [the difficulty Black men face in the labor market] is based on their inability to complete schooling early on, for whatever reason. I don't know a lot of the statistics of the Black race but I do see that. I think that is a good fact. . . . It's going to take unfortunately a heck of a long time to fully eliminate discrimination. We hear about it in the news still, and it's a shame.

. . . Even more important than how employers form their perceptions of Black men is why they hold these views. We argue that the negative views are a complicated combination of stereotypes, realities, and conflicting cultures. The evidence that stereotypes are involved is straightforward. On the one hand, some respondents voiced clearly stereotypical attitudes: "Black men are lazy. . . . They've got no respect for anyone"; and "I'm no doctor . . . but I'm convinced, having dealt with grievance and unrest . . . that Black men, and to some extent Black women, do not deal with stress physically as well as some other races."

On the other hand, certain managers charged that other managers harbored stereotypical views:

> People have a tendency, . . . even if they don't voice it, they deal with their stereotypes. It's easy to identify someone as a female or identify someone as Black male or a White male. And so whatever it is that they have of their expectations of those people, they will project that.

In particular, some managers claimed that their peers engaged in statistical discrimination or generalized from a visible but unrepresentative subset of Black men:

> I think that many employers may feel that because of the large numbers of Black males who are in prison and have problems, that there is a tendency for those who are out and in the workforce to do mischievous things. That's unfortunate.

And one informant, a Black female manager at a Detroit area utility, contended that employers hold Blacks to a different standard than Whites:

> When Blacks and Hispanics and whomever come into the work group, and they are not part of the majority, then one thing they need to know is that they cannot always do what they see others doing, and that is key. That is a lesson

> that needs to be taught. I think the rules aren't always the same, and it may
> not always be intentional. A lot is institutional and lot of it is people just not
> understanding others.

We believe that the perceptions stated by employers contain an element of reality as well. We have no independent way of verifying the statements employers made. However, some comments offered substantial detail and specificity, and some referred to "objective" measures such as absenteeism. . . .

Wilson (1987) argued that soft-skill problems are the product of poor neighborhoods; a number of our informants argued instead that such skills—and particularly motivation—are endogenous to the workplace and labor market. As a Black human resource official of a Detroit-area insurer expressed it,

> I think business drives the work ethic. . . . If business is lax . . . then people
> have casual attitudes about their jobs. . . . You are one thing up to the point
> of entering the business world, but then you are something else. I'm not the
> same person I was 15 years ago. I had to take on certain thoughts and
> attitudes whether I liked it or not.

Several others agreed that motivation is more a function of management than of the workforce. When asked about racial differences in the work ethic, a White male manager of contracted public-sector workers mused,

> I think it's how you motivate each group. Two or three years ago, I would have
> probably said, well, the Black race isn't as motivated as the Oriental or the
> Hispanic. But I've seen that if you motivate, that you have to motivate each
> group differently.

A White male public-sector human resource official added that work ethic may vary by job:

> If I take security, or I take the basic labor jobs, I'm not so sure that when they
> were Caucasian-dominated 20 years ago, that people weren't leaning on a
> shovel and goldbricking. Many times, the classifications we normally associate
> with being more lazy or finding ways to avoid work, are the entry-level, lower
> skilled ones. And now those happen to be dominated by Blacks and to a
> lesser extent Hispanics right now.

A small number of informants also argued that workers can be trained to relate well to customers. Even a store manager who commented that "it does take a certain kind of person" to be "fast, fun, and friendly," added, "but if you work with a person, I think that you could pretty much [get them to] be fast, fun, and friendly." . . .

In addition to stereotype and reality, interviewees spoke of cultural gaps between young Black men and their supervisors, coworkers, and customers—especially as an explanation of difficulties in interaction. A Black human resource manager at an insurance company described the problems of cultural translation:

> I think that, as I attend executive meetings, and in many cases, I'm the only Black man there, the cultural diversity and the strangeness that different people bring to one another—oftentimes people aren't prepared to receive what another person may be prepared to offer. And I think that through that lack of communication, a lot of times things are misunderstood. When problems occur, If I work for you and you had a problem with me, you may not know how to approach me and vice versa, I may not know how to approach you.

White respondents also referred to "a difference in understanding," but were more likely to pose it as a failure of the Black men themselves: "[Young Black men] don't present themselves well to the employer, just because they don't know, they don't realize how they're communicating, or not communicating." This conforms with the view, expressed in focus groups by young, inner-city Black and Latino men, that code switching—being able to present oneself and communicate in ways acceptable to majority White culture—is the most important skill needed to find and keep a job (Jobs for the Future 1995). . . .

CONCLUSIONS

We find that due to competitive pressure, employers are demanding more soft skills, even in low-skill jobs. Soft skills include interaction and motivation, and employers are valuing both more highly. However, many managers perceive Black men as possessing fewer soft skills, along both dimensions. Thus, the same increases in competitive pressure that drive corporate downsizing and restructuring are contributing to widening racial inequality in labor market outcomes.

Employers base their perceptions of Black men on assessments of current or past employees and applicants, as well as interaction outside the workplace, and media images of Blacks. Three factors underlie negative evaluations of Black men as workers: racial stereotypes, cultural differences between employers and young Black men, and actual skill differences. The actual skill differences themselves are in part endogenous to the work situation. Moreover, in a work world characterized by increasing levels of interaction, racially biased attitudes held by customers or coworkers of other racial groups can themselves lead to lower measured productivity—that is, productivity differences can be the direct result of discrimination.

References

Capelli, P. (1995). Is the "skills gap" really about attitudes? *California Management Review, 37*(4), 108–124.

England, P. (1992). *Comparable worth: Theories and evidence.* New York: Aldine de Gruyter.

Hochschild, A. R. (1983). *The managed heart: Commercialization of human feeling.* Berkeley: University of California Press.

Jobs for the Future. (1995). Information from focus groups of young, inner city men of color, for work in progress. Unpublished manuscripts. Boston, MA: Author, for the Annie E. Casey Foundation.

Kilbourne, B. S., England, P., & Beron, K. (1994). Effects of changing individual, occupational, and industrial characteristics on changes in earnings: Intersections of race and gender. *Social Forces, 72,* 1149–1176.

Moss, P., & Tilly, C. (1996). *Growing demand for "soft" skills in four industries: Evidence from in-depth employer interviews* (Working Paper No. 93). New York: Russell Sage Foundation.

Wharton, A. (1993). The affective consequences of service work: Managing emotions on the job. *Work and Occupations, 20*(2), 205–232.

Wilson, W. J. (1987). *The truly disadvantaged.* Chicago: University of Chicago Press.

THE INVISIBLE POOR   27

Katherine S. Newman

Forty years ago, when it first began to dawn on this prosperous nation that the end of the Great Depression had not cured widespread poverty, popular representations of the poor focused on Kentucky hollows where Appalachian children sat on wooden stoops, their bellies swollen and faces blank. Miners whose livelihoods had disappeared with the end of the coal boom were left behind like refugees from another era, their families consigned to a threadbare existence. We understood that these were working folk who had fallen on hard times. The whole country reeled at the thought that such desperate conditions existed in the heartland—among good (white) people—and the war on poverty was born in order to lift up the downtrodden.

In the intervening years, however, as many writers have noted, poverty has been racialized. Even though the majority of the poor are still white and working—as they were in the 1930s and thereafter—the public impression is quite clearly the reverse: poverty wears a black face and is presumed to follow from an unwillingness to enter the labor force. As Herbert Gans explains in his compelling book *The War Against the Poor*, the tendency to racialize the undeserving, the objects of society's contempt, is of long standing. The English poor laws of the mid-nineteenth century distinguished sharply between the shiftless and the working poor, associating the former with the Irish, who were regarded as an undesirable racial caste. In the United States, successive waves of immigrants have taken their place in the unholy category of the undeserving poor, including the Italians, the Slavs, and the Eastern European Jews who filled the boweries of the eastern seaboard in the early twentieth century.

But the Great Migration of African-Americans out of the rural South to the industrialized cities of the North in the twentieth century altered forever the racial dynamics of urban poverty. And urban poverty has long been far more visible, to the media and hence to the public at large, than its rural counterpart, however devastating the latter may be. Initially pulled by war-related job opportunities and pushed by agricultural mechanization that put a damper on southern demands for labor, blacks flooded into cities like Chicago, Milwaukee, Akron, New York, Pittsburgh, and elsewhere, only to be met by intense white hostility and overt policies of containment. They were confined in segregated communities that rapidly became overcrowded. . . .

As the jobs that initially sustained the northern migrants began to dry up, the consequences of extreme segregation became all too evident: places like the Cabrini-Green housing project were devastated and volatile. The underground economy and the welfare system were all that remained as forms of subsistence in some projects, and the association of African-Americans who were not gainfully employed with the damning label of "undeserving poor" took root. It was an easy linkage to make in a country obsessed by race.

It is also misleading because in fact the largest group of poor people in the United States are not those on welfare. They are the working poor whose earnings are so meager that despite their best efforts, they cannot afford decent housing, diets, health care, or child care. The debilitating conditions that impinge upon the working poor—substandard housing, crumbling schools, inaccessible health care—are hardly different from those that surround their nonworking counterparts. For many, indeed, these difficulties are measurably worse because to many of the working poor lack access to government supports that cushion those out of the labor force: subsidized housing, medical care, and food stamps.

DEFINING WORKING POVERTY

Debates over how to define poverty have raged for many years now, and not just because scholars love to argue: federal dollars ebb and flow depending on the definition. The technicalities of these debates need not concern us, but some of the principles that underlie them are relevant because they determine whose lives we are talking about when we use the term "working poor." . . .

We might want to think of the working poor as those employees who receive the minimum wage. This is hardly a perfect measure, for many of these minimum-wage earners are young people who live with their middle-class families. While their earnings are low, they do not live in poor households. Still, this is not universally true: only 36 percent of minimum-wage workers are teenagers; 42 percent are adults twenty-five years old and older. Some scholars dismiss the problems of minimum-wage workers because so many of them are women who work part-time (and are therefore presumed to be bringing in "pin money" rather than the necessities of life). While 65 percent of the nation's low-wage workers are women, the gender picture changes dramatically when we examine only *full-time*, year-round minimum wage workers: about 60 percent of them are men. . . .

When readers picture this low-wage workforce, they are likely to imagine hamburger flippers. Service workers of this kind are indeed among the working poor. Yet a surprisingly varied group of American employees take home wages low enough to pull their households below the poverty line. Household service workers (housekeepers, child care workers, and cooks) had a poverty rate of 21.8 percent during 1996. The rate for dental assistants, bartenders, hairdressers, and waitresses was 13.2 percent, and that for operators, fabricators, and laborers was 8 percent.

AGE, FAMILY STATUS, EDUCATION, AND RACE

With these definitions of the working poor in mind, we might ask who they are in a sociological sense: what kinds of people find themselves in this category? The short answer is that the nation's young, its single parents, the poorly educated, and minorities are more likely than other workers to be poor.

As America's youth enter an increasingly inhospitable labor market, they have found themselves at a disadvantage over those who preceded them in better years. Poverty rates for young workers have nearly doubled in the last twenty years. Family status has a powerful impact on the working poor as well. Not surprisingly, families in which both husband and wife work are the least likely to be poor. Working single mothers, on the other hand, are the most likely to be poor—the poverty rate of families supported by single mothers is almost four times that of married-couple families with at least one worker.

Single-parent families with mothers at the helm are almost twice as likely to be poor as families maintained solely by men, a reflection of the weak position of women in the labor market. But the predicament for single men should not be understated: 10 percent of America's working men who maintain families without a wife suffer poverty as well. Clearly, lacking a second earner—male or female—a low-wage worker is at great risk for falling below the poverty line.

We know that our changing economy favors the well-educated. Their incomes have risen while high school dropouts have seen a precipitous loss in their wages and their employment rates and a concomitant increase in their poverty. . .

Race complicates this picture. In an all-too-familiar story of America's racial divide, we discover that the white population has a significantly lower proportion of working poor (5.4 percent using the BLS* definition) than either African-Americans (13.3 percent) or Hispanics (14.6 percent). This is not just a matter of educational disadvantage concentrated among minorities: African-Americans in the labor force are more likely to be poor than their white counterparts at all levels of the educational continuum. . . .

. . . The clamor over welfare reform that crested in 1996 further tightened the linkage in the public mind between the indigent poor and the whole concept of poverty. The . . . data presented here suggest that this is a relatively minor part of the problem. More important by far are the trends that have increased the size of the *working* population that cannot make ends meet and the millions of children growing up in their households.

THE WORKING POOR IN HARLEM

Central Harlem is one of the poorest parts of New York City. Nearly 30 percent of its households were on public assistance in 1990. Poverty stands at 40 percent. Harlem's walls are laden with graffiti, windows are frequently shattered, and housing projects—magnets for troublemakers—rise above the skyline and dominate the once-elegant brownstones that are now boarded up, their stoops crumbling in decay. Running through the middle of this enclave is a bustling commercial strip, 125th Street (Martin Luther King, Jr. Boulevard), that has a long and glorious history as the crucible of the Harlem Renaissance. The Apollo Theatre, still the grand center of Harlem culture, is on 125th Street, as is a large state office building, named after one of the most famous of the community's native sons, Adam Clayton Powell, Jr.

Harlem residents do much of their shopping along 125th Street, for it is there that the most accessible drugstores, clothing shops, furniture marts, and fast food restaurants are to be found. To buy anything at a fancier store

---

*Editors' note: Bureau of Labor Statistics.

requires a subway trip or bus ride to another neighborhood. Shopping is therefore a community ritual, a constant public parade through the heart of America's best-known African-American enclave.

The main thoroughfare is also the place many Harlem residents turn to when they try to find work. Hence it is also the place where I began to look for the working-poor families. . . . I found them by way of four fast food restaurants that cater to the African-American and Latino clientele in the center of Harlem and on its periphery, where immigrant families from the Dominican Republic and Puerto Rico cluster. Beginning with fast food eateries was not an accident, for the stereotype of hamburger-flipping as the emblematic low-wage job turns out to be quite accurate. *More than one in every fifteen Americans has worked in the fast food industry.* Minority youth are a big part of this labor force: some one in eight of them has worked in the business. Hence the public image of the "McJob" accords fairly well with reality, for the fast food industry is a critical gateway into the labor market for thousands of people who live in inner-city neighborhoods. It is from this starting point that they hope to launch careers that will take them out of the minimum-wage bracket into a job that pays enough to sustain a family.

As is true in any study of a particular occupational group, fast food workers do not necessarily represent the entire universe of low-wage workers. One could just as usefully study people who scan in groceries in supermarkets or clerks in low-priced clothing shops. The question naturally arises, then, how far we can generalize from the experience of Harlem's fast food workers to the working poor elsewhere in the country. . . .

Race/Ethnicity

Nearly all of the residents of central Harlem are African-American. But the workforce we encountered in Harlem's fast food establishments was far more diverse: only slightly over half of its workers are black. A large group, nearly one-quarter, are immigrants who were either born in the Dominican Republic and immigrated to New York as children or first-generation Americans of Dominican parentage. They are members of one of the fastest growing groups of newcomers to New York. . . .

Sons and daughters of Dominican immigrants tend to live in enclaves dominated by friends, family members, and other co-ethnics, communities where most children begin life as monolingual speakers of Spanish. Eventually, of course, they learn the English language and American ways as they move through New York's school system and rub shoulders with outsiders who do not share their culture. Nonetheless, the identification they feel with their fellow immigrants—an identity reinforced by frequent visits and long stays in their "homeland"—is so strong that they think of themselves as Dominican even though many were born in New York.

Sixteen percent of the people whom we encountered in the restaurant workforce were Latinos from elsewhere in the Caribbean, principally Puerto Rico, which has for decades sent workers to the U.S. mainland for jobs, and pulled them back with family ties that run deep. Return migration, or, rather, circular patterns of migration, are so common in this community that a hybrid culture has developed: the Nuyoricans. Puerto Ricans are the largest Hispanic group in New York; one-half of all the Latinos in the city are of Puerto Rican origin. In fact, they are among the largest ethnic groups of any kind, numbering nearly a million and increasing at a rapid rate. . . .

Puerto Ricans and Dominicans represent the two largest groups of Latino immigrants in New York City. They are also among the poorest of the city's residents: Hispanics have median household incomes that are two-thirds of the median for whites or less, and their poverty rates are at least one and a half times greater than those for whites. Nuyoricans and Dominicans, who are important entrants to Harlem's low-wage economy, are also among the poorest of the city's Hispanics, falling well behind the smaller and more prosperous groups of Argentines, Panamanians, and Peruvians. . . .

. . . Although the two groups of Latinos are comparable in many respects (age, fertility, family structure), Puerto Ricans tend to have higher earnings. Dominicans in New York—and in my study—tend to make up for this disadvantage by increasing the number of people per household who are in the labor market. Both groups, however, are at great risk for joining the ranks of the working poor. In this they resemble the native-born African-Americans in Harlem, whose options are similarly limited by the pressures of racial discrimination, an imploding labor market, and a weak educational system that does little to position New York's poverty population for the good jobs that remain.

Age

We usually think of a fast food job as a youngster's first stepping-stone into the labor market, a part-time foray for pocket money and a forestate of an adult lifetime of full-time employment. In most of the nation's suburbs, and in some of its more affluent cities, that is exactly who occupies this employment niche. Not so in central Harlem. . . . While a significant number of high-school-age kids are working in these restaurants, most of the employees are considerably older. Seventy percent of the workers we interviewed are over nineteen. Thirty-five percent are well into adulthood, twenty-three years old and older. This stands in marked contrast to national figures on the fast food workforce. Nationwide, nearly three-quarters of the workers in this industry are twenty years old or younger, with more than one-quarter in the high-school-age group. No doubt the differences reflect the relative paucity of opportunities in a depressed economy like that of central Harlem. Older people have fewer better-paying prospects and are therefore "pushing down" into jobs that would qualify as entry-level way stations elsewhere.

When we look at the local age figures for African-Americans, the contrast with the national average-aged worker grows sharper. . . . One-half of the African-Americans in this workforce are twenty-three years old and older, reflecting once again the more restricted nature of the labor market opportunities open to them. Dominicans, by contrast, look more like the stereotypical young worker in a hamburger flipping job: half of them are in the youngest age group and only 16 percent are as old as the majority of black fast food workers.

Education

Fast food jobs are often thought to act as magnets, pulling young people away from school, distracting their attention from the kind of "human capital" investment that will pay off in the long run. . . . Contrary to prevailing stereotypes of fast food workers as high school dropouts—these inner-city workers are both quite strongly attached to the educational enterprise and better educated as a group than we might have expected. About 70 percent of the youngest workers (fifteen-to-eighteen-year-olds) are in a regular high school program, and another 10 percent are enrolled in an alternative high school. Eight percent are going to college. Less than 10 percent, most of whom are immigrant workers who completed their schooling in Latin America, are completely detached from the world of education. . . .

For many people in these age groups, education is an ongoing process that will continue for many years into adulthood. It is instructive, then, to consider what they hope to accomplish over the long run—what the *aspirations* of these low-wage workers look like. Perhaps in part because of their encounter with the raw world of low-wage employment and their desire to find something better over the long run, Burger Barn employees have very high expectations for their educational futures. Over half of the youngest workers expect to go on to college. African-Americans have higher educational goals than do their Latino counterparts, and part-time workers (many of whom are taking courses while working) expect to do better than those who work full-time.

Whether these aspirations will be realized or not is hard to forecast, but the data on the actual enrollment patterns of these fast food workers are not particularly encouraging. The older the workers, the less likely they are to be enrolled in courses of any kind, and most of those who are in school are taking vocational, technical, or para-professional courses, rather than liberal arts courses of the kind that is typical for the four-year institutions they aspire to attend. Older workers with adult responsibilities may find higher education out of reach.

Still, we see an extraordinary commitment to higher education in aspiration if not in practice, even among those fast food employees who are already quite a bit older than the normal middle-class college student. The numbers reflect what we found in the course of many months of conversation with Harlem's low-wage workers: an enduring belief in education as an essential credential for mobility in the labor market. This portrait could not possibly be farther from the stereotypical picture of the inner-city minority (usually described as a high school dropout with no appreciation of the value of education).

Gender

We have known for some time that low-wage work is often women's work. Nationally, women are concentrated in the jobs that cluster at the bottom of the income distribution. Fast food jobs, prototypical examples of low-wage work, are overwhelmingly held by females in the United States: in 1984, two-thirds of this labor force was female. Not so in Harlem. Nearly half of the workers in these ghetto restaurants were men. This suggests that men face a steeper uphill battle in finding better jobs in communities like Harlem, ultimately "slipping down" into sectors that, elsewhere, would be largely women's employment preserves.

Family/Household Structure

There are many ways to describe . . . Harlem's fast food labor force, and we have already encountered some—race, gender, and age being among the most common demographic descriptors. One other needs to be added, however, if we are to understand who these people are: family structure. In suburban America, we would expect to find that the fast food labor force was composed mainly of young people who are living at home with their parents and passing through this work experience as a way station to a more remunerative future. We already know that Harlem's restaurant workers are older. But what kinds of families do these low-wage workers come from, or more particularly, what do the households they live in look like?

Almost one-third of these Harlem workers live with a single parent (overwhelmingly a mother). Only 13 percent of these workers are children living in a nuclear household (with both parents). Over half of the workers we studied are over the age of eighteen, but still live with their parents or other relatives. These are young adults who cannot earn enough money to launch independent lives and are therefore forced to remain at home, where resources can be pooled and poverty managed collaboratively in the family.

More than one-third of these Harlem workers were parents trying to support families on the strength of these minimum-wage positions. Very few had children under the age of six, most likely because it is so hard to pay for child care out of wages this low. Those who did have young children were older (twenty-five to thirty-four) and worked full-time; one-third were trying to make ends meet alone, as single parents.

The literature on the urban underclass posits a physical separation of welfare recipients and working people, a separation that supposedly underwrites divergent cultures. Aid to Families with Dependent Children, we are told, spawned negative socialization patterns, most especially an unfamiliarity with (or lack of appreciation for) the world of work. When we look at real families, though, we find that what look like separate worlds to the Bureau of Labor Statistics are whole family units that combine work and welfare, and always have.

Twenty-nine percent of central Harlem's households received support from the welfare system in 1995. However, many of the same families also contained members who were in the labor force. Indeed, among restaurant workers, about one-quarter were living in households where someone was receiving AFDC income, though wages remain the overwhelming source of family support for virtually all of these households. Two in five of the Burger Barn workers are the only formally employed person in their household. This is more than designer sneakers and gold chains: the income these young workers receive is a critical source of support in poor households in Harlem.

Rather than paint welfare and work as different worlds, it makes far more sense to describe them as two halves of a single coin, as an integrated economic system at the very bottom of our social structure. Kyesha's family is a clear example of this fusion. Her mother needs the income her working child brings into the house; Kyesha needs the subsidies (housing, medical care, etc.) that state aid provides to her mother. Only because the two domains are linked can this family manage to make ends meet, and then just barely. Of course, many of these restaurant workers are in families that have no contact with the state welfare system at all. Instead, they are wage workers whose family members are also working for a living. Our understanding of their parents' occupational situations is complicated by the fractured family structure that so many are embedded in: 27 percent of their fathers are deceased, and 14 percent of their mothers as well. Since very few of the people we studied were over the age of forty, this suggests a pattern of early mortality among their parents, with all the difficulties this brings about in the lives of young people. Divorce and the incidence of never-married mothers produce enough distance between children and their fathers that many know only that their fathers are alive, and are not aware of what they do for a living.

Still, we do know something about the employment patterns of the parents of our restaurant workers. Over half of their mothers are working, with medical services—hospitals, home care agencies, and nursing homes—providing by far the largest source of employment. The health care industry was one of the few sectors that continued to grow in New York City throughout the relentless recession of the early 1990s; it absorbed many of the inner city's working poor.

What we know about the fathers of these fast food workers is probably less reliable than the information about mothers, simply because the ties between children and their fathers are more tenuous. Nonetheless, 46 percent of the fathers are reported by their sons and daughters to be employed. For many of them, we have no information about their occupation at all because their children do not know where they work. Those for whom we have information are employed as skilled craftsmen and transportation workers. The rest are scattered in janitorial services, factory labor, and retail trades. A number of them are retired and 10 percent are presently unemployed but once worked. . . .

Ongoing changes in the American economy have pulled the rug out from under the low-wage labor market. The continued fiscal instability of cities like New York is spurring cuts in public employment and critical services. Federal government retrenchment has reduced funding for everything from housing subsidies to Medicaid. These trends conspire to make the problems of the working poor more severe than they used to be. That's the bad news. The good news is that despite all of these difficulties, the nation's working poor continue to seek their salvation in the labor market. That such a commitment persists when the economic rewards are so minimal is testimony to the durability of the work ethic, to the powerful reach of mainstream American culture, which has always placed work at the center of our collective moral existence.

Reference

Gans, Herbert. 1995. *The War Against the Poor.* New York: Basic Books.

DOMÉSTICA

28

Pierrette Hondagneu-Sotelo

Particular regional formations have historically characterized the racialization of paid domestic work in the United States. Relationships between domestic employees and employers have always been imbued with racial meanings:

From Pierrette Hondagneu-Sotelo, *Doméstica: Immigrant Workers Caring and Cleaning in America* (Berkeley: University of California Press, 2001) pp. 13–22. Reprinted by permission of the University of California Press.

white "masters and mistresses" have been cast as pure and superior, and "maids and servants," drawn from specific racial-ethnic groups (varying by region), have been cast as dirty and socially inferior. The occupational racialization we see now in Los Angeles or New York City continues this American legacy, but it also draws to a much greater extent on globalization and immigration.

In the United States today, immigrant women from a few non-European nations are established as paid domestic workers. These women—who hail primarily from Mexico, Central America, and the Caribbean and who are perceived as "nonwhite" in Anglo-American contexts—hold various legal statuses. Some are legal permanent residents or naturalized U.S. citizens, many as beneficiaries of the 1986 Immigration Reform and Control Act's amnesty-legalization program. Central American women, most of whom entered the United States after the 1982 cutoff date for amnesty, did not qualify for legalization, so in the 1990s they generally either remained undocumented or held a series of temporary work permits, granted to delay their return to war-ravaged countries. Domestic workers who are working without papers clearly face extra burdens and risks: criminalization of employment, denial of social entitlements, and status as outlaws anywhere in the nation. If they complain about their jobs, they may be threatened with deportation. Undocumented immigrant workers, however, are not the only vulnerable ones. In the 1990s, even legal permanent residents and naturalized citizens saw their rights and privileges diminish, as campaigns against illegal immigration metastasized into more generalized xenophobic attacks on all immigrants, including those here with legal authorization. Immigration status has clearly become an important axis of inequality, one interwoven with relations of race, class, and gender, and it facilitates the exploitation of immigrant domestic workers.

Yet race and immigration are interacting in an important new way, which Latina immigrant domestic workers exemplify: their position as "foreigners" and "immigrants" allows employers, and the society at large, to perceive them as outsiders and thereby overlook the contemporary racialization of the occupation. Immigration does not trump race but, combined with the dominant ideology of a "color-blind" society, manages to shroud it.

With few exceptions, domestic work has always been reserved for poor women, for immigrant women, and for women of color; but over the last century, paid domestic workers have become more homogenous, reflecting the subordinations of both race and nationality/immigration status. In the late nineteenth century, this occupation was the most likely source of employment for U.S.-born women. In 1870, according to the historian David M. Katzman, two-thirds of all nonagricultural female wage earners worked as domestics in private homes. The proportion steadily declined to a little over one-third by 1900, and to one-fifth by 1930. Alternative employment opportunities for women expanded in the mid- and late twentieth century, so by 1990, fewer than 1 percent of employed American women were engaged in domestic work. Census figures, of course, are notoriously unreliable in documenting this increasingly undocumentable, "under-the-table" occupation, but the trend is clear: paid domestic work has gone from being *either* an immigrant woman's job *or* a minority woman's job to one that is now filled by women who, as Latina and Caribbean immigrants, embody subordinate status both racially and as immigrants.

Regional racializations of the occupation were already deeply marked in the late nineteenth and early twentieth centuries, as the occupation re-cruited women from subordinate racial-ethnic groups. In northeastern and midwestern cities of the late nineteenth century, single young Irish, German, and Scandinavian immigrants and women who had migrated from the coun-try to the city typically worked as live-in "domestic help," often leaving the occupation when they married. During this period, the Irish were the main target of xenophobic vilification. With the onset of World War I, European immigration declined and job opportunities in manufacturing opened up for whites, and black migration from the South enabled white employers to recruit black women for domestic jobs in the Northeast. Black women had always predominated as a servant caste in the South, whether in slavery or after, and by 1920 they constituted the single largest group in paid domestic work in both the South and the Northeast. Unlike European immigrant women, black women experienced neither individual nor intergenerational mobility out of the occupation, but they succeeded in transforming the occupation from one characterized by live-in arrangements, with no separation between work and social life, to live-out "day work"—a transformation aided by urbanization, new interurban transportation systems, and smaller urban residences.

In the Southwest and the West of the late nineteenth and early twentieth centuries, the occupation was filled with Mexican American and Mexican im-migrant women, as well as Asian, African American, and Native American women and, briefly, Asian men. Asian immigrant men were among the first recruits for domestic work in the West. California exceptionalism—its Anglo-American conquest from Mexico in 1848, its ensuing rapid development and overnight influx of Anglo settlers and miners, and its scarcity of women—initially created many domestic jobs in the northern part of the territory for Chinese "houseboys," laundrymen, and cooks, and later for Japanese men, followed by Japanese immigrant women and their U.S.-born daughters, the nisei, who remained in domestic work until World War II. Asian American women's experiences . . . provide an intermediate case of intergenerational mobility out of domestic work between that of black and Chicana women who found themselves, generation after generation, stuck in the occupational ghetto of domestic work and that of European immigrant women of the early twentieth century who quickly moved up the mobility ladder.

For Mexican American women and their daughters, domestic work became a dead-end job. From the 1880s until World War II, it provided the largest source of nonagricultural employment for Mexican and Chicana women throughout the Southwest. During this period, domestic vocational training schools, teaching manuals, and Americanization efforts deliberately channeled them into domestic jobs. Continuing well into the 1970s throughout the Southwest, and up to the present in particular regions, U.S.-born Mexican American women have worked as domestics. Over that time, the job has changed. Much as black women helped transform the domestic occupation from live-in to live-out work in the early twentieth century, Chicanas in the Southwest increasingly preferred contractual housecleaning work . . . to live-in or daily live-out domestic work.

While black women dominated the occupation throughout the nation during the 1950s and 1960s, there is strong evidence that many left it during the late 1960s. The 1970 census marked the first time that domestic work did not account for the largest segment of employed black women; and the proportion of black women in domestic work continued to drop dramatically in the 1970s and 1980s, falling from 16.4 percent in 1972 to 7.4 percent in 1980, then to 3.5 percent by the end of the 1980s. By opening up public-sector jobs to black women, the Civil Rights Act of 1964 made it possible for them to leave private domestic service. Consequently, both African American and Mexican American women moved into jobs from which they had been previously barred, as secretaries, sales clerks, and public-sector employees, and into the expanding number of relatively low-paid service jobs in convalescent homes, hospitals, cafeterias, and hotels.

These occupational adjustments and opportunities did not go unnoticed. In a 1973 *Los Angeles Times* article, a manager with thirty years of experience in domestic employment agencies reported, "Our Mexican girls are nice, but the blacks are hostile." Speaking very candidly about her contrasting perceptions of Latina immigrant and African American women domestic workers, she said of black women, "you can feel their anger. They would rather work at Grant's for $1.65 an hour than do housework. To them it denotes a lowering of self." By the 1970s black women in the occupation were growing older, and their daughters were refusing to take jobs imbued with servitude and racial subordination. Domestic work, with its historical legacy in slavery, was roundly rejected. Not only expanding job opportunities but also the black power movement, with its emphasis on self-determination and pride, dissuaded younger generations of African American women from entering domestic work.

It was at this moment that newspaper reports, census data, and anecdotal accounts first register the occupation's demographic shift toward Latina immigrants, a change especially pronounced in areas with high levels of Latino immigration. In Los Angeles, for example, the percentage of African

American women working as domestics in private households fell from 35 percent to 4 percent from 1970 to 1990, while foreign-born Latinas increased their representation from 9 percent to 68 percent. Again, since census counts routinely underestimate the poor and those who speak limited or no English, the women in this group may represent an even larger proportion of private domestic workers.

Ethnographic case studies conducted not only in Los Angeles but also in Washington, D.C., San Francisco, San Diego, Houston, El Paso, suburban areas of Long Island, and New York City provide many details about the experiences of Mexican, Caribbean, and Central American women who now predominate in these metropolitan centers as nanny/housekeepers and housecleaners. Like the black women who migrated from the rural South to northern cities in the early twentieth century, Latina immigrant women newly arrived in U.S. cities and suburbs in the 1970s, 1980s, and 1990s often started as live-ins, sometimes first performing unpaid household work for kin before taking on very low paying live-in jobs for other families. Live-in jobs, shunned by better-established immigrant women, appeal to new arrivals who want to minimize their living costs and begin sending their earnings home. Vibrant social networks channel Latina immigrants into these jobs, where the long hours and the social isolation can be overwhelming. As time passes, many of the women seek live-out domestic jobs. Despite the decline in live-in employment arrangements at the century's midpoint, the twentieth century ended in the United States much as it began, with a resurgence of live-in jobs filled by women of color—now Latina immigrants.

Two factors of the late twentieth century were especially important in creating this scenario. First, . . . globalization has promoted higher rates of immigration. The expansion of U.S. private investment and trade; the opening of U.S. multinational assembly plants (employing mostly women) along the U.S.-Mexico border and in Caribbean and Central American nations, facilitated by government legislative efforts such as the Border Industrialization Program, the North American Free Trade Agreement, and the Caribbean Basin Initiative; the spreading influence of U.S. mass media; and U.S. military aid in Central America have all helped rearrange local economies and stimulate U.S.-bound migration from the Caribbean, Mexico, and Central America. Women from these countries have entered the United States at a propitious time for families looking to employ housecleaners and nannies.

Second, increased immigration led to the racialized xenophobia of the 1990s. The rhetoric of these campaigns shifted focus, from attacking immigrants for lowering wages and competing for jobs to seeking to bar immigrants' access to social entitlements and welfare. In the 1990s, legislation codified this racialized nativism, in large part taking aim at women and children. In 1994 California's Proposition 187, targeting Latina immigrants and their children, won at the polls; and although its denial of all public education and of publicly funded health care was ruled unconstitutional by the courts, the vote helped usher in new federal legislation. In 1996 federal welfare reform, particularly the Immigration Reform Act and Individual Responsibility Act (IRAIRA), codified the legal and social disenfranchisement of legal permanent residents and undocumented immigrants. At the same time, language—and in particular the Spanish language—was becoming racialized; virulent "English Only" and anti-bilingual education campaigns and ballot initiatives spread.

Because Latina immigrants are disenfranchised as immigrants and foreigners, Americans can overlook the current racialization of the job. On the one hand, racial hostilities and fears may be lessened as increasing numbers of Latina and Caribbean nannies care for tow-headed children. As Sau-ling C. Wong suggests in an analysis of recent films, "in a society undergoing radical demographic and economic changes, the figure of the person of color patiently mothering white folks serves to allay racial anxieties."[1] Stereotypical images of Latinas as innately warm, loving, and caring certainly round out this picture. Yet on the other hand, the status of these Latinas as immigrants today serves to legitimize their social, economic, and political subordination and their disproportionate concentration in paid domestic work.

Such legitimation makes it possible to ignore American racism and discrimination. Thus the abuses that Latina domestic workers suffer in domestic jobs can be explained away because the women themselves are foreign and unassimilable. If they fail to realize the American Dream, according to this distorted narrative, it is because they are lazy and unmotivated or simply because they are "illegal" and do not merit equal opportunities with U.S.-born American citizens. Contemporary paid domestic work in the United States remains a job performed by women of color, by black and brown women from the Caribbean, Central America, and Mexico. This racialization of domestic work is masked by the ideology of "a color-blind society" and by the focus on immigrant "foreignness."

GLOBAL TRENDS IN PAID DOMESTIC WORK

Just as paid domestic work has expanded in the United States, so too it appears to have grown in many other postindustrial societies—in Canada and in parts of Europe—in the "newly industrialized countries" (NICs) of Asia, and in the oil-rich nations of the Middle East. Around the globe Caribbean, Mexican, Central American, Peruvian, Sri Lankan, Indonesian, Eastern European, and Filipina women—the latter in disproportionately great numbers—predominate in these jobs. Worldwide, paid domestic work continues its long legacy as a racialized and gendered occupation, but today divisions of nation and citizenship are increasingly salient. Rhacel Parreñas, who has studied Filipina domestic workers, refers to this development as the "international division of reproductive labor," and Anthony Richmond has called it part of a broad, new "global apartheid."[2]

. . . We must remember that the inequality of nations is a key factor in the globalization of contemporary paid domestic work. This inequality has had three results. First, around the globe, paid domestic work is increasingly performed by women who leave their own nations, their communities, and often their families of origin to do it. Second, the occupation draws not only women from the poor socioeconomic classes but also women of relatively high status in their own countries—countries that colonialism made much poorer than those countries where they go to do domestic work. Thus it is not unusual to find middle-class, college-educated women working in other nations as private domestic workers. Third, the development of service-based economies in postindustrial nations favors the international migration of women laborers. Unlike in earlier industrial eras, today the demand for gendered labor favors migrant women's services.

Nations use vastly different methods to "import" domestic workers from other countries. Some countries have developed highly regulated, government-operated, contract labor programs that have institutionalized both the recruitment and working conditions of migrant domestic workers. Canada and Hong Kong exemplify this approach. Since 1981 the Canadian federal government has formally recruited thousands of women to work as live-in nanny/housekeepers for Canadian families. Most come from third world countries, the majority in the 1980s from the Caribbean and in the 1990s from the Philippines; and once in Canada, they must remain in live-in domestic service for two years, until they obtain their landed immigrant status, the equivalent of the U.S. "green card." During this period, they must work in conditions reminiscent of formal indentured servitude and they may not quit their jobs or collectively organize to improve job conditions.

Similarly, since 1973 Hong Kong has relied on the formal recruitment of domestic workers, mostly Filipinas, to work on a full-time, live-in basis for Chinese families. Of the 150,000 foreign domestic workers in Hong Kong in 1995, 130,000 hailed from the Philippines, with smaller numbers drawn from Thailand, Indonesia, India, Sri Lanka, and Nepal. Just as it is now rare to find African American women employed in private domestic work in Los Angeles, so too have Chinese women vanished from the occupation in Hong Kong. . . .

In the larger global context, the United States remains distinctive, as it follows a more laissez-faire approach to incorporating immigrant women into paid domestic work. Unlike in Hong Kong and Canada, here there is no formal government system or policy to legally contract with foreign domestic workers. In the past, private employers in the United States were able to "sponsor" individual immigrant women working as domestics for their green cards, sometimes personally recruiting them while they were vacationing or working in foreign countries, but this route is unusual in Los Angeles today. For such labor certification, the sponsor must document that there is a shortage of labor able to perform a particular, specialized job—and in Los Angeles and many other parts of the country, demonstrating a shortage of domestic workers has become increasingly difficult. And it is apparently unnecessary, as the significant demand for domestic workers in the United States is largely filled not through formal channels but through informal recruitment from the growing number of Caribbean and Latina immigrant women who are already living (legally or illegally) in the United States. The Immigration and Naturalization Service, the federal agency charged with stopping illegal migration, has historically served the interest of domestic employers and winked at the hiring of undocumented immigrant women in private homes.

As we compare the hyperregulated employment systems in Hong Kong and Canada with the U.S. approach to domestic work, we must distinguish between the regulation of labor and the regulation of foreign domestic workers. . . . Here, the United States is again an exception. U.S. labor regulations *do* cover private domestic work—but no one knows about them. . . . Domestic workers' wages and hours are governed by state and federal law, and special regulations cover such details as limits on permissible deductions for breakage and for boarding costs of live-in workers. These regulations did not fall from the sky: they are the result of several important, historic campaigns organized by and for paid domestic workers. Most U.S. employers now know . . . about their obligations for employment taxes—though these obligations are still widely ignored—but few employers and perhaps fewer employees know about the labor laws pertaining to private domestic work. It's almost as though these regulations did not exist. At the same time, the United States does not maintain separate immigration policies for domestic workers, of the sort that mandate live-in employment or decree instant deportation if workers quit their jobs.

This duality has two consequences. On the one hand, both the absence of hyperregulation of domestic workers and the ignorance about existing labor laws further reinforce the belief that paid domestic work is not a real job. Domestic work remains an arrangement that is thought of as private: it remains informal, "in the shadows," and outside the purview of the state and other regulating agencies. On the other hand, the absence of state monitoring of domestic job contracts and of domestic workers' personal movement, privacy, and bodily adornment suggests an opening to upgrade domestic jobs in the United States. Unlike in Hong Kong and Canada, for example, where state regulations prevent Filipina domestic workers from quitting jobs that they find unsatisfactory or abusive, in Los Angeles, Latina immigrant domestic workers can . . . quit their jobs. Certainly they face limited options when they seek jobs outside of private homes, but it is important to note that they are not yoked by law to the same boss and the same job.

The absence of a neocolonialist, state-operated, contractual system for domestic work thus represents an opportunity to seek better job conditions. The chance of success might be improved if existing labor regulations were strengthened, if domestic workers were to work at collective organizing, and if informational and educational outreach to the domestic workers were undertaken. But to be effective, these efforts must occur in tandem with a new recognition that the relationships in paid domestic work are relations of employment.

NOTES

1. Sau-ling C. Wong. 1994. "Diverted Mothering: Representations of Caregivers of Color in the Age of Multiculturalism." pp. 67–91 in *Mothering: Ideology, Experience, and Agency*, edited by Evelyn Nakano Glenn, Grace Chang, and Linda Pennie Forcey. New York: Routledge, p. 69.

2. Rhacel Salazar Parreñas. 2000. "Migrant Filipina Domestic Workers and the International Division of Reproductive Labor." *Gender & Society* 14:560–580; Anthony Richmond. 1994. *Global Apartheid: Refugees, Racism, and the New World Order*. Toronto: Oxford University Press.

Families

OUR MOTHERS' GRIEF

Racial-Ethnic Women
and the Maintenance of Families

Bonnie Thornton Dill

29

REPRODUCTIVE LABOR[1] FOR WHITE WOMEN
IN EARLY AMERICA

In eighteenth- and nineteenth-century America, the lives of white[2] women in the United States were circumscribed within a legal and social system based on patriarchal authority. This authority took two forms: public and private. The social, legal, and economic position of women in this society was controlled through the private aspects of patriarchy and defined in terms of their relationship to families headed by men. The society was structured to confine white wives to reproductive labor within the domestic sphere. At the same time, the formation, preservation, and protection of families among white settlers was seen as crucial to the growth and development of American society. Building, maintaining, and supporting families was a concern of the State and of those organizations that prefigured the State. Thus, while white women had few legal rights as women, they were protected through public forms of patriarchy that acknowledged and supported their family roles of wives, mothers, and daughters because they were vital instruments for building American society. . . .

From *Journal of Family History* 13 (1988): 415–31. Reprinted by permission.

In colonial America, white women were seen as vital contributors to the stabilization and growth of society. They were therefore accorded some legal and economic recognition through a patriarchal family structure. . . .

Throughout the colonial period, women's reproductive labor in the family was an integral part of the daily operation of small-scale family farms or artisan's shops. According to Kessler-Harris (1981), a gender-based division of labor was common, but not rigid. The participation of women in work that was essential to family survival reinforced the importance of their contributions to both the protection of the family and the growth of society.

Between the end of the eighteenth and mid-nineteenth century, what is labeled the "modern American family" developed. The growth of industrialization and an urban middle class, along with the accumulation of agrarian wealth among Southern planters, had two results that are particularly pertinent to this discussion. First, class differentiation increased and sharpened, and with it, distinctions in the content and nature of women's family lives. Second, the organization of industrial labor resulted in the separation of home and family and the assignment to women of a separate sphere of activity focused on childcare and home maintenance. Whereas men's activities became increasingly focused upon the industrial competitive sphere of work, "women's activities were increasingly confined to the care of children, the nurturing of the husband, and the physical maintenance of the home" (Degler 1980, p. 26).

This separate sphere of domesticity and piety became both an ideal for all white women as well as a source of important distinctions between them. As Matthaei (1982) points out, tied to the notion of wife as homemaker is a definition of masculinity in which the husband's successful role performance was measured by his ability to keep his wife in the homemaker role. The entry of white women into the labor force came to be linked with the husband's assumed inability to fulfill his provider role.

For wealthy and middle-class women, the growth of the domestic sphere offered a potential for creative development as homemakers and mothers. Given ample financial support from their husband's earnings, some of these women were able to concentrate their energies on the development and elaboration of the more intangible elements of this separate sphere. They were also able to hire other women to perform the daily tasks such as cleaning, laundry, cooking, and ironing. Kessler-Harris cautions, however, that the separation of productive labor from the home did not seriously diminish the amount of physical drudgery associated with housework, even for middle-class women. . . . In effect, household labor was transformed from economic productivity done by members of the family group to home maintenance; childcare and moral uplift done by an isolated woman who perhaps supervised some servants.

Working-class white women experienced this same transformation but their families' acceptance of the domestic code meant that their labor in the home intensified. Given the meager earnings of working-class men, working-class families had to develop alternative strategies to both survive and keep the wives at home. The result was that working-class women's reproductive labor increased to fill the gap between family need and family income. Women increased their own production of household goods through things such as canning and sewing; and by developing other sources of income, including boarders and homework. A final and very important source of other income was wages earned by the participation of sons and daughters in the labor force. In fact, Matthaei argues that "the domestic homemaking of married women was supported by the labors of their daughters" (1982, p. 130). . . .

Another way in which white women's family roles were socially acknowledged and protected was through the existence of a separate sphere for women. The code of domesticity, attainable for affluent women, became an ideal toward which nonaffluent women aspired. Notwithstanding the personal constraints placed on women's development, the notion of separate spheres promoted the growth and stability of family life among the white middle class and became the basis for working-class men's efforts to achieve a family wage, so that they could keep their wives at home. Also, women gained a distinct sphere of authority and expertise that yielded them special recognition.

During the eighteenth and nineteenth centuries, American society accorded considerable importance to the development and sustenance of European immigrant families. As primary laborers in the reproduction and maintenance of family life, women were acknowledged and accorded the privileges and protections deemed socially appropriate to their family roles. This argument acknowledges the fact that the family structure denied these women many rights and privileges and seriously constrained their individual growth and development. Because women gained social recognition primarily through their membership in families, their personal rights were few and privileges were subject to the will of the male head of the household. Nevertheless, the recognition of women's reproductive labor as an essential building block of the family, combined with a view of the family as the cornerstone of the nation, distinguished the experiences of the white, dominant culture from those of racial ethnics.

Thus, in its founding, American society initiated legal, economic, and social practices designed to promote the growth of family life among European colonists. The reception colonial families found in the United States contrasts sharply with the lack of attention given to the families of racial-ethnics. Although the presence of racial-ethnics was equally as important for the growth of the nation, their political, economic, legal, and social status was quite different.

REPRODUCTIVE LABOR AMONG RACIAL-ETHNICS IN EARLY AMERICA

Unlike white women, racial-ethnic women experienced the oppressions of a patriarchal society but were denied the protections and buffering of a patriarchal family. Their families suffered as a direct result of the organization of the labor systems in which they participated.

Racial-ethnics were brought to this country to meet the need for a cheap and exploitable labor force. Little attention was given to their family and community life except as it related to their economic productivity. Labor, and not the existence or maintenance of families, was the critical aspect of their role in building the nation. Thus they were denied the social structural supports necessary to make *their* families a vital element in the social order. Family membership was not a key means of access to participation in the wider society. The lack of social, legal, and economic support for racial-ethnic families intensified and extended women's reproductive labor, created tensions and strains in family relationships, and set the stage for a variety of creative and adaptive forms of resistance.

AFRICAN-AMERICAN SLAVES

Among students of slavery, there has been considerable debate over the relative "harshness" of American slavery, and the degree to which slaves were permitted or encouraged to form families. It is generally acknowledged that many slaveowners found it economically advantageous to encourage family formation as a way of reproducing and perpetuating the slave labor force. This became increasingly true after 1807 when the importation of African slaves was explicitly prohibited. The existence of these families and many aspects of their functioning, however, were directly controlled by the master. In other words, slaves married and formed families but these groupings were completely subject to the master's decision to let them remain intact. One study has estimated that about 32 percent of all recorded slave marriages were disrupted by sale, about 45 percent by death of a spouse, about 10 percent by choice, with the remaining 13 percent not disrupted at all (Blassingame 1972, pp. 90–92). African slaves thus quickly learned that they had a limited degree of control over the formation and maintenance of their marriages and could not be assured of keeping their children with them. The threat of disruption was perhaps the most direct and pervasive cultural assault[3]on families that slaves encountered. Yet there were a number of other aspects of the slave system which reinforced the precariousness of slave family life.

In contrast to some African traditions and the Euro-American patterns of the period, slave men were not the main provider or authority figure in the family. The mother-child tie was basic and of greatest interest to the slave-owner because it was critical in the reproduction of the labor force.

In addition to the lack of authority and economic autonomy experienced by the husband-father in the slave family, use of the rape of women slaves as a weapon of terror and control further undermined the integrity of the slave family. . . . The slave family, therefore, was at the heart of a peculiar tension in the master-slave relationship. On the one hand, slaveowners sought to encourage familial ties among slaves because, as Matthaei (1982) states: ". . . these provided the basis of the development of the slave into a self-conscious socialized human being" (p. 81). They also hoped and believed that this socialization process would help children learn to accept their place in society as slaves. Yet the master's need to control and intervene in the familial life of the slaves is indicative of the other side of this tension. Family ties had the potential for becoming a competing and more potent source of allegiance than the slavemaster himself. Also, kin were as likely to socialize children in forms of resistance as in acts of compliance.

It was within this context of surveillance, assault, and ambivalence that slave women's reproductive labor took place. She and her menfolk had the task of preserving the human and family ties that could ultimately give them a reason for living. They had to socialize their children to believe in the possibility of a life in which they were not enslaved. The slave woman's labor on behalf of the family was, as Davis (1971) has pointed out, the only labor the slave engaged in that could not be directly appropriated by the slaveowner for his own profit. Yet, its indirect appropriation, as labor crucial to the reproduction of the slaveowner's labor force, was the source of strong ambivalence for many slave women. Whereas some mothers murdered their babies to keep them from being slaves, many sought within the family sphere a degree of autonomy and creativity denied them in other realms of the society. The maintenance of a distinct African-American culture is testimony to the ways in which slaves maintained a degree of cultural autonomy and resisted the creation of a slave family that only served the needs of the master.

Gutman (1976) provides evidence of the ways in which slaves expressed a unique Afro-American culture through their family practices. He provides data on naming patterns and kinship ties among slaves that flies in the face of the dominant ideology of the period. That ideology argued that slaves were immoral and had little concern for or appreciation of family life.

Yet Gutman demonstrated that within a system which denied the father authority over his family, slave boys were frequently named after their fathers, and many children were named after blood relatives as a way of maintaining family ties. Gutman also suggested that after emancipation a number of slaves took the names of former owners in order to reestablish family ties that had been disrupted earlier. On plantation after plantation, Gutman found considerable evidence of the building and maintenance of extensive kinship ties among slaves. In instances where slave families had been disrupted, slaves in new communities reconstituted the kinds of family and kin ties that came to characterize Black family life throughout the South. These patterns included, but were not limited to, a belief in the importance of marriage as a long-term commitment, rules of exogamy that included marriage between first cousins, and acceptance of women who had children outside of marriage. Kinship networks were an important source of resistance to the organization of labor that treated the individual slave, and not the family, as the unit of labor (Caulfield 1974).

Another interesting indicator of the slaves' maintenance of some degree of cultural autonomy has been pointed out by Wright (1981) in her discussion of slave housing. Until the early 1800s, slaves were often permitted to build their housing according to their own design and taste. During that period, housing built in an African style was quite common in the slave quarters. By 1830, however, slaveowners had begun to control the design and arrangement of slave housing and had introduced a degree of conformity and regularity to it that left little room for the slave's personalization of the home. Nevertheless, slaves did use some of their own techniques in construction and often hid it from their masters. . . .

Housing is important in discussions of family because its design reflects sociocultural attitudes about family life. The housing that slaveowners provided for their slaves reflected a view of Black family life consistent with the stereotypes of the period. While the existence of slave families was acknowledged, it certainly was not nurtured. Thus, cabins were crowded, often containing more than one family, and there were no provisions for privacy. Slaves had to create their own. . . .

Perhaps most critical in developing an understanding of slave women's reproductive labor is the gender-based division of labor in the domestic sphere. The organization of slave labor enforced considerable equality among men and women. The ways in which equality in the labor force was translated into the family sphere is somewhat speculative. . . .

. . . We know, for example, that slave women experienced what has recently been called the "double day" before most other women in this society. Slave narratives (Jones 1985; White 1985; Blassingame 1977) reveal that women had primary responsibility for their family's domestic chores. They cooked (although on some plantations meals were prepared for all of the slaves), sewed, cared for their children, and cleaned house, all after completing a full day of labor for the master. Blassingame (1972) and others have pointed out that slave men engaged in hunting, trapping, perhaps some gardening, and furniture making as ways of contributing to the maintenance of their families. Clearly, a gender-based division of labor did exist within the family and it appears that women bore the larger share of the burden for housekeeping and child care. . . .

Black men were denied the male resources of a patriarchal society and therefore were unable to turn gender distinctions into female subordination, even if that had been their desire. Black women, on the other hand, were denied support and protection for their roles as mothers and wives and thus had to modify and structure those roles around the demands of their labor. Thus, reproductive labor for slave women was intensified in several ways: by the demands of slave labor that forced them into the double-day of work; by the desire and need to maintain family ties in the face of a system that gave them only limited recognition; by the stresses of building a family with men who were denied the standard social privileges of manhood; and by the struggle to raise children who could survive in a hostile environment.

This intensification of reproductive labor made networks of kin and quasi-kin important instruments in carrying out the reproductive tasks of the slave community. Given an African cultural heritage where kinship ties formed the basis of social relations, it is not at all surprising that African American slaves developed an extensive system of kinship ties and obligations (Gutman 1976; Sudarkasa 1981). Research on Black families in slavery provides considerable documentation of participation of extended kin in child-rearing, child-birth, and other domestic, social, and economic activities (Gutman 1976; Blassingame 1972; Genovese and Miller 1974). . . .

With individual households, the gender-based division of labor experienced some important shifts during emancipation. In their first real opportunity to establish family life beyond the controls and constraints imposed by a slavemaster, family life among Black sharecroppers changed radically. Most women, at least those who were wives and daughters of able-bodied men, withdrew from field labor and concentrated on their domestic duties in the home. Husbands took primary responsibility for the fieldwork and for relations with the owners, such as signing contracts on behalf of the family. Black women were severely criticized by whites for removing themselves from field labor because they were seen to be aspiring to a model of womanhood that was considered inappropriate for them. This reorganization of female labor, however, represented an attempt on the part of Blacks to protect women from some of the abuses of the slave system and to thus secure their family life. It was more likely a response to the particular set of circumstances that the newly freed slaves faced than a reaction to the lives of their former masters. Jones (1985) argues that these patterns were "particularly significant" because at a time when industrial development was introducing a labor system that divided male and female labor, the freed Black family was establishing a pattern of joint work and complementary tasks between males and females that was reminiscent of the preindustrial American families. Unfortunately, these former slaves had to do this without the institutional supports that white farm families had in the midst of a sharecropping system that deprived them of economic independence.

CHINESE SOJOURNERS

An increase in the African slave population was a desired goal. Therefore, Africans were permitted and even encouraged at times to form families subject to the authority and whim of the master. By sharp contrast, Chinese people were explicitly denied the right to form families in the United States through both law and social practice. Although male laborers began coming to the United States in sizable numbers in the middle of the nineteenth century, it was more than a century before an appreciable number of children of Chinese parents were born in America. Tom, a respondent in Nee and Nee's (1973) book, *Longtime Californ'*, says: "One thing about Chinese men in America was you had to be either a merchant or a big gambler, have lot of side money to have a family here. A working man, an ordinary man, just, can't!" (p. 80).

Working in the United States was a means of gaining support for one's family with an end of obtaining sufficient capital to return to China and purchase land. The practice of sojourning was reinforced by laws preventing Chinese laborers from becoming citizens, and by restrictions on their entry into this country. Chinese laborers who arrived before 1882 could not bring their wives and were prevented by law from marrying whites. Thus, it is likely that the number of Chinese-American families might have been negligible had it not been for two things: the San Francisco earthquake and fire in 1906, which destroyed all municipal records; and the ingenuity and persistence of the Chinese people who used the opportunity created by the earthquake to increase their numbers in the United States. Since relatives of citizens were permitted entry, American-born Chinese (real and claimed) would visit China, report the birth of a son, and thus create an entry slot. Years later the slot could be used by a relative or purchased. The purchasers were called "paper sons." Paper sons became a major mechanism for increasing the Chinese population, but it was a slow process and the sojourner community remained predominantly male for decades.

The high concentration of males in the Chinese community before 1920 resulted in a split-household form of family. . . .

The women who were in the United States during this period consisted of a small number who were wives and daughters of merchants and a larger percentage who were prostitutes. Hirata (1979) has suggested that Chinese prostitution was an important element in helping to maintain the split-household family. In conjunction with laws prohibiting intermarriage, Chinese prostitution helped men avoid long-term relationships with women in the United States and ensured that the bulk of their meager earnings would continue to support the family at home.

The reproductive labor of Chinese women, therefore, took on two dimensions primarily because of the split-household family form. Wives who remained in China were forced to raise children and care for in-laws on the meager remittances of their sojourning husband. Although we know few details about their lives, it is clear that the everyday work of bearing and maintaining children and a household fell entirely on their shoulders. Those women who immigrated and worked as prostitutes performed the more nurturant aspects of reproductive labor, that is, providing emotional and sexual companionship for men who were far from home. Yet their role as prostitute was more likely a means of supporting their families at home in China than a chosen vocation.

The Chinese family system during the nineteenth century was a patriarchal one wherein girls had little value. In fact, they were considered only temporary members of their father's family because when they married, they became members of their husband's families. They also had little social value: girls were sold by some poor parents to work as prostitutes, concubines, or servants. This saved the family the expense of raising them, and their earnings also became a source of family income. For most girls, however, marriages were arranged and families sought useful connections through this process.

With the development of a sojourning pattern in the United States, some Chinese women in those regions of China where this pattern was more prevalent would be sold to become prostitutes in the United States. Most, however, were married off to men whom they saw only once or twice in the 20- or 30-year period during which he was sojourning in the United States. Her status as wife ensured that a portion of the meager wages he earned would be returned to his family in China. This arrangement required considerable sacrifice and adjustment on the part of wives who remained in China and those who joined their husbands after a long separation. . . .

Despite these handicaps, Chinese people collaborated to establish the opportunity to form families and settle in the United States. In some cases it took as long as three generations for a child to be born on United States soil. . . .

CHICANOS

Africans were uprooted from their native lands and encouraged to have families in order to increase the slave labor force. Chinese people were immigrant laborers whose "permanent" presence in the country was denied. By contrast, Mexican-Americans were colonized and their traditional family life was disrupted by war and the imposition of a new set of laws and conditions of labor. The hardships faced by Chicano families, therefore, were the result of the United States colonization of the indigenous Mexican population, accompanied by the beginnings of industrial development in the region. The treaty of Guadalupe Hidalgo, signed in 1848, granted American citizenship to Mexicans living in what is now called the Southwest. The American takeover, however, resulted in the gradual displacement of Mexicans from the land and their incorporation into a colonial labor force (Barrera 1979). In addition, Mexicans who immigrated into the United States after 1848 were also absorbed into the labor force.

Whether natives of Northern Mexico (which became the United States after 1848) or immigrants from Southern Mexico, Chicanos were a largely peasant population whose lives were defined by a feudal economy and a daily struggle on the land for economic survival. Patriarchal families were important instruments of community life and nuclear family units were linked together through an elaborate system of kinship and godparenting. Traditional life was characterized by hard work and a fairly distinct pattern of sex-role segregation. . . .

As the primary caretakers of hearth and home in a rural environment, *Las Chicanas* labor made a vital and important contribution to family survival. . . .

Although some scholars have argued that family rituals and community life showed little change before World War I (Saragoza 1983), the American conquest of Mexican lands, the introduction of a new system of labor, the loss of Mexican-owned land through the inability to document ownership, plus the transient nature of most of the jobs in which Chicanos were employed, resulted in the gradual erosion of this pastoral way of life. Families were uprooted as the economic basis for family life changed. Some immigrated from Mexico in search of a better standard of living and worked in the mines and railroads. Others who were native to the Southwest faced a job market that no longer required their skills and moved into mining, railroad, and agricultural labor in search of a means of earning a living. According to Camarillo (1979), the influx of Anglo[4] capital into the pastoral economy of Santa Barbara rendered obsolete the skills of many Chicano males who had worked as ranchhands and farmers prior to the urbanization of that economy. While some women and children accompanied their husbands to the railroad and mine camps, they often did so despite prohibitions against it. Initially many of these camps discouraged or prohibited family settlement.

The American period (post-1848) was characterized by considerable transiency for the Chicano population. Its impact on families is seen in the growth of female-headed households, which was reflected in the data as early as 1860. Griswold del Castillo (1979) found a sharp increase in female-headed households in Los Angeles, from a low of 13 percent in 1844 to 31 percent in 1880. Camarillo (1979, p. 120) documents a similar increase in Santa Barbara from 15 percent in 1844 to 30 percent by 1880. These increases appear to be due not so much to divorce, which was infrequent in this Catholic population, but to widowhood and temporary abandonment in search of work. Given the hazardous nature of work in the mines and railroad camps, the death of a husband, father or son who was laboring in these sites was not uncommon. Griswold del Castillo (1979) reports a higher death rate among men than women in Los Angeles. The rise in female-headed households, therefore, reflects the instabilities and insecurities introduced into women's lives as a result of the changing social organization of work.

One outcome, the increasing participation of women and children in the labor force was primarily a response to economic factors that required the modification of traditional values. . . .

Slowly, entire families were encouraged to go to railroad workcamps and were eventually incorporated into the agricultural labor market. This was a response both to the extremely low wages paid to Chicano laborers and to the preferences of employers who saw family labor as a way of stabilizing the workforce. For Chicanos, engaging all family members in agricultural work was a means of increasing their earnings to a level close to subsistence for the entire group and of keeping the family unit together. . . .

While the extended family has remained an important element of Chicano life, it was eroded in the American period in several ways. Griswold del Castillo (1979), for example, points out that in 1845 about 71 percent of Angelenos lived in extended families and that by 1880, fewer than half did. This decrease in extended families appears to be a response to the changed economic conditions and to the instabilities generated by the new sociopolitical structure. Additionally, the imposition of American law and custom ignored and ultimately undermined some aspects of the extended family. The extended family in traditional Mexican life consisted of an important set of familial, religious, and community obligations. Women, while valued primarily for their domesticity, had certain legal and property rights that acknowledged the importance of their work, their families of origin and their children. . . .

In the face of the legal, social, and economic changes that occurred during the American period, Chicanas were forced to cope with a series of dislocations in traditional life. They were caught between conflicting pressures to maintain traditional women's roles and family customs and the need to participate in the economic support of their families by working outside the home. During this period the preservation of some traditional customs became an important force for resisting complete disarray. . . .

Of vital importance to the integrity of traditional culture was the perpetuation of the Spanish language. Factors that aided in the maintenance of other aspects of Mexican culture also helped in sustaining the language. However, entry into English-language public schools introduced the children and their families to systematic efforts to erase their native tongue. . . .

Another key factor in conserving Chicano culture was the extended family network, particularly the system of *compadrazgo* or godparenting. Although the full extent of the impact of the American period on the Chicano extended family is not known, it is generally acknowledged that this family system, though lacking many legal and social sanctions, played an important role in the preservation of the Mexican community (Camarillo 1979, p. 13). In Mexican society, godparents were an important way of linking family and community through respected friends or authorities. Named at the important rites of passage in a child's life, such as birth, confirmation, first communion, and marriage, *compadrazgo* created a moral obligation for godparents to act as guardians, to provide financial assistance in times of need, and to substitute in case of the death of a parent. Camarillo (1979) points out that in traditional society these bonds cut across class and racial lines. . . .

The extended family network—which included godparents—expanded the support groups for women who were widowed or temporarily abandoned and for those who were in seasonal, part-, or full-time work. It suggests, therefore, the potential for an exchange of services among poor people whose income did not provide the basis for family subsistence. . . . This family form is important to the continued cultural autonomy of the Chicano community.

CONCLUSION: OUR MOTHERS' GRIEF

Reproductive labor for Afro-American, Chinese-American, and Mexican-American women in the nineteenth century centered on the struggle to maintain family units in the face of a variety of cultural assaults. Treated primarily as individual units of labor rather than as members of family groups, these women labored to maintain, sustain, stabilize, and reproduce their families while working in both the public (productive) and private (reproductive) spheres. Thus, the concept of reproductive labor, when applied to women of color, must be modified to account for the fact that labor in the productive sphere was required to achieve even minimal levels of family subsistence. Long after industrialization had begun to reshape family roles among middle-class white families, driving white women into a cult of domesticity, women of color were coping with an extended day. This day included subsistence labor outside the family and domestic labor within the family. For slaves, domestics, migrant farm laborers, seasonal factory-workers, and prostitutes, the distinctions between labor that reproduced family life and which economically sustained it were minimized. The expanded workday was one of the primary ways in which reproductive labor increased.

Racial-ethnic families were sustained and maintained in the face of various forms of disruption. Yet racial-ethnic women and their families paid a high price in the process. High rates of infant mortality, a shortened life span, the early onset of crippling and debilitating disease provided some insight into the costs of survival.

The poor quality of housing and the neglect of communities further increased reproductive labor. Not only did racial-ethnic women work hard outside the home for a mere subsistence, they worked very hard inside the home to achieve even minimal standards of privacy and cleanliness. They were continually faced with disease and illness that directly resulted from the absence of basic sanitation. The fact that some African women murdered their children to prevent them from becoming slaves is an indication of the emotional strain associated with bearing and raising children while participating in the colonial labor system.

We have uncovered little information about the use of birth control, the prevalence of infanticide, or the motivations that may have generated these or other behaviors. We can surmise, however, that no matter how much children were accepted, loved, or valued among any of these groups of people, their futures in a colonial labor system were a source of grief for their mothers. For those children who were born, the task of keeping them alive, of helping them to understand and participate in a system that exploited them, and the challenge of maintaining a measure—no matter how small—of cultural integrity, intensified reproductive labor.

Being a racial-ethnic woman in nineteenth-century American society meant having extra work both inside and outside the home. It meant having a contradictory relationship to the norms and values about women that were being generated in the dominant white culture. As pointed out earlier, the notion of separate spheres of male and female labor had contradictory outcomes for the nineteenth-century whites. It was the basis for the confinement of women to the household and for much of the protective legislation that subsequently developed. At the same time, it sustained white families by providing social acknowledgment and support to women in the performance of their family roles. For racial-ethnic women, however, the notion of separate spheres served to reinforce their subordinate status and became, in effect, another assault. As they increased their work outside the home, they were forced into a productive labor sphere that was organized for men and "desperate" women who were so unfortunate or immoral that they could not confine their work to the domestic sphere. In the productive sphere, racial-ethnic women faced exploitative jobs and depressed wages. In the reproductive sphere, however, they were denied the opportunity to embrace the dominant ideological definition of "good" wife or mother. In essence, they were faced with a double-bind situation, one that required their participation in the labor force to sustain family life but damned them as women, wives, and mothers because they did not confine their labor to the home. Thus, the conflict between ideology and reality in the lives of racial-ethnic women during the nineteenth century sets the stage for stereotypes, issues of self-esteem, and conflicts around gender-role prescriptions that surface more fully in the twentieth century. Further, the tensions and conflicts that characterized their lives during this period provided the impulse for community activism to jointly address the inequities, which they and their children and families faced.

NOTES

1. The term *reproductive labor* is used to refer to all of the work of women in the home. This includes but is not limited to: the buying and preparation of food and clothing, provision of emotional support and nurturance for all family members, bearing children, and planning, organizing, and carrying out a wide variety of tasks associated with their socialization. All of these activities are necessary for the growth of patriarchal capitalism because they maintain, sustain, stabilize, and *reproduce* (both biologically and socially) the labor force.

2. The term *white* is a global construct used to characterize peoples of European descent who migrated to and helped colonize America. In the seventeenth century, most of these immigrants were from the British Isles. However, during the time period covered by this article, European immigrants became increasingly diverse. It is a limitation of this article that time and space does not permit a fuller discussion of the variations in the white European immigrant experience. For the purposes of the argument made herein and of the contrast it seeks to draw between the experiences of mainstream (European) cultural groups and that of racial/ethnic minorities, the differences among European settlers are joined and the broad similarities emphasized.

3. Cultural assaults, according to Caulfield (1974), are benign and systematic attacks on the institutions and forms of social organization that are fundamental to the maintenance and flourishing of a group's culture.

4. This term is used to refer to white Americans of European ancestry.

References

Barrera, Mario. 1979. *Race and Class in the Southwest.* South Bend, IN: Notre Dame University Press.

Blassingame, John. 1972. *The Slave Community: Plantation Life in the Antebellum South.* New York: Oxford University Press.

————. 1977. *Slave Testimony: Two Centuries of Letters, Speeches, Interviews, and Autobiographies.* Baton Rouge, LA: Louisiana State University Press.

Camarillo, Albert. 1979. *Chicanos in a Changing Society.* Cambridge, MA: Harvard University Press.

Caulfield, Mina Davis. 1974. "Imperialism, the Family, and Cultures of Resistance." *Socialist Review* 4(2)(October): 67–85.

Davis, Angela. 1971. "The Black Woman's Role in the Community of Slaves." *Black Scholar* 3(4)(December): 2–15.

Degler, Carl. 1980. *At Odds.* New York: Oxford University Press.

Genovese, Eugene D., and Elinor Miller, eds. 1974. *Plantation, Town, and County: Essays on the Local History of American Slave Society.* Urbana: University of Illinois Press.

Griswold del Castillo, Richard. 1979. *The Los Angeles Barrio: 1850–1890.* Los Angeles: The University of California Press.

Gutman, Herbert. 1976. *The Black Family in Slavery and Freedom: 1750–1925.* New York: Pantheon.

Hirata, Lucie Cheng. 1979. "Free, Indentured, Enslaved: Chinese Prostitutes in Nineteenth-Century America." *Signs* 5 (Autumn): 3–29.

Jones, Jacqueline. 1985. *Labor of Love, Labor of Sorrow*. New York: Basic Books.

Kessler-Harris, Alice. 1981. *Women Have Always Worked*. Old Westbury: The Feminist Press.

Matthaei, Julie. 1982. *An Economic History of Women in America*. New York: Schocken Books.

Nee, Victor G., and Brett de Bary Nee. 1973. *Longtime Californ'*. New York: Pantheon Books.

Saragoza, Alex M. 1983. "The Conceptualization of the History of the Chicano Family: Work, Family, and Migration in Chicanos." Research Proceedings of the Symposium on Chicano Research and Public Policy. Stanford, CA: Stanford University, Center for Chicano Research.

Wright, Deborah Gray. 1985. *Ar'n't I a Woman?: Female Slaves in the Plantation South*. New York: W. W. Norton.

Wright, Gwendolyn. 1981. *Building the Dream: A Social History of Housing in America*. New York: Pantheon Books.

THE DIVERSITY OF AMERICAN FAMILIES   30

Eleanor Palo Stoller and Rose Campbell Gibson

Our understanding of family, perhaps more than any other institution, is clouded by myth. Masking diversity among contemporary U.S. families is an ideology that describes the American family as "one man who goes off to earn the bacon, one woman who waits at home to fry it, and roughly 2.5 children poised to eat it" (Cole 1986, p. 11). This image of family-dominated U.S. culture during the decades following World War II, a period during which the majority of older people today experienced at least part of their childrearing years. As described by Betty Friedan (1964) in *The Feminine Mystique*, this ideology told women that caring for their home and family was the path to fulfillment and gratification (Andersen 1997). The sociologist Talcott Parsons (1955) described the American family as an isolated, nuclear unit in which the husband, wife, and their dependent children live geographically and economically independent from other relatives. Families were viewed as safe havens from the competitiveness of the workplace, and both popular images and

From Eleanor Palo Stoller and Rose Campbell Gibson, eds., *Worlds of Difference: Inequality in the Aging Experience*, 3d ed. (Thousand Oaks, CA: Pine Forge Press, 2000). Reprinted by permission.

social scientific treatises emphasized the split between the public sphere of employment and the private sphere of home (Coontz 1992).

. . . We will explore the consequences of this idealized image of the American family on our efforts to understand the diversity of family experiences among older people today. Our exploration of older families will ask these questions:

1. How does the myth of the American family mask the diversity of families in which today's cohorts of elderly people grew up and grew old?
2. How do interlocking hierarchies based on gender, race, and class structure the family experiences of today's older Americans?
3. How do definitions of family based on the experiences of the dominant groups in society bias research on other older families?

HOW THE MYTH OF THE AMERICAN FAMILY MASKS THE DIVERSITY OF FAMILIES IN WHICH TODAY'S COHORTS OF ELDERLY PEOPLE GREW UP AND GREW OLD

There are a number of problems with the ideological view of the nuclear family as a key to understanding the family experiences of older people today. First, not all older people married, and not all married people had children. Of the people who are 65 and older today, 5% never married, 6% divorced or separated, and 20% do not have adult children. Marriage and divorce rates also vary by gender and ethnicity. Older American Indian and Asian American women are slightly less likely to have remained single throughout their lives than are other older people. Rates of divorce and separation for elderly Blacks, Hispanic Americans, American Indians, and Asian Americans are higher than rates for elderly Whites.

Second, the idealized view of the family glorified the role of the full-time housewife, wife, and mother. Married women who did not enter the paid labor force and worked full time at home often found that their lives did not correspond to the image fostered by popular culture. Many full-time mothers were troubled by the isolation of their daily lives, the repetitiveness of household tasks, and the invisibility of their accomplishments (Oakley 1974). Betty Friedan (1964) labeled the vague, unspoken anxiety experienced by the full-time housewives in her study as "the problem that has no name." According to popular ideology, these women should have been among the most happy and fulfilled. Their families came closest to the idealized American family. Yet Friedan's (1964) subjects shared doubts and dissatisfactions. Moreover, they felt responsible for this unease, because the ideology of the feminine mystique shifted the blame to women themselves if they were unsatisfied with their lives. Women were taught that

> [they] could desire no greater destiny than to glory in their femininity. . . .
> If a woman had a problem in the 1950s and 1960s, she knew that something
> must be wrong with her marriage and with herself. . . . What kind of woman
> was she if she did not feel this mysterious fulfillment waxing the kitchen floor?
> (1964, p. 19)

Not all women who opted for full-time work at home over paid work in the labor market were dissatisfied with their lives. Caring for a home and family offered opportunities for creativity and satisfaction. Because work at home provides freedom from supervision and some flexibility in organizing tasks, many women preferred it to the more alienating conditions they would have encountered in the paid positions available to them (Andersen 1997). Nevertheless, as the 1960s came to a close, the vague discontent described by Friedan began to crystallize into a feminist analysis that defined motherhood as a serious obstacle to women's fulfillment. This strain of feminist thought reflects the race and class position of its creators. In gathering data for *The Feminine Mystique*, Friedan interviewed her former classmates at Smith College, and her results reflect the experiences of these White, middle-class, college-educated women. bell hooks (1984) describes the bias of an analysis developed from the standpoint of this particular group:

> [They] argued that motherhood was a serious obstacle to women's liberation,
> a trap confining women to the home, keeping them tied to cleaning, cooking,
> and child care. . . . Had black women voiced their views on motherhood,
> [motherhood] would not have been named a serious obstacle to our freedom
> as women. Racism, availability of jobs, lack of skills or education, and a num-
> ber of other issues would have been at the top of the list—but not mother-
> hood. Black women would not have said motherhood prevented us from en-
> tering the world of paid work, because we have always worked. (p. 133)

. . . Not all of today's older women were able to choose whether or not they would stay out of the paid labor force when their children were young, despite the popular and scholarly rhetoric advocating full-time motherhood. Women rearing children alone because of single motherhood, divorce, or widowhood had little choice but to combine the nurturing roles assigned to mothers with the breadwinning responsibilities supposedly reserved for fathers. Furthermore, poor and working-class two-parent families often needed two incomes to survive. Women of color were more likely than White women to find themselves in this situation, as racial oppression denied many of these women sufficient economic resources to maintain nuclear family households (Collins 1990). Elderly Black women, for example, typically combined childrearing with paid work at a time when employment opportunities for women and for people of color were even more restricted than they are today.

The rigid gender-based division of labor that characterized popular constructions of the American family of the 1950s did not apply across all combinations of race and class. For example, the strict gender-role segregation within the domestic sphere applied more to White families than to Black families. Black couples have traditionally been more flexible in allocating family obligations, with parents sharing household tasks and childrearing as well as responsibility for earning a living (McAdoo 1986). Upper-class women, most of whom were White, retained personal responsibility for managing children and household, but rarely did the actual work themselves (Ostrander 1984). Rather, they supervised the work of paid domestic workers, who were usually poor or working-class women and often women of color (Cole 1986).

Upper-class women also had more resources with which to exercise their family responsibilities than less affluent women. Susan Ostrander's (1984) wealthy subjects wanted their children "to develop to their fullest potential, . . . to be 'the best'" (p. 75). They expressed concerns about their children's personal happiness and their future success. Most important, they had resources to facilitate these goals for their children. For example, if their children encountered academic difficulties or adjustment problems, the parents often sent them to private schools. Mrs. Miles, a woman now in her 70s, whose husband had been chairman of the board of a major bank, describes her decision to enroll her teenage daughter in a private boarding school: "She was a very shy child . . . so I thought . . . going away to school would be good for her, would give her some confidence and put her on her feet" (p. 85). As Ostrander explains, "Upper-class children . . . are not allowed to fail academically or personally. This gives them strong advantages" (p. 84). Parents from other social classes share many of the same dreams and concerns for their children expressed by the women Ostrander interviewed, but they do not share the same resources for helping their children realize these dreams.

Another aspect of the mythic American family that clouds our understanding of the family experiences of today's elderly people is the assumption of the nuclear family as isolated from other relatives. Contrary to this image of independent nuclear households, women have traditionally maintained family relationships across households through visiting, writing letters, organizing holiday gatherings, and remembering birthdays and anniversaries — activities that have been described as "kinship work" (DiLeonardo 1987; Rosenthal 1985). The work of "kinkeeping" can produce complex networks of female kin, who entrust confidences, pool resources, and share social activities. Furthermore, as social gerontologists have demonstrated, women are the primary caregivers of frail elderly people in their families.

Strong female-centered networks have also linked families and households among Blacks. Mothering responsibilities have traditionally been shared among women in Black communities. Boundaries distinguishing biological mothers from other women who care for children are less rigid within these communities, and women often feel a sense of obligation for "our children," a term that includes all of the children in their community. Patricia Hill Collins (1991) explains that

> Black communities recognized that vesting one person with full responsibility for mothering a child may not be wise or possible. As a result, "othermothers," women who assist bloodmothers by sharing mothering responsibilities, traditionally have been a central part of the institution of Black motherhood. (p. 47)

When children are orphaned, when parents are ill or at work, or when biological mothers are too young to care for their children alone, other women in the community take on child care responsibilities, sometimes temporarily but other times permanently.

These women-centered networks of bloodmothers and othermothers within Black communities have often been described as a reaction to the legacy of slavery and to generations of poverty and oppression. Some scholars have challenged this interpretation. The impact of racial oppression on Black families must be acknowledged, but it is also important to recognize the ways in which today's elderly Blacks evolved new definitions of family from their everyday experiences and cultural legacies. Networks of community-based child care, for example, provided supervision for children while biological parents worked to provide economic support. Alternative definitions of motherhood, emphasizing both emotional support and physical provision of care, were adapted from West African culture, which had been retained as a culture of resistance since slavery (Sudarkasa 1981). Building on this cultural heritage, Black women devised strategies for ensuring their children's physical survival within a racist society—ways of ensuring their physical survival without compromising their sense of self-worth (Collins 1991). The historian Elsa Barkley Brown (1991) describes this approach to mothering by explaining how her mother understood "the need to socialize me to live my life one way and, at the same time, to provide me with the tools I would need to live it differently" (p. 231). Like other Black women writing about family, Brown stresses the contributions of mothers and grandmothers in teaching the "delicate balance between conformity and resistance. . . . Black daughters must learn how to survive in interlocking structures of race, class, and gender oppression while rejecting and transcending these very same structures" (Collins 1991, p. 54).

Other examples also challenge the myth of self-reliance of isolated, nuclear households. Coontz (1992) points out that working-class and ethnic subcommunities "evolved mutual aid in finding jobs, surviving tough times, and pooling money for recreation" (p. 71). Immigrant groups developed infrastructures of churches, temperance societies, workers' associations, fraternal orders, and cooperatives that provided instrumental and emotional aid beyond the confines of the nuclear household. Godparenting created extrafamilial bonds among Catholic populations, bonds that often cut across social class boundaries, and as Coontz points out, "the notion of 'going for sisters' has long and still thriving roots in black communities" (p. 72). Extended networks of kin also characterize Latino families, in which child care responsibilities are often shared with older siblings, aunts, uncles, and grandparents. Ceremonial events frequently involve close friends linked to the family through *compadrazgo*, a network of kinlike ties among very close friends who exchange tangible assistance and social support (Heyck 1994).

The idealized image of the American family also assumes a heterosexual couple. Despite more recent attention to gay and lesbian rights, it is important to remember that today's cohorts of elderly gay men and lesbians grew up in a social environment in which they encountered strong pressures to hide their sexual orientation. They lived most of their lives during a historical period characterized by hostility toward homosexuality (Fullmer 1995). Many kept their relationships secret to avoid discriminatory treatment by their families, their employers, and their communities (D'Augelli & Hart 1987). Lesbians and gay men who revealed their relationships to others often needed to develop strategies for coping with negative attitudes and discriminatory treatment (Hooyman & Kiyak 1999). Indeed, there is some evidence that the experience of "coming out" strengthens the adaptive resources for meeting challenges of late life (Quan & Whitford 1992). Although marriage between gay and lesbian partners is not legally sanctioned, many couples establish long-term relationships (Peplau 1991). The absence of legal recognition creates a number of obstacles for these couples in late life. Many employers do not extend insurance or retirement benefits to unmarried partners, and social security does not extend survivor's benefits to gay men or lesbians whose partners have died. Although provision of domestic partnership benefits is increasing, these benefits were not available for current cohorts of gay or lesbian elders. Locating assisted living or nursing home environments receptive to gay or lesbian couples can also be difficult. When one partner is hospitalized, the other partner may be denied access to an intensive care unit or other contexts reserved for "next of kin." Without a medical power of attorney, hospital personnel do not recognize the authority of one partner to make health care decisions for his or her incapacitated partner. Gay or lesbian couples can try to protect each other's financial situations in their wills, but other family members sometimes contest the right of the surviving partner to an inheritance. . . .

Generalizing the image of the American family to all couples masks the complexity of family experiences and resources with which older Americans face the challenges of aging. Uncritically accepting dominant stereotypes of older families is like wearing blinders. Our understanding is limited, because we don't think to ask questions about dimensions of social reality that we overlook. We forget to ask how people who never married develop other relationships to meet their needs for companionship in old age. We forget to ask how gay and lesbian couples activate a support network to care for them when they are ill. We forget to ask how more fluid definitions of family and more flexible divisions of labor both within and beyond nuclear-family households influence strategies used by older Blacks in coping with disability in old age. These oversights are not intentional. They reflect limited views of social reality from particular standpoints. The poet Adrienne Rich (as quoted in West & Fenstermaker 1995) used the term *White solipsism* to refer to the process of thinking, imagining, and speaking "as if Whiteness described the world," resulting in a "tunnel-vision which simply does not see nonwhite experience or existence as precious or significant" (p. 306).

References

Andersen, M. L. (1997). *Thinking about women: Sociological perspectives on sex and gender* (4th ed.). Boston: Allyn & Bacon.

Brown, E. B. (1991). Mothers of mind. In P. Bell-Scott, B. Guy-Sheftall, J. Jones Royster, J. Sims-Wood, M. DeCosta-Willis, & L. Fultz (Eds.), *Double stitch: Black women write about mothers and daughters* (pp. 74–93). Boston: Beacon.

Cole, J. (1986). *All American woman: Lives that divide, ties that bind.* New York: Free Press.

Collins, P. H. (1990). *Black feminist thought: Knowledge, consciousness, and the politics of empowerment.* Boston: Unwin Hyman.

Collins, P. H. (1991). The meaning of motherhood in black culture and black mother-daughter relationships. In P. Bell-Scott, B. Guy-Sheftall, J. Jones Royster, J. Sims-Wood, M. DeCosta-Willis, & L. Fultz (Eds.), *Double stitch: Black women write about mothers and daughters* (pp. 42–60). Boston: Beacon.

Coontz, S. (1992). *The way we never were: American families and the nostalgia trap.* New York: Basic Books.

D'Augelli, A., & Hart, M. M. (1987). Gay women, men and families in rural settings: Toward the development of helping communities. *American Journal of Community Psychology, 15*(1), 79–93.

DiLeonardo, M. (1987). The female world of cards and holidays: Women, families, and the work of kinship. *Signs: Journal of Women in Culture and Society, 12*, 440–453.

Friedan, B. (1964). *The feminine mystique.* New York: Norton.

Fullmer, E. M. (1995). Challenging biases against families of older gays and lesbians. In G. C. Smith, S. Tobin, E. A. Robertson-Tchabo, & P. Power (Eds.), *Strengthening aging families: Diversity in practice and power.* Thousand Oaks, CA: Sage.

Heyck, D. L. D. (1994). *Barrios and borderlands: Cultures of Latinos and Latinas in the United States.* New York: Routledge.

hooks, b. (1984). *Feminist theory from margin to center.* Boston: South End Press.

Hooyman, N., & Kiyak, H. A. (1999). *Social gerontology: A multidisciplinary perspective* (5th ed.). Boston: Allyn & Bacon.

McAdoo, H. (1986). Societal stress: The black family. In J. Cole (Ed.), *All American women: Lines that divide, ties that bind* (pp. 187–197). New York: Free Press.

Oakley, A. (1974). *The sociology of housework*. Oxford: Martin Robertson.

Ostrander, S. (1984). Community volunteer. In *Women of the upper class* (pp. 111–139). Philadelphia: Temple University Press.

Parsons, T. (1955). The American family: Its relations to personality and the social structure. In T. Parsons & R. Bales (Eds.), *Family socialization and interaction process* (pp. 3–21). Glencoe, IL: Free Press.

Peplau, L. A. (1991). Lesbian and gay relationships. In J. C. Gonsiorek & J. D. Weinrich (Eds.), *Homosexuality: Research implications for public policy*. Newbury Park, CA: Sage.

Quan, J. K., & Whitford, G. (1992). Adaptation and age-related expectations of older gay and lesbian adults. *The Gerontologist, 32*, 367–374.

Rosenthal, C. J. (1985). Kinkeeping in the familial division of labor. *Journal of Marriage and the Family, 47*, 965–974.

Sudarkasa, N. (1981). Interpreting the African heritage in Afro-American family organizations. In H. McAdoo (Ed.), *Black families* (pp. 37–53). Beverly Hills, CA: Sage.

West, C., & Fenstermaker, S. (1995). Doing difference. *Gender & Society, 9*(1), 8–37.

COUNTERING THE CONSPIRACY TO IGNORE BLACK GIRLS

31

Robin D. G. Kelley

"Okay. Pretend I'm Kimberly and you're Tommy and, oh, you have to be Billy and Rocky, too. Anyway, pretend I'm the waitress in a restaurant and you see me for the first time and ask for the menu and then look up and then say, 'Kimberly, is that you?' Okay? and then you say, 'would you like to come to my pool party? All the Rangers will be there, and we'll have cake and ice cream.' And then Trini will get really mad because you didn't invite her and she'll be jealous, okay?"

This is how my five-year-old daughter, Elleza, imagines the lifestyles of *The Mighty Morphin Power Rangers*, a horrible, campy, excessively violent television show that has kids mesmerized and parents going broke on Power Rangers toys, clothes, and coloring books. We forbid her to watch the show and refuse to take her to see *Power Rangers—The Movie*, although she does own a couple of Power Rangers dolls and accessories. The Power Ranger craze is simply unavoidable; virtually every kid in the country has a general idea who these Morphin teens are and what color outfits each one wears.* However, Elleza has no detailed knowledge of the Power Rangers, besides the basic fact that this multicultural team of super adolescents fight "evil," derive their power from dinosaurs, and are experts in karate.

Because the Power Rangers are largely a mystery to her, she has to invent their world, their crises, their adventures. Although she is able to create fresh, highly imaginative tales about these dynamic teens dressed in colorful body suits, all of her narratives ultimately lead in the same direction: romance. Usually Billy, Tommy, Jason, or Rocky get in trouble, Kimberly comes to the rescue, and they get married. Occasionally Kimberly is in a bind and one of the boys saves her, but the outcome is usually the same—first love, then marriage, then Kimberly pushing a baby carriage. Romance is the dominant theme, yet in her mind the Power Rangers are not always teens; sometimes they're infants, other times they're toddlers or little kids who magically turn into adults with a snap of a finger. Those who are not romantically involved are usually siblings.

Of course, we try to do the P.C. thing. My wife, Diedra, and I encourage those narratives in which Kimberly rescues the boys rather than vice versa, and we constantly challenge her understanding of heterosexual relationships, which she derives from daily observation, daycare, fairy tales, Walt Disney, and perhaps a bit of instinct. We believe these are very serious issues, and we do the best we can to raise her with a healthy gender and sexual identity, to help her understand and challenge the sexist world she will inherit.

And yet I must confess that, for me at least, Elleza's style of play is far more enjoyable than being kicked in the face by an out-of-control "round-house" or karate-chopped in the stomach. Elleza is smart, strong, and outspoken when she's not feeing shy, but she is also strongly female-identified and takes pleasure in her girlhood. Yes, I know "girlhood" and "femininity" are social constructions, and we are always reminding her that boys are not necessarily stronger, faster, more heroic, better leaders, or that they should be barred from wearing dresses, and that girls can have short hair, tattoos, big muscles, positions of power, or wear short skirts without inviting sexual assault. But having witnessed my nephews and cousins in action, it is hard for me to accept the notion that aggressive behavior among boys is entirely a

---

*For the uninitiated, I should point out that in the early version of Power Rangers, Zack, the only black member, wore a black suit and Trini, the only other minority, wore yellow. Yes, you guessed it: she is Asian American.

product of socialization. Little boys, for example, will hurt themselves repeat-edly by jumping off furniture or running into walls. It's as if they cannot com-prehend the connection between their pain and their actions. On the other hand, most girls I've witnessed, Elleza included, only have to hurt themselves once in order to figure out that flying off the coffee table head first will result in excruciating pain. Indeed, Elleza was among that elite group of toddlers who came to that conclusion via a priori reasoning rather than experience.

Elleza has been described by friends as the classic girl, mainly because she tends to be nonviolent, cooperative, more verbal than physical, and deeply ro-mantic to the point of bordering on utopianism. She has an astounding grasp of language that allows her to employ complex analogies and similes as well as create stunning descriptions. When she was barely four years old, for ex-ample, she drew a small crowd at the Guggenheim Museum in SoHo when she started giving alternate names to works by Antoni Tapies, an abstract painter who usually titles his pieces by the materials or colors he uses. Elleza came up with evocative titles such as "Midnight Angel," "Rainy Night," "The Day the Sun Escaped," and "Peace."

Perhaps, as social psychologist Carol Gilligan suggests, this is the nature of girls' play.* But these were also the elements of my own boyhood I enjoyed most. My mother taught me to abhor violence, enjoy reading, and to use my powers of imagination to create a beautiful world rather than reproduce the reality surrounding us. I tried to do this while living on 156th Street between Amsterdam and Broadway in New York City, a neighborhood famous for male posturing and escalating violence. Although I spent a lot of time in the library, I made frequent visits to the local playgrounds and fire hydrants. Besides, my older sister's friends protected me from the baby thugs and saved me from the unmanly label of being a "pussy" ("sissy" simply wasn't in our vocabulary).

Yet, if the truth be told, I was never crazy about hanging out at the play-ground. Not that I disliked the playground: I still have fond memories of those hot summer days when I thought it couldn't get better than this, days when the music was blasting and everyone was laughing and joking on the sun-drenched concrete, stealing some shade on the other side of the handball court, using rocks, ballpoint pens, and matches to pop ten-cent rolls of "caps" (in our day, the phrase "bustin' caps" had a more benign meaning). But except for basketball and (later) tennis, I dreaded playing sports and felt extremely uncomfortable in gatherings of nappy-headed eight-year-old virgins bragging about the length of their "dicks" or which fourth-grader they claimed to have "fucked" in the coat room.

When I moved to Seattle to live with my father, "boy culture" (which some misguided social scientists wrongly label "street culture") was thrust

---

* See Carol Gilligan, *In a Different Voice: Psychological Theory and Women's Development* (Cambridge, Mass.: Harvard University Press, 1982).

upon me with a vengeance. My dad took it upon himself to turn me into a "man," by any means necessary. He insisted that I, not my sister, help him work on the car, he gave me clinics on "da mentals" of basketball (he took all the "fun" out of it), made me field fly balls, and sent me to a summer camp where boxing was strongly encouraged. My most traumatic memory was in the summer of 1974, when he signed me up to play for the Central Area Youth Association football team. I really didn't want to play football but was too afraid to tell him so. I went through the motions, expecting to get cut, but due to an unfortunate shortage of twelve-year-old players that year, everyone made the team. That the coming months would be spent in utter pain and agony hit me when my father took me to an athletic store to pick out my first jock strap and a protective cup.

To be fair, my father was not wholly responsible for my short-lived football career; I also succumbed to peer pressure from friends and relatives who also regarded organized contact sports as a male rite of passage. However, he did establish a climate at home in which becoming a man (i.e., conforming to established gender norms) and learning to respect authority was the raison d'etre for growing up. There was no give and take or open discussion about what we wanted to do or how we wanted to live our lives. He told us what to do and we did it, period. And like his father, he believed in discipline and followed the adage that to spare the rod (or the belt, the Hot Wheels track, the backhand slap) is to spoil the child. Rather than respect authority, we learned to fear it. Indeed, if I had to divide my life into neat historical epochs, I would call my five years in my father's house the "age of terror." To this day, I am afraid to tell him what I really think and dread the thought that he might accidentally discover this book while looking to buy me another one of those "inspirational" Christian fundamentalist paperbacks on how men should lead their families.

Although I lived with my father only a grand total of eight years,* I'm pretty sure his parenting style has had some impact on my own—what exactly, I'm not sure. Because I am raising a girl, most of his little maxims about what it means to be a man have no direct bearing on my parenting problems. On the other hand, it is entirely possible that his utter paranoia about making me a man, manifested in his extreme misogynist and homophobic views, rendered his version of masculinity less than stable, if not downright comical. In other words, maybe his obsession with me becoming a man (i.e., not gay, not weak, not "pussy-whipped") exposed deep insecurities about his own masculinity and therefore revealed just how precarious "manhood" really is. A simpler, though more likely explanation is that my father, through his own example, taught me how not to raise a child. Instead of teaching us to respect authority, for example, his punishments merely demonstrated to me the

---

*My mother and father split up when I was three years old. My sister and I ended up living with him and his second wife from 1971 to 1976, after which time we moved back with my mother.

futility of violence as a method of disciplining children. And the hierarchical manner in which he ran the household reinforced in my mind the limitations of patriarchy as a model for family life. I really do believe that my commitment to family based on shared authority, mutuality, and collective decision-making is largely an outgrowth of having lived—briefly—in a household where we were expected to follow orders and keep silent unless spoken to. I don't want my daughter to grow up like I did, feeling helpless and powerless, afraid to say what she thinks or fight for what she believes in. I don't want her to grow up in a family where men are the leaders, women and the children are followers, and all transgressors deviants who must be punished. Thus, long before I cracked open a book on the sociology of the family or on the case against patriarchy, I had concluded that my father's understanding of fatherhood was extremely dysfunctional.

Yet, within the current political discourse in African-American communities, which places a premium on saving black boys and puts the responsibility for doing so squarely on the shoulders of black men, my father should probably be regarded as a hero. According to certain popular schools of thought, the delicate balance of corporal punishment, steady work, family responsibilities, and father-son interaction on the playing field was the key to my success, not the "feminine" values my mother instilled into me when I lived with her.

I find this line of reasoning not only questionable but potentially damaging and shortsighted. It not only proposed a limited vision of masculinity, but more important, it signals that the only progressive, useful role black fathers ought to play is raising black boys. We see this discourse clear as day in films like John Singleton's *Boyz N the Hood* and in books such as Baba Zak A. Kondo's *For Homeboys Only: Arming and Strengthening Young Brothers for Black Manhood* and Jawanza Kunjufu's three-volume *Countering the Conspiracy to Destroy Black Boys.** We see it in the mushrooming of black male academies in innercity communities, where the curriculum places a heavy emphasis on promoting male role models in the classroom as well as in their reading assignments, preparing boys for "leadership" roles, and building students' self-esteem and sense of masculinity through discipline and responsibility.†

---

*Baba Zak A. Kondo, *For Homeboys Only: Arming and Strengthening Young Brothers for Black Manhood* (Washington, D.C.: Nubia Press, 1991); Jawanza Kunjufu, *Countering the Conspiracy to Destroy Black Boys* (Chicago: African American Images, 1985–86).

†Clearly, the purpose of these institutions is not to eliminate or reduce misogyny among young black males, for if that was the goal one might expect to find a curriculum that emphasizes women's history, female role models, and issues of gender and sexism. Ironically, one of the most vicious of these attacks on black women generally and feminism in particular was written by a woman: Shahrazad Ali. Titled *The Blackman's Guide to Understanding the Blackwoman* (1990), this highly controversial and vastly popular book caricatured black women as selfish, power-hungry, aggressive, manipulating, and even dirty. She argued that true liberation required that black women return to their traditional roles as child-care givers and supporters of black men.

Some of the same Afrocentric critics who support black male academies and insist that saving our sons and reestablishing black manhood is the key to liberation are also nudging women into more traditional roles or ignoring them altogether. Thus, while criticizing the media for what they believe are unduly negative images of black women, many contemporary black nationalists have invested more of their energy into reinstituting the traditional family and chastising black women for failing to support black men. Besides being anti-feminist, they tend to place the blame for the behavior of young black males squarely on the shoulders of single mothers, whom they characterize as irresponsible and incapable of disciplining their sons.

By implication, this limited focus on men raising boys not only renders my daughter invisible but makes me an inauthentic black father. There are far, far fewer guides on how black fathers ought to raise their daughters or popular texts on nurturing healthy, strong black girls. Likewise, books specifically for black mothers raising sons are also few and far between. And in my personal case, I'm worse than an inauthentic black father. I'm a traitor to my race and my gender. Why? Because I'm not raising a son, and I have no intention or desire to become a surrogate father to wayward boyz in the 'hood. Not that I'm suggesting that this sort of work isn't important. On the contrary, respect is due to the brothers out there trying to deal with these kids—boys who live their lives as if on a suicide mission.

Even if I had the desire and drive to take up the manly challenge of black fatherhood in the age of black male endangerment, I don't have a whole lot of confidence in my ability to bring a boy up in the manner to which our defenders of manhood are accustomed. I'm not good with my hands, I don't like baseball, fighting, or toy guns, and my wife can vouch for the fact that as a patriarch I couldn't rule myself out of a paper bag. Every time I'm asked to lecture at a boys detention center or neighborhood youth organization, the poets seem to identify with me much more than the regular homies. And during every family picnic hosted by my in-laws on Staten Island, I either end up playing with Elleza the whole time or swapping baby stories with Diedra's female cousins. I've never taken a seat inside the men's circle, where beer and sports are the order of the day. In short, to most black folks who subscribe to the notion that the role of black men is to save black boys, I'm probably not a very good male role model. Three decades ago, Senator Daniel Patrick Moynihan would have characterized me as one of those "dysfunctional" black men who lacked a healthy masculine self-image and needed to enlist in the army in order to attain one.

But the question remains: where is the concern for raising our daughters, and what are the issues and challenges fathers have to face? After all, even in today's world black women not only experience both race and gender discrimination but are potential victims of sexual harassment, rape, battery, incest, etc.—acts often committed by loved ones. Most African-American working women—not unlike other women—are still expected to put in double days, balancing their time between a full-time job, domestic work, and childcare. How should fathers help prepare their daughters for a world where black women are generally paid less than white men and white women, and in some circumstances less than black men, for the same job? How do we explain the underrepresentation of women in positions of authority, or help our daughters navigate a popular culture in which women are often portrayed as passive and weak or as sex objects for male gratification?

These are extremely critical issues whose political import ought to match that of, say, the rising black male incarceration rate in the U.S. Fathers (and mothers) simply cannot afford to ignore the complex problems confronting their offspring who are young, black, and female. Black girls, much more than boys, for example, constantly have to maintain their self-esteem and sense of beauty in the face of a racialist aesthetic that values white skin and straight hair. While this fact is certainly no revelation to African-Americans, we nonetheless unwittingly pass value-laden signals to our children about skin and hair. Besides the media, the point is sometimes made explicit by the way classmates decide who is ugly and who is beautiful, or family members making reference to "good hair" or to the "pretty babies" interracial couples supposedly produce. More often than not, the message is more subtle: a mother complaining about how difficult it is to do her daughter's hair, a father who is always admiring light-skinned women. Indeed, before Elleza turned three years old, she went through a brief phase when she wanted to be a little white girl. The phase passed quickly (at least for the time being), partly because we had always told her she was beautiful, partly because we never tried to compare her with anyone else. At the same time, we would not allow her to speak of someone else's physical appearance in a degrading fashion.

Battling the aesthetic assault on black girls and boosting their self-esteem in a racist world does carry a certain price. To be told incessantly that you're beautiful or cute or gorgeous while boys are told that they're strong, big, and smart reinforces the subordination of women while conveying the message that the only worthwhile possession girls have is their looks. With Elleza we try to teach her to love herself (including her body) while simultaneously playing down her appearance in order to emphasize her strength and intellect.

As most parents know, actions often speak louder than words. If we want to raise girls to break out of traditional gender conventions that reproduce female subordination and male dominance, dads need to be more cognizant of how our behavior and attitudes as men shape our children's understanding of the world. The way we speak to and interact with women and other men, the way decisions are made in the household, the way parenting responsibilities get defined—all of these factors affect our kids' identities. Daughters, for instance, need to see their fathers breaking traditional gender boundaries, but doing so without fanfare. Making a big deal out of transgressing the bounds of gender only reinforces the idea that traditional roles assigned to men and women are part of the "natural" order of things. The standard example is household work, although to make an impact men need to do at least half of all the work and should resist whining or making a big production out of it. Aside from household chores, dads should spend a lot more time with their daughters playing their games on *their* terms. Don't think for a minute that playing with dolls or makeup, hopscotch or jacks, ought to be Mom's domain, or that every Barbie needs a GI Joe. Those dads who feel uncomfortable playing girls' games ought to get over it or get some counseling. By refusing to play "girls' games," men and boys sometimes unwittingly degrade or ridicule girls' activities, which could have a damaging effect on their self-esteem. Besides, these intimate play moments are opportunities to listen to what your daughter has to say, to get a clearer sense of what she thinks about, and for her to know the many faces of Daddy.

Obviously, these brief and underdeveloped insights represent only the tip of an iceberg I'm only just beginning to explore myself. If there is any single lesson here, it is that arming our daughters for the twenty-first century is just as important as arming our sons. And the responsibility for doing so should not fall entirely on the shoulders of mothers (just as the task for turning boys into men should never be relegated to men alone). But what do I give my daughter to help prepare her for the future? What does she need to survive? Of course, there are the obvious things: unconditional love, affection, honesty, a sense of safety and security, independence, responsibility, and a right to her own private world. Hopefully, she'll leave the nest with enormous self-confidence; an unwavering love for herself as a woman and as an African-American; and a critical appreciation for her cultural heritage—again, as a woman, an African-descended person, and as a product of the modern world. I hope she grows up with a healthy and positive attitude toward sexuality as well as an analysis of society that will enable her to embrace and critically engage feminism(s). She needs to recognize that the sex-based limits placed in front of her are not natural but discriminatory barriers designed to maintain male dominance. She also must understand that these barriers are penetrable through struggle, and that a just society cannot exist until they are removed entirely. Finally, she should learn that valuing herself does not mean devaluing others, and that equality does not mean sameness.

Ironically, if Elleza ends up with a little brother I would probably arm him with the same principles. I will love him the same way I love his sister. I will flood him with affection, try to instill the same kinds of values, and place a great deal of emphasis on cooperation and nonviolence. He will learn to love his unique blackness, his cultural heritage with all of its multicultural roots, without apologetics or chauvinism. Better yet, through his big sister as well as his mother, he will learn to admire and take leadership from women, and vigilantly reject the idea that men are "superior" or that male dominance is the natural order of things. If we're successful, he too will grow up as a feminist. And like Elleza, I will teach him to love literature (from Toni Morrison to Oscar Ameringer), art (from Akan sculpture to Flemish painting), and music (from Monk to Debussy). If he wants a clinic in basketball or baseball, I'll certainly do everything in my power to encourage him. Given my limited athletic abilities, however, I'll have to send him to a summer program.

But if for some reason Power Rangers are still the rage by the time he hits three or four, I may have to bite the bullet and improve my karate skills.

RACIAL SAFETY AND CULTURAL MAINTENANCE

32

The Child Care Concerns of Employed Mothers of Color

Lynet Uttal

In the 1930s, doctors professed a single model of infant care that was promoted as superior to traditional ethnic infant care practices.[1] A similar movement is currently taking place in the 1990s in the field of paid child care services. Child care advocates are pressing for the professionalization of child care work and the practice of a single model of developmentally appropriate care. Underlying this proposal is the assumption that child rearing can be stripped of cultural values and practices, and that the type of care a child receives can be offered independent of the social and cultural location of the child's family.

This model ignores how membership in historically subordinated racial ethnic groups creates a different experience of child care for people of color than for the White population. Child care research has identified systematic differences in preferences by racial ethnic groups. One difference frequently noted is that African American parents view child care as an educational setting more so than do White parents. White parents, especially middle-class ones, are more likely to view child care as an opportunity for their children to have social interactions with other children.[2] African American parents express a greater preference for child care that provides structured academic programs for preschool-aged children, whereas middle-class White parents

From *Ethnic Studies Review* 19 (February 1996). Reprinted by permission.

prefer loosely structured activities that expose their children to different concepts through play. This high valuation of education is rooted in beliefs that early education will prepare children for kindergarten and create a stronger foundation for social mobility through education.[3]

One study found that African American parents expected the day care center's staff to be aware of and sensitive to racial issues and objected when the daycare center's programming violated this expectation.[4] In another study, Chinese American parents expressed concern about the conflicting messages children get when what is taught at home differs from what is taught at their daycares,[5] such as differing beliefs about how to address elders and differing eating practices (e.g., whether picking up a bowl and eating from it is acceptable). These concerns are important to take into account because early childhood education research has shown that presentations of positive ethnic images are important in the formulation of children's self-images and for the transmission of cultural values.[6] Yet, when child care advocates propose a single model of developmentally appropriate child care, they ignore the significance of membership in a historically subordinated racial ethnic group and cultural values in how child care arrangements are viewed.

Racial group membership and cultural practices are important because they create a lens, or historical consciousness, through which child care is assessed. Historical consciousness, according to poet and scholar Janice Gould, is the awareness of one's historical identity.[7] In her discussion of Native American women, Gould states that historical consciousness is the collective awareness of five hundred years of internal colonialism and genocide. Although individual Native American women come from many different tribes and lead very different lives, Gould argues that their historical consciousness informs how individual women live out their lives on a daily basis. Their historical consciousness reflects their social histories as members of particular gender, race, ethnic, and class groups.

The historical consciousness of their status as members of historically subordinated racial ethnic groups informs the types of concerns employed mothers of color have about leaving their children in other people's care. . . . I explore two expressions of this historical consciousness in employed mothers' views of their child care arrangements: racial safety and cultural maintenance. The introduction of these two concepts is central to the understanding of why parents of color seek out child care providers who are members of their own racial ethnic group and of how they view child care provided by persons who do not share their racial ethnic group membership or knowledge of their racial ethnic histories. Furthermore, this chapter not only identifies child care problems related to overt forms of racism but also discusses the problems that occur when well-intentioned White child care providers lack experience with caring for children of different racial ethnic groups and with negotiating multicultural interactions. . . .

THE CONCERN FOR RACIAL SAFETY

. . . Awareness of racism in U.S. society was a common topic when mothers of color talked about their child care arrangements. Because of their own experiences with racism, they were concerned about how their children would be treated when the child care providers were White. Often times, mothers discovered these problems only after they had established child care arrangements. For example, Gloria Thomas, an African American waitress and mother of two children, observed behaviors that she defined as racist. Gloria said:

> I don't know if she was used to [Black people]. I think she was kind of narrow-minded. I didn't feel comfortable, me being Black. [And] she looked like she put more energy into the White kids than the Black kids. I think she felt that I was on to her, because she said in a couple days, or actually I said, "this isn't going to work," and she pretty much knew also that it wasn't going to work.

Gloria expected White child care providers to have knowledge of how to negotiate cross-racial interactions. She said,

> If you are dealing with my kids, I hope you do have some cultural skills. I don't like prejudiceness at all. . . . You have to be not dumb. Some white people can be really stupid. They say the stupidest things.

When it was clear that child care providers lacked these skills, mothers removed their children from the child care setting. Frances Trudeau, an African American lawyer and the mother of two children, responded this way when the school's teachers and administrators at her five-year-old son's school failed to acknowledge and address that the namecalling and chasing of African American children was racism. When she and other parents spoke to the director, they were told that the school could not develop a policy to address the problem because families came from so many different walks of life and the school did not want to tell people how to behave. This response reduced cross-racial interactions to individual interactions and personal disagreements and failed to acknowledge the more systemic nature of racism. Mothers of color found this kind of response inadequate and frustrating. Because they are aware of the pervasiveness of the problem of racism, they do not define unpleasant cross-racial interactions as occasional, individual disagreements even in child care settings.

Mothers of color also experienced racism when they used predominantly White child care settings. Gloria Thomas described one such encounter.

> This one woman was pretty annoying. She asked me this question and to this day I still want to ask her what did she mean by it. She said, "Oh, are you a

single parent?" And I said, "yes." And she goes, "Oh, do you live around here?" And I said, "Yes, I live right around the corner." You could tell her mind was [thinking], "she goes to this day care? She's a single parent, Black, and she lives up here. How can she afford it?" It's really weird.

One of the strategies that mothers of color used to protect their children from racism was to find child care within their own racial ethnic communities. The use of kin and community networks protected the children and the mothers from having to deal with cross-racial interactions. When mothers of color were able to make child care arrangements with child care providers of their same racial ethnic group, the concerns about racial safety and cultural maintenance were eliminated. Yet, care within one's racial ethnic community did not guarantee a fit between the values and child rearing practices of mothers and child care providers. Often times, mothers had several relatives and acquaintances from which to choose. When this was the case, the mothers carefully discriminated between their choices based on what they considered to be a good environment and good care. After insuring their child's racial safety and exposure to traditional cultural practices and values, they invoked additional criteria to decide which child care setting was the best. Sylvia Rodriquez, an office manager and the mother of two children, chose a Mexican immigrant woman over her sister-in-law. She explained:

It depends on who the relatives are. Like for example, you know, financially [my husband's sister] could have used watch[ing for pay] my son and my daughter at her house. She's real good about feeding them and things like that. But she has a lot of marital problems that I wouldn't want my kids to be around, watching the arguments and fights. I know they use bad language and that's another thing I don't like.

Like Sylvia, Lupe Gonzalez, an administrative assistant and the mother of an eleven-month-old, was discriminating in terms of which relative she chose to watch her young baby. She had two options: an elderly grandmother and an aunt who was the same age as herself. She was pleased that her aunt was available to care for her baby, though she would have left her baby in her grandmother's care if necessary. But her grandmother was elderly and was already watching several other grandchildren. Because of her grandmother's age, Lupe felt that she would not be as attentive or as physically able to pick up her baby. The advantage of care by the aunt was that her infant son would also be the only child for whom the aunt provided care.

Although Gloria Thomas had left a White child care provider because she felt the White provider was unable to negotiate the cross-racial interactions, she found that simply finding an African American child care provider did not necessarily create satisfactory child care arrangements. Gloria had found a family daycare run by an African American woman, yet other factors prevented her from feeling comfortable with this arrangement. . . . Gloria and her provider had disagreements about what were appropriate disciplining practices. Gloria talked with her provider about these issues but felt that her preferences were not validated by the child care provider. She said:

> Well, I did, I said, I don't believe in hitting. And she said, "what do you mean by hitting." I said, "just swatting," and she said, "I do, you know, a slap on the hand." And I said, "pretty much even that, I don't want." But I could feel like that she didn't want to hear that.

Since Gloria also wanted something different from what she perceived as traditional African American child rearing practices, she moved her children into a day care center where she was the only African American parent, as well as of the lowest socioeconomic status and background. She often found herself irritated with what she perceived as a White style of interaction, yet she felt the social, educational, and environmental advantages of the day care center outweighed the need to have her child cared for by her previous African American child care provider.

Young mothers often opposed some of the traditional child rearing practices used by the older and more traditional women in their communities. For example, Maria Hernandez, a Mexican American office manager and the mother of a four year old, expressed dissatisfaction with the care provided by her Mexican American mother-in-law. She said:

> I don't really like the idea of them being yelled at or spanked. I think if there is a behavior problem, they should be able to tell [the parents] and for us to deal with it. Luckily, I have been in the situation where my kids is pretty mellow, but I've seen her spanking her other grandchildren. I wouldn't like that.

Occasionally, Maria would consider moving her child to a day care center. Yet, when Maria weighed out all factors (e.g., convenience, location, flexibility, cost, quality of care, being within the family for child care), she decided that this care by her mother-in-law was the best choice, in spite of the differences about disciplining practices. . . .

Another consideration was that simply being of the same race did not guarantee racial safety. Gloria felt that her African American child care provider was uncomfortable with the fact that her children's father was White. Being biracial located her children in a different category of race than simply "Black." Similarly, other mothers found that their searches for child care were complicated by having mixed-race children. Julie Lopez described how her background complicated her search for child care.

> I'm bilingual, but I'm not bicultural. My father was Black, my mother was White, my husband is Mexican. My child is half Mexican, Chicano. My grandparents

> are Jewish. We had all these different types of people all there and I picked
> parts of different cultures. . . . My child is going to get a different concept of
> different people. . . .

Several of the middle-class, predominantly White daycares had made a formal commitment to diversify their ethnic composition as well as to developing a multicultural curriculum. They offered full scholarships to children of color in order to diversify race and ethnicity. Yet, even when the day care center had a formal commitment to a multiculturalism, child care providers' behaviors and attitudes often demonstrated a lack of cultural competency that resulted in racially unsafe environments for children of color and their mothers. Racism ranged from outright hostile relations with child care staff and other parents at the day care to the interactions with well-intentioned White child care providers who lacked experience with caring for children of color and negotiating cross-cultural interactions. Aurora Garcia, a Mexican American mother, explained how this happens:

> They're all White, and they come from that perspective. . . . And they have
> blind spots. I don't know how else to put it. They're coming from their perspec-
> tives and their reality, their experiences, and so to change that, you have to
> ask them to. You have to help them do it, too.

And indeed, one of the consequences of being a parent of a child of color in a majority White daycare was the increased need for parental involvement. Aurora negotiated her child's racial safety by becoming, informally, the day care center's multicultural consultant. She intervened when the staff at her daughter's day care center did not interrupt behavior that was racist and stereotypical, such as when a White child pretended to be an Indian and came to school stereotypically dressed in feathers and headbands, wielding a toy tomahawk and whooping war cries. First, she brought to their awareness that certain behaviors and practices were racist and stereotypical. In the case of the White boy who came to school dressed as a stereotypical American Indian, she told them that she objected to the child's practice as well as to the staff's encouragement of it by painting stereotypical American Indian war paint on him. When the day care center was responsive to her concerns and asked her to work with them on it, she talked to the children and staff and recommended multicultural readings. Aurora gave the day care staff credit for trying to improve, but at the same time she was aware that cross-cultural gaffs were going to be a regular part of being in a predominantly White day care. She added:

> They are very actively trying to deal with some of these issues, and to me that
> felt good, culturally, you know. They made some boo-boos. [Like] at one point
> one of these teachers was talking to one of the [Latino] kids in Spanish, and

she said, "She's bilingual, right?" [She wasn't.] Then you have to decode what you are and [let them know that] not all Chicanos speak Spanish. So, on one level, it was like you could ignore it. But I had to talk to her and explain who I am, and this has been my experience, and people assume that if you're a particular ethnicity then you're going to do what they perceive are the stereotypical things of that ethnicity.

Because of their awareness of racism in U.S. society, mothers of color were acutely aware of whether their children would be racially safe in child care settings. When one is a member of a historically subordinated racial ethnic group, finding child care that provides children with racial safety is an important concern. Yet, the search is complicated by other racial-ethnic factors.

THE SEARCH FOR CULTURAL MAINTENANCE

Many of the mothers expressed interest in child care by racially and ethnically similar caregivers. For some, this was motivated by the desire to protect their children and have them in racially safe situations. For others, staying within one's own community was an explicit strategy to ensure that their children would learn about their cultural heritage and histories. Many of the mothers had been young adults at a time in history when racial ethnic groups began to take pride in claiming their cultural histories. Prior to the 1960s, historically subordinated racial ethnic groups were expected to socialize their children to the dominant Anglo Saxon Protestant values of U.S. society. As far back as the 1920s, child care services were used to "Americanize" immigrant children and their parents.[8] Mothers were aware of this bias and purposefully sought out culturally similar providers because they saw child care as a site that would influence their children's understandings of their cultural heritage.

Several Mexican American mothers sought Spanish-speaking Mexican or Mexican American caregivers for this reason. For example, when Elena Romero, a Mexican American nutritionist, first needed child care, she used this strategy and found a Spanish-speaking Mexican American child care provider. She said:

We found out about this [family] day care that was run by a preschool teacher that had decided to open up her own day care. And she was Chicana and . . . I really wanted him to know Spanish. Since birth I had [talked] to him [in Spanish]. . . . Anyway, so I went to this day care and I really was impressed with the day care center because . . . she was really organized . . . and I liked her right away, you know. Then she had like senoras mexicanas . . . come in and cook for her and like they would make a big ol' pot of albóndigas . . . a meatball soup, you know. So like they would make really good Mexican food.

Similarly, Aurora Garcia said:

> I was hoping that, given that my child would be in the household for a sig-
> nificant number of hours during the day, that there be some [ethnic] similarity,
> you know, not that I'm traditional, I don't consider myself traditional, but those
> values I wanted, kind of implanted, you know, issues of discipline, you know,
> being really caring and nurturing and her being familiar with Spanish.

In describing what she looked for in child care, Julie Lopez, an African American mother whose ex-husband and stepfather are Latino, said:

> There's a cultural thing . . . one of the things for me, and our family, it has
> been really important to have [my child] in a bilingual place where she can
> sit down with other kids and speak Spanish and have a teacher that speaks
> Spanish and they sit down at lunch and they speak Spanish together. And
> the writing they do is both in English and Spanish and the pictures on the
> walls and stuff, because that cultural thing to me is really important . . .
> I'm always more comfortable if they're bicultural as well, versus just being
> bilingual.

Thus, their concerns were not simply about the skills and types of food that
their children would be eating but also addressed a broader understanding
that shared cultural practices were expressions of shared cultural values.

Another issue that confronted employed mothers of color was whether to
foster cultural maintenance and racial safety at the cost of educational opportunities. In particular, mothers who had been raised working class and were
now middle class grappled with this problem. When Aurora Garcia switched
her daughter from a family day care home with a Mexican American caregiver
to a predominantly White day care center, she felt as if she had to make compromises. She said:

> I'm not getting the ideal. I can't find the ideal . . . there are very few children
> of color there. I think diversity to them is Jewish. That's being diverse cultur-
> ally . . . I mean, the ideal to me would be that she be in a school where she
> would be learning Spanish, she would be learning those things. . . . And that's
> just a tradeoff for me right now . . . I think of all the skills she's learning right
> now, but there's a cultural context to them that would be nice to have.

Aurora acknowledged that because she used a predominantly White day care
center she was raising her child in a White environment. However, she pointed
out that her daughter was exposed to traditional cultural values because of who
her parents were. She said, "I'm very much entrenched in who I am and what
my cultural values are and my experience, and my partner is in his." Similarly,

Elena Romero reconciled herself to the fact that her children would learn about their culture and their history at home. She said:

> [My husband and I] are both real proud of being Mexicanos, Chicanos, you know. And we're both constantly involved in the Movement kind of things. And we both have friends who are bilingual and that have kids, and, you, our families. If we have a birthday party, we have a piñata and all that stuff. So we decided well, that they would get it from us.

Yet, Aurora and Elena both realized that placing children in White daycares removed them from being fully immersed in their traditional ethnic community. Aurora said:

> It's not the same for a child. I mean, it really is how you play, who you play with, what you play, it's what you eat, it's how people treat you, what they say to you. . . . [Her teachers] are going to present it from a white perspective because they don't have bilingual teachers. They don't have African American teachers. So for me, it's a trade off.

Clearly, choosing to move outside of one's culture into a predominantly White daycare was not an easy decision. When they made the decision to place their children in a predominantly White daycare, they continued to be conflicted about not being able to find a daycare for their children which would provide cultural maintenance and exposure to traditional cultural practices and values. Although on the surface it may appear that mothers of color who place their children in predominantly White daycare settings are rejecting their own cultural practices and turning their back on their racial ethnic group, this was not the case. They were highly self-conscious that their children's child care was not fulfilling one of their major criteria for their child care arrangements. By providing their children with the social opportunities and formal education which they had come to expect for any well-educated child of the middle class, they had to work harder at home to ensure that their children learned about their cultures and histories. Furthermore, by placing themselves in predominantly White settings, they more frequently encountered racism and, more frequently and at a younger age, had to explain to their young children about race relations with White society and how to navigate them.

Both of these concerns—racial safety and cultural maintenance—reflect how membership in specific racial ethnic groups influences views of what constitutes appropriate caregiving. The concern of mothers of color for racial safety addresses their awareness that their children can be the targets of racism by a society that has historically devalued their racial ethnic group. The concern for cultural maintenance reflects their preference to retain and/or retrieve traditional cultural practices and values. . . .

NOTES

1. Jacquelyn Litt, "Mothering, Medicalization, and Jewish Identity, 1928–1940," *Gender & Society* 10, no. 2 (1996): 185–198.

2. Carole E. Joffe, *Friendly Intruders: Childcare Professionals and Family Life* (Berkeley: University of California Press, 1977).

3. Mary Larner and Anne Mitchell, "Meeting the Child Care Needs of Low-Income Families," *Child and Youth Care Forum* 21, no. 5 (1992): 317–334. Katherine Brown Rosier and William A. Corsario, "Competent Parents, Complex Lives: Managing Parenthood in Poverty," *Journal of Contemporary Ethnography* 22, no. 2 (1993): 171–204.

4. Joffe, *Friendly Intruders.*

5. Stevanne Auerbach, "What Parents Want From Day Care," in *Child Care: A Comprehensive Guide: Philosophy, Programs, and Practices for the Creation of Quality Services for Children*, vol. 1, *Rationale for Child Care Services: Programs vs. Politics*, ed. Stevanne Auerbach with James A. Rivaldo (New York: Human Sciences Press, 1975), 137–152.

6. Janice Hale, "The Transmission of Cultural Values to Young African American Children," *Young Children* (September 1991): 7–14.

7. Janice Gould, "American Indian Women's Poetry: Strategies of Rage and Hope," *Signs* 20, no. 4 (1995): 797–817.

8. Julia Wrigley, "Different Care for Different Kids: Social Class and Child Care Policy," *Educational Policy* 3, no. 4 (1989): 421–439.

STRAIGHT IS TO GAY
AS FAMILY IS TO NO FAMILY

33

Kath Weston

IS "STRAIGHT" TO "GAY"
AS "FAMILY" IS TO "NO FAMILY"?

For years, and in an amazing variety of contexts, claiming a lesbian or gay identity has been portrayed as a rejection of "the family" and a departure from kinship. In media portrayals of AIDS, Simon Watney . . . observes that "we are

From Kath Weston, *Families We Choose: Lesbians, Gays, Kinship* (New York: Columbia University Press, 1991), pp. 22–29. Reprinted by permission of Columbia University Press.

invited to imagine some absolute divide between the two domains of 'gay life' and 'the family,' as if gay men grew up, were educated, worked and lived our lives in total isolation from the rest of society." Two presuppositions lend a dubious credence to such imagery: the belief that gay men and lesbians do not have children or establish lasting relationships, and the belief that they invariably alienate adoptive and blood kin once their sexual identities become known. By presenting "the family" as a unitary object, these depictions also imply that everyone participates in identical sorts of kinship relations and subscribes to one universally agreed-upon definition of family.

Representations that exclude lesbians and gay men from "the family" invoke what Blanche Wiesen Cook . . . has called "the assumption that gay people do not love and do not work," the reduction of lesbians and gay men to sexual identity, and sexual identity to sex alone. In the United States, sex apart from heterosexual marriage tends to introduce a wild card into social relations, signifying unbridled lust and the limits of individualism. If heterosexual intercourse can bring people into enduring association via the creation of kinship ties, lesbian and gay sexuality in these depictions isolates individuals from one another rather than weaving them into a social fabric. To assert that straight people "naturally" have access to family, while gay people are destined to move toward a future of solitude and loneliness, is not only to tie kinship closely to procreation, but also to treat gay men and lesbians as members of a nonprocreative species set apart from the rest of humanity. . . .

It is but a short step from positioning lesbians and gay men somewhere beyond "the family"—unencumbered by relations of kinship, responsibility, or affection—to portraying them as a menace to family and society. A person or group must first be outside and other in order to invade, endanger, and threaten. My own impression from fieldwork corroborates Frances FitzGerald's . . . observation that many heterosexuals believe not only that gay people have gained considerable political power, but also that the absolute number of lesbians and gay men (rather than their visibility) has increased in recent years. Inflammatory rhetoric that plays on fears about the "spread" of gay identity and of AIDS finds a disturbing parallel in the imagery used by fascists to describe syphilis at mid-century, when "the healthy" confronted "the degenerate" while the fate of civilization hung in the balance. . . .

A long sociological tradition in the United States of studying "the family" under siege or in various states of dissolution lent credibility to charges that this institution required protection from "the homosexual threat." . . .

. . . By shifting without signal between reproduction's meaning of physical procreation and its sense as the perpetuation of society as a whole, the characterization of lesbians and gay men as nonreproductive beings links their supposed attacks on "the family" to attacks on society in the broadest sense. Speaking of parents who had refused to accept her lesbian identity, a Jewish woman explained, "They feel like I'm finishing off Hitler's job." The plausibility of the contention that gay people pose a threat to "the family" (and, through the family, to ethnicity) depends upon a view of family grounded in heterosexual relations, combined with the conviction that gay men and lesbians are incapable of procreation, parenting, and establishing kinship ties.

Some lesbians and gay men . . . had embraced the popular equation of their sexual identities with the renunciation of access to kinship, particularly when first coming out. "My image of gay life was very lonely, very weird, no family," Rafael Ortiz recollected. "I assumed that my family was gone now—that's it." After Bob Korkowski began to call himself gay, he wrote a series of poems in which an orphan was the central character. Bob said the poetry expressed his fear of "having to give up my family because I was queer." When I spoke with Rona Bren after she had been home with the flu, she told me that whenever she was sick, she relived old fears. That day she had remembered her mother's grim prediction: "You'll be a lesbian and you'll be alone the rest of your life. Even a dog shouldn't be alone."

Looking backward and forward across the life cycle, people who equated their adoption of a lesbian or gay identity with a renunciation of family did so in the double-sided sense of fearing rejection by the families in which they had grown up, and not expecting to marry or have children as adults. Although few in numbers, there were still those who had considered "going straight" or getting married specifically in order to "have a family." Vic Kochifos thought he understood why:

> It's a whole lot easier being straight in the world than it is being gay. . . .
> You have built-in loved ones: wife, husband, kids, extended family. It just works
> easier. And when you want to do something that requires children, and you
> want to have a feeling of knowing that there's gonna be someone around
> who cares about you when you're 85 years old, there are thoughts that go
> through your head, sure. There must be. There's a way of doing it gay, but
> it's a whole lot harder, and it's less secure.

Bernie Margolis had been sexually involved with men since he was in his teens, but for years had been married to a woman with whom he had several children. At age 67 he regretted having grown to adulthood before the current discussion of gay families, with its focus on redefining kinship and constructing new sorts of parenting arrangements.

> I didn't want to give up the possibility of becoming a family person. Of having
> kids of my own to carry on whatever I built up. . . . My mother was always talk-
> ing about she's looking forward to the day when she would bring her children
> under the canopy to get married. It never occurred to her that I wouldn't be
> married. It probably never occurred to me either.

The very categories "good family person" and "good family man" had seemed to Bernie intrinsically opposed to a gay identity. In his fifties at the time

I interviewed him, Stephen Richter attributed never having become a father to "not having the relationship with the woman." Because he had envisioned parenting and procreation only in the context of a heterosexual relationship, regarding the two as completely bound up with one another, Stephen had never considered children an option.

Older gay men and lesbians were not the only ones whose adult lives had been shaped by ideologies that banish gay people from the domain of kinship. Explaining why he felt uncomfortable participating in "family occasions," a young man who had no particular interest in raising a child commented, "When families get together, what do they talk about? Who's getting married, who's having children. And who's not, okay? Well, look who's not." Very few of the lesbians and gay men I met believed that claiming a gay identity automatically requires leaving kinship behind. In some cases people described this equation as an outmoded view that contrasted sharply with revised notions of what constitutes a family.

Well-meaning defenders of lesbian and gay identity sometimes assert that gays are not inherently "anti-family," in ways that perpetuate the association of heterosexual identity with exclusive access to kinship. Charles Silverstein . . . , for instance, contends that lesbians and gay men may place more importance on maintaining family ties than heterosexuals do because gay people do not marry and raise children. Here the affirmation that gays and lesbians are capable of fostering enduring kinship ties ends up reinforcing the implication that they cannot establish "families of their own," presumably because the author regards kinship as unshakably rooted in heterosexual alliance and procreation. In contrast, discourse on gay families cuts across the politically loaded couplet of "pro-family" and "anti-family" that places gay men and lesbians in an inherently antagonistic relation to kinship solely on the basis of their nonprocreative sexualities. "Homosexuality is not what is breaking up the Black family," declared Barbara Smith . . . , a black lesbian writer, activist, and speaker at the 1987 Gay and Lesbian March on Washington. "Homophobia is. My Black gay brothers and my Black lesbian sisters are members of Black families, both the ones we were born into and the ones we create."

At the height of gay liberation, activists had attempted to develop alternatives to "the family," whereas by the 1980s many lesbians and gay men were struggling to legitimate gay families as a form of kinship. . . . Gay or chosen families might incorporate friends, lovers, or children, in any combination. Organized through ideologies of love, choice, and creation, gay families have been defined through a contrast with what many gay men and lesbians . . . called "straight," "biological," or "blood" family. If families we choose were the families lesbians and gay men created for themselves, straight family represented the families in which most had grown to adulthood.

What does it mean to say that these two categories of family have been defined through contrast? One thing it emphatically does *not* mean is that heterosexuals share a single coherent form of family (although some of the lesbians and gay men doing the defining believed this to be the case). I am not arguing here for the existence of some central, unified kinship system vis-à-vis which gay people have distinguished their own practice and understanding of family. In the United States, race, class, gender, ethnicity, regional origin, and context all inform differences in household organization, as well as differences in notions of family and what it means to call someone kin.

In any relational definition, the juxtaposition of two terms gives meaning to both. Just as light would not be meaningful without some notion of darkness, so gay or chosen families cannot be understood apart from the families lesbians and gay men call "biological," "blood," or "straight." Like others in their society, most gay people . . . considered biology a matter of "natural fact." When they applied the terms "blood" and "biology" to kinship, however, they tended to depict families more consistently organized by procreation, more rigidly grounded in genealogy, and more uniform in their conceptualization than anthropologists know most families to be. For many lesbians and gay men, blood family represented not some naturally given unit that provided a base for all forms of kinship, but rather a procreative principle that organized only one possible *type* of kinship. In their descriptions they situated gay families at the opposite end of a spectrum of determination, subject to no constraints beyond a logic of "free" choice that ordered membership. To the extent that gay men and lesbians mapped "biology" and "choice" onto identities already opposed to one another (straight and gay, respectively), they polarized these two types of family along an axis of sexual identity.

The following chart recapitulates the ideological transformation generated as lesbians and gay men began to inscribe themselves within the domain of kinship.

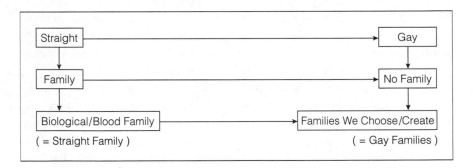

What this chart presents is not some static substitution set, but a historically motivated succession. To move across or down the chart is to move through time. Following along from left to right, time appears as process, periodized with reference to the experience of coming out. In the first opposition, coming out defines the transition from a straight to a gay identity. For the person who maintains an exclusively biogenetic notion of kinship, coming out can mark the renunciation of kinship, the shift from "family" to "no family" portrayed in the second opposition. In the third line, individuals who accepted the possibility of gay families after coming out could experience themselves making a transition from the biological or blood families in which they had grown up to the establishment of their own chosen families.

Moving from top to bottom, the chart depicts the historical time that inaugurated contemporary discourse on gay kinship. "Straight" changes from a category with an exclusive claim on kinship to an identity allied with a specific kind of family symbolized by biology or blood. Lesbians and gay men, originally relegated to the status of people without family, later lay claim to a distinctive type of family characterized as families we choose or create. While dominant cultural representations have asserted that straight is to gay as family is to no family (lines 1 and 2), at a certain point in history gay people began to contend that straight is to gay as blood family is to chosen families (lines 1 and 3).

What provided the impetus for this ideological shift? Transformations in the relation of lesbians and gay men to kinship are inseparable from sociohistorical developments: changes in the context for disclosing a lesbian or gay identity to others, attempts to build urban gay "community," cultural inferences about relationships between "same-gender" partners, and the lesbian baby boom associated with alternative (artificial) insemination. . . . If . . . kinship is something people use to act as well as to think, then its transformations should have unfolded not only on the "big screen" of history, but also on the more modest stage of day-to-day life, where individuals have actively engaged novel ideological distinctions and contested representations that would exclude them from kinship.

Cultural Institutions and the Production of Ideas

RACIST STEREOTYPING IN THE ENGLISH LANGUAGE

34

Robert B. Moore

LANGUAGE AND CULTURE

An integral part of any culture is its language. Language not only develops in conjunction with a society's historical, economic and political evolution; it also reflects that society's attitudes and thinking. Language not only *expresses* ideas and concepts but actually *shapes* thought.[1] If one accepts that our dominant white culture is racist, then one would expect our language—an indispensable transmitter of culture—to be racist as well. Whites, as the dominant group, are not subjected to the same abusive characterization by our language that people of color receive. Aspects of racism in the English language that will be discussed in this essay include terminology, symbolism, politics, ethnocentrism, and context.

Before beginning our analysis of racism in language we would like to quote part of a TV film review which shows the connection between language and culture.[2]

> Depending on one's culture, one interacts with time in a very distinct fashion. One example which gives some cross-cultural insights into the concept of time is language. In Spanish, a watch is said to "walk." In English, the watch "runs." In German, the watch "functions." And in French, the watch "marches." In the Indian culture of the Southwest, people do not refer to time

From Paula S. Rothenberg, ed., *Racism and Sexism: An Integrated Study* (New York: St. Martin's Press, 1988), pp. 269–79. Reprinted by permission of the Council on Interracial Books for Children c/o Lawrence Jordan Literary Agency, 345 West 121st Street, NY, NY 10027.

in this way. The value of the watch is displaced with the value of "what time it's getting to be." Viewing these five cultural perspectives of time, one can see some definite emphasis and values that each culture places on time. For example, a cultural perspective may provide a clue to why the negative stereotype of the slow and lazy Mexican who lives in the "Land of Manana" exists in the Anglo value system, where time "flies," the watch "runs" and "time is money."

A SHORT PLAY ON "BLACK" AND "WHITE" WORDS

Some may blackly (angrily) accuse me of trying to blacken (defame) the English language, to give it a black eye (a mark of shame) by writing such black words (hostile). They may denigrate (to cast aspersions; to darken) me by accusing me of being blackhearted (malevolent), of having a black outlook (pessimistic, dismal) on life, of being a blackguard (scoundrel)—which would certainly be a black mark (detrimental fact) against me. Some may black-brow (scowl at) me and hope that a black cat crosses in front of me because of this black deed. I may become a black sheep (one who causes shame or embarrassment because of deviation from the accepted standards), who will be blackballed (ostracized) by being placed on a blacklist (list of undesirables) in an attempt to blackmail (to force or coerce into a particular action) me to retract my words. But attempts to blackjack (to compel by threat) me will have a Chinaman's chance of success, for I am not a yellow-bellied Indian-giver of words, who will whitewash (cover up or gloss over vices or crimes) a black lie (harmful, inexcusable). I challenge the purity and innocence (white) of the English language. I don't see things in black and white (entirely bad or entirely good) terms, for I am a white man (marked by upright firmness) if there ever was one. However, it would be a black day when I would not "call a spade a spade," even though some will suggest a white man calling the English language racist is like the pot calling the kettle black. While many may be niggardly (grudging, scanty) in their support, others will be honest and decent— and to them I say, that's very white of you (honest, decent).

The preceding is of course a white lie (not intended to cause harm), meant only to illustrate some examples of racist terminology in the English language.

OBVIOUS BIGOTRY

Perhaps the most obvious aspect of racism in language would be terms like "nigger," "spook," "chink," "spic," etc. While these may be facing increasing social disdain, they certainly are not dead. Large numbers of white Americans continue to utilize these terms. "Chink," "gook," and "slant-eyes" were in common usage among U.S. troops in Vietnam. An NBC nightly news broadcast, in

February 1972, reported that the basketball team in Pekin, Illinois, was called the "Pekin Chinks" and noted that even though this had been protested by Chinese Americans, the term continued to be used because it was easy, and meant no harm. Spiro Agnew's widely reported "fat Jap" remark and the "little Jap" comment of lawyer John Wilson, during the Watergate hearings, are surface indicators of a deep-rooted Archie Bunkerism.

Many white people continue to refer to Black people as "colored," as for instance in a July 30, 1975, *Boston Globe* article on a racist attack by whites on a group of Black people using a public beach in Boston. One white person was quoted as follows:

> We've always welcomed good colored people in South Boston but we will not tolerate radical blacks or Communists. . . . Good colored people are welcome in South Boston, black militants are not.

Many white people may still be unaware of the disdain many African Americans have for the term "colored," but it often appears that whether used intentionally or unintentionally, "colored" people are "good" and "know their place," while "Black" people are perceived as "uppity" and "threatening" to many whites. Similarly, the term "boy" to refer to African American men is now acknowledged to be a demeaning term, though still in common use. Other terms such as "the pot calling the kettle black" and "calling a spade a spade" have negative racial connotations but are still frequently used, as for example when President Ford was quoted in February 1976 saying that even though Daniel Moynihan had left the U.N., the U.S. would continue "calling a spade a spade."

COLOR SYMBOLISM

The symbolism of white as positive and black as negative is pervasive in our culture, with the black/white words used in the beginning of this essay only one of many aspects. "Good guys" wear white hats and ride white horses, "bad guys" wear black hats and ride black horses. Angels are white, and devils are black. The definition of *black* includes "without any moral light or goodness, evil, wicked, indicating disgrace, sinful," while that of *white* includes "morally pure, spotless, innocent, free from evil intent."

A children's TV cartoon program, *Captain Scarlet*, is about an organization called Spectrum, whose purpose is to save the world from an evil extraterrestrial force called the Mysterons. Everyone in Spectrum has a color name—Captain Scarlet, Captain Blue, etc. The one Spectrum agent who has been mysteriously taken over by the Mysterons and works to advance their evil aims is Captain Black. The person who heads Spectrum, the good organization out to defend the world, is Colonel White.

Three of the dictionary definitions of white are "fairness of complexion, purity, innocence." These definitions affect the standards of beauty in our culture, in which whiteness represents the norm. "Blondes have more fun" and "Wouldn't you really rather be a blonde" are sexist in their attitudes toward women generally, but are racist white standards when applied to third world women. A 1971 *Mademoiselle* advertisement pictured a curly-headed, ivory-skinned woman over the caption, "When you go blonde go all the way," and asked: "Isn't this how, in the back of your mind, you always wanted to look? All wide-eyed and silky blonde down to there, and innocent?" Whatever the advertising people meant by this particular woman's innocence, one must remember that "innocent" is one of the definitions of the word white. This standard of beauty when preached to all women is racist. The statement "Isn't this how, in the back of your mind, you always wanted to look?" either ignores third world women or assumes they long to be white.

Time magazine in its coverage of the Wimbledon tennis competition between the black Australian Evonne Goolagong and the white American Chris Evert described Ms. Goolagong as "the dusky daughter of an Australian sheepshearer," while Ms. Evert was "a fair young girl from the middle-class groves of Florida." *Dusky* is a synonym of "black" and is defined as "having dark skin; of a dark color; gloomy; dark; swarthy." Its antonyms are "fair" and "blonde." *Fair* is defined in part as "free from blemish, imperfection, or anything that impairs the appearance, quality, or character: pleasing in appearance, attractive; clean; pretty; comely." By defining Evonne Goolagong as "dusky," *Time* technically defined her as the opposite of "pleasing in appearance; attractive; clean; pretty; comely."

The studies of Kenneth B. Clark, Mary Ellen Goodman, Judith Porter and others indicate that this persuasive "rightness of whiteness" in U.S. culture affects children before the age of four, providing white youngsters with a false sense of superiority and encouraging self-hatred among third world youngsters.

ETHNOCENTRISM OR FROM A WHITE PERSPECTIVE

Some words and phrases that are commonly used represent particular perspectives and frames of reference, and these often distort the understanding of the reader or listener. David R. Burgest[3] has written about the effect of using the terms "slave" or "master." He argues that the psychological impact of the statement referring to "the master raped his slave" is different from the impact of the same statement substituting the words: "the white captor raped an African woman held in captivity."

Implicit in the English usage of the "master-slave" concept is ownership of the "slave" by the "master," therefore, the "master" is merely abusing his property (slave). In reality, the captives (slave) were African individuals with human worth, right and dignity and the term "slave" denounces that human quality thereby making the mass rape of African women by white captors more acceptable in the minds of people and setting a mental frame of reference for legitimizing the atrocities perpetuated against African people.

The term slave connotes a less than human quality and turns the captive person into a thing. For example, two McGraw-Hill Far Eastern Publishers textbooks (1970) stated, "At first it was the slaves who worked the cane and they got only food for it. Now men work cane and get money." Next time you write about slavery or read about it, try transposing all "slaves" into "African people held in captivity," "Black people forced to work for no pay" or "African people stolen from their families and societies." While it is more cumbersome, such phrasing conveys a different meaning.

PASSIVE TENSE

Another means by which language shapes our perspective has been noted by Thomas Greenfield,[4] who writes that the achievements of Black people—and Black people themselves—have been hidden in

> the linguistic ghetto of the passive voice, the subordinate clause, and the "understood" subject. The seemingly innocuous distinction (between active/passive voice) holds enormous implications for writers and speakers. When it is effectively applied, the rhetorical impact of the passive voice—the art of making the creator or instigator of action totally disappear from a reader's perception—can be devastating.

For instance, some history texts will discuss how European immigrants came to the United States seeking a better life and expanded opportunities, but will note that "slaves *were brought* to America." Not only does this omit the destruction of African societies and families, but it ignores the role of northern merchants and southern slaveholders in the profitable trade in human beings. Other books will state that "the continental railroad *was built*," conveniently omitting information about the Chinese laborers who built much of it or the oppression they suffered.

Another example. While touring Monticello, Greenfield noted that the tour guide

> made all the black people of Monticello disappear through her use of the passive voice. While speaking of the architectural achievements of Jefferson in the active voice, she unfailingly shifted to passive when speaking of the work performed by Negro slaves and skilled servants.

Noting a type of door that after 166 years continued to operate without need for repair, Greenfield remarks that the design aspect of the door was much simpler than the actual skill and work involved in building and installing it. Yet his guide stated: "Mr. Jefferson designed these doors . . ." while "the doors were installed in 1809." The workers who installed those doors were African people whom Jefferson held in bondage. The guide's use of the passive tense enabled her to dismiss the reality of Jefferson's slaveholding. It also meant that she did not have to make any mention of the skills of those people held in bondage.

POLITICS AND TERMINOLOGY

"Culturally deprived," "economically disadvantaged" and "underdeveloped" are other terms which mislead and distort our awareness of reality. The application of the term "culturally deprived" to third world children in this society reflects a value judgment. It assumes that the dominant whites are cultured and all others without culture. In fact, third world children generally are bicultural, and many are bilingual, having grown up in their own culture as well as absorbing the dominant culture. In many ways, they are equipped with skills and experiences which white youth have been deprived of, since most white youth develop in a monocultural, monolingual environment. Burgest[5] suggests that the term "culturally deprived" be replaced by "culturally dispossessed," and that the term "economically disadvantaged" be replaced by "economically exploited." Both these terms present a perspective and implication that provide an entirely different frame of reference as to the reality of the third world experience in U.S. society.

Similarly, many nations of the third world are described as "underdeveloped." These less wealthy nations are generally those that suffered under colonialism and neo-colonialism. The "developed" nations are those that exploited their resources and wealth. Therefore, rather than referring to these countries as "underdeveloped," a more appropriate and meaningful designation might be "over exploited." Again, transpose this term next time you read about "underdeveloped nations" and note the different meaning that results.

Terms such as "culturally deprived," "economically disadvantaged" and "underdeveloped" place the responsibility for their own conditions on those being so described. This is known as "Blaming the Victim."[6] It places responsibility for poverty on the victims of poverty. It removes the blame from those in power who benefit from, and continue to permit, poverty.

Still another example involves the use of "non-white," "minority" or "third world." While people of color are a minority in the U.S., they are part of the vast majority of the world's population, in which white people are a distinct minority. Thus, by utilizing the term minority to describe people of color in the U.S., we can lose sight of the global majority/minority reality—a fact of some importance in the increasing and interconnected struggles of people of color inside and outside the U.S.

To describe people of color as "non-white" is to use whiteness as the standard and norm against which to measure all others. Use of the term "third world" to describe all people of color overcomes the inherent bias of "minority" and "non-white." Moreover, it connects the struggles of third world people in the U.S. with the freedom struggles around the globe.

The term third world gained increasing usage after the 1955 Bandung Conference of "non-aligned" nations, which represented a third force outside of the two world superpowers. The "first world" represents the United States, Western Europe and their sphere of influence. The "second world" represents the Soviet Union and its sphere. The "third world" represents, for the most part, nations that were, or are, controlled by the "first world" or West. For the most part, these are nations of Africa, Asia and Latin America.

"LOADED" WORDS AND NATIVE AMERICANS

Many words lead to a demeaning characterization of groups of people. For instance, Columbus, it is said, "discovered" America. The word *discover* is defined as "to gain sight or knowledge of something previously unseen or unknown; to discover may be to find some existent thing that was previously unknown." Thus, a continent inhabited by millions of human beings cannot be "discovered." For history books to continue this usage represents a Eurocentric (white European) perspective on world history and ignores the existence of, and the perspective of, Native Americans. "Discovery," as used in the Euro-American context, implies the right to take what one finds, ignoring the rights of those who already inhabit or own the "discovered" thing.

Eurocentrism is also apparent in the usage of "victory" and "massacre" to describe the battles between Native Americans and whites. *Victory* is defined in the dictionary as "a success or triumph over an enemy in battle or war; the decisive defeat of an opponent." *Conquest* denotes the "taking over of control by the victor, and the obedience of the conquered." *Massacre* is defined as "the unnecessary, indiscriminate killing of a number of human beings, as in barbarous warfare or persecution, or for revenge or plunder." *Defend* is described as "to ward off attack from; guard against assault or injury; to strive to keep safe by resisting attack."

Eurocentrism turns these definitions around to serve the purpose of distorting history and justifying Euro-American conquest of the Native American homelands. Euro-Americans are not described in history books as invading Native American lands, but rather as defending *their* homes against "Indian" attacks. Since European communities were constantly encroaching on land already occupied, then a more honest interpretation would state that it was the Native Americans who were "warding off," "guarding" and "defending" their homelands.

Native American victories are invariably defined as "massacres," while the indiscriminate killing, extermination and plunder of Native American nations by Euro-Americans is defined as "victory." Distortion of history by the choice of "loaded" words used to describe historical events is a common racist practice. Rather than portraying Native Americans as human beings in highly defined and complex societies, cultures and civilizations, history books use such adjectives as "savages," "beasts," "primitive," and "backward." Native people are referred to as "squaw," "brave," or "papoose" instead of "woman," "man," or "baby."

Another term that has questionable connotations is *tribe*. The *Oxford English Dictionary* defines this noun as "a race of people; now applied especially to a primary aggregate of people in a primitive or barbarous condition, under a headman or chief." Morton Fried,[7] discussing "The Myth of Tribe," states that the word "did not become a general term of reference to American Indian society until the nineteenth century. Previously, the words commonly used for Indian populations were 'nation' and 'people.'" Since "tribe" has assumed a connotation of primitiveness or backwardness, it is suggested that the use of "nation" or "people" replace the term whenever possible in referring to Native American peoples.

The term *tribe* invokes even more negative implications when used in reference to American peoples. As Evelyn Jones Rich[8] has noted, the term is "almost always used to refer to third world people and it implies a stage of development which is, in short, a put-down."

"LOADED" WORDS AND AFRICANS

Conflicts among diverse peoples within African nations are often referred to as "tribal warfare," while conflicts among the diverse peoples within European countries are never described in such terms. If the rivalries between the Ibo and the Hausa and Yoruba in Nigeria are described as "tribal," why not the rivalries between Serbs and Slavs in Yugoslavia, or Scots and English in Great Britain, Protestants and Catholics in Ireland, or the Basques and the Southern Spaniards in Spain? Conflicts among African peoples in a particular nation have religious, cultural, economic and/or political roots. If we can analyze the roots of conflicts among European peoples in terms other than "tribal warfare," certainly we can do the same with African peoples, including correct reference to the ethnic groups or nations involved. For example, the terms "Kaffirs," "Hottentot" or "Bushmen" are names imposed by white Europeans. The correct names are always those by which a people refer to themselves. (In these instances Xhosa, Khoi-Khoin and San are correct.[9])

The generalized application of "tribal" in reference to Africans—as well as the failure to acknowledge the religious, cultural and social diversity of African peoples—is a decidedly racist dynamic. It is part of the process whereby Euro-Americans justify, or avoid confronting, their oppression of third world peoples. Africa has been particularly insulted by this dynamic, as witness the pervasive "darkest Africa" image. This image, widespread in Western culture, evokes an Africa covered by jungles and inhabited by "uncivilized," "cannibalistic," "pagan," "savage" peoples. This "darkest Africa" image avoids the geographical reality. Less than 20 percent of the African continent is wooded savanna, for example. The image also ignores the history of African cultures and civilizations. Ample evidence suggests this distortion of reality was developed as a convenient rationale for the European and American slave trade. The Western powers, rather than exploiting, were civilizing and Christianizing "uncivilized" and "pagan savages" (so the rationalization went). This dynamic also served to justify Western colonialism. From Tarzan movies to racist children's books like *Doctor Dolittle* and *Charlie and the Chocolate Factory*, the image of "savage" Africa and the myth of "the white man's burden" has been perpetuated in Western culture.

A 1972 *Time* magazine editorial lamenting the demise of *Life* magazine, stated that the "lavishness" of *Life*'s enterprises included "organizing safaris into darkest Africa." The same year, the *New York Times*' C. L. Sulzberger wrote that Africa has "a history as dark as the skins of many of its people." Terms such as "darkest Africa," "primitive," "tribe" ("tribal") or "jungle," in reference to Africa, perpetuate myths and are especially inexcusable in such large circulation publications.

Ethnocentrism is similarly reflected in the term "pagan" to describe traditional religions. A February 1973 *Time* magazine article on Uganda stated, "Moslems account for only 500,000 of Uganda's 10 million people. Of the remainder, 5,000,000 are Christians and the rest pagan." *Pagan* is defined as "Heathen, a follower of a polytheistic religion; one that has little or no religion and that is marked by a frank delight in and uninhibited seeking after sensual pleasures and material goods." *Heathen* is defined as "Unenlightened; an unconverted member of a people or nation that does not acknowledge the God of the Bible. A person whose culture or enlightenment is of an inferior grade, especially an irreligious person." Now, the people of Uganda, like almost all Africans, have serious religious beliefs and practices. As used by Westerners, "pagan" connotes something wild, primitive and inferior— another term to watch out for.

The variety of traditional structures that African people live in are their "houses," not "huts." A *hut* is "an often small and temporary dwelling of simple construction." And to describe Africans as "natives" (noun) is derogatory terminology—as in, "the natives are restless." The dictionary definition of *native* includes: "one of a people inhabiting a territorial area at the time of its discovery or becoming familiar to a foreigner; one belonging to a people having a less complex civilization." Therefore, use of "native," like use of "pagan" often implies a value judgment of white superiority.

QUALIFYING ADJECTIVES

Words that would normally have positive connotations can have entirely different meanings when used in a racial context. For example, C. L. Sulzberger, the columnist of the *New York Times*, wrote in January 1975, about conversations he had with two people in Namibia. One was the white South African administrator of the country and the other a member of SWAPO, the Namibian liberation movement. The first is described as "Dirk Mudge, who as senior elected member of the administration is a kind of acting Prime Minister. . . ." But the second person is introduced as "Daniel Tijongarero, an intelligent Herero tribesman who is a member of SWAPO. . . ." What need was there for Sulzberger to state that Daniel Tijongarero is "intelligent"? Why not also state that Dirk Mudge was "intelligent"—or do we assume he wasn't?

A similar example from a 1968 *New York Times* article reporting on an address by Lyndon Johnson stated, "The President spoke to the well-dressed Negro officials and their wives." In what similar circumstances can one imagine a reporter finding it necessary to note that an audience of white government officials was "well-dressed"?

Still another word often used in a racist context is "qualified." In the 1960s white Americans often questioned whether Black people were "qualified" to hold public office, a question that was never raised (until too late) about white officials like Wallace, Maddox, Nixon, Agnew, Mitchell, et al. The question of qualifications has been raised even more frequently in recent years as white people question whether Black people are "qualified" to be hired for positions in industry and educational institutions. "We're looking for a qualified Black" has been heard again and again as institutions are confronted with affirmative action goals. Why stipulate that Blacks must be "qualified," when for others it is taken for granted that applicants must be "qualified"?

SPEAKING ENGLISH

Finally, the depiction in movies and children's books of third world people speaking English is often itself racist. Children's books about Puerto Ricans or Chicanos often connect poverty with a failure to speak English or to speak it well, thus blaming the victim and ignoring the racism which affects third world people regardless of their proficiency in English. Asian characters speak a stilted English ("Honorable so and so" or "Confucius say") or have a speech impediment ("roots or ruck," "very solly," "flied lice"). Native American characters speak another variation of stilted English ("Boy not hide. Indian take boy."), repeat certain Hollywood-Indian phrases ("Heap big" and "Many moons") or simply grunt out "Ugh" or "How." The repeated use of these language characterizations functions to make third world people seem less intelligent and less capable than the English-speaking white characters.

WRAP-UP

A *Saturday Review* editorial[10] on "The Environment of Language" stated that language

> . . . has as much to do with the philosophical and political conditioning of a society as geography or climate. . . . people in Western cultures do not realize the extent to which their racial attitudes have been conditioned since early childhood by the power of words to ennoble or condemn, augment or detract, glorify or demean. Negative language infects the subconscious of most Western people from the time they first learn to speak. Prejudice is not merely imparted or superimposed. It is metabolized in the bloodstream of society. What is needed is not so much a change in language as an awareness of the power of words to condition attitudes. If we can at least recognize the underpinnings of prejudice, we may be in a position to deal with the effects.

To recognize the racism in language is an important first step. Consciousness of the influence of language on our perceptions can help to negate much of that influence. But it is not enough to simply become aware of the effects of racism in conditioning attitudes. While we may not be able to change the language, we can definitely change our usage of the language. We can avoid using words that degrade people. We can make a conscious effort to use terminology that reflects a progressive perspective, as opposed to a distorting perspective. It is important for educators to provide students with opportunities to explore racism in language and to increase their awareness of it, as well as learning terminology that is positive and does not perpetuate negative human values.

NOTES

1. Simon Podair, "How Bigotry Builds Through Language," *Negro Digest*, March 1967.
2. Jose Armas, "Antonia and the Mayor: A Cultural Review of the Film," *The Journal of Ethnic Studies*, Fall 1975.

3. David R. Burgest, "The Racist Use of the English Language," *Black Scholar*, Sept. 1973.

4. Thomas Greenfield, "Race and Passive Voice at Monticello," *Crisis*, April 1975.

5. David R. Burgest, "Racism in Everyday Speech and Social Work Jargon," *Social Work*, July 1973.

6. William Ryan, *Blaming the Victim*, Pantheon Books, 1971.

7. Morton Fried, "The Myth of Tribe," *National History*, April 1975.

8. Evelyn Jones Rich, "Mind Your Language," *Africa Report*, Sept./Oct. 1974.

9. Steve Wolf, "Catalogers in Revolt Against LC's Racist, Sexist Headings," *Bulletin of Interracial Books for Children*, Vol. 6, Nos. 3&4, 1975.

10. "The Environment of Language," *Saturday Review*, April 8, 1967.

Also see:

Roger Bastide, "Color, Racism and Christianity," *Daedalus*, Spring 1967.

Kenneth J. Gergen, "The Significance of Skin Color in Human Relations," *Daedalus*, Spring 1967.

Lloyd Yabura, "Towards a Language of Humanism," *Rhythm*, Summer 1971.

UNESCO, "Recommendations Concerning Terminology in Education on Race Questions," June 1968.

CRIMES AGAINST HUMANITY   **35**

Ward Churchill

If nifty little "pep" gestures like the "Indian Chant" and the "Tomahawk Chop" are just good clean fun, then let's spread the fun around, shall we?

During the past couple of seasons, there has been an increasing wave of controversy regarding the names of professional sports teams like the Atlanta "Braves," Cleveland "Indians," Washington "Redskins," and Kansas City "Chiefs." The issue extends to the names of college teams like Florida State University "Seminoles," University of Illinois "Fighting Illini," and so on, right on down to high school outfits like the Lamar (Colorado) "Savages." Also involved have been team adoption of "mascots," replete with feathers, buckskins, beads, spears, and "warpaint" (some fans have opted to adorn

From *Z Magazine* 6 (March 1993): 43–47. Reprinted by permission of the author.

themselves in the same fashion), and nifty little "pep" gestures like the "Indian Chant" and "Tomahawk Chop."

A substantial number of American Indians have protested that use of native names, images, and symbols as sports team mascots and the like is, by definition, a virulently racist practice. Given the historical relationship between Indians and non-Indians during what has been called the "Conquest of America," American Indian Movement leader (and American Indian Anti-Defamation Council founder) Russell Means has compared the practice to contemporary Germans naming their soccer teams the "Jews," "Hebrews," and "Yids," while adorning their uniforms with grotesque caricatures of Jewish faces taken from the Nazis' anti-Semitic propaganda of the 1930s. Numerous demonstrations have occurred in conjunction with games—most notably during the November 15, 1992 match-up between the Chiefs and Redskins in Kansas City—by angry Indians and their supporters.

In response, a number of players—especially African Americans and other minority athletes—have been trotted out by professional team owners like Ted Turner, as well as university and public school officials, to announce that they mean not to insult but to honor native people. They have been joined by the television networks and most major newspapers, all of which have editorialized that Indian discomfort with the situation is "no big deal," insisting that the whole thing is just "good, clean fun." The country needs more such fun, they've argued, and "a few disgruntled Native Americans" have no right to undermine the nation's enjoyment of its leisure time by complaining. This is especially the case, some have argued, "in hard times like these." It has even been contended that Indian outrage at being systematically degraded—rather than the degradation itself—creates "a serious barrier to the sort of intergroup communication so necessary in a multicultural society such as ours."

Okay, let's communicate. We are frankly dubious that those advancing such positions really believe their own rhetoric, but, just for the sake of argument, let's accept the premise that they are sincere. If what they say is true, then isn't it time we spread such "inoffensiveness" and "good cheer" around among *all* groups so that *everybody* can participate *equally* in fostering the round of national laughs they call for? Sure it is—the country can't have too much fun or "intergroup involvement"—so the more, the merrier. Simple consistency demands that anyone who thinks the Tomahawk Chop is a swell pastime must be just as hearty in their endorsement of the following ideas—by the logic used to defend the defamation of American Indians—should help us all really start yukking it up.

First, as a counterpart to the Redskins, we need an NFL team called "Niggers" to honor Afro-Americans. Half-time festivities for fans might include a simulated stewing of the opposing coach in a large pot while players and cheerleaders dance around it, garbed in leopard skins and wearing fake bones in their noses. This concept obviously goes along with the kind of gaiety attending the Chop, but also with the actions of the Kansas City Chiefs, whose team members—prominently including black team members—lately appeared on a poster looking "fierce" and "savage" by way of wearing Indian regalia. Just a bit of harmless "morale boosting," says the Chiefs' front office. You bet.

So that the newly formed Niggers sports club won't end up too out of sync while expressing the "spirit" and "identity" of Afro-Americans in the above fashion, a baseball franchise—let's call this one the "Sambos"—should be formed. How about a basketball team called the "Spearchuckers"? A hockey team called the "Jungle Bunnies"? Maybe the "essence" of these teams could be depicted by images of tiny black faces adorned with huge pairs of lips. The players could appear on TV every week or so gnawing on chicken legs and spitting watermelon seeds at one another. Catchy, eh? Well, there's "nothing to be upset about," according to those who love wearing "war bonnets" to the Super Bowl or having "Chief Illiniwik" dance around the sports arenas of Urbana, Illinois.

And why stop there? There are plenty of other groups to include. Hispanics? They can be "represented" by the Galveston "Greasers" and San Diego "Spics," at least until the Wisconsin "Wetbacks" and Baltimore "Beaners" get off the ground. Asian Americans? How about the "Slopes," "Dinks," "Gooks," and "Zipperheads?" Owners of the latter teams might get their logo ideas from editorial page cartoons printed in the nation's newspapers during World War II: slant-eyes, buck teeth, big glasses, but nothing racially insulting or derogatory, according to the editors and artists involved at the time. Indeed, this Second World War–vintage stuff can be seen as just another barrel of laughs, at least by what current editors say are their "local standards" concerning American Indians.

Let's see. Who's been left out? Teams like the Kansas City "Kikes," Hanover "Honkies," San Leandro "Shylocks," Daytona "Dagos," and Pittsburgh "Polacks" will fill a certain social void among white folk. Have a religious belief? Let's all go for the gusto and gear up the Milwaukee "Mackerel Snappers" and Hollywood "Holy Rollers." The Fighting Irish of Notre Dame can be rechristened the "Drunken Irish" or "Papist Pigs." Issues of gender and sexual preferences can be addressed through creation of teams like the St. Louis "Sluts," Boston "Bimbos," Detroit "Dykes," and the Fresno "Fags." How about the Gainesville "Gimps" and Richmond "Retards," so the physically and mentally impaired won't be excluded from our fun and games?

Now, don't go getting "overly sensitive" out there. None of this is demeaning or insulting, at least not when it's being done to Indians. Just ask the folks who are doing it, or their apologists like Andy Rooney in the national media. They'll tell you—as in fact they *have* been telling you—that there's been no harm done, regardless of what their victims think, feel, or say. The situation is exactly the same as when those with precisely the same mentality used to insist that Step 'n' Fetchit was okay, or Rochester on the *Jack Benny Show*, or Amos and Andy, Charlie Chan, the Frito Bandito, or any of the other cutesy symbols making up the lexicon of American racism. Have we communicated yet?

Let's get just a little bit real here. The notion of "fun" embodied in rituals like the Tomahawk Chop must be understood for what it is. There's not a single non-Indian example used above which can be considered socially acceptable in even the most marginal sense. The reasons are obvious enough. So why is it different where American Indians are concerned? One can only conclude that, in contrast to the other groups at issue, Indians are (falsely) perceived as being too few, and therefore too weak, to defend themselves effectively against racist and otherwise offensive behavior.

Fortunately, there are some glimmers of hope. A few teams and their fans have gotten the message and have responded appropriately. Stanford University, which opted to drop the name "Indians" from Stanford, has experienced no resulting drop-off in attendance. Meanwhile, the local newspaper in Portland, Oregon, recently decided its long-standing editorial policy prohibiting use of racial epithets should include derogatory team names. The Redskins, for instance, are now referred to as "the Washington team," and will continue to be described in this way until the franchise adopts an inoffensive moniker (newspaper sales in Portland have suffered no decline as a result).

Such examples are to be applauded and encouraged. They stand as figurative beacons in the night, proving beyond all doubt that it is quite possible to indulge in the pleasure of athletics without accepting blatant racism into the bargain.

NUREMBERG PRECEDENTS

On October 16, 1946, a man named Julius Streicher mounted the steps of a gallows. Moments later he was dead, the sentence of an international tribunal composed of representatives of the United States, France, Great Britain, and the Soviet Union having been imposed. Streicher's body was then cremated, and—so horrendous were his crimes thought to have been—his ashes dumped into an unspecified German river so that "no one should ever know a particular place to go for reasons of mourning his memory."

Julius Streicher had been convicted at Nuremberg, Germany, of what were termed "Crimes Against Humanity." The lead prosecutor in his case—Justice Robert Jackson of the United States Supreme Court—had not argued that the defendant had killed anyone, nor that he had personally committed any especially violent act. Nor was it contended that Streicher had held any particularly important position in the German government during the period in which the so-called Third Reich had exterminated some 6,000,000 Jews, as well as several million Gypsies, Poles, Slavs, homosexuals, and other unter-menschen (subhumans).

The sole offense for which the accused was ordered put to death was in having served as publisher/editor of a Bavarian tabloid entitled *Der Sturmer* during the early-to-mid 1930s, years before the Nazi genocide actually began. In this capacity, he had penned a long series of virulently anti-Semitic editorials and "news" stories, usually accompanied by cartoons and other images graphically depicting Jews in extraordinarily derogatory fashion. This, the prosecution asserted, had done much to "dehumanize" the targets of his distortion in the mind of the German public. In turn, such dehumanization had made it possible—or at least easier—for average Germans to later indulge in the outright liquidation of Jewish "vermin." The tribunal agreed, holding that Streicher was therefore complicit in genocide and deserving of death by hanging.

During his remarks to the Nuremberg tribunal, Justice Jackson observed that, in implementing its sentences, the participating powers were morally and legally binding themselves to adhere forever after to the same standards of conduct that were being applied to Streicher and the other Nazi leaders. In the alternative, he said, the victorious allies would have committed "pure murder" at Nuremberg—no different in substance from that carried out by those they presumed to judge—rather than establishing the "permanent benchmark for justice" which was intended.

Yet in the United States of Robert Jackson, the indigenous American Indian population had already been reduced, in a process which is ongoing to this day, from perhaps 12.5 million in the year 1500 to fewer than 250,000 by the beginning of the 20th century. This was accomplished, according to official sources, "largely through the cruelty of [Euro-American] settlers," and an informal but clear governmental policy which had made it an articulated goal to "exterminate these red vermin," or at least whole segments of them.

Bounties had been placed on the scalps of Indians—any Indians—in places as diverse as Georgia, Kentucky, Texas, the Dakotas, Oregon, and California, and had been maintained until resident Indian populations were decimated or disappeared altogether. Entire peoples such as the Cherokee had been reduced to half their size through a policy of forced removal from their homelands east of the Mississippi River to what were then considered less preferable areas in the West.

Others, such as the Navajo, suffered the same fate while under military guard for years on end. The United States Army had also perpetrated a long series of wholesale massacres of Indians at places like Horseshoe Bend, Bear River, Sand Creek, the Washita River, the Marias River, Camp Robinson, and Wounded Knee.

Through it all, hundreds of popular novels—each competing with the next to make Indians appear more grotesque, menacing, and inhuman—were sold in the tens of millions of copies in the U.S. Plainly, the Euro-American public was being conditioned to see Indians in such a way as to allow their eradication to continue. And continue it did until the Manifest Destiny of the U.S.—a direct precursor to what Hitler would subsequently call Lebensraumpolitik (the politics of living space)—was consummated.

By 1900, the national project of "clearing" Native Americans from their land and replacing them with "superior" Anglo-American settlers was complete: the indigenous population had been reduced by as much as 98 percent while approximately 97.5 percent of their original territory had "passed" to the invaders. The survivors had been concentrated, out of sight and mind of the public, on scattered "reservations," all of them under the self-assigned "plenary" (full) power of the federal government. There was, of course, no Nuremberg-style tribunal passing judgment on those who had fostered such circumstances in North America. No U.S. official or private citizen was ever imprisoned—never mind hanged—for implementing or propagandizing what had been done. Nor had the process of genocide afflicting Indians been completed. Instead, it merely changed form.

Between the 1880s and the 1980s, nearly half of all Native American children were coercively transferred from their own families, communities, and cultures to those of the conquering society. This was done through compulsory attendance at remote boarding schools, often hundreds of miles from their homes, where native children were kept for years on end while being systematically "deculturated" (indoctrinated to think and act in the manner of Euro-Americans rather than Indians). It was also accomplished through a pervasive foster home and adoption program—including "blind" adoptions, where children would be permanently denied information as to who they were/are and where they'd come from—placing native youths in non-Indian homes.

The express purpose of all this was to facilitate a U.S. governmental policy to bring about the "assimilation" (dissolution) of indigenous societies. In other words, Indian cultures as such were to be caused to disappear. Such policy objectives are directly contrary to the United Nations 1948 Convention on Punishment and Prevention of the Crime of Genocide, an element of international law arising from the Nuremberg proceedings. The forced "transfer of the children" of a targeted "racial, ethnical, or religious group" is explicitly prohibited as a genocidal activity under the Convention's second article.

Article II of the Genocide Convention also expressly prohibits involuntary sterilization as a means of "preventing births among" a targeted population. Yet, in 1975, it was conceded by the U.S. government that its Indian Health Service (IHS), then a subpart of the Bureau of Indian Affairs (BIA), was even then conducting a secret program of involuntary sterilization that had affected approximately 40 percent of all Indian women. The program was allegedly discontinued, and the IHS was transferred to the Public Health Service, but no one was punished. In 1990, it came out that the IHS was inoculating Inuit children in Alaska with Hepatitis-B vaccine. The vaccine had already been banned by the World Health Organization as having a demonstrated correlation with the HIV-Syndrome which is itself correlated to AIDS. As this is written, a "field test" of Hepatitis-A vaccine, also HIV-correlated, is being conducted on Indian reservations in the northern plains region.

The Genocide Convention makes it a "crime against humanity" to create conditions leading to the destruction of an identifiable human group, as such. Yet the BIA has utilized the government's plenary prerogatives to negotiate mineral leases "on behalf of" Indian peoples paying a fraction of standard royalty rates. The result has been "super profits" for a number of preferred U.S. corporations. Meanwhile, Indians, whose reservations ironically turned out to be in some of the most mineral-rich areas of North America, which makes us, the nominally wealthiest segment of the continent's population, live in dire poverty.

By the government's own data in the mid-1980s, Indians received the lowest annual and lifetime per capita incomes of any aggregate population group in the United States. Concomitantly, we suffer the highest rate of infant mortality, death by exposure and malnutrition, disease, and the like. Under such circumstances, alcoholism and other escapist forms of substance abuse are endemic in the Indian community, a situation which leads both to a general physical debilitation of the population and a catastrophic accident rate. Teen suicide among Indians is several times the national average.

The average life expectancy of a reservation-based Native American man is barely 45 years; women can expect to live less than three years longer.

Such itemizations could be continued at great length, including matters like the radioactive contamination of large portions of contemporary Indian Country, the forced relocation of traditional Navajos, and so on. But the point should be made: genocide, as defined in international law, is a continuing fact of day-to-day life (and death) for North America's native peoples. Yet there has been—and is—only the barest flicker of public concern about, or even consciousness of, this reality. Absent any serious expression of public outrage, no one is punished and the process continues.

A salient reason for public acquiescence before the ongoing holocaust in Native North America has been a continuation of the popular legacy, often through more effective media. Since 1925, Hollywood has released more than 2,000 films, many of them rerun frequently on television, portraying Indians as strange, perverted, ridiculous, and often dangerous things of the past. Moreover, we are habitually presented to mass audiences one-dimensionally, devoid of recognizable human motivations and emotions; Indians thus serve as props, little more. We have thus been thoroughly and systematically dehumanized.

Nor is this the extent of it. Everywhere, we are used as logos, as mascots, as jokes: "Big-Chief" writing tablets, "Red Man" chewing tobacco, "Winnebago" campers, "Navajo" and "Cherokee" and "Pontiac" and "Cadillac" pickups and automobiles. There are the Cleveland "Indians," the Kansas City "Chiefs," the Atlanta "Braves" and the Washington "Redskins" professional sports teams—not to mention those in thousands of colleges, high schools, and elementary schools across the country—each with their own degrading caricatures and parodies of Indians and/or things Indian. Pop fiction continues in the same vein, including an unending stream of New Age manuals purporting to expose the inner works of indigenous spirituality in everything from pseudo-philosophical to do-it-yourself styles. Blond yuppies from Beverly Hills amble about the country claiming to be reincarnated 17th century Cheyenne Ushamans ready to perform previously secret ceremonies.

In effect, a concerted, sustained, and in some ways accelerating effort has gone into making Indians unreal. It is thus of obvious importance that the American public begin to think about the implications of such things the next time they witness a gaggle of face-painted and war-bonneted buffoons doing the "Tomahawk Chop" at a baseball or football game. It is necessary that they think about the implications of the grade-school teacher adorning their child in turkey feathers to commemorate Thanksgiving. Think about the significance of John Wayne or Charlton Heston killing a dozen "savages" with a single bullet the next time a western comes on TV. Think about why Land-o-Lakes finds it appropriate to market its butter with the stereotyped image of an "Indian princess" on the wrapper. Think about what it means when non-Indian academics profess—as they often do—to "know more about Indians than Indians do themselves." Think about the significance of charlatans like Carlos Castaneda and Jamake Highwater and Mary Summer Rain and Lynn Andrews churning out "Indian" bestsellers, one after the other, while Indians typically can't get into print.

Think about the real situation of American Indians. Think about Julius Streicher. Remember Justice Jackson's admonition. Understand that the treatment of Indians in American popular culture is not "cute" or "amusing" or just "good, clean fun."

Know that it causes real pain and real suffering to real people. Know that it threatens our very survival. And know that this is just as much a crime against humanity as anything the Nazis ever did. It is likely that the indigenous people of the United States will never demand that those guilty of such criminal activity be punished for their deeds. But the least we have the right to expect—indeed, to demand—is that such practices finally be brought to a halt.

MEDIA MAGIC   36
Making Class Invisible

Gregory Mantsios

Of the various social and cultural forces in our society, the mass media is arguably the most influential in molding public consciousness. Americans spend an average twenty-eight hours per week watching television. They also spend an undetermined number of hours reading periodicals, listening to the radio, and going to the movies. Unlike other cultural and socializing institutions, ownership and control of the mass media is highly concentrated. Twenty-three corporations own more than one-half of all the daily newspapers, magazines, movie studios, and radio and television outlets in the United States.[1] The number of media companies is shrinking and their control of the industry is expanding. And a relatively small number of media outlets is producing and packaging the majority of news and entertainment programs. For the most part, our media is national in nature and single-minded (profit-oriented) in purpose. This media plays a key role in defining our cultural tastes, helping us locate ourselves in history, establishing our national identity, and ascertaining the range of national and social possibilities. In this essay, we will examine the way the mass media shapes how people think about each other and about the nature of our society.

The United States is the most highly stratified society in the industrialized world. Class distinctions operate in virtually every aspect of our lives, determining the nature of our work, the quality of our schooling, and the health and safety of our loved ones. Yet remarkably, we, as a nation, retain illusions about living in an egalitarian society. We maintain these illusions, in large part, because the media hides gross inequities from public view. In those instances when inequities are revealed, we are provided with messages that obscure the nature of class realities and blame the victims of class-dominated society for their own plight. Let's briefly examine what the news media, in particular, tells us about class.

ABOUT THE POOR

The news media provides meager coverage of poor people and poverty. The coverage it does provide is often distorted and misleading.

From Paula Rothenberg, ed., *Race, Class, and Gender in the United States: An Integrated Study*, 4th ed. (New York: St. Martin's Press, 1998). Reprinted with permission of the author.

The Poor Do Not Exist

For the most part, the news media ignores the poor. Unnoticed are forty million poor people in the nation—a number that equals the entire population of Maine, Vermont, New Hampshire, Connecticut, Rhode Island, New Jersey, and New York combined. Perhaps even more alarming is that the rate of poverty is increasing twice as fast as the population growth in the United States. Ordinarily, even a calamity of much smaller proportion (e.g., flooding in the Midwest) would garner a great deal of coverage and hype from a media usually eager to declare a crisis, yet less than one in five hundred articles in the *New York Times* and one in one thousand articles listed in the *Readers Guide to Periodic Literature* are on poverty. With remarkably little attention to them, the poor and their problems are hidden from most Americans.

When the media does turn its attention to the poor, it offers a series of contradictory messages and portrayals.

The Poor Are Faceless

Each year the Census Bureau releases a new report on poverty in our society and its results are duly reported in the media. At best, however, this coverage emphasizes annual fluctuations (showing how the numbers differ from previous years) and ongoing debates over the validity of the numbers (some argue the number should be lower, most that the number should be higher). Coverage like this desensitizes us to the poor by reducing poverty to a number. It ignores the human tragedy of poverty—the suffering, indignities, and misery endured by millions of children and adults. Instead, the poor become statistics rather than people.

The Poor Are Undeserving

When the media does put a face on the poor, it is not likely to be a pretty one. The media will provide us with sensational stories about welfare cheats, drug addicts, and greedy panhandlers (almost always urban and Black). Compare these images and the emotions evoked by them with the media's treatment of middle-class (usually white) "tax evaders," celebrities who have a "chemical dependency," or wealthy businesspeople who use unscrupulous means to "make a profit." While the behavior of the more affluent offenders is considered an "impropriety" and a deviation from the norm, the behavior of the poor is considered repugnant, indicative of the poor in general, and worthy of our indignation and resentment.

The Poor Are an Eyesore

When the media does cover the poor, they are often presented through the eyes of the middle class. For example, sometimes the media includes a story about community resistance to a homeless shelter or storekeeper annoyance with panhandlers. Rather than focusing on the plight of the poor, these stories are about middle-class opposition to the poor. Such stories tell us that the poor are an inconvenience and an irritation.

The Poor Have Only Themselves to Blame

In another example of media coverage, we are told that the poor live in a personal and cultural cycle of poverty that hopelessly imprisons them. They routinely center on the Black urban population and focus on perceived personality or cultural traits that doom the poor. While the women in these stories typically exhibit an "attitude" that leads to trouble or a promiscuity that leads to single motherhood, the men possess a need for immediate gratification that leads to drug abuse or an unquenchable greed that leads to the pursuit of fast money. The images that are seared into our mind are sexist, racist, and classist. Census figures reveal that most of the poor are white not Black or Hispanic, that they live in rural or suburban areas not urban centers, and hold jobs at least part of the year.[2] Yet, in a fashion that is often framed in an understanding and sympathetic tone, we are told that the poor have inflicted poverty on themselves.

The Poor Are Down on Their Luck

During the Christmas season, the news media sometimes provides us with accounts of poor individuals or families (usually white) who are down on their luck. These stories are often linked to stories about soup kitchens or other charitable activities and sometimes call for charitable contributions. These "Yule time" stories are as much about the affluent as they are about the poor: they tell us that the affluent in our society are a kind, understanding, giving people—which we are not.* The series of unfortunate circumstances that have led to impoverishment are presumed to be a temporary condition that will improve with time and a change in luck.

Despite appearances, the messages provided by the media are not entirely disparate. With each variation, the media informs us what poverty is not (i.e., systemic and indicative of American society) by informing us what it is. The media tells us that poverty is either an aberration of the American way of life

---

*American households with incomes of less than $10,000 give an average of 5.5 percent of their earning to charity or to a religious organization, while those making more than $100,000 a year give only 2.9 percent. After changes in the 1986 tax code reduced the benefits of charitable giving, taxpayers earning $500,000 or more slashed their average donation by nearly one-third. Furthermore, many of these acts of benevolence do not help the needy. Rather than provide funding to social service agencies that aid the poor, the voluntary contributions of the wealthy go to places and institutions that entertain, inspire, cure, or educate wealthy Americans—art museums, opera houses, theaters, orchestras, ballet companies, private hospitals, and elite universities. (Robert Reich, "Secession of the Successful," *New York Times Magazine*, February 17, 1991, p. 43.)

(it doesn't exist, it's just another number, it's unfortunate but temporary) or an end product of the poor themselves (they are a nuisance, do not deserve better, and have brought their predicament upon themselves).

By suggesting that the poor have brought poverty upon themselves, the media is engaging in what William Ryan has called "blaming the victim."[3] The media identifies in what ways the poor are different as a consequence of deprivation, then defines those differences as the cause of poverty itself. Whether blatantly hostile or cloaked in sympathy, the message is that there is something fundamentally wrong with the victims—their hormones, psychological makeup, family environment, community, race, or some combination of these—that accounts for their plight and their failure to lift themselves out of poverty.

But poverty in the United States is systemic. It is a direct result of economic and political policies that deprive people of jobs, adequate wages, or legitimate support. It is neither natural nor inevitable: there is enough wealth in our nation to eliminate poverty if we chose to redistribute existing wealth or income. The plight of the poor is reason enough to make the elimination of poverty the nation's first priority. But poverty also impacts dramatically on the non-poor. It has a dampening effect on wages in general (by maintaining a reserve army of unemployed and underemployed anxious for any job at any wage) and breeds crime and violence (by maintaining conditions that invite private gain by illegal means and rebellion-like behavior, not entirely unlike the urban riots of the 1960s). Given the extent of poverty in the nation and the impact it has on us all, the media must spin considerable magic to keep the poor and the issue of poverty and its root causes out of the public consciousness.

ABOUT EVERYONE ELSE

Both the broadcast and the print news media strive to develop a strong sense of "we-ness" in their audience. They seek to speak to and for an audience that is both affluent and like-minded. The media's solidarity with affluence, that is, with the middle and upper class, varies little from one medium to another. Benjamin DeMott points out, for example, that the *New York Times* understands affluence to be intelligence, taste, public spirit, responsibility, and a readiness to rule and "conceives itself as spokesperson for a readership awash in these qualities."[4] Of course, the flip side to creating a sense of "we," or "us," is establishing a perception of the "other." The other relates back to the faceless, amoral, undeserving, and inferior "underclass." Thus, the world according to the news media is divided between the "underclass" and everyone else. Again the messages are often contradictory.

The Wealthy Are Us

Much of the information provided to us by the news media focuses attention on the concerns of a very wealthy and privileged class of people. Although the concerns of a small fraction of the populace, they are presented as though they were the concerns of everyone. For example, while relatively few people actually own stock, the news media devotes an inordinate amount of broadcast time and print space to business news and stock market quotations. Not only do business reports cater to a particular narrow clientele, so do the fashion pages (with $2,000 dresses), wedding announcements, and the obituaries. Even weather and sports news often have a class bias. An all news radio station in New York City, for example, provides regular national ski reports. International news, trade agreements, and domestic policies issues are also reported in terms of their impact on business climate and the business community. Besides being of practical value to the wealthy, such coverage has considerable ideological value. Its message: the concerns of the wealthy are the concerns of us all.

The Wealthy (as a Class) Do Not Exist

While preoccupied with the concerns of the wealthy, the media fails to notice the way in which the rich as a class of people create and shape domestic and foreign policy. Presented as an aggregate of individuals, the wealthy appear without special interests, interconnections, or unity in purpose. Out of public view are the class interests of the wealthy, the interlocking business links, the concerted actions to preserve their class privileges and business interests (by running for public office, supporting political candidates, lobbying, etc.). Corporate lobbying is ignored, taken for granted, or assumed to be in the public interest. (Compare this with the media's portrayal of the "strong arm of labor" in attempting to defeat trade legislation that is harmful to the interests of working people.) It is estimated that two-thirds of the U.S. Senate is composed of millionaires.[5] Having such a preponderance of millionaires in the Senate, however, is perceived to be neither unusual nor antidemocratic; these millionaire senators are assumed to be serving "our" collective interests in governing.

The Wealthy Are Fascinating and Benevolent

The broadcast and print media regularly provide hype for individuals who have achieved "super" success. These stories are usually about celebrities and superstars from the sports and entertainment world. Society pages and gossip columns serve to keep the social elite informed of each others' doings, allow the rest of us to gawk at their excesses, and help to keep the American dream alive. The print media is also fond of feature stories on corporate empire builders. These stories provide an occasional "insider's" view of the private and corporate life of industrialists by suggesting a rags to riches account of corporate success. These stories tell us that corporate success is a series of smart moves, shrewd acquisitions, timely mergers, and well thought out executive suite shuffles. By painting the upper class in a positive light, innocent of any wrongdoing (labor leaders and union organizations usually get the opposite treatment), the media assures us that wealth and power are benevolent. One person's capital accumulation is presumed to be good for all. The elite, then, are portrayed as investment wizards, people of special talent and skill, who even their victims (workers and consumers) can admire.

The Wealthy Include a Few Bad Apples

On rare occasions, the media will mock selected individuals for their personality flaws. Real estate investor Donald Trump and New York Yankees owner George Steinbrenner, for example, are admonished by the media for deliberately seeking publicity (a very un-upper class thing to do); hotel owner Leona Helmsley was caricatured for her personal cruelties; and junk bond broker Michael Milkin was condemned because he had the audacity to rob the rich. Michael Parenti points out that by treating business wrongdoings as isolated deviations from the socially beneficial system of "responsible capitalism," the media overlooks the features of the system that produce such abuses and the regularity with which they occur. Rather than portraying them as predictable and frequent outcomes of corporate power and the business system, the media treats abuses as if they were isolated and atypical. Presented as an occasional aberration, these incidents serve not to challenge, but to legitimate, the system.[6]

The Middle Class Is Us

By ignoring the poor and blurring the lines between the working people and the upper class, the news media creates a universal middle class. From this perspective, the size of one's income becomes largely irrelevant: what matters is that most of "us" share an intellectual and moral superiority over the disadvantaged. As *Time* magazine once concluded, "Middle America is a state of mind."[7] "We are all middle class," we are told, "and we all share the same concerns": job security, inflation, tax burdens, world peace, the cost of food and housing, health care, clean air and water, and the safety of our streets. While the concerns of the wealthy are quite distinct from those of the middle class (e.g., the wealthy worry about investments, not jobs), the media convinces us that "we [the affluent] are all in this together."

The Middle Class Is a Victim

For the media, "we" the affluent not only stand apart from the "other"—the poor, the working class, the minorities, and their problems—"we" are also victimized by the poor (who drive up the costs of maintaining the welfare roles), minorities (who commit crimes against us), and by workers (who are greedy and drive companies out and prices up). Ignored are the subsidies to the rich, the crimes of corporate America, and the policies that wreak havoc on the economic well-being of middle America. Media magic convinces us to fear, more than anything else, being victimized by those less affluent than ourselves.

The Middle Class Is Not a Working Class

The news media clearly distinguishes the middle class (employees) from the working class (i.e., blue collar workers) who are portrayed, at best, as irrelevant, outmoded, and a dying breed. Furthermore, the media will tell us that the hardships faced by blue collar workers are inevitable (due to progress), a result of bad luck (chance circumstances in a particular industry), or a product of their own doing (they priced themselves out of a job). Given the media's presentation of reality, it is hard to believe that manual, supervised, unskilled, and semiskilled workers actually represent more than 50 percent of the adult working population.[8] The working class, instead, is relegated by the media to "the other."

In short, the news media either lionizes the wealthy or treats their interests and those of the middle class as one in the same. But the upper class and the middle class do not share the same interests or worries. Members of the upper class worry about stock dividends (not employment), they profit from inflation and global militarism, their children attend exclusive private schools, they eat and live in a royal fashion, they call on (or are called upon by) personal physicians, they have few consumer problems, they can escape whenever they want from environmental pollution, and they live on streets and travel to other areas under the protection of private police forces.*[9]

The wealthy are not only a class with distinct life-styles and interests, they are a ruling class. They receive a disproportionate share of the country's yearly income, own a disproportionate amount of the country's wealth, and contribute a disproportionate number of their members to governmental bodies and decision-making groups—all traits that William Domhoff, in his classic work *Who Rules America*, defined as characteristic of a governing class.[10]

This governing class maintains and manages our political and economic structures in such a way that these structures continue to yield an amazing proportion of our wealth to a minuscule upper class. While the media is not above referring to ruling classes in other countries (we hear, for example, references to Japan's ruling elite),[11] its treatment of the news proceeds as though there were no such ruling class in the United States.

---

*The number of private security guards in the United States now exceeds the number of public police officers. (Robert Reich, "Secession of the Successful," *New York Times Magazine*, February 17, 1991, p. 42.)

Furthermore, the news media inverts reality so that those who are working class and middle class learn to fear, resent, and blame those below, rather than those above them in the class structure. We learn to resent welfare, which accounts for only two cents out of every dollar in the federal budget (approximately $10 billion) and provides financial relief for the needy,* but learn little about the $11 billion the federal government spends on individuals with incomes in excess of $100,000 (not needy),[12] or the $17 billion in farm subsidies, or the $214 billion (twenty times the cost of welfare) in interest payments to financial institutions.

Middle-class whites learn to fear African Americans and Latinos, but most violent crime occurs within poor and minority communities and is neither interracial[†] nor interclass. As horrid as such crime is, it should not mask the destruction and violence perpetrated by corporate America. In spite of the fact that 14,000 innocent people are killed on the job each year, 100,000 die prematurely, 400,000 become seriously ill, and 6 million are injured from work-related accidents and diseases, most Americans fear government regulation more than they do unsafe working conditions.

Through the media, middle-class—and even working-class—Americans learn to blame blue collar workers and their unions for declining purchasing power and economic security. But while workers who managed to keep their jobs and their unions struggled to keep up with inflation, the top 1 percent of American families saw their average incomes soar 80 percent in the last decade.[13] Much of the wealth at the top was accumulated as stockholders and corporate executives moved their companies abroad to employ cheaper labor (56 cents per hour in El Salvador) and avoid paying taxes in the United States. Corporate America is a world made up of ruthless bosses, massive layoffs, favoritism and nepotism, health and safety violations, pension plan losses, union busting, tax evasions, unfair competition, and price gouging, as well as fast buck deals, financial speculation, and corporate wheeling and dealing that serve the interests of the corporate elite, but are generally wasteful and destructive to workers and the economy in general.

It is no wonder Americans cannot think straight about class. The mass media is neither objective, balanced, independent, nor neutral. Those who own and direct the mass media are themselves part of the upper class, and neither they nor the ruling class in general have to conspire to manipulate public opinion. Their interest is in preserving the status quo, and their view of society as fair and equitable comes naturally to them. But their ideology

---

*A total of $20 billion is spent on welfare when you include all state funding. But the average state funding also comes to only two cents per state dollar.

†In 92 percent of the murders nationwide the assailant and the victim are of the same race (46 percent are white/white, 46 percent are black/black), 5.6 percent are black on white, and 2.4 percent are white on black. (FBI and Bureau of Justice Statistics, 1985–1986, quoted in Raymond S. Franklin, *Shadows of Race and Class*, University of Minnesota Press, Minneapolis, 1991, p. 108.)

dominates our society and justifies what is in reality a perverse social order—one that perpetuates unprecedented elite privilege and power on the one hand and widespread deprivation on the other. A mass media that did not have its own class interests in preserving the status quo would acknowledge that inordinate wealth and power undermines democracy and that a "free market" economy can ravage a people and their communities.

NOTES

1. Martin Lee and Norman Solomon, *Unreliable Sources*, Lyle Stuart (New York, 1990), p. 71. See also Ben Bagdikian, *The Media Monopoly*, Beacon Press (Boston, 1990).

2. Department of Commerce, Bureau of the Census, "Poverty in the United States: 92," *Current Population Reports, Consumer Income*, Series P60-185, pp. xi, xv, 1.

3. William Ryan, *Blaming the Victim*, Vintage (New York, 1971).

4. Benjamin Demott, *The Imperial Middle*, William Morrow (New York, 1990), p. 123.

5. Fred Barnes, "The Zillionaires Club," *The New Republic*, January 29, 1990, p. 24.

6. Michael Parenti, *Inventing Reality*, St. Martin's Press (New York, 1986), p. 109.

7. *Time*, January 5, 1979, p. 10.

8. Vincent Navarro, "The Middle Class—A Useful Myth," *The Nation*, March 23, 1992, p. 1.

9. Charles Anderson, *The Political Economy of Social Class*, Prentice Hall (Englewood Cliffs, N.J., 1974), p. 137.

10. William Domhoff, *Who Rules America*, Prentice Hall (Englewood Cliffs, N.J., 1967), p. 5.

11. Lee and Solomon, *Unreliable Sources*, p. 179.

12. *Newsweek*, August 10, 1992, p. 57.

13. *Business Week*, June 8, 1992, p. 86.

THE MYTH OF THE LATIN WOMAN

37

I Just Met a Girl Named María

Judith Ortiz Cofer

On a bus trip to London from Oxford University, where I was earning some graduate credits one summer, a young man, obviously fresh from a pub,

From Judith Ortiz Cofer, *The Latin Deli: Prose & Poetry* (Athens: University of Georgia Press, 1993). Copyright © 1993 by Judith Ortiz Cofer. Reprinted by permission of the publisher.

spotted me and as if struck by inspiration went down on his knees in the aisle. With both hands over his heart he broke into an Irish tenor's rendition of "María" from *West Side Story*. My politely amused fellow passengers gave his lovely voice the round of gentle applause it deserved. Though I was not quite as amused, I managed my version of an English smile: no show of teeth, no extreme contortions of the facial muscles—I was at this time of my life practicing reserve and cool. Oh, that British control, how I coveted it. But María had followed me to London, reminding me of a prime fact of my life: you can leave the Island, master the English language, and travel as far as you can, but if you are a Latina, especially one like me who so obviously belongs to Rita Moreno's gene pool, the Island travels with you.

This is sometimes a very good thing—it may win you that extra minute of someone's attention. But with some people, the same things can make *you* an island—not so much a tropical paradise as an Alcatraz, a place nobody wants to visit. As a Puerto Rican girl growing up in the United States and wanting like most children to "belong," I resented the stereotype that my Hispanic appearance called forth from many people I met.

Our family lived in a large urban center in New Jersey during the sixties, where life was designed as a microcosm of my parents' casas on the island. We spoke in Spanish, we ate Puerto Rican food bought at the bodega, and we practiced strict Catholicism complete with Saturday confession and Sunday mass at a church where our parents were accommodated into a one-hour Spanish mass slot, performed by a Chinese priest trained as a missionary for Latin America.

As a girl I was kept under strict surveillance, since virtue and modesty were, by cultural equation, the same as family honor. As a teenager I was instructed on how to behave as a proper señorita. But it was a conflicting message girls got, since the Puerto Rican mothers also encouraged their daughters to look and act like women and to dress in clothes our Anglo friends and their mothers found too "mature" for our age. It was, and is, cultural, yet I often felt humiliated when I appeared at an American friend's party wearing a dress more suitable to a semiformal than to a playroom birthday celebration. At Puerto Rican festivities, neither the music nor the colors we wore could be too loud. I still experience a vague sense of letdown when I'm invited to a "party" and it turns out to be a marathon conversation in hushed tones rather than a fiesta with salsa, laughter, and dancing—the kind of celebration I remember from my childhood.

I remember Career Day in our high school, when teachers told us to come dressed as if for a job interview. It quickly became obvious that to the barrio girls, "dressing up" sometimes meant wearing ornate jewelry and clothing that would be more appropriate (by mainstream standards) for the company Christmas party than as daily office attire. That morning I had agonized in front of my closet, trying to figure out what a "career girl" would wear because, essentially, except for Marlo Thomas on TV, I had no models on which to base my decision. I knew how to dress for school: at the Catholic school I attended we all wore uniforms; I knew how to dress for Sunday mass, and I knew what dresses to wear for parties at my relatives' homes. Though I do not recall the precise details of my Career Day outfit, it must have been a composite of the above choices. But I remember a comment my friend (an Italian-American) made in later years that coalesced my impressions of that day. She said that at the business school she was attending the Puerto Rican girls always stood out for wearing "everything at once." She meant, of course, too much jewelry, too many accessories. On that day at school, we were simply made the negative models by the nuns who were themselves not credible fashion experts to any of us. But it was painfully obvious to me that to the others, in their tailored skirts and silk blouses, we must have seemed "hopeless" and "vulgar." Though I now know that most adolescents feel out of step much of the time, I also know that for the Puerto Rican girls of my generation that sense was intensified. The way our teachers and classmates looked at us that day in school was just a taste of the culture clash that awaited us in the real world, where prospective employers and men on the street would often misinterpret our tight skirts and jingling bracelets as a come-on.

Mixed cultural signals have perpetuated certain stereotypes—for example, that of the Hispanic woman as the "Hot Tamale" or sexual firebrand. It is a one-dimensional view that the media have found easy to promote. In their special vocabulary, advertisers have designated "sizzling" and "smoldering" as the adjectives of choice for describing not only the foods but also the women of Latin America. From conversations in my house I recall hearing about the harassment that Puerto Rican women endured in factories where the "boss men" talked to them as if sexual innuendo was all they understood and, worse, often gave them the choice of submitting to advances or being fired.

It is custom, however, not chromosomes, that leads us to choose scarlet over pale pink. As young girls, we were influenced in our decisions about clothes and colors by the women—older sisters and mothers who had grown up on a tropical island where the natural environment was a riot of primary colors, where showing your skin was one way to keep cool as well as to look sexy. Most important of all, on the island, women perhaps felt freer to dress and move more provocatively, since, in most cases, they were protected by the traditions, mores, and laws of a Spanish/Catholic system of morality and machismo whose main rule was: *You may look at my sister, but if you touch her I will kill you.* The extended family and church structure could provide a young woman with a circle of safety in her small pueblo on the Island; if a man "wronged" a girl, everyone would close in to save her family honor.

This is what I have gleaned from my discussions as an adult with older Puerto Rican women. They have told me about dressing in their best party clothes on Saturday nights and going to the town's plaza to promenade with their girlfriends in front of the boys they liked. The males were thus given an opportunity to admire the women and to express their admiration in the form of *piropos:* erotically charged street poems they composed on the spot. I have been subjected to a few piropos while visiting the Island, and they can be outrageous, although custom dictates that they must never cross into obscenity. This ritual, as I understand it, also entails a show of studied indifference on the woman's part; if she is "decent," she must not acknowledge the man's impassioned words. So I do understand how things can be lost in translation. When a Puerto Rican girl dressed in her idea of what is attractive meets a man from the mainstream culture who has been trained to react to certain types of clothing as a sexual signal, a clash is likely to take place. The line I first heard based on this aspect of the myth happened when the boy who took me to my first formal dance leaned over to plant a sloppy overeager kiss painfully on my mouth, and when I didn't respond with sufficient passion said in a resentful tone: "I thought you Latin girls were supposed to mature early"—my first instance of being thought of as a fruit or vegetable—I was supposed to *ripen*, not just grow into womanhood like other girls.

It is surprising to some of my professional friends that some people, including those who should know better, still put others "in their place." Though rarer, these incidents are still commonplace in my life. It happened to me most recently during a stay at a very classy metropolitan hotel favored by young professional couples for their weddings. Late one evening after the theater, as I walked toward my room with my new colleague (a woman with whom I was coordinating an arts program), a middle-aged man in a tuxedo, a young girl in satin and lace on his arm, stepped directly into our path. With his champagne glass extended toward me, he exclaimed, "Evita!"

Our way blocked, my companion and I listened as the man half-recited, half-bellowed "Don't Cry for Me, Argentina." When he finished, the young girl said: "How about a round of applause for my daddy?" We complied, hoping this would bring the silly spectacle to a close. I was becoming aware that our little group was attracting the attention of the other guests. "Daddy" must have perceived this too, and he once more barred the way as we tried to walk past him. He began to shout-sing a ditty to the tune of "La Bamba"—except the lyrics were about a girl named María whose exploits all rhymed with her name and gonorrhea. The girl kept saying "Oh, Daddy" and looking at me with pleading eyes. She wanted me to laugh along with the others. My companion and I stood silently waiting for the man to end his offensive song. When he finished, I looked not at him but at his daughter. I advised her calmly never to ask her father what he had done in the army. Then I walked between them and to my room. My friend complimented me on my cool handling of the situation. I confessed to her that I really had wanted to push the jerk into the swimming pool. I knew that this same man—probably a corporate executive, well educated, even worldly by most standards—would not have been likely to regale a white woman with a dirty song in public. He would perhaps have checked his impulse by assuming that she could be somebody's wife or mother, or at least *somebody* who might take offense. But to him, I was just an Evita or a María: merely a character in his cartoon-populated universe.

Because of my education and my proficiency with the English language, I have acquired many mechanisms for dealing with the anger I experience. This was not true for my parents, nor is it true for the many Latin women working at menial jobs who must put up with stereotypes about our ethnic group such as: "They make good domestics." This is another facet of the myth of the Latin women in the United States. Its origin is simple to deduce. Work as domestics, waitressing, and factory jobs are all that's available to women with little English and few skills. The myth of the Hispanic menial has been sustained by the same media phenomenon that made "Mammy" from *Gone with the Wind* America's idea of the black woman for generations; María, the housemaid or counter girl, is now indelibly etched into the national psyche. The big and the little screens have presented us with the picture of the funny Hispanic maid, mispronouncing words and cooking up a spicy storm in a shiny California kitchen.

This media-engendered image of the Latina in the United States has been documented by feminist Hispanic scholars, who claim that such portrayals are partially responsible for the denial of opportunities for upward mobility among Latinas in the professions. I have a Chicana friend working on a Ph.D. in philosophy at a major university. She says her doctor still shakes his head in puzzled amazement at all the "big words" she uses. Since I do not wear my diplomas around my neck for all to see, I too have on occasion been sent to that "kitchen," where some think I obviously belong.

One such incident that has stayed with me, though I recognize it as a minor offense, happened on the day of my first public poetry reading. It took place in Miami in a boat-restaurant where we were having lunch before the event. I was nervous and excited as I walked in with my notebook in hand. An older woman motioned me to her table. Thinking (foolish me) that she wanted me to autograph a copy of my brand new slender volume of verse, I went over. She ordered a cup of coffee from me, assuming that I was the waitress. Easy enough to mistake my poems for menus, I suppose. I know that it wasn't an intentional act of cruelty, yet with all the good things that happened that day, I remember that scene most clearly, because it reminded me of what I had to overcome before anyone would take me seriously. In retrospect I understand that my anger gave my reading fire, that I have almost always taken doubts in my abilities as a challenge—and that the result is, most times, a feeling of satisfaction at having won a convert when I see the cold, appraising eyes warm to my words, the body language change, the smile that indicates that I have opened some avenue for communication. That day I read to that woman and her lowered eyes told me that she was embarrassed at her little faux pas, and when I willed her to look up to me, it was my victory, and she graciously allowed me to punish her with my full attention. We shook hands at the end of the reading, and I never saw her again. She has probably forgotten the whole thing but maybe not.

Yet I am one of the lucky ones. My parents made it possible for me to acquire a stronger footing in the mainstream culture by giving me the chance at an education. And books and art have saved me from the harsher forms of ethnic and racial prejudice that many of my Hispanic *compañeras* have had to endure. I travel a lot around the United States, reading from my books of poetry and my novel, and the reception I most often receive is one of positive interest by people who want to know more about my culture. There are, however, thousands of Latinas without the privilege of an education or the entrée into society that I have. For them life is a struggle against the misconceptions perpetuated by the myth of the Latina as whore, domestic, or criminal. We cannot change this by legislating the way people look at us. The transformation, as I see it, has to occur at a much more individual level. My personal goal in my public life is to try to replace the old pervasive stereotypes and myths about Latinas with a much more interesting set of realities. Every time I give a reading, I hope the stories I tell, the dreams and fears I examine in my work, can achieve some universal truth which will get my audience past the particulars of my skin color, my accent, or my clothes.

I once wrote a poem in which I called us Latinas "God's brown daughters." This poem is really a prayer of sorts, offered upward, but also, through the human-to-human channel of art, outward. It is a prayer for communication, and for respect. In it, Latin women pray "in Spanish to an Anglo God / with a Jewish heritage," and they are "fervently hoping / that if not omnipotent, / at least He be bilingual."

GLADIATORS, GAZELLES, AND GROUPIES   38
Basketball Love and Loathing

Julianne Malveaux

The woman in me loves the sheer physical impact of a basketball game. I don't dwell on the fine points, the three-point shot or the overtime, and can't even

From Todd Boyd and Kenneth L. Shropshire, eds., *Basketball Jones: America Above the Rim* (New York: New York University Press, 2000), pp. 51–58. Reprinted by permission of New York University Press.

tell you which teams make me sweat (though I could have told you twenty years ago). I can tell you, though, how much I enjoy watching mostly black men engage in an elegantly (and sometimes inelegantly) skilled game, and how engrossing the pace and physicality can be. In my younger days, I spent hours absorbed by street basketball games, taking the subway to watch the Rucker Pros in upper Manhattan, hanging on a bench near the basketball courts on West Fourth Street in the Village. I didn't consider myself a basketball groupie—I was a nerd who liked the game. Still, there is something about muscled, scantily clad men (the shorts have gotten longer in the 1990s) that spoke to me. Once I even joked that watching basketball games was like safe sex in the age of AIDS. Yes, the woman in me, even in my maturity, loves the physical impact of a basketball game, a game where muscles strain, sweat flies, and intensity demands attention.

The feminist in me abhors professional basketball and the way it reinforces gender stereotypes. Men play, women watch. Men at the center, women at the periphery, and eagerly so. Men making millions, women scheming to get to some of the millions through their sex and sexuality. The existence of the Women's National Basketball Association hardly ameliorates my loathing for the patriarchal underbelly of the basketball sport, since women players are paid a scant fraction of men's pay and attract a fraction of the live and television audience. To be sure, women's basketball will grow and develop and perhaps even provide men with role models of what sportsmanship should be. I am also emphatically clear that basketball isn't the only patriarchy in town. Despite women's participation in politics it, too, is a patriarchy, with too many women plying sex and sexuality as their stock in trade, as the impeachment imbroglio of 1998–99 reminds us.

The race person in me has mixed feelings about basketball. On one hand, I enjoy seeing black men out there making big bank, the kind of bank they can't make in business, science, or more traditional forms of work. On the other hand, I'm aware of the minuscule odds that any high school hoopster will become a Michael Jordan. If some young brothers spent less time on the hoops and more in the labs, perhaps the investment of time in chemistry would yield the same mega-millions that professional basketball does. This is hardly the sole fault of the youngsters with NBA aspirations. It is also the responsibility of coaches to introduce a reality check to those young people whose future focus is exclusively on basketball. And it is society's responsibility to make sure there are opportunities for young men, especially young African American men, outside basketball. The athletic scholarship should not be the sole passport to college for low-income young men. Broader access to quality education must be a societal mandate; encouraging young African American men to focus on higher education is critical to our nation's fullest development.

Professional basketball bears an unfortunate similarity to an antebellum plantation, with some coaches accustomed to barking orders and uttering racial expletives to get maximum performance from their players. While few condone the fact that Latrell Sprewell, then of the Golden State Warriors, choked his coach P. J. Carlesimo in 1997, many understand that rude, perhaps racist, but certainly dismissive treatment motivated his violent action. The race of coaches, writers, and commentators, in contrast to that of the players, often has broader racial implications. White players whose grammar needs a boost often get it from sportswriters, who make them sound far more erudite than they are. The same writers quote black players with exaggerated, cringe-producing broken English—"dis," "dat," "dese," and "dose"—almost a parody of language. White commentators frequently remark on the "natural talent" of black players, compared to the skill of white ones. Retired white players are more likely to get invitations to coach and manage than are their black counterparts.

Despite this plantation tension, there are those who tout the integration in basketball as something lofty and desirable and the sport itself as one that teaches discipline, teamwork, and structure. In *Values of the Game*, former New Jersey senator and 2000 presidential candidate Bill Bradley describes the game as one of passion, discipline, selflessness, respect, perspective, courage, and other virtues. Though Bradley took a singular position as a senator in lifting up the issue of police brutality around the Rodney King beating, his record on race matters is otherwise relatively unremarkable. But Bradley gets credit for being far more racially progressive than he actually is because he, a white man, played for the New York Knicks for a decade, living at close quarters with African American men.

As onerous as I find the basketball plantation, the race person in me wonders why I am so concerned about the gladiators. Few if any of them exhibit anything that vaguely resembles social consciousness. Michael Jordan, for example, passed on the opportunity to make a real difference in the 1990 North Carolina Senate race between the former Charlotte mayor and Democratic candidate, the African American architect Harvey Gantt, and the ultraconservative and racially manipulative Republican senator Jesse Helms, preferring to save his endorsements for Nike. Charles Barkley once arrogantly crowed that he was not a role model, ignoring the biblical adage that much is expected of those to whom much is given. Few of the players, despite their millions, invest in the black community and in black economic development. Magic Johnson is a notable exception, with part ownership in a Los Angeles–based black bank and development of a theater chain that has the potential for revitalizing otherwise abandoned inner-city communities. Johnson's example notwithstanding, basketball players are more likely to make headlines for their shenanigans than for giving back to the community.

With my mixed feelings, love and loathing, why pay attention to basketball at all? Why not tune it out as forcefully as I tune out anything else I'm disinterested in? The fact is that it is nearly impossible to tune out, turn off, or ignore basketball. It is a cultural delimiter, a national export, a medium through which messages about race, gender, and power are transmitted not only nationally but also internationally. As Walter LaFeber observes in his book, *Michael Jordan and the New Global Capitalism*, Mr. Jordan's imagery has been used not only to sell the basketball game and Nike shoes but to make hundreds of millions of dollars for both Mr. Jordan and the companies he represents. Jordan is not to be faulted for brokering his skill into millions of dollars. Still, the mode and methods of his enrichment make the game of basketball a matter of intellectual curiosity as one explores the way in which capitalism and popular culture intersect. Jordan's millions, interesting as they may be, are less interesting than the way in which media have created his iconic status in both domestic and world markets and offered him up as a role model of the reasonable, affable African American man, the antithesis of "bad boy" trash talkers who so frequently garner headlines.

From a gender perspective, too, there is another set of questions. What would a woman have to do to achieve the same influence, iconic status, and bankability as Jordan? Is it even conceivable, in a patriarchal world, that a woman could earn, gain, or be invented in as iconic a status as a Michael Jordan? A woman might sing—but songbirds come and go and aren't often connected with international marketing of consumer products. She might dance—but that, too, would not turn into international marketability. She might possess the dynamic athleticism of the tennis-playing Williams sisters, Venus and Serena, or she might have the ethereal beauty of a Halle Berry. In either case, her product identification would solely be connected with "women's" products—hair, beauty, and women's sports items. The commercial heavy lifting has been left to the big boys, or to one big boy in particular, Michael Jordan.

To be sure, the public acceptance of women's sports has changed. Thanks to Title IX, women's sports get better funding and more attention at the college level than they did only decades ago. After several fits and starts, the professional women's basketball league seems to be doing well, although not as well as the NBA. In time this may change, but my informal survey of male basketball fans suggests that some find women's basketball simply less entertaining than the NBA. "When women start dunking and trash talking, more people will start watching," said a man active in establishing Midnight Basketball teams in urban centers. If this is the case, then it raises questions both about women's basketball and about the tension between male players and officials about on-court buffoonery. Do officials kill the goose that laid the golden egg when they ask a Dennis Rodman to "behave himself"?

My ire is not about gender envy; it is about imagery and the replication of gender-oppressive patterns by using basketball, especially, to connect an image of sportsmanship, masculinity, with product identification. In a commercial context that promotes this connection, there is both a subtle and a not-so-subtle message about the status of women.

There is also a set of subtle messages about race and race relations that emerges when African American men are used in the same way that African American women were used in the "trade cards" of the early twentieth century. In the patriarchal context, hoopsters become hucksters while the invisible (white) male role as power broker is reinforced. In too many ways the rules and profit of the game reinforce the rules, profit, and history of American life, with iconic gladiators serving as symbols and servants of multinational interests. Black men who entertain serve as stalkers for white men who measure profits. Neutered black men can join their white colleagues in chachinging cash registers but can never unlock the golden handcuffs and the platinum muzzles that limit their ability to generate independent opinions. Women? Always seen, never heard. Ornamental or invisible assets. Pawns in an unspoken game.

A subtext of the basketball culture relegates women, especially African American women, to a peripheral, dependent, and soap-operatic role. These women are sometimes seen as sources of trouble (as in the imbroglio with the Washington then Bullets, now Wizards, where apparently false rape charges spotlighted the lives of two players and perhaps changed the long-term composition of the team). Or they are depicted (for example, in a 1998 *Sports Illustrated* feature) as predatory sperm collectors, whose pregnancies are part of a "plot" to collect child support from high-earning hoopsters.

There are aspects to some women's behavior in relationship with basketball stars that are hardly laudable. (Some of this is detailed in the fictional *Homecourt Advantage*, written by two basketball significant others.) At the same time, it is clear that this behavior is more a symptom than a cause of predatory patriarchy. After all, no one forces a player into bed or into a relationship with an unscrupulous woman. But conditions often make a player irresistible to women whose search for legitimacy comes through attachment to a man.

The disproportionate attention and influence that the basketball lifestyle gives some men makes them thoroughly irresistible to some women. This begs the question: In a patriarchy, just what can women do to attain the same influence and irresistibility? In a patriarchy, male games are far more intriguing than female games. Women, spectators, can watch and attach themselves to high-achieving gladiators. There is little that they can do, in the realm of sport, capitalism, and imagery, to gain the same access that men have. Again, this is not restricted to basketball but is a reflection of gender roles in our society. One might describe this as the "Hillary Rodham Clinton dilemma." Does a woman gain more power and influence by developing her own career or by attaching herself to a powerful man?

This dilemma may be a short-term one, given changes in the status of women in our society. Still, a range of scholars have noted that while women have come a long way, with higher incomes, higher levels of labor-force participation, and increased representation in executive ranks, at the current pace, income equality (which may not be the equivalent of the breakdown of patriarchy) will occur in the middle of the twenty-first century.

Those who depict the basketball culture have been myopic in their focus on the men who shoot hoops, because their coverage has ignored the humanity of the women who hover around hoopsters like moths drawn to a flame. Those writers who are eager to write about the "paternity ward" or about "Darwin's athletes" might also ask how the gender relations in basketball replicate or differ from gender relations in our society. Unfortunately, gender relations in basketball are too often the norm in society. Men play, women watch. To the extent that the basketball culture is elevated, women's roles are denigrated. Why are we willing to endow male hoopsters with a status that no woman can attain in our society? Sedentary men sing the praises of male hoopsters, sit enthralled and engrossed by their games, thus elevating a certain form of achievement in our society. This behavior is seen as normal, even laudable. Yet it takes women, often the daughters or sisters of these enthralled men, out of the iconic, high-achievement mix and pushes them to the periphery of a culture that reveres the athletic achievement derived from the combination of basketball prowess and patriarchy.

The combination of prowess and patriarchy denigrates the groupie who is attracted by the bright lights, but it also dehumanizes the swift gazelle, the gladiator, and the hoopster, whose humanity is negated by his basketball identity. Media coverage depicts young men out of control, trash-talking, wild-walking, attention-grabbing icons. Men play, women watch. Were there other center stages, this would be of limited interest. But the fact is that this stage is one from which other stages reverberate. Men play, women watch, in politics, economics, technology, and sports. And it is "blown up" into international cultural supremacy in basketball.

This woman watches, with love and loathing, the way basketball norms reinforce those that exist in our society. It is a triumph of patriarchy, this exuberance of masculine physicality. It is a reminder to women that feminism notwithstanding, we have yet to gain economic and cultural equivalency with Michael Jordan's capitalist dominance. While we should not, perhaps, ask men to walk away from arenas in which they can dominate, we must ask ourselves why there is no equivalent space for us; why the basketball tenet that men play, women watch, reverberates in so many other sectors of our society.

References

Bradley, Bill. *Values of the Game.* New York: Artisan Press, 1998.

Ewing, Rita, and Crystal McCrary. *Homecourt Advantage.* New York: Avon Books, 1998.

Frey, Darcy. *The Last Shot: City Street, Basketball Dreams.* Boston: Houghton Mifflin, 1994.

Hoberman, John. *Darwin's Athletes: How Sport Has Damaged Black America and Preserved the Myth of Race.* New York: Mariner Books, 1997.

LaFeber, Walter. *Michael Jordan and the New Global Capitalism.* New York: W. W. Norton & Co., 1999.

Wahl, Grant, and L. Jon Wertheim. "Paternity Ward." *Sports Illustrated,* May 4, 1998, at 62.

TAKING MULTICULTURAL, ANTIRACIST EDUCATION SERIOUSLY

39

An Interview with Enid Lee

Barbara Miner

What do you mean by a multicultural education?   The term *multicultural education* has a lot of different meanings. The term I use most often is *antiracist education.*

Multicultural or antiracist education is fundamentally a perspective. It's a point of view that cuts across all subject areas, and addresses the histories and experiences of people who have been left out of the curriculum. Its purpose is to help us deal equitably with all the cultural and racial differences that you find in the human family. It's also a perspective that allows us to get at explanations for why things are the way they are in terms of power relationships, in terms of equality issues.

So when I say multicultural or antiracist education, I am talking about equipping students, parents, and teachers with the tools needed to combat racism and ethnic discrimination, and to find ways to build a society that includes all people on an equal footing.

It also has to do with how the school is run in terms of who gets to be involved with decisions. It has to do with parents and how their voices are heard or not heard. It has to do with who gets hired in the school.

If you don't take multicultural education or antiracist education seriously, you are actually promoting a monocultural or racist education. There is not neutral ground on this issue.

From David Levine, Robert Lowe, Bob Peterson, and Rita Tenorio, eds., *Rethinking Schools: An Agenda for Change* (New York: New Press, 1995), pp. 9–16. Reprinted by permission.

Why do you use the term antiracist education instead of multicultural education?
Partly because in Canada, multicultural education often has come to mean
something that is quite superficial: the dances, the dress, the dialect, the
dinners. And it does so without focusing on what those expressions of culture
mean: the values, the power relationships that shape the culture.

I also use the term *antiracist education* because a lot of multicultural education hasn't looked at discrimination. It has the view, "People are different
and isn't that nice," as opposed to looking at how some people's differences are
looked upon as deficits and disadvantages. In antiracist education, we attempt
to look at—and change—those things in school and society that prevent
some differences from being valued.

Oftentimes, whatever is white is treated as normal. So when teachers
choose literature that they say will deal with a universal theme or story, like
childhood, all the people in the stories are of European origin; it's basically
white culture and civilization. That culture is different from others, but it
doesn't get named as different. It gets named as normal.

Antiracist education helps us move that European perspective over to the
side to make room for other cultural perspectives that must be included.

*What are some ways your perspective might manifest itself in a kindergarten
classroom, for example?* It might manifest itself in something as basic as the
kinds of toys and games that you select. If all the toys and games reflect the
dominant culture and race and language, then that's what I call a monocultural
classroom even if you have kids of different backgrounds in the class.

I have met some teachers who think that just because they have kids from
different races and backgrounds, they have a multicultural classroom. Bodies
of kids are not enough.

It also gets into issues such as what kind of pictures are up on the wall?
What kinds of festivals are celebrated? What are the rules and expectations
in the classroom in terms of what kinds of languages are acceptable? What
kinds of interactions are encouraged? How are the kids grouped? These are
just some of the concrete ways in which a multicultural perspective affects a
classroom.

How does one implement a multicultural or antiracist education? It usually happens in stages. Because there's a lot of resistance to change in schools, I don't
think it's reasonable to expect to move straight from a monocultural school to
a multiracial school.

First there is this surface stage in which people change a few expressions
of culture in the school. They make welcome signs in several languages and
have a variety of foods and festivals. My problem is not that they start there.
My concern is that they often stop there. Instead, what they have to do is
move very quickly and steadily to transform the entire curriculum. For

example, when we say classical music, whose classical music are we talking about? European? Japanese? And what items are on the tests? Whose culture do they reflect? Who is getting equal access to knowledge in the school? Whose perspective is heard? Whose is ignored?

The second stage is transitional and involves creating units of study. Teachers might develop a unit on Native Americans, or Native Canadians, or people of African background. And they have a whole unit that they study from one period to the next. But it's a separate unit and what remains intact is the main curriculum, the main menu. One of the ways to assess multicultural education in your school is to look at the school organization. Look at how much time you spend on which subjects. When you are in the second stage you usually have a two- or three-week unit on a group of people or an area that's been omitted in the main curriculum.

You're moving into the next stage of structural change when you have elements of that unit integrated into existing units. Ultimately, what is at the center of the curriculum gets changed in its prominence. For example, civilizations. Instead of just talking about Western civilization, you begin to draw on what we need to know about India, Africa, China. We also begin to ask different questions about why and what we are doing. Whose interest is it in that we study what we study? Why is it that certain kinds of knowledge get hidden? In mathematics, instead of studying statistics with sports and weather numbers, why not look at employment in light of ethnicity?

Then there is the social change stage, when the curriculum helps lead to changes outside of the school. We actually go out and change the nature of the community we live in. For example, kids might become involved in how the media portray people and start a letter-writing campaign about news that is negatively biased. Kids begin to see this as a responsibility that they have to change the world.

I think about a group of elementary school kids who wrote to the manager of the store about the kinds of games and dolls that they had. That's a long way from having some dinner and dances that represent an "exotic" form of life.

In essence, in antiracist education we use knowledge to empower people and to change their lives.

Teachers have limited money to buy new materials. How can they begin to incorporate a multicultural education even if they don't have a lot of money?   We do need money and it is a pattern to underfund antiracist intiatives so that they fail. We must push for funding for new resources because some of the information we have is downright inaccurate. But if you have a perspective, which is really a set of questions that you ask about your life, and you have the kids ask, then you can begin to fill in the gaps.

Columbus is a good example. It turns the whole story on its head when you have the children try to find out what the people who were on this continent might have been thinking and doing and feeling when they were being "discovered," tricked, robbed, and murdered. You might not have that information on hand, because that kind of knowledge is deliberately suppressed. But if nothing else happens, at least you shift your teaching, to recognize the native peoples as human beings, to look at things from their view.

There are other things you can do without new resources. You can include, in a sensitive way, children's backgrounds and life experiences. One way is through interviews with parents and with community people, in which they can recount their own stories, especially their interactions with institutions like schools, hospitals, and employment agencies. These are things that often don't get heard.

I've seen schools inviting grandparents who can tell stories about their own lives, and these stories get to be part of the curriculum later in the year. It allows excluded people, it allows humanity, back into the schools. One of the ways that discrimination works is that it treats some people's experiences, lives, and points of view as though they don't count, as though they are less valuable than other people's.

I know we need to look at materials. But we can also take some of the existing curriculum and ask kids questions about what is missing, and whose interest is being served when things are written in the way they are. Both teachers and students must alter that material.

How can a teacher who knows little about multiculturalism be expected to teach multiculturally?   I think the teachers need to have the time and encouragement to do some reading and to see the necessity to do so. A lot has been written about multiculturalism. It's not like there's no information.

You also have to look around at what people of color are saying about their lives and draw from those sources. You can't truly teach this until you reeducate yourself from a multicultural perspective. But you can begin. It's an ongoing process.

Most of all, you have to get in touch with the fact that your current education has a cultural bias, that it is an exclusionary, racist bias and that it needs to be purged. A lot of times people say, "I just need to learn more about those other groups." And I say, "No, you need to look at how the dominant culture and biases affect your view of nondominant groups in society." You don't have to fill your head with little details about what other cultural groups eat and dance. You need to take a look at your culture, what your idea of normal is, and realize it is quite limited and is in fact just reflecting a particular experience. You have to realize that what you recognize as universal is, quite often, exclusionary. To be really universal, you must begin to learn what Africans,

Asians, Latin Americans, the aboriginal peoples, and all silenced groups of Americans have had to say about the topic.

How can one teach multiculturally without making white children feel guilty or threatened?   Perhaps a sense of being threatened or feeling guilty will occur. But I think it is possible to have kids move beyond that.

First of all, recognize that there have always been white people who have fought against racism and social injustice. White children can proudly identify with these people and join in that tradition of fighting for social justice.

Second, it is in their interest to be opening their minds and finding out how things really are. Otherwise, they will constantly have an incomplete picture of the human family.

The other thing is, if we don't make clear that some people benefit from racism, then we are being dishonest. What we have to do is talk about how young people can use that from which they benefit to change the order of things so that more people will benefit.

If we say we are all equally discriminated against on the basis of racism, or sexism, that's not accurate. We don't need to be caught up in the guilt of our benefit, but should use our privilege to help change things.

I remember a teacher telling me last summer that after she listened to me on the issue of racism, she felt ashamed of who she was. And I remember wondering if her sense of self was founded on a sense of superiority. Because if that's true, then she is going to feel shaken. but if her sense of self is founded on working with people of different colors to change things, then there is no need to feel guilt or shame.

Where does an antisexist perspective fit into a multicultural perspective?   In my experience, when you include racism as just another of the "isms," it tends to get sidetracked or omitted. That's because people are sometimes uncomfortable with racism, although they may be comfortable with class and gender issues. I like to put racism in the foreground, and then include the others by example and analysis.

I certainly believe that sexism—and ageism and heterosexism and class issues—have to be taken up. But in my way of thinking I don't list them under multicultural education.

For me, the emphasis is race. I've seen instances where teachers have replaced a really sexist set of materials with nonsexist materials—but the new resources included only white people. In my judgment, more has been done in the curriculum in terms of sexism that in terms of racism.

Of course, we must continue to address sexism in all its forms, no question about that. But we cannot give up on the fight against racism either. As a black woman, both hurt my heart. . . .

What can school districts do to further multicultural education?   Many teachers will not change the curriculum if they have no administrative support. Sometimes, making these changes can be scary. You can have parents on your back, kids who can be resentful. You can be told you are making the curriculum too political.

What we are talking about here is pretty radical; multicultural education is about challenging the status quo and the basis of power. You need administrative support to do that.

In the final analysis, multicultural or antiracist education is about allowing educators to do the things they have wanted to do in the name of their profession: to broaden the horizons of the young people they teach, to give them skills to change a world in which the color of their skin defines their opportunities, where some human beings are treated as if they are just junior children.

Maybe teachers don't have this big vision all the time. But I think those are the things that a democratic society is supposed to be about.

When you look at the state of things in the United States and Canada, it's almost as if many parts of the society have given up on decency, doing the right thing, and democracy in any serious way. I think that antiracist education gives us an opportunity to try again.

State Institutions and Social Policy

THE FIRST AMERICANS **40**

American Indians

C. Matthew Snipp

By the end of the nineteenth century, many observers predicted that American Indians were destined for extinction. Within a few generations, disease, warfare, famine, and outright genocide had reduced their numbers from millions to less than 250,000 in 1890. Once a self-governing, self-sufficient people, American Indians were forced to give up their homes and their land, and to subordinate themselves to an alien culture. The forced resettlement to reservation lands or the Indian Territory (now Oklahoma) frequently meant a life of destitution, hunger, and complete dependency on the federal government for material needs.

Today, American Indians are more numerous than they have been for several centuries. While still one of the most destitute groups in American society, tribes have more autonomy and are now more self-sufficient than at any time since the last century. In cities, modern pan-Indian organizations have been successful in making the presence of American Indians known to the larger community, and have mobilized to meet the needs of their people (Cornell 1988; Nagel 1986; Weibel-Orlando 1991). In many rural areas, American Indians and especially tribal governments have become increasingly more important and increasingly more visible by virtue of their growing political and economic power. The balance of this [reading] is devoted to explaining their unique place in American society.

From Silvia Pedraza and Rubén G. Rumbaut, eds., *Origins and Destinies: Immigration, Race, and Ethnicity in America.* (Belmont, CA: Wadsworth, 1996), pp. 390–403. Reprinted by permission.

THE INCORPORATION OF AMERICAN INDIANS

The current political and economic status of American Indians is the result of the process by which they were incorporated into Euro-American society (Hall 1989). This amounts to a long history of efforts aimed at subordinating an otherwise self-governing and self-sufficient people that eventually culminated in widespread economic dependency. The role of the U.S. government in this process can be seen in the five major historical periods of federal Indian relations: removal, assimilation, the Indian New Deal, termination and relocation, and self-determination.

Removal

In the early nineteenth century, the population of the United States expanded rapidly at the same time that the federal government increased its political and military capabilities. The character of Indian-American relations changed after the War of 1812. The federal government increasingly pressured tribes settled east of the Appalachian Mountains to move west to the territory acquired in the Louisiana Purchase. Numerous treaties were negotiated by which the tribes relinquished most of their land and eventually were forced to move west.

Initially the federal government used bargaining and negotiation to accomplish removal, but many tribes resisted (Prucha 1984). However, the election of Andrew Jackson by a frontier constituency signaled the beginning of more forceful measures to accomplish removal. In 1830 Congress passed the Indian Removal Act, which mandated the eventual removal of the eastern tribes to points west of the Mississippi River, in an area which was to become the Indian Territory and is now the state of Oklahoma. Dozens of tribes were forcibly removed from the eastern half of the United States to the Indian Territory and newly created reservations in the west, a long process ridden with conflict and bloodshed.

As the nation expanded beyond the Mississippi River, tribes of the plains, southwest, and west coast were forcibly settled and quarantined on isolated reservations. This was accompanied by the so-called Indian Wars—a bloody chapter in the history of Indian-White relations (Prucha 1984; Utley 1963). This period in American history is especially remarkable because the U.S. government was responsible for what is unquestionably one of the largest forced migrations in history.

The actual process of removal spanned more than a half-century and affected nearly every tribe east of the Mississippi River. Removal often meant extreme hardships for American Indians, and in some cases this hardship reached legendary proportions. For example, the Cherokee removal has become known as the "Trail of Tears." In 1838, nearly 17,000 Cherokees were ordered to leave their homes and assemble in military stockades (Thornton 1987, p. 117). The march to the Indian Territory began in October and continued through the winter months. As many as 8,000 Cherokees died from cold weather and diseases such as influenza (Thornton 1987, p. 118).

According to William Hagan (1979), removal also caused the Creeks to suffer dearly as their society underwent a profound disintegration. The contractors who forcibly removed them from their homes refused to do anything for "the large number who had nothing but a cotton garment to protect them from the sleet storms and no shoes between them and the frozen ground of the last stages of their hegira. About half of the Creek nation did not survive the migration and the difficult early years in the West" (Hagan 1979, pp. 77–81). In the West, a band of Nez Perce men, women, and children, under the leadership of Chief Joseph, resisted resettlement in 1877. Heavily outnumbered, they were pursued by cavalry troops from the Wallowa valley in eastern Oregon and finally captured in Montana near the Canadian border. Although the Nez Perce were eventually captured and moved to the Indian Territory, and later to Idaho, their resistance to resettlement has been described by one historian as "one of the great military movements in history" (Prucha 1984, p. 541).

Assimilation

Near the end of the nineteenth century, the goal of isolating American Indians on reservations and the Indian Territory was finally achieved. The Indian population also was near extinction. Their numbers had declined steadily throughout the nineteenth century, leading most observers to predict their disappearance (Hoxie 1984). Reformers urged the federal government to adopt measures that would humanely ease American Indians into extinction. The federal government responded by creating boarding schools and the allotment acts—both were intended to "civilize" and assimilate American Indians into American society by Christianizing them, educating them, introducing them to private property, and making them into farmers. American Indian boarding schools sought to accomplish this task by indoctrinating Indian children with the belief that tribal culture was an inferior relic of the past and that Euro-American culture was vastly superior and preferable. Indian children were forbidden to wear their native attire, to eat their native foods, to speak their native language, or to practice their traditional religion. Instead, they were issued Euro-American clothes, and expected to speak English and become Christians. Indian children who did not relinquish their culture were punished by school authorities. The curriculum of these schools taught vocational arts along with "civilization" courses.

The impact of allotment policies is still evident today. The 1887 General Allotment Act (the Dawes Severalty Act) and subsequent legislation mandated that tribal lands were to be allotted to individual American Indians, . . . and the surplus lands left over from allotment were to be sold on the open market. Indians who received allotted tribal lands also received citizenship, farm implements, and encouragement from Indian agents to adopt farming as a livelihood (Hoxie 1984, Prucha 1984).

For a variety of reasons, Indian lands were not completely liquidated by allotment, many Indians did not receive allotments, and relatively few changed their lifestyles to become farmers. Nonetheless, the allotment era was a disaster because a significant number of allottees eventually lost their land. Through tax foreclosures, real estate fraud, and their own need for cash, many American Indians lost what for most of them was their last remaining asset (Hoxie 1984).

Allotment took a heavy toll on Indian lands. It caused about 90 million acres of Indian land to be lost, approximately two-thirds of the land that had belonged to tribes in 1887 (O'Brien 1990). This created another problem that continues to vex many reservations: "checkerboarding." Reservations that were subjected to allotment are typically a crazy quilt composed of tribal lands, privately owned "fee" land, and trust land belonging to individual Indian families. Checkerboarding presents reservation officials with enormous administrative problems when trying to develop land use management plans, zoning ordinances, or economic development projects that require the construction of physical infrastructure such as roads or bridges.

The Indian New Deal

The Indian New Deal was short-lived but profoundly important. Implemented in the early 1930s along with the other New Deal programs of the Roosevelt administration, the Indian New Deal was important for at least three reasons. First, signaling the end of the disastrous allotment era as well as a new respect for American Indian tribal culture, the Indian New Deal repudiated allotment as a policy. Instead of continuing its futile efforts to detribalize American Indians, the federal government acknowledged that tribal culture was worthy of respect. Much of this change was due to John Collier, a long-time Indian rights advocate appointed by Franklin Roosevelt to serve as Commissioner of Indian Affairs (Prucha 1984).

Like other New Deal policies, the Indian New Deal also offered some relief from the Great Depression and brought essential infrastructure development to many reservations, such as projects to control soil erosion and to build hydroelectric dams, roads, and other public facilities. These projects created jobs in New Deal programs such as the Civilian Conservation Corps and the Works Progress Administration.

An especially important and enduring legacy of the Indian New Deal was the passage of the Indian Reorganization Act (IRA) of 1934. Until then,

Indian self-government had been forbidden by law. This act allowed tribal governments, for the first time in decades, to reconstitute themselves for the purpose of overseeing their own affairs on the reservation. Critics charge that this law imposed an alien form of government, representative democracy, on traditional tribal authority. On some reservations, this has been an on-going source of conflict (O'Brien 1990). Some reservations rejected the IRA for this reason, but now have tribal governments authorized under different legislation.

Termination and Relocation

After World War II, the federal government moved to terminate its long-standing relationship with Indian tribes by settling the tribes' outstanding legal claims, by terminating the special status of reservations, and by helping reservation Indians relocate to urban areas (Fixico 1986). The Indian Claims Commission was a special tribunal created in 1946 to hasten the settlement of legal claims that tribes had brought against the federal government. In fact, the Indian Claims Commission became bogged down with prolonged cases, and in 1978 the commission was dissolved by Congress. At that time, there were 133 claims still unresolved out of an original 617 that were first heard by the commission three decades earlier (Fixico 1986, p. 186). The unresolved claims that were still pending were transferred to the Federal Court of Claims.

Congress also moved to terminate the federal government's relationship with Indian tribes. House Concurrent Resolution (HCR) 108, passed in 1953, called for steps that eventually would abolish all reservations and abolish all special programs serving American Indians. It also established a priority list of reservations slated for immediate termination. However, this bill and subsequent attempts to abolish reservations were vigorously opposed by Indian advocacy groups such as the National Congress of American Indians. Only two reservations were actually terminated, the Klamath in Oregon and the Menominee in Wisconsin. The Menominee reservation regained its trust status in 1975 and the Klamath reservation was restored in 1986.

The Bureau of Indian Affairs (BIA) also encouraged reservation Indians to relocate and seek work in urban job markets. This was prompted partly by the desperate economic prospects on most reservations, and partly because of the federal government's desire to "get out of the Indian business." The BIA's relocation programs aided reservation Indians in moving to designated cities, such as Los Angeles and Chicago, where they also assisted them in finding housing and employment. Between 1952 and 1972, the BIA relocated more than 100,000 American Indians (Sorkin 1978). However, many Indians returned to their reservations (Fixico 1986). For some American Indians, the return to the reservation was only temporary; for example, during periods when seasonal employment such as construction work was hard to find.

Self-Determination

Many of the policies enacted during the termination and relocation era were steadfastly opposed by American Indian leaders and their supporters. As these programs became stalled, critics attacked them for being harmful, ineffective, or both. By the mid-1960s, these policies had very little serious support. Perhaps inspired by the gains of the Civil Rights movement, American Indian leaders and their supporters made "self-determination" the first priority on their political agendas. For these activists, self-determination meant that Indian people would have the autonomy to control their own affairs, free from the paternalism of the federal government.

The idea of self-determination was well received by members of Congress sympathetic to American Indians. It also was consistent with the "New Federalism" of the Nixon administration. Thus, the policies of termination and relocation were repudiated in a process that culminated in 1975 with the passage of the American Indian Self-Determination and Education Assistance Act, a profound shift in federal Indian policy. For the first time since this nation's founding, American Indians were authorized to oversee the affairs of their own communities, free of federal intervention. In practice, the Self-Determination Act established measures that would allow tribal governments to assume a larger role in reservation administration of programs for welfare assistance, housing, job training, education, natural resource conservation, and the maintenance of reservation roads and bridges (Snipp and Summers 1991). Some reservations also have their own police forces and game wardens, and can issue licenses and levy taxes. The Onondaga tribe in upstate New York have taken their sovereignty one step further by issuing passports that are internationally recognized. Yet there is a great deal of variability in terms of how much autonomy tribes have over reservation affairs. Some tribes, especially those on large and well-organized reservations have nearly complete control over their reservations, while smaller reservations with limited resources often depend heavily on BIA services. . . .

CONCLUSION

Though small in number, American Indians have an enduring place in American society. Growing numbers of American Indians occupy reservation and other trust lands, and equally important has been the revitalization of tribal governments. Tribal governments now have a larger role in reservation affairs than ever in the past. Another significant development has been the urbanization of American Indians. Since 1950, the proportion of American Indians in cities has grown rapidly. These American Indians have in common with reservation Indians many of the same problems and disadvantages, but they also face other challenges unique to city life.

The challenges facing tribal governments are daunting. American Indians are among the poorest groups in the nation. Reservation Indians have substantial needs for improved housing, adequate health care, educational opportunities, and employment, as well as developing and maintaining reservation infrastructure. In the face of declining federal assistance, tribal governments are assuming an ever-larger burden. On a handful of reservations, tribal governments have assumed completely the tasks once performed by the BIA.

As tribes have taken greater responsibility for their communities, they also have struggled with the problems of raising revenues and providing economic opportunities for their people. Reservation land bases provide many reservations with resources for development. However, these resources are not always abundant, much less unlimited, and they have not always been well managed. It will be yet another challenge for tribes to explore ways of efficiently managing their existing resources. Legal challenges also face tribes seeking to exploit unconventional resources such as gambling revenues. Their success depends on many complicated legal and political contingencies.

Urban American Indians have few of the resources found on reservations, and they face other difficult problems. Preserving their culture and identity is an especially pressing concern. However, urban Indians have successfully adapted to city environments in ways that preserve valued customs and activities—powwows, for example, are an important event in all cities where there is a large Indian community. In addition, pan-Indianism has helped urban Indians set aside tribal differences and forge alliances for the betterment of urban Indian communities.

These alliances are essential, because unlike reservation Indians, urban American Indians do not have their own form of self-government. Tribal governments do not have jurisdiction over urban Indians. For this reason, urban Indians must depend on other strategies for ensuring that the needs of their community are met, especially for those new to city life. Coping with the transition to urban life poses a multitude of difficult challenges for many American Indians. Some succumb to these problems, especially the hardships of unemployment, economic deprivation, and related maladies such as substance abuse, crime, and violence. But most successfully overcome these difficulties, often with help from other members of the urban Indian community.

Perhaps the greatest strength of American Indians has been their ability to find creative ways for dealing with adversity, whether in cities or on reservations. In the past, this quality enabled them to survive centuries of oppression and persecution. Today this is reflected in the practice of cultural traditions that Indian people are proud to embrace. The resilience of American Indians is an abiding quality that will no doubt ensure that they will remain part of the ethnic mosaic of American society throughout the twenty-first century and beyond.

References

Cornell, Stephen. 1988. *The Return of the Native: American Indian Political Resurgence.* New York: Oxford University Press.

Fixico, Donald L. 1986. *Termination and Relocation: Federal Indian Policy, 1945–1960.* Albuquerque, NM: University of New Mexico Press.

Hagan, William T. 1979. *American Indians.* Chicago, IL: University of Chicago Press.

Hall, Thomas D. 1989. *Social Change in the Southwest, 1350–1880.* Lawrence, KS: University Press of Kansas.

Hoxie, Frederick E. 1984. *A Final Promise: The Campaign to Assimilate the Indians, 1880–1920.* Lincoln, NE: University of Nebraska Press.

Nagel, Joanne. 1986. "American Indian Repertoires of Contention." Paper presented at the annual meeting of the American Sociological Association, San Francisco, CA.

O'Brien, Sharon. 1990. *American Indian Tribal Governments.* Norman, OK: University of Oklahoma Press.

Prucha, Francis Paul. 1984. *The Great Father.* Lincoln, NE: University of Nebraska Press.

Snipp, C. Matthew and Gene F. Summers. 1991. "American Indian Development Policies," pp. 166–180 in *Rural Policies for the 1990s,* edited by Cornelia Flora and James A. Christenson. Boulder, CO: Westview Press.

Sorkin, Alan L. 1978. *The Urban American Indian.* Lexington, MA: Lexington Books.

Thornton, Russell. 1987. *American Indian Holocaust and Survival: A Population History since 1942.* Norman, OK: University of Oklahoma Press.

Utley, Robert M. 1963. *The Last Days of the Sioux Nation.* New Haven: Yale University Press.

Weibel-Orlando, Joan. 1991. *Indian Country, L.A.* Urbana, IL: University of Illinois Press.

CAN EDUCATION ELIMINATE RACE, CLASS, AND GENDER INEQUALITY?

41

Roslyn Arlin Mickelson and Stephen Samuel Smith

INTRODUCTION

Parents, politicians, and educational policy makers share the belief that a "good education" is *the* meal ticket. It will unlock the door to economic

Printed by permission of the authors.

opportunity and thus enable disadvantaged groups or individuals to improve their lot dramatically.[1] This belief is one of the assumptions that has long been part of the American Dream. According to the putative dominant ideology, the United States is basically a meritocracy in which hard work and individual effort are rewarded, especially in financial terms.[2] Related to this central belief are a series of culturally enshrined misconceptions about poverty and wealth. The central one is that poverty and wealth are the result of individual inadequacies or strengths rather than the results of the distributive mechanisms of the capitalist economy. A second misconception is the belief that everyone is the master of her or his own fate. The dominant ideology assumes that American society is open and competitive, a place where an individual's status depends on talent and motivation, not inherited position, connections, or privileges linked to ascriptive characteristics like gender or race. To compete fairly, everyone must have access to education free of the fetters of family background, gender, and race. Since the middle of this century, the reform policies of the federal government have been designed, at least officially, to enhance individuals' opportunities to acquire education. The question we will explore in this essay is whether expanding educational opportunity is enough to reduce the inequalities of race, social class, and gender which continue to characterize U.S. society.

We begin by discussing some of the major educational policies and programs of the past forty-five years that sought to reduce social inequality through expanding equality of educational opportunity. This discussion highlights the success and failures of programs such as school desegregation, compensatory education, Title IX, and job training. We then focus on the barriers these programs face in actually reducing social inequality. Our point is that inequality is so deeply rooted in the structure and operation of the U.S. political economy, that, at best, educational reforms can play only a limited role in ameliorating such inequality. In fact, there is considerable evidence that indicates that, for poor and many minority children, education helps legitimate, if not actually reproduce, significant aspects of social inequality in their lives. Finally, we speculate about education's potential role in individual and social transformation.

First, it is necessary to distinguish among equality, equality of opportunity, and equality of educational opportunity. The term *equality* has been the subject of extensive scholarly and political debate, much of which is beyond the scope of this essay. Most Americans reject equality of life conditions as a goal, because it would require a fundamental transformation of our basic economic and political institutions, a scenario most are unwilling to accept. As Ralph Waldo Emerson put it, "The genius of our country has worked out our true policy—opportunity."

The distinction between equality of opportunity and equality of outcome is important. Through this country's history, equality has most typically been understood in the former way. Rather than a call for the equal distribution of money, property, or many other social goods, the concern over equality has been with equal opportunity in pursuit of these goods. In the words of Jennifer Hochschild, "So long as we live in a democratic capitalist society—that is, so long as we maintain the formal promise of political and social equality while encouraging the practice of economic inequality—we need the idea of equal opportunity to bridge that otherwise unacceptable contradiction."[3] To use a current metaphor: If life is a game, the playing field must be level; if life is a race, the starting line must be in the same place for everyone. For the playing field to be level, many believe education is crucial because it gives individuals the wherewithal to compete in the allegedly meritocratic system. In America, then, *equality* is really understood to mean *equality of opportunity*, which itself hinges on *equality of educational opportunity*.

THE SPOTTY RECORD OF FEDERAL EDUCATIONAL REFORMS

In the past [fifty] years, a series of educational reforms initiated at the national level has been introduced into local school systems. All of the reforms aimed to move education closer to the ideal of equality of educational opportunity. Here we discuss several of these reforms, and how the concept of equality of educational opportunity has evolved. Given the importance of race and racism in U.S. history, many of the federal education policies during this period attempted to redress the most egregious forms of inequality based on race.

School Desegregation

Although American society has long claimed to be based on equality of opportunity, the history of race relations suggests the opposite. Perhaps the most influential early discussion of this disparity was Gunnar Myrdal's *An American Dilemma*, published in 1944. The book vividly exposed the contradictions between the ethos of freedom, justice, equality of opportunity and the actual experiences of African Americans in the United States.[4]

The links among desegregation, expanded educational opportunity, and the larger issue of equality of opportunity are very clear from the history of the desegregation movement. This movement, whose first phase culminated in the 1954 *Brown* decision outlawing *de jure* segregation in school, was the first orchestrated attempt in U.S. history to directly address inequality of educational opportunity. The NAACP strategically chose school segregation to be the camel's nose under the tent of the Jim Crow (segregated) society. That one of the nation's foremost civil rights organizations saw the attack on segregated schools as the opening salvo in the battle against society-wide inequality is a powerful example of the American belief that education has a pivotal role in promoting equality of opportunity.

Has desegregation succeeded? This is really three questions: First, to what extent are the nation's schools desegregated? Second, have desegregation efforts enhanced students' academic outcomes? Third, what are the long-term outcomes of desegregated educational experiences?

Since 1954, progress toward the desegregation of the nation's public schools has been uneven and limited. Blacks experienced little progress in desegregation until the mid-1960s when, in response to the civil rights movement, a series of federal laws, executive actions, and judicial decisions resulted in significant gains, especially in the South. Progress continued until 1988, when the effects of a series of federal court decisions and various local and national political developments precipitated marked trends toward the resegregation of Black students. Nationally, in 1994–1995, 33 percent of Black students attended majority White schools compared with the approximately 37 percent who attended majority White schools for much of the 1980s.[5]

Historically, Latinos were relatively less segregated than African Americans. However, from the mid-1960s to the mid-1990s there was a steady increase in the percentage of the Latino students who attended segregated schools. As a result, education for Latinos is now more segregated than it is for Blacks.

Given the long history of legalized segregation in the south, it is ironic that the South's school systems are now generally the country's most *desegregated*, while those in the northeast are the most intensely segregated. However, even desegregated schools are often resegregated at the classroom level by tracking or ability grouping. There is a strong relationship between race and social class, and racial isolation is often an outgrowth of residential segregation and socioeconomic background.

Has desegregation helped to equalize educational outcomes? A better question might be which desegregation programs under what circumstances accomplish which goals? Evidence from recent desegregation research suggests that, overall, children benefit academically and socially from well-run programs. Black students enjoy modest academic gains, while the academic achievement of White children is not hurt, and in some cases is helped, by desegregation. In school systems which have undergone desegregation efforts, the racial gap in educational outcomes has generally been reduced, but not eliminated.

More important than short-term academic gains are the long-term consequence of desegregation for Black students. Compared to those who attended racially isolated schools, Black adults who experienced desegregated education as children are more likely to attend multiracial colleges and graduate from them, work in higher-status jobs, live in integrated neighborhoods, assess their abilities more realistically when choosing an occupation, and to report interracial friendships.[6]

Despite these modest, but positive, outcomes, in the last decade of the twentieth century, most American children attend schools segregated by race, ethnicity, and social class. Consequently, [fifty] years of official federal interventions aimed at achieving equality of educational opportunity through school desegregation have only made small steps toward achieving that goal; children from different race and class backgrounds continue to receive segregated and, in many respects, unequal educations. . . .

Title IX

Title IX of the 1972 Higher Education Act is the primary federal law prohibiting sex discrimination in education. It states, "No person in the United States shall, on the basis of sex, be excluded from participation in, be denied benefit of, or be subjected to discrimination under any program or activity receiving Federal financial assistance." Until Title IX's passage, gender inequality in educational opportunity received minimal legislative attention. The act mandates gender equality of treatment in admission, courses, financial aid, counseling services, employment, and athletics.

The effect of Title IX upon college athletics has been especially controversial. While women constitute 53 percent of undergraduates, they are only 37 percent of college athletes. This is undoubtedly due to the complex interaction between institutional practices and gender-role socialization over the life course. Certainly, the fact that the vast majority of colleges spend much more money on recruiting and scholarships for male athletes contributes to the disparities.

In spring 1997, the U.S. Supreme Court refused to review a lower court's ruling in *Brown v. Cohen* that states in essence that Title IX requires universities to provide equal athletic opportunities for male and female students regardless of cost. Courts have generally upheld the following three-pronged test for compliance: (1) the percentage of athletes who are females should reflect the percentage of students who are female; (2) there must be a continuous record of expanding athletic opportunities for females athletes; and (3) schools must accommodate the athletic interests and abilities of female students. As of the ruling, very few universities were in compliance with the law.

Gender discrimination exists in other areas of education where it takes a variety of forms. For example, in K–12 education official curricular materials frequently feature a preponderance of male characters. Male and female characters typically exhibit traditional gender roles. Vocational education at the high school and college level remains gender-segregated to some degree. School administrators at all levels are overwhelmingly male although most teachers in elementary and secondary schools are female. In higher education, the situation is more complex. Faculty women in academia are found disproportionately in the lower ranks, are less likely to be promoted, and continue to earn less than their male colleagues.

Like the laws and policies aimed at eliminating race differences in school processes and outcomes, those designed to eliminate gender differences in educational opportunities have, at best, only narrowed them. Access to educational opportunity in the United States remains unequal for people of different gender, race, ethnic, and socioeconomic backgrounds.

EQUALITY OF EDUCATIONAL OPPORTUNITY AND EQUALITY OF INCOME

Despite the failures of . . . many programs to eliminate the inequality of educational opportunity over the past 45 years, there is one indicator of substantial progress: measured in median years, the gap in educational attainment between Blacks and Whites, and between males and females, has all but disappeared. In 1997, the median educational attainment of most groups was slightly more than twelve years. In the 1940s, by contrast, White males and females had a median educational attainment of just under nine years, African American males about five years, and African American women about six.

However, the main goal of educational reform is not merely to give all groups the opportunity to receive the same quality and quantity of education. According to the dominant ideology, the ultimate goal of these reforms is to provide equal educational opportunity in order to facilitate equal access to jobs, housing, and various other aspects of the American dream. It thus becomes crucial to examine whether the virtual elimination of the gap in educational attainment has been accompanied by a comparable decrease in other measures of inequality.

Of the various ways inequality can be measured, income is one of the most useful. Much of a person's social standing and access to the good things in life depends on his or her income.[7] Unfortunately, the dramatic progress in narrowing the gap in educational attainment has not been matched by a comparable narrowing of the gap in income inequality. Median individual earnings by race and gender indicate that White men still earn significantly more than any other group. Black men trail White men, and all women earn significantly less than all men. Even when occupation, experience, and level of education are controlled, women earn less than men, and Black men earn less than White men. It is only Black and White women with comparable educational credentials in similar jobs who earn about the same.

The discrepancy between the near elimination of the gap in median educational attainment and the ongoing gaps in median income is further evidence that addressing the inequality of educational opportunity is woefully insufficient for addressing broader sources of inequality throughout society.

This discrepancy can be explained by the nature of the U.S. political economy. The main cause of income inequality is the structure and operation of U.S. capitalism, a set of institutions which scarcely have been affected by the educational reforms discussed earlier. Greater equality of educational opportunity has not led to a corresponding decrease in income inequality because educational reforms do not create more good-paying jobs, affect gender-segregated and racially segmented occupational structures, or limit the mobility of capital either between regions of the country or between the United States and other countries. For example, no matter how good an education White working-class or minority youth may receive, it does nothing to alter the fact that thousands of relatively good paying manufacturing jobs have left northern inner cities for northern suburbs, the sunbelt, or foreign countries.

Many argue that numerous service jobs remain or that new manufacturing positions have been created in the wake of this capital flight. But these pay less than the departed manufacturing jobs, are often part-time or temporary, and frequently do not provide benefits. Even middle-class youth are beginning to fear the nature of the jobs which await them once they complete their formal education. Without changes in the structure and operation of the capitalist economy, educational reforms alone cannot markedly improve the social and economic position of disadvantaged groups. This is the primary reason that educational reforms do little to affect the gross social inequalities that inspired them in the first place.

BEYOND ATTAINMENT: THE PERSISTENCE OF EDUCATIONAL INEQUALITY

Educational reforms have not led to greater overall equality for several additional reasons. While race and gender gaps in educational attainment have narrowed considerably, educational achievement remains highly differentiated by social class, gender, and race. Many aspects of school processes and curricular content are deeply connected to race, social class, and gender inequality. But gross measures of educational outputs, such as median years of schooling completed, mask these indicators of inequality.

Not all educational experiences are alike. Four years of public high school in Beverly Hills are quite different from four years in an inner-city school. Family background, race, and gender have a great deal to do with whether a person goes to college and which institution of higher education she or he attends. The more privileged the background, the more likely a person is to attend an elite private university.

For example, according to Jacobs, women trail men slightly in representation in high status institutions of higher education because women are less likely to attend engineering programs and are more likely to be part-time students (who are themselves more likely to attend lower status institutions such as community colleges). Gender segregation in fields of study remains marked, with women less likely than men to study in scientific and mathematical fields. Furthermore, there is substantial race and ethnic segregation between institutions of higher education. Asian-Americans and Latinos are more segregated from Whites than are African-Americans. Whites and Asian-Americans are more likely to attend higher status universities than are African-Americans and Latinos.[8]

These patterns of race and gender segregation in higher education have direct implications for gender and race gaps in occupational and income attainment. Math and science degree recipients are more likely to obtain more lucrative jobs. A degree from a state college is not as competitive as one from an elite private university. Part of the advantage of attending more prestigious schools comes from the social networks to which a person has access and can join.

Another example of persistent inequality of educational opportunity is credential inflation. Even though women, minorities, and members of the working class now obtain higher levels of education than they did before, members of more privileged social groups gain even higher levels of education. At the same time, the educational requirements for the best jobs (those with the highest salaries, benefits, agreeable working conditions, autonomy, responsibility) are growing. Those with the most education from the best schools tend to be the top candidates for the best jobs. Because people from more privileged backgrounds are almost always in a better position to gain these desirable educational credentials, members of the working class, women, and minorities are still at a competitive disadvantage. Due to the dynamics of credential inflation, educational requirements previously necessary for the better jobs and now within the reach of many dispossessed groups are inadequate and insufficient in today's labor market. The credential inflation process keeps the already privileged one step (educational credential) ahead of the rest of the job seekers.[9]

One additional aspect of the persistent inequalities in educational opportunities concerns what sociologists of education call the hidden curriculum. This concept refers to two separate but related processes. The first is that the content and process of education differ for children according to their race, gender, and class. The second is that these differences reflect and thus help reproduce the inequalities based on race, gender, and class that characterize U.S. society as a whole.

One aspect of the hidden curriculum is the formal curriculum's ideological content. Anyon's work on U.S. history texts demonstrates that children from more privileged backgrounds are more likely to be exposed to rich, sophisticated, and complex materials than are their working-class counterparts.[10] Another aspect of the hidden curriculum concerns the social organization of the school and the classroom. Some hidden curriculum theorists suggest that tracking, ability grouping, and conventional teacher-centered classroom interactions contribute to the reproduction of the social relations of production at the workplace. Lower-track classrooms are disproportionately filled with working-class and minority students. Students in lower tracks are more likely than those in higher tracks to be assigned repetitive exercises with low levels of cognitive challenge. Lower-track students are likely to work individually and to lack classroom experience with problem solving or other independent, creative activities. Such activities are more conducive to preparing students for working-class jobs than for professional and managerial positions. Correspondence principle theorists argue that educational experiences from preschool to high school are designed to differentially prepare students for their ultimate positions in the work force, and that a student's placement in various school programs is strongly related to her or his race and class origin. Critics charge that the correspondence principle has been applied in too deterministic and mechanical a fashion. Evidence abounds of student resistance to class, gender, and race differentiated education.[11] This is undoubtedly why so many students drop out or graduate from high school with minimal levels of literacy and formal skills. Nonetheless, hidden curriculum theory offers a compelling contribution to explanations of how and why school processes and outcomes are so markedly different according to the race, gender, and social class of students.

CONCLUSION

In this [essay] we have argued that educational reforms alone cannot reduce inequality. Nevertheless, education remains important to any struggle to reduce inequality. Moreover, education is more than a meal ticket; it is intrinsically worthwhile and crucially important for the survival of democratic society. Many of the programs discussed in this essay contribute to the enhancement of individuals' cognitive growth and thus promote important nonsexist, nonracist attitudes and practices. Many of these programs also make schools somewhat more humane places for adults and children. Furthermore, education, even reformist liberal education, contains the seeds of individual and social transformation. Those of us committed to the struggle against inequality cannot be paralyzed by the structural barriers that make it impossible for education to eliminate inequality. We must look upon the schools as arenas of struggle against race, gender, and social class inequality.

NOTES

1. This essay draws on an article by Roslyn Arlin Mickelson that appeared as "Education and the Struggle Against Race, Class and Gender Inequality," *Humanity and Society* 11(4) (1987): 440–64.

2. Ascertaining whether a set of beliefs constitutes the dominant ideology in a particular society involves a host of difficult theoretical and empirical questions. For this reason we use the term *putative dominant ideology*. For discussion of these questions, see Nicholas Abercrombie et al., *The Dominant Ideology Thesis* (London: George Allen & Unwin, 1980); James C. Scott, *Weapons of the Weak* (New Haven: Yale University Press, 1985); Stephen Samuel Smith, "Political Acquiescence and Beliefs About State Coercion" (unpublished Ph.D. dissertation, Stanford University, 1990).

3. Jennifer Hochschild, "The Double-Edged Sword of Equal Educational Opportunity." Paper presented at the meeting of the American Education Research Association, Washington, D.C., April 22, 1987.

4. Gunnar Myrdal, *An American Dilemma: The Negro Problem and Modern Democracy* (New York: Harper & Row, 1944).

5. Gary Orfield, Mark D. Bachmeier, David R. James, and Tamela Eitle, "Deepening Segregation in American Public Schools" (Cambridge, MA: Harvard Project on School Desegregation, 1997).

6. Amy Stuart Wells and Robert L. Crain, "Perpetuation Theory and the Long-Term Effects of School Desegregation," *Review of Educational Research* 64(4) (1994): 531–55.

7. To be sure, income does not measure class-based inequality, but there is a positive correlation between income and class. Income has the additional advantage of being easily quantifiable. Were we to use another measure of inequality, e.g., wealth, the disjuncture between it and increases in educational attainment would be even larger. Although the distribution of wealth in U.S. society has remained fairly stable since the Depression, the gap between rich and poor increased in the 1980s and 1990s. Although accurate data are difficult to obtain, a 1992 study by the Federal Reserve found that in 1989 the top one-half of one percent of households held 29 percent of the wealth held by all households.

8. Jerry A. Jacobs, "Gender and Race Segregation Between and Within Colleges" (paper presented at the Eastern Sociological Society, Boston, MA, April 1996).

9. Randall Collins, *The Credential Society* (New York: Academic Press, 1979).

10. Jean Anyon, "Social Class and the Hidden Curriculum of Work," *Journal of Education* 162(1) (1980): 67–92; Jean Anyon, "Social Class and School Knowledge," *Curriculum Inquiry* 10 (1981): 3–42.

11. Samuel Bowles and Herbert Gintis, *Schooling in Capitalist America: Educational Reform and the Contradictions of Economic Life* (New York: Basic Books, 1976); Roslyn Arlin Mickelson, "The Case of the Missing Brackets: Teachers and Social Reproduction," *Journal of Education* 169(2) (1987): 78–88.

WELFARE REFORM, FAMILY HARDSHIP, AND WOMEN OF COLOR

42

Linda Burnham

Tens of thousands of women's and human rights activists gathered in September 1995 at the United Nations Fourth World Conference on Women, held in Beijing, China, to focus their attention on improving the condition and status of women worldwide. Working through cultural, religious, political, economic, and regional differences, women from the nations of the world produced a comprehensive document, the Beijing Platform for Action, that detailed actions to be taken by governments, nongovernmental organizations, and multilateral financial and development institutions to improve women's conditions. The platform for action called on governments to take action to relieve "the persistent and increasing burden of poverty on women" and address gender "inequality in economic structures and policies, in all forms of productive activities and in access to resources" (United Nations 1995).

Yet . . . since Beijing, in a time of unparalleled national prosperity, policies contradictory to the spirit and intent of the platform for action were promulgated in the United States, targeting the most vulnerable citizens and, rather than assisting women onto the path of economic security, driving many deeper into poverty. While U.S. officials pledged in international forums to uphold women's human rights, those rights were substantially undermined by the 1996 passage of the Personal Responsibility and Work Opportunities Reconciliation Act (PRWORA).

INCREASING FAMILY HARDSHIP

There are many studies that document how much worse off women are due to welfare reform. Those who remain in the welfare system, those who leave for employment, and those who might have used Aid to Families with Dependent Children (AFDC) are in worse shape, with less support than the woefully inadequate earlier system provided. . . .

The stated intent of welfare reform was at least twofold: to reduce the welfare rolls and to move women toward economic self-sufficiency. The first objective has been achieved: welfare rolls have declined dramatically since 1996.

From Randy Albelda and Ann Withorn, eds., *Lost Ground: Welfare Reform, Poverty, and Beyond* (Cambridge, MA: South End Press, 2002), pp. 43–56. Reprinted by permission of the South End Press.

Welfare reform has stripped single mothers of any sense that they are entitled to government support during the years when they are raising their children.

Despite the "success" of welfare reform, research has repeatedly found that many women who move from welfare to work do not achieve economic independence. Instead, most find only low-paid, insecure jobs that do not lift their families above the poverty line. They end up worse off economically than they were on welfare: they work hard and remain poor. Others are pushed off welfare and find no employment. They have no reported source of income.

Women in transition from welfare to work—or to no work—face particular difficulties and crises related to housing insecurity and homelessness and food insecurity and hunger.

Low-income people in the United States faced a housing crisis long before the passage of the PRWORA. In most states, the median fair-market cost of housing for a family of three is considerably higher than total income from a Temporary Aid to Needy Families (TANF) grant (Dolbeare 1999). Further, as a consequence of two decades of declining federal support for public and subsidized housing, the great majority of both current and former TANF recipients are at the mercy of an unforgiving private housing market. . . .

Utility payment problems are another important indicator of housing insecurity because they reveal that many families, while they may have a roof over them, spend at least some time without heat and light. And utility problems are often a prelude to inability to pay the rent. A 1998 survey of social service clients who had left welfare within the previous six months found that 25 percent had had their heat cut off (Sherman et al. 1998, 13). A recent Illinois study found that 61 percent of TANF recipients who were not working could not pay their utility bills. But former recipients who were working were also struggling with their budgets, and 48 percent were unable to meet their utility payments (Work, Welfare and Families 2000, 25). . . .

Welfare reform has also put severe pressures on an already strained shelter system. The U.S. Conference of Mayors reported that requests for emergency shelter increased by 12 percent between 1998 and 1999 in the 26 cities surveyed and were at their highest levels since 1994 (U.S. Conference of Mayors 1999, 94). "When I started here three years ago, we had plenty of family space. Since welfare reform, I don't have a bed," said a social service worker in a Salvation Army Shelter in New Orleans (Cobb 1999, 1). . . .

Although the PRWORA was trumpeted as a step toward strengthening families, increased housing insecurity and homelessness have led to families being split apart. Most family shelters do not take men, so the fathers of two-parent families that become homeless must either go to a single men's shelter or make other housing arrangements. Many shelters also do not accommodate adolescent boys or older male teens. Family breakup may be required for a shelter stay.

The housing instability of poor women and their children has profound consequences, both for them and for society as a whole. Homelessness compromises the emotional and physical health of women and children, disrupts schooling, and creates a substantial barrier to employment. It widens the chasm between those who are prospering in a strong economy and those who fall ever farther behind. In the six years since the United States made its Beijing commitments to improving women's lives, welfare policy, rather than widening poor women's access to safe and affordable housing, has created higher levels of housing instability and homelessness.

Like homelessness, the problems of food insecurity predate welfare reform. Low-income workers and welfare recipients alike have struggled for years to provide adequate food for themselves and their families. The robust economy of the late 1990s did not fundamentally alter this reality. Of families headed by single women, one in three experiences food insecurity and one in ten experiences hunger (Work, Welfare and Families 2000, 25).

Welfare reform has made women's struggles to obtain food for themselves and their families more difficult. Several studies document that former recipients cannot pay for sufficient food and that their families skip meals, go hungry, and/or use food pantries or other emergency food assistance.

The figures are astoundingly high. In New Jersey, 50.3 percent of former recipients who were not working reported an inability to sufficiently feed themselves or their children. Former recipients who were working were no better off: Almost 50 percent reported the very same problem (Work, Poverty and Welfare Evaluation Project 1999, 58). . . . . The higher costs associated with participating in the labor force, combined with reduction or elimination of the food stamp allotment, meant women's access to adequate food became more precarious rather than less so as they moved from welfare to work. Entering the workforce came at a very high price.

The Food Stamp Program is intended to ensure that no family goes hungry, but many families do not receive the food stamps to which they are entitled. Even before welfare reform, the rate of participation in the Food Stamp Program was declining more rapidly than the poverty rate. The number of people receiving food stamps dropped even more steeply later, from 25.5 million average monthly recipients in 1996 to 18.5 million in the first half of 1999 (U.S. General Accounting Office 1999, 46). The rate of participation is the lowest it has been in two decades, with a growing gap between the need for food assistance and families' use of food stamps.

Welfare reform has itself contributed to the underutilization of food stamps. Many families that leave the welfare system do not know that as long as their income remains below a certain level, they are still eligible for food stamps. Believing that termination of TANF benefits disqualifies them from receiving food stamps as well, they fail to apply or to reconfirm eligibility. Confusion and misinformation on the part of eligibility workers, or their withholding of information, are also factors in the low participation of former recipients. Additional contributing factors include the lack of bilingual staff and burdensome application and recertification processes (Venner, Sullivan, and Seavey 2000, 17). Among families who had left welfare, only 42 percent of those who were eligible for food stamps were receiving them (Zedlewski and Brauner 1999, 1–6). . . .

WOMEN OF COLOR AND IMMIGRANT WOMEN

Welfare reform is a nominally race-neutral policy suffused with racial bias, both in the politics surrounding its promulgation and in its impact. It may not have been the intent to racially target women of color for particular punishment, yet women of color and immigrant women have nonetheless been particularly hard hit in ways that were highly predictable.

Feminist theory has for some time recognized that the social and economic circumstances women of color must negotiate are shaped by the intersection of distinct axes of power—in this case primarily race, class, and gender. The relationships of subordination and privilege that define these axes generate multiple social dynamics that influence, shape, and transform each other, creating, for women of color, multiple vulnerabilities and intensified experiences of discrimination.

Welfare reform might legitimately be regarded as a class-based policy intended to radically transform the social contract with the poor. Poverty in the United States, however, is powerfully structured by racial and gender inequities. It is not possible, therefore, to institute poverty policy of any depth that does not also reconfigure other relations, either augmenting or diminishing race and gender inequalities. By weakening the social safety net for the poor, PRWORA necessarily has its greatest effect on those communities that are disproportionately represented among the poor. Communities of color and immigrant communities, already characterized by significantly higher levels of minimum-wage work, homelessness, hunger, and poor health, are further jeopardized by the discriminatory impact of welfare reform.

As a consequence of the historical legacy and current practices of, among other things, educational inequity and labor market disadvantage, patterns of income and wealth in the United States are strongly skewed along racial lines, for example, the disproportionate burden of poverty carried by people of color. While the white non-Hispanic population constituted 72.3 percent of the total population in 1998, it made up only 45.8 percent of the population living below the poverty line. In stark contrast, blacks made up just over 12 percent of the general population but 26.4 percent of the U.S. population in poverty. People of Hispanic origin, of all races, constitute 23.4 percent of the people below the poverty line, while making up 11.2 percent of the total population.

While 8.2 percent of the white non-Hispanic population lives in poverty, 12.5 percent of Asian and Pacific Islanders do (U.S. Census Bureau 1999, 2000).[1]

Economic vulnerabilities due to race and ethnicity may be further compounded by disadvantages based on gender and immigration/citizenship status. Thus, for households headed by single women, the poverty rates are also stark. Over 21 percent of such white non-Hispanic households were below the poverty line in 1998, as compared to over 46 percent of black and 48 percent of Hispanic female-headed households (U.S. Census Bureau 1999). Immigrants, too, are disproportionately poor, with 18 percent below the poverty line as compared to 12 percent of the native born (U.S. Census Bureau 1999). Given the disproportionate share of poverty experienced by people of color, and the significant poverty of single-mother households, it is no surprise that the welfare rolls are racially unbalanced, with women of color substantially overrepresented (see Table 1).

This racial imbalance has been cynically used for decades in the ideological campaign to undermine support for welfare—a crude but ultimately effective interweaving of race, class, gender, and anti-immigrant biases that prepared the consensus to "end welfare as we know it." Having been maligned as lazy welfare cheats and something-for-nothing immigrants, Latinas, African American women, and Asian women of particular nationality groups are now absorbing a punishing share of welfare reform's negative impacts.

Much of the data on welfare reform are not disaggregated by race. We will not know the full impact of welfare reform on women of color until we have county, statewide, and national studies of women transitioning from welfare to work that consistently include race as a variable. However, to the extent that communities of color experience some of the most devastating effects of poverty at exceptionally high rates, and to the extent that welfare reform has rendered these communities more, rather than less, vulnerable, we may expect that the policy will deepen already entrenched inequalities.

Table 1
TANF Recipients by Race, 1998
(In percentages)

White	32.7
Black	39.0
Hispanic	22.2
Native	1.5
Asian	3.4
Other	0.6
Unknown	0.7

Source: U.S. Department of Health and Human Services, 1999.

For example, African American women are massively overrepresented in the urban homeless population. Their particular vulnerability to homelessness has been shaped by, among many factors, high rates of reliance on welfare in a period in which the value of the welfare grant plummeted and housing costs climbed steeply; low marriage rates and, therefore, lack of access to a male wage; overconcentration on the bottom rungs of the wage ladder; and high unemployment rates, especially for young women with less than a high school education.

Beyond intensified impact due to disproportionate representation in the affected population, additional factors compound the disadvantages of women of color and immigrant women. One Virginia study found noteworthy differences in how caseworkers interact with black and white welfare recipients. A substantial 41 percent of white recipients were encouraged to go to school to earn their high school diplomas, while no black recipients were. A much higher proportion of whites than blacks found their caseworkers to be helpful in providing information about potential jobs (Gooden 1997). Other studies showed that blacks were removed from welfare for noncompliance with program rules at considerably higher rates than white recipients, while a higher proportion of the cases of white recipients were closed because they earned too much to qualify for welfare (Savner 2000). . . .

Some of the most punitive provisions of PRWORA are directed at immigrants. The 1996 legislation banned certain categories of legal immigrants from a wide array of federal assistance programs, including TANF, food stamps, Supplementary Security Income, and Medicaid. In the year following passage, 940,000 legal immigrants lost their food stamp eligibility. Strong advocacy reversed some of the cuts and removed some restrictions, but legal immigrants arriving in the United States after the passage of the legislation are ineligible for benefits for five years. States have the right to bar pre-enactment legal immigrants from TANF and nonemergency Medicaid as well (National Immigration Law Center 1999).

These restrictions have had profound effects on immigrant communities. First of all, many immigrant women who are on welfare face significant barriers to meeting TANF work requirements. Perhaps the most formidable obstacles are limited English proficiency and low educational levels. A study of immigrant recipients in California found that 87 percent of the Vietnamese women and 48 percent of the Mexican American women had limited or no proficiency in English. Many of these women were also not literate in their native languages, with the Mexican Americans averaging 6.5 years in school and the Vietnamese 8.7 years (Equal Rights Advocates 1999, 7). A study of Hmong women found 90 percent with little or no English proficiency, 70 percent with no literacy in the Hmong language, and 62 percent with no formal education whatsoever (Moore and Selkowe 1999).

Limited English, lack of education, and limited job skills severely restrict immigrant women's options in the job market, making it very difficult for them to comply with welfare-to-work requirements. Language problems also impede their ability to negotiate the welfare bureaucracy, which provides very limited or no translation services. These women lack information about programs to which they are entitled, and they worry about notices that come to them in English. When immigrant women recipients are able to find work, it is most often in minimum-wage or low-wage jobs without stability or benefits (Center for Urban Research and Learning 1999, 5; Equal Rights Advocates 1999, 31). . . .

Immigrant women recipients are also likely to experience severe over-crowding and to devote a huge portion of their income to housing. They share housing with relatives or with unrelated adults; live in garages or other makeshift, substandard dwellings; and worry constantly about paying the rent.

A more hidden, but still pernicious, impact of welfare reform has been the decline in applications for aid from immigrants who are eligible to receive it. One report documents PRWORA's "chilling effect on immigrants" who mistakenly believe they are no longer eligible for any benefits. Reporting on the numbers of TANF applications approved each month, this study showed a huge drop—71 percent—in the number of legal immigrant applicants approved for TANF and MediCal between January 1996 and January 1998. That number fell from 1,545 applicants in January 1996 to only 450 in January 1998 (Zimmerman and Fix 1998, 5). The intensive anti-immigrant propaganda that accompanied the passage of PRWORA and statewide anti-immigrant initiatives appears to have discouraged those who need and are entitled to aid from applying for it, surely undermining the health and welfare of immigrant women and their families.

WELFARE REFORM IS INCOMPATIBLE WITH WOMEN'S HUMAN RIGHTS

One of the chief accomplishments of the Beijing conference and the Platform for Action was to position women's issues squarely within the context of human rights. Building on the foundational work of activists worldwide, Beijing became the first U.N. women's conference in which "women's rights are human rights" was articulated not as a platitude but as a strategic assertion. Indeed, the phrase was taken up by former First Lady Hillary Rodham Clinton, who, in her September 5, 1995, speech to the conference, asserted that "women will never gain full dignity until their human rights are respected and protected."

PRWORA is wholly incompatible with the strategic objectives of the Beijing Platform for Action and profoundly compromises the exercise of women's human rights. Rather than improving the status of poor women, the legislation has deepened the misery of tens of thousands of women and their children. By undermining women's access to a stable livelihood, welfare reform constructs barriers to their exercise of political, civil, cultural, and social rights.

Undoing the damage of welfare reform—and bringing U.S. policy in line with its stated commitments to the world community—will require the promulgation and implementation of policies that restore and strengthen the social safety net for women and children while funding programs that support women along the path to economic self-sufficiency. In the absence of the political will for such a comprehensive reworking of U.S. social welfare policy, advocates for poor women and families face an extended, defensive battle to ameliorate the cruelest and most discriminatory effects of this radically regressive policy.

NOTE

1. In citing Census Bureau statistics, I use their terminology. Elsewhere, I use the term "Latino" to refer to immigrants from Mexico, Central and South America, and the Spanish-speaking Caribbean and their descendants in the United States.

References

Center for Urban Research and Learning. 1999. *Cracks in the System: Conversations with People Surviving Welfare Reform.* Chicago: Center for Urban Research and Learning, Loyola University, Howard Area Community Center, Organization of the NorthEast.

Clinton, Hillary. 1995. Remarks by First Lady Hillary Rodham Clinton for the United Nations Fourth World Conference on Women, 5 Sept.

Cobb, Kim. 1999. Homeless Kids Problem Worst in Louisiana; Welfare Reform, Housing Crunch Are Among Reasons. *Houston Chronicle,* 15 Aug.

Dolbeare, Cushing. 1999. *Out of Reach: The Gap Between Housing Costs and Income of Poor People in the United States.* Washington, DC: National Low-Income Housing Coalition.

Equal Rights Advocates. 1999. *From War on Poverty to War on Welfare: The Impact of Welfare Reform on the Lives of Immigrant Women.* San Francisco.

Gooden, Susan. 1997. Examining Racial Differences in Employment Status Among Welfare Recipients. In *Race and Welfare Report.* Oakland, CA: Grass Roots Innovative Policy Program.

Moore, Thomas and Vicky Selkowe. 1999. *The Impact of Welfare Reform on Wisconsin's Hmong Aid Recipients.* Milwaukee: Institute for Wisconsin's Future.

National Immigration Law Center. 1999. *Immigrant Eligibility for Public Benefits.* Washington, DC.

Savner, Steve. 2000. Welfare Reform and Racial/Ethnic Minorities: The Questions to Ask. *Poverty & Race* 9(4):3–5.

Sherman, Arloc, Cheryl Amey, Barbara Duffield, Nancy Ebb, and Deborah Weinstein. 1998. *Welfare to What: Early Findings on Family Hardship and Well-Being.* Washington, DC: Children's Defense Fund and National Coalition for the Homeless.

United Nations. 1995. *Fourth World Conference on Women Platform for Action.* Geneva, Switzerland.

U.S. Census Bureau. 1999. *Poverty Thresholds in 1998 by Size of Family and Number of Related Children Under 18 Years.* Washington, DC.

U.S. Conference of Mayors. 1999. *A Status Report on Hunger and Homelessness in America's Cities.* Washington, DC.

U.S. Department of Health and Human Services, Administration for Children and Families. 1999. *Characteristics and Financial Circumstances of TANF Recipients.* Washington, DC.

U.S. General Accounting Office. 1999. *Food Stamp Program: Various Factors Have Led to Declining Participation.* Washington, DC.

Venner, Sandra H., Ashley F. Sullivan, and Dorie Seavey. 2000. *Paradox of Our Times: Hunger in a Strong Economy.* Medford, MA: Tufts University, Center on Hunger and Poverty.

Work, Poverty and Welfare Evaluation Project. 1999. Assessing Work First: What Happens After Welfare? Report for the Study Group on Work, Poverty and Welfare. Legal Services of New Jersey, New Jersey Poverty Research Institute, Edison, NJ.

Work, Welfare and Families. 2000. *Living with Welfare Reform: A Survey of Low Income Families in Illinois.* Chicago: Chicago Urban League and UIC Center for Urban Economic Development.

Zedlewski, Sheila R. and Sarah Brauner. 1999. *Are the Steep Declines in Food Stamp Participation Linked to Falling Welfare Caseloads?* Washington, DC: Urban Institute.

Zimmerman, Wendy and Michael Fix. 1998. *Declining Immigrant Applications for Medi-Cal and Welfare Benefits in Los AnGeles County.* Washington, DC: Urban Institute.

AID TO DEPENDENT CORPORATIONS*

43

Exposing Federal Handouts to the Wealthy

Chuck Collins

In 1992 rancher J. R. Simplot of Grandview, Idaho, paid the U.S. government $87,000 for grazing rights on federal lands, about one-quarter the rate charged by private landowners. Simplot's implicit subsidy from U.S. taxpayers, $261,000, would have covered the welfare costs of about 60 poor families.

---

*Editors' note: This is a play on words regarding Aid to Families with Dependent Children—the federal welfare program that was eliminated in 1996 by the Personal Responsibility and Work Opportunities Reconciliation Act.

From *Dollars & Sense* 1 (May/June 1995): 15–17, 40. Reprinted by permission.

With a net worth exceeding $500 million, it's hard to argue that Simplot needed the money.

Since 1987, American Barrick Resources Corporation has pocketed $8.75 billion by extracting gold from a Nevada mine owned by the U.S. government. But Barrick has paid only minimal rent to the Department of the Interior. In 1992 Barrick's founder was rewarded for his business acumen with a $32 million annual salary.

Such discounts are only one form of corporate welfare, dubbed "wealthfare" by some activists, that U.S. taxpayers fund. At a time when Congress is attempting to slash or eliminate the meager benefits received by the poor, we are spending far more to subsidize wealthy corporations and individuals. Wealthfare comes in five main varieties: discounted user fees for public resources; direct grants; corporate tax reductions and loopholes; giveaways of publicly funded research and development (R&D) to private profit-making companies; and tax breaks for wealthy individuals. . . .

TAX AVOIDANCE

The largest, yet most invisible, part of wealthfare is tax breaks for corporations and wealthy individuals. The federal Office of Management and Budget (OMB) estimates that these credits, deductions, and exemptions, called "tax expenditures," will cost $440 billion in fiscal 1996. This compares, for example, to the $16 billion annual federal cost of child support programs.

Due both to lower basic tax rates and to myriad loopholes, corporate taxes fell from one-third of total federal revenues in 1953 to less than 10% today (see "Disappearing Corporate Taxes," *Dollars & Sense*, July 1994).* Were corporations paying as much tax now as they did in the 1950s, the government would take in another $250 billion a year—more than the entire budget deficit.

The tax code is riddled with tax breaks for the natural resource, construction, corporate agri-business, and financial industries. Some serve legitimate purposes, or did at one time. Others have been distorted to create tax shelters and perpetuate bad business practices. . . .

One particularly generous tax break is the foreign tax credit, which allows U.S.-based multinational corporations to deduct from their U.S. taxes the income taxes they pay to other nations. Donald Bartlett and James Steele, authors of *America: Who Really Pays the Taxes*, say that by 1990 this writeoff was worth $25 billion a year.

While in many cases this credit is a valid method of preventing double taxation on profits earned overseas, the oil companies have used it to avoid

---

*Editors' note: 9.9% in fiscal year 2000.

most of their U.S. tax obligations. Until 1950, Saudi Arabia had no income tax, but charged royalties on all oil taken from their wells. Such royalties are a payment for use of a natural resource. They are a standard business expense, payable *before* a corporation calculates the profits on which it will pay taxes.

These royalties were a major cost to ARAMCO, the oil consortium operating there (consisting of Exxon, Mobil, Chevron, and Texaco). But since royalties are not income taxes, they could not be used to reduce Exxon and friends' tax bills back home.

When King Saud decided to increase the royalty payments, ARAMCO convinced him to institute a corporate income tax and to substitute this for the royalties. The tax was a sham, since it applied only to ARAMCO, not to any other business in Saudi Arabia's relatively primitive economy. The result was that the oil companies avoided hundreds of millions of dollars in their American taxes. Eventually the other oil-producing nations, including Kuwait, Iraq, and Nigeria, followed suit, at huge cost to the U.S. Treasury.

In contrast to the ARAMCO problem, many corporate executive salaries should not be counted as deductible expenses. These salaries and bonuses are often so large today that they constitute disguised profits. Twenty years ago the average top executive made 34 times the wages of the firm's lowest paid workers. Today the ratio is 140 to 1.* . . .

OTHER VITAL MATTERS

Taxes are but one form of wealthfare. . . . Many corporations also receive direct payments from the federal government. The libertarian Cato Institute argues that every cabinet department "has become a conduit for government funding of private industry. Within some cabinet agencies, such as the U.S. Department of Agriculture and the Department of Commerce, almost every spending program underwrites private business."

Agriculture subsidies typically flow in greater quantities the larger is the recipient firm. Of the $1.4 billion in annual sugar price supports, for example, 40% of the money goes to the largest 1% of firms, with the largest ones receiving more than $1 million each. . . .

The Progressive Policy Institute estimates that taxpayers could save $114 billion over five years by eliminating or restricting such direct subsidies. Farm subsidies, for example, could be limited to only small farmers.

The government also pays for scientific research and development, then allows the benefits to be reaped by private firms. This occurs commonly in

---

*Editors' note: The AFL-CIO reports that since 1980 the average pay of working people has increased 74 percent; CEO pay, 1,884 percent. The AFL-CIO maintains a Web site, "Executive Pay Watch." See www.afl-cio.org

medical research. One product, the anti-cancer drug Taxol, cost the U.S. government $32 million to develop as part of a joint venture with private industry. But in the end the government gave its share to Bristol-Myers Squibb, which now charges cancer patients almost $1,000 for a three-week supply of the drug.

WHO IS ENTITLED?

Beyond corporate subsidies, the government also spends far more than necessary to help support the lifestyles of wealthy individuals. This largess pertains to several of the most expensive and popular "entitlements" in the federal budget, such as Social Security, Medicare, and the deductibility of interest on home mortgages. As the current budget-cutting moves in Congress demonstrate, such universal programs have much greater political strength than do programs targeted solely at low-income households.

While this broad appeal is essential to maintain, billions of dollars could be saved by restricting the degree to which the wealthy benefit from universal programs. If Social Security and Medicare payments were denied to just the richest 3% of households this would reduce federal spending by $30 to $40 billion a year—more than the total federal cost of food stamps.

Similarly, mortgage interest is currently deductible up to $1 million per home, justifying the term "mansion subsidy" for its use by the rich. The government could continue allowing everyone to use this deduction, but limit it to $250,000 per home. This would affect only the wealthiest 5% of Americans, but would save taxpayers $10 billion a year.

Progressive organizations have mounted a renewed focus on the myriad handouts to the corporate and individual rich. One effort is the Green Scissors coalition, an unusual alliance of environmental groups such as Friends of the Earth, and conservative taxcutters, such as the National Taxpayers Union. Last January Green Scissors proposed cutting $33 billion over the next ten years in subsidies that they contend are wasteful and environmentally damaging. These include boondoggle water projects, public land subsidies, highways, foreign aid projects, and agricultural programs.

Another new organization, Share the Wealth, is a coalition of labor, religious, and economic justice organizations. It recently launched the "Campaign for Wealth-Fare Reform," whose initial proposal targets over $35 billion in annual subsidies that benefit the wealthiest 3% of the population. The campaign rejects the term "corporate welfare" because it reinforces punitive anti-welfare sentiments. Welfare is something a humane society guarantees to people facing poverty, unemployment, low wages, and racism. "Wealthfare," in contrast, is the fees and subsidies extracted from the public by the wealthy and powerful—those who are least in need.

Today's Congress is not sympathetic to such arguments. But the blatant anti-poor, pro-corporate bias . . . has already begun to awaken a dormant public consciousness. This will leave more openings, not less, for progressives to engage in public education around the true nature of government waste.

References

Bartlett, Donald, and James Steele. 1994. *America: Who Really Pays the Taxes?* New York: Simon & Schuster.

Crittendon, Anne. 1993. *Killing the Sacred Cows.* New York: Penguin Books.

Friends of the Earth. 1995. *Green Scissors Report: Cutting Wasteful and Environmentally Harmful Spending and Subsidies.* Washington, DC: Green Scissors Campaign of Citizens United to Terminate Subsidies.

Shapiro, Robert. 1994. *Cut-and-Invest to Compete and Win: A Budget Strategy for American Growth.* Washington, DC: Progressive Policy Institute.

Shields, Janice C. 1994. *Aid to Dependent Corporations.* Washington, DC: Essential Information.

POLICING THE NATIONAL BODY   44
Sex, Race, and Criminalization

Jael Silliman

American politicians, eager to garner support from the large middle-class, promote "family values," endlessly debate issues of abortion, and outdo each other as champions of "working families" (read "middle-class" families). . . .

Essentially, what we have is an America deeply divided across class and race lines. This makes it possible for mainstream America—its politicians and media—to ignore or rarely address issues of poverty, criminalization, and race that are pressing for communities of color. Incarceration rates for people of color are disproportionately high and assaults and searches by police, the

From Jael Silliman and Anannya Bhattacharjee, eds., *Policing the National Body: Race, Gender, and Criminalization* (Boston: South End Press, 2002), pp. x–xxvi. Reprinted by permission of the South End Press.

Immigration and Naturalization Service (INS), and border patrol forces are daily occurrences in communities of color. This aggressive law enforcement regime is increasingly accepted by the mainstream as the price to be paid for law and order. A lead article in the February 2001 edition of the *New York Times Magazine* marks this decisive shift in public attitude in New York City from a more libertarian, turbulent, and nonconformist city towards a greater acceptance of aggressive law enforcement.

. . . The new wisdom in New York City—one of the bastions of liberalism in the country—is "We no longer believe that to solve crime we have to deal with the root causes of poverty and racism; we now believe that we can reduce crime through good policing."[1]

Aggressive law enforcement policies and actions are devastating women of color and their communities. Though there is a strong and growing law enforcement accountability movement,[2] the women's movement in general has not seen state violence as a critical concern.[3] The mainstream reproductive rights movement, consumed with protecting the right to abortion, has failed to respond adequately to the policing, criminalization, and incarceration of large numbers of poor people and people of color. It has not sufficiently addressed cuts in welfare and immigrant services that have made one of the most fundamental reproductive rights—the right to have a child and to rear a family—most tenuous for a large number of people.

The mainstream reproductive rights movement, largely dominated by white women, is framed around choice: the choice to determine whether or not to have children, the choice to terminate a pregnancy, and the ability to make informed choices about contraceptive and reproductive technologies. This conception of choice is rooted in the neoliberal tradition that locates individual rights at its core, and treats the individual's control over her body as central to liberty and freedom. This emphasis on individual choice, however, obscures the social context in which individuals make choices, and discounts the ways in which the state regulates populations, disciplines individual bodies, and exercises control over sexuality, gender, and reproduction.[4]

The state regulates and criminalizes reproduction for many poor women through mandatory or discriminatory promotion of long-acting contraceptives and sterilization, and by charging pregnant women on drugs with negligence or child abuse. An examination of the body politics, the state's power of "regulation, surveillance and control of bodies (individual and collective)," elucidates the scope and venues through which the state regulates its populations "in reproduction and sexuality, in work and in leisure, in sickness and other forms of deviance and human difference."[5] . . .

Women of color have independently articulated a broad reproductive rights agenda embedded in issues of equality and social justice, while keenly tuned to the state's role in the reproduction and regulation of women's bodies. In an effort to protect their reproductive rights, they have challenged coercive population policies, demanded access to safe and accessible birth control, and asserted their right to economic and political resources to maintain healthy children. Through these demands, they move from an emphasis on individual rights to rights that are at once politicized and collectivized. African-American women leaders articulate the barriers to exercising their reproductive rights:

> Hunger and homelessness. Inadequate housing and income to provide for themselves and their children. Family instability. Rape. Incest. Abuse. Too young, too old, too sick, too tired. Emotional, physical, mental, economic, social—the reasons for not carrying a pregnancy to term are endless and varied.[6]

These leaders remind us that a range of individual and social concerns must be engaged to realize reproductive rights for all women. . . . A key concern among women of color and poor communities today is: the difficulty of maintaining families and sustaining community in the face of increasing surveillance and criminalization. . . . Particular communities and women within them are conceived and reproduced as threats to the national body, imagined as white and middle-class. . . .

. . . Poor women and women of color are criminalized. Bhattacharjee elaborates upon the prison system, police, INS, and border patrol forces to illustrate how they routinely undermine and endanger women's caretaking, caregiving, and reproductive functions. She shows how systemic and frequent abuses of reproductive rights and threats to bodily integrity are often overlooked by narrow definitions of reproductive rights and single-issue movements. . . . Greater coalition-building efforts between the women's movement, the reproductive rights movement, the immigrant rights movement, the violence against women movement, and the enforcement account-ability movement are needed to break down barriers and to ensure the safety and self-determination of women of color. . . .

CRIME AND PUNISHMENT IN THE UNITED STATES

The criminal justice system has become a massive machine for arrest, detention, and incarceration. The events of September 11, 2001 have intensified this trend. Citing "national security" and the "war against terrorism," President Bush has furthered the power of the criminal justice system to arrest noncitizens and to circumvent the court system. Immigrants and communities of color will bear the brunt of the intensified assault on civil liberties. In 1998, on any given day, there were approximately six million people under some form of correctional authority. The number of people in American prisons is expected to surpass two million by late 2001. Federal Judge U. W. Clemon, after a visit to the Morgan County Jail in Alabama, wrote in a blistering ruling, "To say the Morgan County Jail is overcrowded is an understatement. The sardine-can appearance of its cell units more nearly resemble the holding units of slave ships during the Middle Passage of the eighteenth century than anything in the twenty-first century."[7] Imprisonment is the solution currently proffered for drug offenses in minority communities and for the other social problems spawned by poverty. As welfare and service programs are gutted, the only social service available to many of America's poor is jail![8]

Despite this surge in incarceration rates, it is widely accepted that prisons encourage recidivism, transform the occasional offender into a habitual delinquent,[9] fail to eliminate crime, and ignore the social problems that drive individuals to engage in illegal actions. French historian and social critic Michel Foucault explains the political rationale behind what the terms the "production of delinquency," and its usefulness for those in power:

> [T]he existence of a legal prohibition creates around it a field of illegal practices, which one manages to supervise, while extracting from it an illicit profit through elements, themselves illegal, but rendered manipulable by their organization in delinquency. . . . Delinquency represents a diversion of illegality for the illicit circuits of profit and power of the dominant class.[10]

The building and maintenance of policing and prison systems is politically expedient and highly profitable. Prisons boost local economies. Fremont County, Colorado, home to thirteen prisons, promotes itself as the Corrections Capital of the World. In *Going Up the River*, Joseph Hallinan explains how prisons have become public works projects that require a steady flow of inmates to sustain them.[11] Corporations engage in bidding wars to run prisons, and the federal government boasts about saving money by contracting out prison management to the private sector.

People of color are disproportionately represented in the prison industrial complex. Bureau of Justice statistics indicate that in 1999, 46 percent of prison inmates were Black and 18 percent were Hispanic. In "Killing the Black Community," Judith Scully argues that the war on drugs is used to justify and exercise control over the Black community. She contends that the United States government has historically maintained control over the Black community by selectively enforcing the law, arbitrarily defining criminal behavior and incarceration, and failing to punish white people engaged in lawless acts against the Black community, as exemplified in the infamous Rodney King and Amadou Diallo verdicts. The war on drugs demonstrates how government officials employ legal tools as well as racial rhetoric and criminal theory to criminalize and destroy Black communities. Scully explores how the creation of drug-related crimes such as the "crack baby" demonizes Black motherhood and undermines Black childbearing. Her essay exposes the institutional links between Blackness, suspicion, and criminality.

As strongly as class and race biases determine the criminal justice system, so does gender bias. Since 1980, the number of women in state and federal correctional facilities has tripled.[12] Amnesty International figures indicate that the majority of the over 140,000 women in the American penitentiary system are Black, Latina, and poor women, incarcerated largely for petty crimes.[13] For the same offense, Black and Latina women are respectively eight and four times more likely to be incarcerated than white women. The United Nations special report on violence against women in US state and federal prisons noted the trend in prison management that emphasizes punishment rather than rehabilitation and a widespread reduction in welfare and support services within the criminal justice system. . . .

The criminal justice system works collaboratively with government, corporate, and professional institutions to perform and carry out disciplinary functions deemed necessary to uphold the system of injustice. The recent exposés on racial profiling, discriminatory sentencing, and the compliance of hospitals, medical professionals, and private citizens in administering drug tests or reporting substance abuse among pregnant women are a few examples of the ways in which a range of actors are drawn in (sometimes reluctantly) to the surveillance and disciplinary system. For example, feminist lawyer Lynn Paltrow, executive director of National Advocates for Pregnant Women, describes how the Medical University Hospital in Charleston instituted a policy of reporting and facilitating the arrest of pregnant patients, overwhelmingly African-Americans, who tested positive for cocaine. African-American women were dragged out of the hospital in chains and shackles where the medical staff worked in collaboration with the prosecutor and police to see if the threat of arrest would deter drug use among pregnant women.[14]

Such violations are not going unchallenged. In *Ferguson vs. City of Charleston* (March 2001), a lawsuit engineered by Lynn Paltrow, the US Supreme Court agreed that Americans have the right, when they seek medical help, to expect that their doctor will examine them to provide diagnosis and treatment, and not search them to facilitate their arrest.[15] . . .

It is essential that we do not separate the more blatant forms of policing—videocameras within prisons that track every move of a prisoner, racial profiling, drug tests disproportionately administered in poor communities, and the raids on illegal immigrants crossing over to the United States—from the disciplinary apparatus being deployed across society. The wide use of differential forms of control and discipline is apparent in the ways in which the public has acquiesced to the policing and surveillance increasingly employed in everyday lives.

As a society we are no longer outraged when we hear that public schools in poor neighborhoods are routinely policed and equipped with metal detectors, and that students are hauled away in handcuffs for petty misdemeanors. The bodies being patrolled, segregated by race, determine the form of discipline applied. Perhaps this differential treatment explains why, in response to the spate of shootings in public schools across the country, parents in affluent neighborhoods have rallied and called for greater policing to ensure their children's safety. Middle-class parents invite policing to "protect their children." This contrasts sharply with the policing imposed in schools in poor communities and communities of color that criminalizes rather than protects. Though surveillance and policing differ according to whether they are there to protect or to criminalize, both kinds of interventions further extend state control over individual and collective bodies. The overt and insidious intrusions consolidate power in state and corporate entities.

BIOLOGICAL CONTROL

Women of color have been the target of biological control ideologies since the founding of America. In her book *Killing the Black Body,* Dorothy Roberts traces the history of reproductive rights abuses perpetrated on the Black community from slavery to the present. Roberts shows how control over reproduction is systematically deployed as a form of racial oppression and argues that the denial of Black reproductive autonomy serves the interests of white supremacy.[16] Others have documented the history of sterilization abuse in the Latina and Native American communities, indicating that population control has a long history in the United States. A Native American reproductive activist reports that on the Pine Ridge reservation today, pregnant women with drinking problems are put in jail, as it is the only holding place for them.[17] . . .

The potential to extend biological control has expanded exponentially. Advances in data collection and storage through the computer, Internet, and genetic revolutions have made surveillance systems more efficient and invasive. The nineteenth-century "panoptical gaze" made it possible for a prisoner to be seen at all times, and through that process the prisoner internalized surveillance.[18] The new technologies radically expand the ability to collect, process, and encode large amounts of information on ever smaller surfaces. This makes it possible for the human body to be manipulated and controlled in radically new ways—from within the body itself. This intensification takes control from a mental to a physiological realm.

Like policing, corporate intrusion into the private sphere is increasingly naturalized. A great deal of data on individuals is bought, sold, and traded. In this instance, it is information on the rich and middle-class that is particularly coveted. The Internet revolution has made it possible for corporate and security interests to track every move made by an individual on the web to determine a person's consumer preferences, interests, and purchases, in addition to getting credit card information. Such intrusive data collecting techniques regarding an individual's tastes and preferences for corporate niche marketing is rarely framed as a surveillance problem. Critics have discussed it sometimes as a privacy concern, but by and large these incursions are accepted as a market-driven intrusion into our life.

Emerging reproductive technologies, such as cloning, have the potential to blur the distinctions between genetically distinct and genetically determined individuals. This raises ethical questions regarding who would count as fully human with the attendant civil rights and liberties. Valerie Hartouni points to how standards of humanity get "partialized" (making some less human than others) in this process.[19] She fears that, in the current social context, such technologies will be used to manage and contain diversity and the proliferation of difference.[20] Other critics are concerned with the eugenic possibilities of cloning and similar practices that work on humans from the inside out. The commodification of human life and the disruption that such technologies could have on kinship structures and human relations are a source of grave concern.

SURVEILLANCE AND NATIONAL (IN)SECURITY

Whereas the prison and policing systems are supposed to protect the nation from dangers within, the military, Border Patrols, and INS are ostensibly designed to protect the public from danger and threats that emanate from outside the national body. The discretionary authority and budgets of these institutions have expanded exponentially since September 11, 2001. . . . An influx of immigrants, the rhetoric of explosive population growth in the Third World, and angry young men inside and outside the United States are manufactured as a threat to national security.[21]

Immigrants in the United States are constructed as a source of danger. This threat has been used to justify the allocation of billions of dollars to the enforcement programs of the INS that patrol the nation. At present, the INS has more armed agents with arrest power than any other federal law enforcement agency.[22] Mandatory detention provisions have made immigrants the fastest growing incarcerated population in the United States. Stringent controls and border security forces are positioned at strategic places along the United States-Mexico border. Military-style tactics and equipment result in immigrants undertaking more dangerous, isolated routes to cross over where the risks of death, dehydration, and assault are exponentially higher.

Immigrant rights organizations and the press record detailed accounts of immigrants risking their lives to make their way through tortuous terrain to find work in the United States. The Mojave Desert in Southern California with its inhospitable terrain has become one such death trap for immigrants.[23] This dangerous crossing is a dramatic example of the extreme risks that immigrants take to escape INS agents and provide adequately for their families.

Vigilante groups and private citizens in Arizona have taken it upon themselves to "help" patrol ranchers "hunt" Mexican undocumented immigrants. Jose Palafox reports on the roundup of over 3,000 undocumented immigrants on Roger and Don Barnett's 22,000-acre property in Douglas, Arizona, near the United States-Mexico border. Ranchers circulated a leaflet asking for volunteers to help patrol their land while having "Fun in the Sun." Immigrants were considered "fair game."[24]

While border patrol forces and private citizens tighten their grip on illegal immigration and regulate the mobility of poor workers, social services are being slashed for legal immigrants. The latter are often portrayed as a drain on national resources and a direct threat to low-wage workers in the United States, and female immigrants of color are particularly targeted for their family settlement and community building roles. . . .

Even though many studies have shown that immigrant labor contributes to the economy, there is a widely held perception that they represent a financial drain. Despite the proverbial anti-immigrant sentiment, the United States economy depends on cheap immigrant labor. Very often undocumented workers perform jobs or are made to engage in activities usually considered too low-paying or too risky for citizens and residents. It has been estimated that 45,000 to 50,000 women and children are trafficked annually into and across the United States for the sex industry, sweatshops, domestic labor, and agricultural work. The INS compares the trafficking in women and children with the drug and weapons smuggling industry. . . .

BEYOND ANALYSIS: BUILDING A MOVEMENT

The . . . Committee on Women, Population, and the Environment (CWPE) [is] a multiracial alliance of feminist activists, health practitioners, and scholars committed to promoting the social and economic empowerment of women in a context of global peace and justice. We work toward eliminating poverty, inequality, racism, and environmental degradation. A crucial feature of our work . . . is our identification and cultivation of a political common ground, given our widely ranging ethnic and national identities. . . .

CWPE asserts that the oppression of women, not their reproductive capacities, needs to be eliminated and calls for drug treatment for women with substance abuse problems, decent jobs, educational opportunities, and mental health and child-care services. It is the lack of these services that deny human dignity and exacerbate conditions of poverty, social status, and gender discrimination. . .

This [essay] is directed at activists, students, policy-makers, and scholars who seek to understand the connections between the criminalization of people of color and the poor, and social and population control. We seek to [foster] a dialogue that builds collaborations between social movements to move beyond single-issue and identity-based politics towards an inclusive political agenda across progressive movements.

NOTES

1. Quoting Myron Magnet, editor of *City Journal*.

2. The law enforcement agencies referred to include local and state police agencies; prison systems at the local, state and federal levels; the United States Border Patrol and Interior Enforcement Operations of the Immigration and Naturalization Service (INS); and the rapidly expanding INS detention system.

3. This is not true in other parts of the world. For example, state violence has been a critical issue for the contemporary women's movement in India. Radha Kumar in *The History of Doing* writes: "The issue of rape has been one that most contemporary feminist movements internationally have focused on, firstly because sexual assault is one of the ugliest and most brutal expressions of masculine violence towards women, because rape and the historical discourse around it reveal a great deal about the social relations of reproduction, and thirdly because of what it shows about the way in which the woman's body is seen as representing the community. In India, it has been the latter reason which has been the most dominant in the taking up of campaigns against rape." (Delhi: Kali for Women, 1993), 128.

4. Foucault, in *The History of Sexuality*, refers to this form of state control as "bio-power."

5. Nancy Scheper-Hughes and Margaret Lock, "The Mindful Body: A Prolegomenon to Future Work in Medical Anthropology," *Medical Anthropology Quarterly*, Vol. 1 (1987), 6–41.

6. "We Remember: African American Women are for Reproductive Freedom" (1998); statement signed and distributed by leaders in the African-American community including Byllye Avery, Reverend Willie Barro, Donna Brazil, Shirley Chisholm, and Dorothy Heights in support of keeping abortion safe and legal.

7. David Firestone, "Alabama's Packed Jails Draw Ire of Courts, Again," *New York Times*, May 1, 2001, A1.

8. Eve Goldberg and Linda Evans, "The Prison Industrial Complex and the Global Economy," *Political Environments* (Fall 1999/Winter 2000), 47.

9. Dr. James Gilligan, a Harvard psychotherapist who has worked on prisons and recidivism, argues that punishing violence with imprisonment does not stop violence because it continues to replicate the patriarchal code. See *Violence: Reflections on a National Epidemic* (New York: Vintage, 1997).

10. Michel Foucault, *Discipline and Punish* (New York: Vintage, 1995), 280.

11. Joseph Hallinan, *Going Up The River: Travels in a Prison Nation* (New York: Random House, 2001).

12. Jennifer Yanco, "Breaking the Silence: Women and the Criminal Justice System," *Political Environments*, No. 7 (Fall 1999/Winter 2000), 20.

13. This document, "The UN Special Report on Violence Against Women" by Special Rapporteur Radhika Coomaraswamy, is available from United Nations Publications at www.un.org.

14. Lynn Paltrow, "Pregnant Drug Users, Fetal Persons, and the Threat to *Roe v. Wade*," *Albany Law Review*, Vol. 62, No. 3 (1999), 1024.

15. This decision affirms the Fourth Amendment to the US Constitution that protects every American, including those who are pregnant and those with substance abuse problems, from warrantless, unreasonable searches.

16. Dorothy Roberts, *Killing the Black Body* (New York: Vintage, 1998).

17. Native American Health Education Resource Center, SisterSong Native Women's Reproductive Health and Rights Roundtable Report (Lake Andes, SD: January 2001), 17.

18. Foucault, *Discipline and Punish*, 195–228.

19. For a rich discussion of this set of issues, see Valerie Hartouni, "Replicating the Singular Self," in *Cultural Conceptions: On Reproduction Technologies and the Remaking of Life* (Minneapolis, MN: University of Minnesota Press, 1997), 110–132.

20. Ibid., 119.

21. For more on the subject see Betsy Hartmann, "Population, Environment, and Security: A New Trinity" in Jael Silliman and Ynestra King, eds., *Dangerous Intersections* (Cambridge, MA: South End Press, 1999), 1–23.

22. Maria Jimenez, "Legalization Then and Now: An Eighty Year History," *Network News* (Oakland, CA: National Network for Immigrant and Refugee Rights, Summer 2000), 6.

23. Ginger Thompson, "The Desperate Risk of Death in a Desert," *New York Times*, October 31, 2000, A12.

24. Jose Palafox, "Welcome to America: Arizona Ranchers Hunt Mexicans," *Network News* (Summer 2000), 4.

Thinking Further

After reading the articles in this section, you will find it helpful to think about the following:

For Part III, "Rethinking Institutions," we could have emphasized any number of social institutions. Health, education, sports, and religion are also important social institutions where race, class, and gender matter. But in searching for materials, we wondered why we were able to find many articles in some areas and virtually none in others. For example, education is not identified here as a section, but we located more material written by students, faculty, administrators, and community members that grapple with intersections of race, class, and gender than we could include. In contrast, writings about religion that use a race, class, and gender intersectional framework are much harder to find. We anticipated finding reflective essays on how religion is shaped by and influences intersections of race, class, and gender. Why do these patterns exist and what might be done to change them?

Suggested Readings

Coontz, Stephanie. 1992. *The Way We Never Were: American Families and the Nostalgia Trap*. New York: Basic Books.

Kelley, Robin D. G. 1994. *Race Rebels: Culture, Politics, and the Black Working Class*. New York: Free Press.

Kumashiro, Kevin K., ed. 2001. *Troubling Intersections of Race and Sexuality: Queer Students of Color and Anti-Oppressive Education*. Lanham, MD: Rowman & Littlefield.

Neal, Mark Anthony. 2002. *Soul Babies: Black Popular Culture and the Post-Soul Aesthetic*. New York: Routledge.

Quadagno, Jill. 1994. *The Color of Welfare: How Racism Undermined the War on Poverty*. New York: Oxford University Press.

Squires, Gregory D. 1994. *Capital and Communities in Black and White: The Intersections of Race, Class, and Uneven Development*. Albany: State University of New York Press.

InfoTrac College Edition: Search Terms

Use your access to the online library, InfoTrac College Edition, to learn more about subjects covered in this section. The following are some suggested terms for searches:

bilingual education
Bureau of Indian Affairs

comparable worth
feminization of poverty
Jim Crow laws
minimum wage
North American Free Trade Agreement (NAFTA 1994)
Personal Responsibility and Work Opportunities Reconciliation Act
    (PRWORA 1996)
Title IX (Higher Education Act of 1972)

 ## InfoTrac College Edition: Bonus Reading

You can use the InfoTrac College Edition feature to find an additional reading pertinent to this section and can also search on the author's name or keywords to locate the following reading:

Pintado-Vertner, Ryan. "From Sweatshop to Hip-Hop: Once Ignored by Fashion, Youth of Color Become the Focus of Marketing." *ColorLines* (Summer 2002): 35–41.

Ryan Pintado-Vertner reports on the changing role of the hip-hop generation in the global political economy. Ten years ago, youth of color often provided the cheap labor that made gym shoes, t-shirts, and high fashion apparel. Now that hip-hop culture is in vogue, youth of color have become the focus of marketing campaigns. Race, class, and gender all interact in this production process that exploits workers and that sells goods made with exploited youth labor back to youth, often using their own images. How does Pintado-Vertner's article show how major social institutions of the economy and culture interconnect in producing these race, class, and gender outcomes?

Wadsworth Sociology Resource Center: Virtual Society

For additional links, quizzes, and learning tools related to this section, see the Wadsworth Sociology Resource Center, Virtual Society, at www.wadsworth.com/sociology_d/

Applying the
Framework

Once you understand the race/class/gender framework that we have presented, you can use it to shed light on a wide array of social issues. Many of the social issues that are of public concern—violence, immigration, the influence of the mass media, sexuality, economic inequality—are framed by race, class, and gender, even though the race, class, and gender dimensions of social issues are not readily apparent in the facile interpretations that characterize typical media and public discussions. Yet, discussion about such issues as education reform, health care, or any number of other issues would be better informed if race, class, and gender were included in the analysis. Having such an analysis would also better inform efforts for social change.

We have seen that race, class, and gender are systemic features of U.S. society. Although their form changes as society evolves, understanding how they are interrelated is critical to understanding the issues of the day. When you understand the connections among race, class, and gender, what new questions and interpretations might you have about current events? With the intersection of race, class, and gender in mind, you see things differently. You both think about different groups *and* you think about groups differently. Race, class, and gender also create many contemporary social issues and are reflected in how people learn about and understand them. How can a race/class/gender framework be used to understand social issues?

The social issues that arise in society are products of social institutions that are based on a certain race, class, and gender configuration. Health care provides an example. Different groups have widely varying patterns of health and illness—patterns that are tremendously influenced by race, class, and gender. Things as fundamental as who lives, how long, and how people die reflect race, class, and gender arrangements in society. Infant mortality is two and a half times more likely among African American babies than White babies. Native American infant mortality is one and a half times the rate for White Americans. Latino groups have comparable infant mortality rates to Whites, except among Cuban Americans, where the rate of infant mortality is slightly higher than for Whites; for Mexicans, Puerto Ricans, and Central/South Americans, it is comparable to Whites (National Center for Health Statistics 2002). Infant mortality is exacerbated by poor nutrition and inadequate health care for many mothers, particularly poor, young women who are less likely than other groups to have a usual source of health care (National Center for Health Statistics 2002).

Access to health care also reflects the realities of race, class, and gender. Having no health insurance is a problem for millions in the United States; one third have no health insurance, but Latinos, African Americans, and Native Americans are much less likely than Whites to have health insurance (National Center for Health Statistics 2002).

Race, class, and gender also shape health research and health care delivery. Middle-class White couples and individuals may seek infertility treatment, funded—at least in part—through insurance, whereas the medical and pharmaceutical industries invest in efforts to control the fertility of poor women via sterilization (Roberts 1997).

It is important to realize that race, class, and gender arrangements are not just about poverty, though. Corporate boards, institutions of higher education, and the mass media, along with other social institutions, all have a particular race, class, and gender character that influences how the institution works, who it serves, and how people experience the institution, as we saw in the prior section. The point here is to understand how these institutional structures also lie behind public issues, such as health care.

How social issues are understood is typically also the result of interlocking race, class, and gender ideologies. Ideologies distort social realities and make existing social arrangements seem legitimate. Media representations of violence, for example, give the impression that violence is rampant. The media associate violence primarily with people of color by nightly broadcasting shootings and other forms of violence that are routinely depicted as happening in communities of color. Violence within middle- and upper-class households is rarely, if ever, seen. Similarly, corporate crime, though revealed occasionally through scandals like Enron's malfeasance, are not routinely reported, nor is the toll that corporate crime takes on human lives ever tallied in the way that street crime is reported—even though the magnitude of corporate crime in terms of dollars is much greater. Likewise, violence against women, though more visible than in the past, is rarely presented in the media and is more often silenced—or whispered about. Violence is also depicted as the work of deranged and crazy individuals, seldom as the method by which social control is administered, such as in who is most likely to get the death penalty or who is most likely to be shot by police. If we look beyond these facile interpretations of violence, we can see clear race, class, and gender dimensions to these important social issues.

Despite what race, class, and gender reveal about violence, their importance typically remains hidden in taken-for-granted ways of thinking about social issues. This is because the same social institutions that frame our discussion of social issues—particularly the mass media—are themselves organized along race, class, and gender lines. As a result, ideologies about social issues reflect the understandings on which such arrangements rest.

If we use a more inclusive framework to understand social issues, we are better positioned to take action to address the issues before us. As an example, consider the environment. Thinking narrowly about the environment leaves many people with the impression that the environmental movement is more concerned with protecting animals and other wildlife than with protecting people. How does an analysis of race, class, and gender change the questions and actions one would consider in analyzing environmental issues? Using this perspective, you might first ask what communities are most affected by pollution and toxic dumping. Where is the least desirable land, and who lives there? When you ask these questions, you can see that communities with high concentrations of poor people or people of color incur more of the negative costs associated with industrial development and environmental degradation than do more affluent communities (Pellow 2002). An environmental movement centered in race, class, and gender politics recognizes that poor, working-class, and minority communities are most affected by toxic pollution of various forms. These communities have often organized in the environmental justice movement—a national coalition of grassroots organizations from diverse communities—to protest this form of human and environmental abuse.

In this section of the book, we include readings on three significant social issues that are currently important in the United States: ethnicity and migration, sexuality, and violence. Although these three issues are not the only ones meriting attention, these are prominent issues of public debate in the United States. How can we apply an analysis of race, class, and gender to these important social issues? As we discuss each, we will see how these issues also increase our understanding of how race, class, and gender connect.

ETHNICITY AND MIGRATION

The view of America as a "white country" is increasingly challenged by the changing contours of the U.S. population. The overall percentage of the U.S. population consisting of people of color is increasing. Racial-ethnic groups are a larger proportion of the U.S. population, and their numbers are expected to rise over the years ahead. One quarter of the U.S. population is now Black, Hispanic, Asian American, Pacific Islander, or Native American; by 2050, non-Hispanic Whites are predicted to make up only slightly more than half the total population.

At the same time that the nation is becoming more diverse, changes are occurring in which groups predominate. Latinos have recently exceeded African Americans as the largest minority group in the population. Hispanic, Native American, Asian American, and African American populations are also growing more rapidly than White American populations, with the greatest growth among Hispanics and Asian Americans (Grieco and Cassidy 2001; U.S. Census Bureau 2001).

Concurrently, categories of race and ethnic designation are changing. The 2000 U.S. census allowed people for the first time to indicate that they were of more than one racial identity. Two and a half percent of the population chose to do so; being "Hispanic" was treated as an additional category. Thus, a person could identify herself as one race, for example "Black," but also check Hispanic; such a person would not have been considered as identified with two races. But, a person who selected "Black" and "White"—a biracial person—would be counted as having two races. These population categories reveal the complexity of defining race and ethnicity in a nation marked by so much diversity. They also show the socially constructed nature of race and ethnicity, because these can be matters of self-identification, as well as categories about how others perceive and treat you.

The nation's diversity stems both from the complex system of race and ethnic relations, as well as from the increased immigration that has characterized recent years. Throughout the 1990s and into the new century, immigration rates have been high—although not as high as at the beginning of the twentieth century (1901–1910) when 10 of every 1,000 people were immigrants. In the early 1990s, 7 of every 1,000 people in the United States were immigrants; now, the numbers have returned to levels typical of most years in the twentieth century (2.4 immigrants for every 1,000 people in the population). Of course, these numbers do not account for undocumented immigrants, so the rate of immigration is actually higher than these numbers indicate (U.S. Census Bureau 2001).

Such transformations in the character of the population are challenging long-standing perceptions of American national identity and require us to think carefully about how ethnicity and migration can be seen through a framework of race, class, and gender. For example, the majority of immigrants (two-thirds) are "family-sponsored" immigrants; that is, spouses, parents, or children of U.S. citizens. Many people also probably think of immigrant workers as largely poor, perhaps agricultural workers, but the truth is that a significant number of immigrants arrive with advanced educational degrees and are highly skilled or professional workers (Immigration and Naturalization Service 2002). The largest number of immigrants come from Mexico and the different Caribbean nations, though China, India, the Philippines, Central American nations, and Russia make up the next largest groups, in respective order (U.S. Census Bureau 2001). Furthermore, immigrants do not typically come from the poorest segments of society, as it takes substantial resources to relocate (Portes and Rumbaut 1996). Despite stereotypes to the contrary, there is no single image of immigration that captures the diverse national, cultural, and class identities that immigrants bring to the United States.

Immigration into the United States is only one facet of the increasing importance that ethnicity and migration play in contemporary social life. Changes in ethnicity and migration are being enacted on a world stage—a global context that increasingly links nations together in a global social order where diversity is more typical than not. You need only consult an older world atlas to get a sense of the widespread changes that characterize the current period. *World cities* have been formed that are linked through international systems of commerce; within these cities, migrants play an important role in providing the labor that is needed to keep pace with the global economy. Migrant (or "guest") workers may provide the labor for multinational corporations; or, they may provide the service work that increasingly characterizes postindustrial society. Some may provide agricultural labor for multinational producers of food; others do domestic labor for middle- and upper-class families whose own lives are being transformed by changes in the world economic system. Wherever you look, the world is being changed by the increasingly global basis of modern life.

Such changes indicate that globalization is not just about what is happening elsewhere in the world—as if global studies were just a matter of comparing societies. Globalization is everywhere. No nation, including the United States, can really be understood without seeing how globalization is affecting life even at home. Patterns of life in any one society are now increasingly shaped by the connections between societies (Andersen and Taylor 2003), and this is evident by looking at how ethnicity and migration are changing life in the United States. *Transnational families* are more common—that is, families whose members live and reside in different nations (often at a great distance from each other). Cultural features associated with different cultural traditions are increasingly evident, even in communities in the United States that have been thought of as "All-American" towns. Youth cultures embrace common features whether in Russia, Mexico, Japan, or the United States—because of the penetration of world capitalist markets. Thus, hip-hop plays on the streets of Mexico City and salsa is heard on radio stations in the United States.

What do such changes mean for understanding ethnicity and migration in a transnational, global context? Within the formal borders of the United States, changes associated with migration are raising new questions about national identity. How will the United States define itself as a nation-state in the changing global context? How will the increased visibility and vocality of people of color in the United States, in Latin America, in Africa, in Asia, and within the borders of former European colonial powers shape the future? How are cultural representations and institutions changed in a world context marked by such an ethnic mix—both within and between countries? In this context, what does race mean? Who gets defined as a race, and how does that map onto concepts of citizenship and the division of labor in different nations? What will it mean to be "American" in a nation where racial-ethnic diversity is so much a part of the national fabric?

The events of September 11, 2001, give these questions additional meaning. In response to terrorism, most in the United States were catapulted into a strong sense of national identity. But, just as racial profiling was becoming increasingly criticized by the public, fears of terrorism have made it suddenly permissible to many. In the aftermath of 9/11, more restrictive immigration practices and policies have been employed; international and "ethnic" students are increasingly under surveillance; civil liberties have been restricted in the interests of national security. The events of 9/11 and the international tensions that have followed bring increasing urgency to grappling with questions of how we can maintain a society of such great diversity within a framework of social justice.

Thus, several themes emerge in rethinking ethnicity and migration via the lens of race, class, and gender. Despite the ideology of the "melting pot," national identity in the United States has been closely linked to a history of racial privilege. As Lillian Rubin points out in "Is This a White Country, or What?" the term "American" is assumed to mean White. Other types of "Americans," such as African Americans and Asian Americans, become distinguished from the "real" Americans by virtue of their race. "To be an 'all-American' is, by definition, not to be an Asian American, Pacific American, American Indian, Latino, Arab American, or African American," observes political theorist Manning Marable (Marable 1995: 114). More importantly, certain benefits are reserved for those deemed to be "deserving Americans."

Many Americans never "melted" into the melting pot. Many people think that race is like ethnicity and that the failure of people of color to assimilate into the mainstream as White ethnic groups purportedly have done represents an unwillingness to shed their culture. But as Mary Waters points out in "Optional Ethnicities: For Whites Only?" this view seriously misreads the meaning of both race and ethnicity in shaping American national identity. Waters suggests that White Americans of European ancestry have *symbolic ethnicity* in that their ethnic identity does not influence their lives unless they want it to. Waters contrasts this symbolic ethnic identity among many Whites with the socially enforced and imposed racial identity among African Americans. Because race operates as a physical marker in the United States, intersections of race and ethnicity operate differently for Whites and for people of color. White ethnics can thus have "ethnicity" without cost, but people of color pay the price for their ethnic identity.

We see this in Robert Smith's discussion of the increasing presence of Mexicans in the Untied States ("'Mexicanness' in New York"). Smith uses the concept of *racialization* to describe how groups become defined as having racial characteristics that are also the basis for their mistreatment. Although we might imagine Mexicans and Mexican Americans as an ethnic group—in that their identity comes from a commonly shared cultural background—in the context of migration, they have been redefined within the context of racial, social, and class hierarchies in the United States. As Smith argues, Mexican immigrants confront the racial segregation and discrimination that have characterized Black American experience. They are also juxtaposed against both White ethnics and other groups, such as Puerto Ricans, by social structures that place different values on groups exploited for their labor. His article shows how race and ethnicity intertwine—both between dominant and subordinate groups and among subordinate groups—those "racialized" as Black.

Despite these structural forces, however, people are not just passive victims. Nazli Kibria's research ("Migration and Vietnamese American Women") shows how Vietnamese American women and men negotiate with each other to adjust to the specific experiences they face as immigrants. By examining gender roles in Vietnamese American families, Kibria shows how women rebuild families to provide economic and social support during the transition to a new community. Her analysis reminds us that immigration and migration are not just about ethnicity and race, but are also gendered processes. Families are sites where gender is both *created* and also *contested*, as people face circumstances that require reconstruction of prior relationships. Kibria's research also emphasizes that people actively shape viable family networks and support systems that enable them to survive even in the face of cultural, economic, and social assaults.

The experiences of groups migrating to the United States show both the tenacity of race, class, and gender structures *and* how people work to change these systems. In "Chappals and Gym Shorts" Almas Sayeed depicts living as an Indian Muslim woman in Kansas—a place historically seen as quintessentially "American." She describes the complexity of negotiating the multiple and different cultural spheres that, as she says, "interlink" in her experience. Her narrative shows how her ethnic status, her gender, her religion, and her community all shape her identity as a feminist Islamic woman.

Though experienced on a personal and immediate level, patterns of migration stem from long-term historical and global change. Migration is embedded in patterns of social, economic, and political organization where world capitalism plays a key role. For centuries people of color have provided the work that has enabled the domination of others and that has been the foundation for the development of the class system. Slavery, forced servitude, and the exploitation of women's labor have all been mechanisms through which the power of dominant groups has been created. Together, the articles in this section reveal the complex and interconnected ways that race, class, and gender operate in the construction of racial, ethnic, and gender identities and in the creation of a social order based on these social realities.

SEXUALITY

As an important part of structures of race, class, and gender, sexuality is another social issue critical in contemporary society. The linkage between race, class, and gender is revealed within studies of sexuality, just as sexuality is a dimension of each. For example, constructing images about Black sexuality is central to maintaining institutional racism. Similarly, beliefs about women's sexuality structure gender oppression. At the same time, sexuality operates as a system of oppression comparable to the systems of race, class, and gender. *Heterosexism*, the institutionalized structures and beliefs that define and enforce heterosexual behavior as the only natural and permissible form of sexual expression, is a system of oppression that affects everyone.

Understanding this rests on seeing sexuality as a social construction, the point argued by Pepper Schwartz and Virginia Rutter in "The Gender of Sexuality." Schwartz and Rutter see that sexuality has a biological context, because bodies make sex pleasurable. But the real significance of sexuality lies in its social dimensions. Thus, ideas about gender in society influence how we experience sexuality and how sexuality is controlled. At the same time, ideas about sexuality also construct gender. Thus, homophobia is used as a mechanism to enforce the construction of "masculinity." The hatred directed toward lesbians, gays, and bisexuals is thus part of the system by which gender is created and maintained. In this regard, sexuality and gender are deeply linked.

Cornel West ("Black Sexuality: The Taboo Subject") adds another social dimension to sexuality, showing how sexuality has been used as the vehicle to support racial fears and racial subordination. Racial subordination was built on the exploitation of Black bodies—both as labor and as sexual objects.

West's essay untangles multiple dimensions of racial-sexual politics. Strictures against certain interracial, sexual relationships—but not these between White men and African American women—are a way to maintain White patriarchy. And, as West points out, race mixes with both gender and sex. Without examining the ways that sexuality, race, and gender—and, we should add, class—entangle, we remain shrouded in myths and distortions about human sexuality and human relationships.

How do race, class, gender, and sexuality interrelate? This is a complex question, explored in the interview by Amy Gluckman and Betsy Reed with longtime activist Barbara Smith ("Where Has Gay Liberation Gone?"). Many have argued that race, class, gender, and sexuality are similar in social relations and processes of society. Can we collapse them as similar categories of difference? Smith carefully identifies some of the ways in which race and sexuality are not similar. She notes that racial oppression is historically and structurally embedded in the founding of the United States, whereas lesbian and gay oppression was not. Differences in historical paths and varying structural arrangements should not blind us, however, to the similarities that link race, class, and gender oppression to homophobia and heterosexism. As Smith notes, similar strategies are used to oppress people by race and by sexuality. As she succinctly puts it, "Our enemies are the same. That to me is the major thing that should be pulling us together."

When you use a race/class/gender framework to think about sexuality, you will see both common patterns and different realities in race, class, gender, and sexual oppression. Sexuality, for example, has not been used as an explicit category to organize the division of labor, as have race, class, and gender. Nonetheless, sexuality has been a key part of the division of labor, primarily in the form of heterosexual households where the White middle and upper classes rely on male breadwinners. But, this family form, as we have seen, has not been available to all and is currently undergoing great change. Female-headed and female-supported families have long been characteristics of families of color and are now increasingly common among White, middle-class families, too. Thus, the idea of the male breadwinner—supported as it was through heterosexism—is more an ideological ideal than a reality.

Kamala Kempadoo in "Globalizing Sex Workers' Rights" raises another dimension to the discussion of sexuality—sex as work. Sex workers sell their bodies or images of their bodies for money. Kempadoo emphasizes how sex workers have organized worldwide to address issues of human rights, working conditions, and decriminalization. Her analysis links sex work to other struggles for women's rights and social justice. She also brings a global perspective to the discussion of sexuality. Sex work on a global scale is also linked to world politics about race and class, given the specific position of women of color in international sex work. Furthermore, the class politics of international relations are also apparent in the place of women in the international traffic in sex, where women's sexuality is used to promote tourism and where images of the "exotic other" are used to attract more affluent classes to various regions of the world. Kempadoo also links sexuality to processes of migration, showing how race, class, gender, and sex are part of the international traffic in women.

Jason Schultz ("Getting Off on Feminism") questions sexuality and gender, specifically in the form of White heterosexual male sexualities. He points to the narrow range of sexualities that are available to men, especially middle-class White men, under heterosexism. "Just as hetero women are often forced to choose between the images of the virgin and the whore, modern straight men are caught in a cultural tug-of-war between the Marlboro Man and the Wimp," Schultz observes. Rather than bemoan this dilemma, Schultz describes how he and his male friends tried to explore new ways of thinking about heterosexual masculinity.

VIOLENCE

The violence permeating the United States is a pressing social issue. School shootings, snipers, and street criminals dominate the news and project the idea that violence is rampant. Usually violence is depicted as the action of crazed individuals who are socially maladjusted, angry, or desperate. Although this may be the case for some, deeper questions about violence arise when we think about violence in the context of race, class, and gender relations. Who is perceived as violent shifts, as does who we see as victims of violence. The consequences of violence in society are numerous, not the least of which is the fear that images of violence create and the action of the state toward those perceived as violent. How can we apply our knowledge of race, class, and gender to understanding violence?

Many of the current forms of violence are directly linked to race, class, and gender inequality. Although not the form of violence most usually depicted in the evening news, violence against women is common in society—

even more common than the street crimes that are nightly shown on television. Most women live with some fears of violence—one of the reasons that feminists have argued that violence against women is a form of social control—making women fearful of being alone, being out at night, being in public (Madriz 1997). Pornography is also a form of violence against women, representing the link between violence and sexuality as a commodity. Both sell products, fueling the social-class system's ongoing search for new consumer markets.

Inequality is also found in victimization by crime. Property crime is far more common among those with the least, even higher among renters than home owners and twice as likely to occur against those with incomes under $15,000 per year. Black Americans and Hispanics are more likely than Whites to be victims of violent crimes. Crime victimization is highest among young people (U.S. Bureau of Justice Statistics 2001).

Although we might abhor violence, it is widely marketed in American culture. Films, video games, and other forms of popular culture enhance their share of market "power" with violent images meant to entertain and captivate the public. Gun shows and sales offer another example of the marketing of violence. Unfortunately, violence often hurts the most vulnerable segments of society. For example, violence against children has reached alarming proportions. Homicide by guns is the second leading cause of death for Black males under seventeen years of age (Violence Policy Center 2000). In a sense, the violence within the United States is an outcome of the conditions of inequality. When people are wronged, they may use violence as a weapon. Conflict that escalates into violence can characterize relationships among groups disadvantaged within existing race, class, and gender arrangements—sometimes with tragic consequences.

It is important to see violence not as acts of individual social deviance but as one outcome of how the politics of race, class, and gender permeates a range of social institutions. Violent acts, the threat of violence, and more generalized policies based on the use of force find organizational homes in designated institutions of social control—primarily the police, the military, and the criminal justice system. Violence simultaneously permeates a range of other social institutions. Some social institutions are known for their use of force, while others are less peaceful than they appear. Men's violence against women and children (whether physical, emotional, or sexual) in families of all social classes and racial-ethnic compositions contradicts the dominant belief that the home is a place of tranquility and love. Sexual harassment on the job as a dimension of social control belies the belief that men and women encounter a similar work environment. In the media, depicting rape as pleasurable to women and portraying violence against Native Americans, Asian Americans, and African Americans in numerous movies as justified contributes to a generalized belief system condoning violence. Thus, violence is both hated and condoned in a society where the meaning, commission, and consequences of violence are deeply tied to race, class, and gender inequality.

Deirdre Davis (in "The Harm That Has No Name") analyzes the daily violence that millions of women experience by simply being out in public. Street harassment, Davis writes, is a form of sexual terrorism. Though many might think of street harassment as a minor nusiance, Davis argues that it is entangled with women's subjective awareness of rape and with the gender and racial meanings that are associated with sexual fears. "Spirit-murder," she calls it, revealing how racism and sexism co-mingle to define women—and women of color, in particular—as targets for social control and harm.

Understanding violence in the context of systems of social control unveils the interconnected nature of dominant ideologies of masculinity, the individualized and systemic forms of violence they justify, and the mechanisms of social control that support masculinity. "More Power Than We Want: Masculine Sexuality and Violence," by Bruce Kokopeli and George Lakey, investigates these links. Violence against women enforces interlocking belief systems about masculinity and violence. Violence based on race, gender, and heterosexism are parts of one system of social control: All of these types of violence aim to reinforce systems of privilege.

The everyday realities of violence are also shown in a different vein by Brent Staples ("Just Walk on By"). Read in the context of racial profiling, Staples's narrative is indicative of the threat of violence in the everyday life of African American men and, now, others as well. One prominent theme in the readings on violence is the link between individual acts of violence and more routinized, systemic violence. To illustrate historically, lynchings of African American men can be seen as the random acts of unruly mobs; yet, the failure to arrest, prosecute, and convict those who did the lynching is a powerful testament to public endorsement of these seemingly individual acts of violence. Likewise, rape appears to be a private act, but it occurs as part of a generalized climate that condones violence against women. Although violence may be experienced individually, it occurs in specific organizational and institutional contexts.

Helen Zia in "Where Race and Gender Meet" shows how racism, hate crimes, and pornography enforce the subordination of oppressed groups. She applies her analysis to Asian Americans, but you can use it to understand how homophobic hate crimes also work. Homophobic violence is a way—through extreme means—that dominant groups try to maintain control. As we link gender, race, class, and homophobia, we see how violence is a means by which dominant groups assert their power.

In the section on ethnicity and migration, we ask how the United States can build a national community that is founded on a community of justice even with a population diversified by race, ethnicity, class, gender, sexuality, and citizenship status. Thinking about violence brings a new dimension to this complex question. Most in the United States feel fearful about violence—now because of fears about terrorism and international war. For many in the United States who have been less touched in the past by threats of violence, these are new fears, but, as Desiree Taylor points out ("How Safe Is America?"), feeling unsafe is not new for many—those who because of their class, their race, their sexual orientation, or their gender have never been safe within the United States. For Taylor, cries to defend freedom and democracy in the face of terrorist threats must be accompanied with the realization that justice and freedom have long been denied to many living within the United States. The point is not to dismiss people's fears about new threats but to create a new nation where race, class, and gender justice within the United States are as important to us as our position in the world.

Further Resources

For additional materials relating to this section, see the features on pages 509–10 at the conclusion of this section.

References

Andersen, Margaret L., and Howard F. Taylor. 2003. *Sociology: The Essentials*. Belmont, CA: Wadsworth.

Grieco, Elizabeth M., and Rachel C. Cassidy. 2001. *Overview of Race and Hispanic Origin: Census 2000 Brief*. Washington, DC: U.S. Census Bureau.

Immigration and Naturalization Service. 2002. *Legal Immigration, Fiscal Year 2001. Office of Policy and Planning Annual Report*. Washington, DC: Immigration and Naturalization Service, U.S. Department of Justice. Web site: www.ins.gov

Madriz, Esther. 1997. *Nothing Bad Happens to Good Girls: Fear of Crime in Women's Lives.* Berkeley: University of California Press.

Marable, Manning. 1995. "Beyond Racial Identity Politics: Toward a Liberation Theory for Multicultural Democracy." *Race & Class* 35 (July/September): 113–30.

National Center for Health Statistics. 2002. *Health United States 2002.* Rockville, MD: U.S. Department of Health and Human Services.

Portes, Alejandro, and Rubén G. Rumbaut. 1996. *Immigrant America.* Berkeley: University of California Press.

Pellow, David Naguib. 2002. *Garbage Wars: The Struggle for Environmental Justice in Chicago.* Cambridge, MA: M.I.T. Press.

Roberts, Dorothy. 1997. *Killing the Black Body: Race, Reproduction, and the Meaning of Liberty.* New York: Vintage.

U.S. Bureau of Justice Statistics. 2001. *Crime Victimization 2001.* Washington, DC: U.S. Department of Justice. Web page: www.ojp.usdoj.gov/bis/

U.S. Census Bureau. 2001. *Statistical Abstract of the United States 2001.* Washington, DC: U.S. Department of Commerce.

Violence Policy Center. 2000. *Fact Sheet on Firearms Violence against Black Children and Youth.* Washington, DC: Violence Policy Center. Web site: www.vpc.org

Ethnicity and Migration

"IS THIS A WHITE COUNTRY, OR WHAT?"

45

Lillian Rubin

"They're letting all these coloreds come in and soon there won't be any place left for white people," broods Tim Walsh, a thirty-three-year-old white construction worker. "It makes you wonder: Is this a white country, or what?"

It's a question that nags at white America, one perhaps that's articulated most often and most clearly by the men and women of the working class. For it's they who feel most vulnerable, who have suffered the economic contractions of recent decades most keenly, who see the new immigrants most clearly as direct competitors for their jobs.

It's not whites alone who stew about immigrants. Native-born blacks, too, fear the newcomers nearly as much as whites—and for the same economic reasons. But for whites the issue is compounded by race, by the fact that the newcomers are primarily people of color. For them, therefore, their economic anxieties have combined with the changing face of America to create a profound uneasiness about immigration—a theme that was sounded by nearly 90 percent of the whites I met, even by those who are themselves first-generation, albeit well-assimilated, immigrants.

Sometimes they spoke about this in response to my questions; equally often the subject of immigration arose spontaneously as people gave voice to their concerns. But because the new immigrants are dominantly people of color, the discourse was almost always cast in terms of race as well as immigration, with the talk slipping from immigration to race and back again as if these are not two separate phenomena. "If we keep letting all them foreign-ers in, pretty soon there'll be more of them than us and then what will this country be like?" Tim's wife, Mary Anne, frets. "I mean, this is *our* country, but the way things are going, white people will be the minority in our own country. Now does that make any sense?"

Such fears are not new. Americans have always worried about the strangers who came to our shores, fearing that they would corrupt our soci-ety, dilute our culture, debase our values. So I remind Mary Anne, "When your ancestors came here, people also thought we were allowing too many foreigners into the country. Yet those earlier immigrants were successfully in-tegrated into the American society. What's different now?"

"Oh, it's different, all right," she replies without hesitation. "When my people came, the immigrants were all white. That makes a big difference." . . .

Listening to Mary Anne's words I was reminded again how little we Americans look to history for its lessons, how impoverished is our historical memory. For, in fact, being white didn't make "a big difference" for many of those earlier immigrants. The dark-skinned Italians and the eastern European Jews who came in the late nineteenth and early twentieth centuries didn't look very white to the fair-skinned Americans who were here then. Indeed, the same people we now call white—Italians, Jews, Irish—were seen as another race at that time. Not black or Asian, it's true, but an alien other, a race apart, although one that didn't have a clearly defined name. Moreover, the racist fears and fantasies of native-born Americans were far less contained then than they are now, largely because there were few social constraints on their expression.

When, during the nineteenth century, for example, some Italians were taken for blacks and lynched in the South, the incidents passed virtually un-noticed. And if Mary Anne and Tim Walsh, both of Irish ancestry, had come to this country during the great Irish immigration of that period, they would have found themselves defined as an inferior race and described with the same language that was used to characterize blacks: "low-browed and savage, grov-elling and bestial, lazy and wild, simian and sensual."[1] Not only during that period but for a long time afterward as well, the U.S. Census Bureau counted the Irish as a distinct and separate group, much as it does today with the cat-egory it labels "Hispanic."

But there are two important differences between then and now, differ-ences that can be summed up in a few words: the economy and race. Then, a growing industrial economy meant that there were plenty of jobs for both im-migrant and native workers, something that can't be said for the contracting economy in which we live today. True, the arrival of the immigrants, who were more readily exploitable than native workers, put Americans at a disad-vantage and created discord between the two groups. Nevertheless, work was available for both.

Then, too, the immigrants—no matter how they were labeled, no matter how reviled they may have been—were ultimately assimilable, if for no other reason than that they were white. As they began to lose their alien ways, it became possible for native Americans to see in the white ethnics of yesteryear a reflection of themselves. Once this shift in perception occurred, it was possible for the nation to incorporate them, to take them in, chew them up, digest them, and spit them out as Americans—with subcultural variations not always to the liking of those who hoped to control the manners and mores of the day, to be sure, but still recognizably white Americans.

Today's immigrants, however, are the racial other in a deep and profound way. . . . And integrating masses of people of color into a society where race conciousness lies at the very heart of our central nervous system raises a whole new set of anxieties and tensions. . . .

The increased visibility of other racial groups has focused whites more self-consciously than ever on their own racial identification. Until the new immigration shifted the complexion of the land so perceptibly, whites didn't think of themselves as white in the same way that Chinese know they're Chinese and African-Americans know they're black. Being white was simply a fact of life, one that didn't require any public statement, since it was the definitive social value against which all others were measured. "It's like everything's changed and I don't know what happened," complains Marianne Bardolino. "All of a sudden you have to be thinking all the time about these race things. I don't remember growing up thinking about being white like I think about it now. I'm not saying I didn't know there was coloreds and whites; it's just that I didn't go along thinking, *Gee, I'm a white person.* I never thought about it at all. But now with all the different colored people around, you have to think about it because they're thinking about it all the time."

"You say you feel pushed now to think about being white, but I'm not sure I understand why. What's changed?" I ask.

"I told you," she replies quickly, a small smile covering her impatience with my question. "It's because they think about what they are, and they want things their way, so now I have to think about what I am and what's good for me and my kids." She pauses briefly to let her thoughts catch up with her tongue, then continues. "I mean, if somebody's always yelling at you about being black or Asian or something, then it makes you think about being white. Like, they want the kids in school to learn about their culture, so then I think about being white and being Italian and say: What about my culture? If they're going to teach about theirs, what about mine?"

To which America's racial minorities respond with bewilderment. "I don't understand what white people want," says Gwen Tomalson. "They say if black kids are going to learn about black culture in school, then white people want their kids to learn about white culture. I don't get it. What do they think kids have been learning about all these years? It's all about white people and how they live and what they accomplished. When I was in school you wouldn't have thought black people existed for all our books ever said about us."

As for the charge that they're "thinking about race all the time," as Marianne Bardolino complains, people of color insist that they're forced into it by a white world that never lets them forget. "If you're Chinese, you can't forget it, even if you want to, because there's always something that reminds you," Carol Kwan's husband, Andrew, remarks tartly. "I mean, if Chinese kids get good grades and get into the university, everybody's worried and you read about it in the papers."

While there's little doubt that racial anxieties are at the center of white concerns, our historic nativism also plays a part in escalating white alarm. The new immigrants bring with them a language and an ethnic culture that's vividly expressed wherever they congregate. And it's this also, the constant reminder of an alien presence from which whites are excluded, that's so troublesome to them.

The nativist impulse isn't, of course, given to the white working class alone. But for those in the upper reaches of the class and status hierarchy— those whose children go to private schools, whose closest contact with public transportation is the taxi cab—the immigrant population supplies a source of cheap labor, whether as nannies for their children, maids in their households, or workers in their businesses. They may grouse and complain that "nobody speaks English anymore," just as working-class people do. But for the people who use immigrant labor, legal or illegal, there's a payoff for the inconvenience—a payoff that doesn't exist for the families in this study but that sometimes costs them dearly. For while it may be true that American workers aren't eager for many of the jobs immigrants are willing to take, it's also true that the presence of a large immigrant population—especially those who come from developing countries where living standards are far below our own—helps to make these jobs undesirable by keeping wages depressed well below what most American workers are willing to accept. . . .

It's not surprising, therefore, that working-class women and men speak so angrily about the recent influx of immigrants. They not only see their jobs and their way of life threatened, they feel bruised and assaulted by an environment that seems suddenly to have turned color and in which they feel like strangers in their own land. So they chafe and complain: "They come here to take advantage of us, but they don't really want to learn our ways," Beverly Sowell, a thirty-three-year old white electronics assembler, grumbles irritably. "They live different than us; it's like another world how they live. And they're so clannish. They keep to themselves, and they don't even *try* to learn English. You go on the bus these days and you might as well be in a foreign country; everybody's talking some other language, you know, Chinese or Spanish or something. Lots of them have been here a long time, too, but they don't care; they just want to take what they can get."

But their complaints reveal an interesting paradox, an illuminating glimpse into the contradictions that beset native-born Americans in their relations with those who seek refuge here. On the one hand, they scorn the immigrants; on the other, they protest because they "keep to themselves." It's the same contradiction that dominates black-white relations. Whites refuse to integrate blacks but are outraged when they stop knocking at the door, when they move to sustain the separation on their own terms—in black theme houses on campuses, for example, or in the newly developing black middle-class suburbs.

I wondered, as I listened to Beverly Sowell and others like her, why the same people who find the lifeways and languages of our foreign-born population offensive also care whether they "keep to themselves."

"Because like I said, they just shouldn't, that's all," Beverly says stubbornly. "If they're going to come here, they should be willing to learn our ways—you know what I mean, be real Americans. That's what my grandparents did, and that's what they should do."

"But your grandparents probably lived in an immigrant neighborhood when they first came here, too," I remind her.

"It was different," she insists. "I don't know why; it was. They wanted to be Americans; these here people now, I don't think they do. They just want to take advantage of this country. . . .

"Everything's changed, and it doesn't make sense. Maybe you get it, but I don't. We can't take care of our own people and we keep bringing more and more foreigners in. Look at all the homeless. Why do we need more people here when our own people haven't got a place to sleep?"

"Why do we need more people here?"—a question Americans have asked for two centuries now. Historically, efforts to curb immigration have come during economic downturns, which suggests that when times are good, when American workers feel confident about their future, they're likely to be more generous in sharing their good fortune with foreigners. But when the economy falters, as it did in the 1990s, and workers worry about having to compete for jobs with people whose standard of living is well below their own, resistance to immigration rises. "Don't get me wrong; I've got nothing against these people," Tim Walsh demurs. "But they don't talk English, and they're used to a lot less, so they can work for less money than guys like me can. I see it all the time; they get hired and some white guy gets left out."

It's this confluence of forces—the racial and cultural diversity of our new immigrant population; the claims on the resources of the nation now being made by those minorities who, for generations, have called America their home; the failure of some of our basic institutions to serve the needs of our people; the contracting economy, which threatens the mobility aspirations of working-class families—all these have come together to leave white workers feeling as if everyone else is getting a piece of the action while they get nothing. "I feel like white people are left out in the cold," protests Diane Johnson, a twenty-eight-year-old white single mother who believes she lost a job as a bus driver to a black woman. "First it's the blacks; now it's all those other colored people, and it's like everything always goes their way. It seems like a white person doesn't have a chance anymore. It's like the squeaky wheel gets the grease, and they've been squeaking and we haven't," she concludes angrily.

Until recently, whites didn't need to think about having to "squeak"—at least not specifically as whites. They have, of course, organized and squeaked at various times in the past—sometimes as ethnic groups, sometimes as workers. But not as whites. As whites they have been the dominant group, the favored ones, the ones who could count on getting the job when people of color could not. Now suddenly there are others—not just individual others but identifiable groups, people who share a history, a language, a culture, even a color—who lay claim to some of the rights and privileges that formerly had been labeled "for whites only." And whites react as if they've been betrayed, as if a sacred promise has been broken. They're white, aren't they? They're *real* Americans, aren't they? This is their country, isn't it?

The answers to these questions used to be relatively unambiguous. But not anymore. Being white no longer automatically assures dominance in the politics of a multiracial society. Ethnic group politics, however, has a long and fruitful history. As whites sought a social and political base on which to stand, therefore, it was natural and logical to reach back to their ethnic past. Then they, too, could be "something"; they also would belong to a group; they would have a name, a history, a culture, and a voice. "Why is it only the blacks or Mexicans or Jews that are 'something'?" asks Tim Walsh. "I'm Irish, isn't that something, too? Why doesn't that count?"

In reclaiming their ethnic roots, whites can recount with pride the tribulations and transcendence of their ancestors and insist that others take their place in the line from which they have only recently come. "My people had a rough time, too. But nobody gave us anything, so why do we owe them something? Let them pull their share like the rest of us had to do," says Al Riccardi, a twenty-nine-year-old white taxi driver.

From there it's only a short step to the conviction that those who don't progress up that line are hampered by nothing more than their own inadequacies or, worse yet, by their unwillingness to take advantage of the opportunities offered them. "Those people, they're hollering all the time about discrimination," Al continues, without defining who "those people" are. "Maybe once a long time ago that was true, but not now. The problem is that a lot of those people are lazy. There's plenty of opportunities, but you've got to be willing to work hard."

He stops a moment, as if listening to his own words, then continues, "Yeah, yeah, I know there's a recession on and lots of people don't have jobs. But it's different with some of those people. They don't really want to work, because if they did, there wouldn't be so many of them selling drugs and getting in all kinds of trouble."

"You keep talking about 'those people' without saying who you mean," I remark.

"Aw c'mon, you know who I'm talking about," he says, his body shifting uneasily in his chair. "It's mostly the black people, but the Spanish ones, too."

In reality, however, it's a no-win situation for America's people of color, whether immigrant or native born. For the industriousness of the Asians comes in for nearly as much criticism as the alleged laziness of other groups. When blacks don't make it, it's because, whites like Al Riccardi insist, their culture doesn't teach respect for family; because they're hedonistic, lazy, stupid, and/or criminally inclined. But when Asians demonstrate their ability to overcome the obstacles of an alien language and culture, when the Asian family seems to be the repository of our most highly regarded traditional values, white hostility doesn't disappear. It just changes its form. Then the accomplishments of Asians, the speed with which they move up the economic ladder, aren't credited to their superior culture, diligence, or intelligence—even when these are granted—but to the fact that they're "single minded," "untrustworthy," "clannish drones," "narrow people" who raise children who are insufficiently "well rounded."[2] . . .

Not surprisingly, as competition increases, the various minority groups often are at war among themselves as they press their own particular claims, fight over turf, and compete for an ever-shrinking piece of the pie. In several African-American communities, where Korean shopkeepers have taken the place once held by Jews, the confrontations have been both wrenching and tragic. A Korean grocer in Los Angeles shoots and kills a fifteen-year-old black girl for allegedly trying to steal some trivial item from the store.[3] From New York City to Berkeley, California, African-Americans boycott Korean shop owners who, they charge, invade their neighborhoods, take their money, and treat them disrespectfully.[4] But painful as these incidents are for those involved, they are only symptoms of a deeper malaise in both communities—the contempt and distrust in which the Koreans hold their African-American neighbors, and the rage of blacks as they watch these new immigrants surpass them.

Latino-black conflict also makes headlines when, in the aftermath of the riots in South Central Los Angeles, the two groups fight over who will get the lion's share of the jobs to rebuild the neighborhood. Blacks, insisting that they're being discriminated against, shut down building projects that don't include them in satisfactory numbers. And indeed, many of the jobs that formerly went to African-Americans are now being taken by Latino workers. In an article entitled "Black vs. Brown," Jack Miles, an editorial writer for the *Los Angeles Times*, reports that "janitorial firms serving downtown Los Angeles have almost entirely replaced their unionized black work force with non-unionized immigrants."[5] . . .

But the disagreements among America's racial minorities are of little interest or concern to most white working-class families. Instead of conflicting groups, they see one large mass of people of color, all of them making claims that endanger their own precarious place in the world. It's this perception that has led some white ethnics to believe that reclaiming their ethnicity alone is not enough, that so long as they remain in their separate and distinct groups, their power will be limited. United, however, they can become a formidable countervailing force, one that can stand fast against the threat posed by minority demands. But to come together solely as whites would diminish their impact and leave them open to the charge that their real purpose is simply to retain the privileges of whiteness. A dilemma that has been resolved, at least for some, by the birth of a new entity in the history of American ethnic groups—the "European-Americans."[6] . . .

At the University of California at Berkeley, for example, white students and their faculty supporters insisted that the recently adopted multicultural curriculum include a unit of study of European-Americans. At Queens College in New York City, where white ethnic groups retain a more distinct presence, Italian-American students launched a successful suit to win recognition as a disadvantaged minority and gain the entitlements accompanying that status, including special units of Italian-American studies.

White high school students, too, talk of feeling isolated and, being less sophisticated and wary than their older sisters and brothers, complain quite openly that there's no acceptable and legitimate way for them to acknowledge a white identity. "There's all these things for all the different ethnicities, you know, like clubs for black kids and Hispanic kids, but there's nothing for me and my friends to join," Lisa Marshall, a sixteen-year-old white high school student, explains with exasperation. "They won't let us have a white club because that's supposed to be racist. So we figured we'd just have to call it something else, you know, some ethnic thing, like Euro-Americans. Why not? They have African-American clubs."

Ethnicity, then, often becomes a cover for " white," not necessarily because these students are racist but because racial identity is now such a prominent feature of the discourse in our social world. In a society where racial consciousness is so high, how else can whites define themselves in ways that connect them to a community and, at the same time, allow them to deny their racial antagonisms?

Ethnicity and race—separate phenomena that are now inextricably entwined. Incorporating newcomers has never been easy, as our history of controversy and violence over immigration tells us.[7] But for the first time, the new immigrants are also people of color, which means that they tap both the nativist and racist impulses that are so deeply a part of American life. As in the past, however, the fear of foreigners, the revulsion against their strange customs and seemingly unruly ways, is only part of the reason for the anti-immigrant attitudes that are increasingly being expressed today. For whatever xenophobic suspicions may arise in modern America, economic issues play a critical role in stirring them up.

NOTES

1. David R. Roediger, *The Wages of Whiteness* (New York: Verso, 1991), p. 133.

2. These were, and often still are, the commonly held stereotypes about Jews. Indeed, the Asian immigrants are often referred to as "the new Jews."

3. Soon Ja Du, the Korean grocer who killed fifteen-year-old Latasha Harlins, was found guilty of voluntary manslaughter and sentenced to four hundred hours of community service, a $500 fine, reimbursement of funeral costs to the Harlins family, and five years' probation.

4. The incident in Berkeley didn't happen in the black ghetto, as most of the others did. There, the Korean grocery store is near the University of California campus, and the woman involved in the incident is an African-American university student who was Maced by the grocer after an argument over a penny.

5. Jack Miles, "Blacks vs. Browns," *Atlantic Monthly* (October 1992), pp. 41–68.

6. For an interesting analysis of what he calls "the transformation of ethnicity," see Richard D. Alba, *Ethnic Identity* (New Haven, CT: Yale University Press, 1990).

7. In the past, many of those who agitated for a halt to immigration were immigrants or native-born children of immigrants. The same often is true today. As anti-immigrant sentiment grows, at least some of those joining the fray are relatively recent arrivals. One man in this study, for example—a fifty-two-year-old immigrant from Hungary— is one of the leaders of an anti-immigration group in the city where he lives.

References

Alba, Richard D. *Ethnic Identity*. New Haven: Yale University Press, 1990.
Roediger, David R. *The Wages of Whiteness*. New York: Verso, 1991.

OPTIONAL ETHNICITIES

For Whites Only?

46

Mary C. Waters

What does it mean to talk about ethnicity as an option for an individual? To argue that an individual has some degree of choice in their ethnic identity flies in the face of the commonsense notion of ethnicity many of us believe in—

From Silvia Pedraza and Rubén G. Rumbaut, eds., *Origins and Destinies: Immigration, Race and Ethnicity in America* (Belmont, CA: Wadsworth, 1996), pp. 444–54. Reprinted by permission.

that one's ethnic identity is a fixed characteristic, reflective of blood ties and given at birth. However, social scientists who study ethnicity have long concluded that while ethnicity is based on a *belief* in a common ancestry, ethnicity is primarily a *social* phenomenon, not a biological one (Alba 1985, 1990; Barth 1969; Weber [1921] 1968, p. 389). The belief that members of an ethnic group have that they share a common ancestry may not be a fact. There is a great deal of change in ethnic identities across generations through intermarriage, changing allegiances, and changing social categories. There is also a much larger amount of change in the identities of individuals over their lives than is commonly believed. While most people are aware of the phenomenon known as "passing"—people raised as one race who change at some point and claim a different race as their identity—there are similar life course changes in ethnicity that happen all the time and are not given the same degree of attention as "racial passing."

White Americans of European ancestry can be described as having a great deal of choice in terms of their ethnic identities. The two major types of options White Americans can exercise are (1) the option of whether to claim any specific ancestry, or to just be "White" or American, [Lieberson (1985) called these people "unhyphenated Whites"] and (2) the choice of which of their European ancestries to choose to include in their description of their own identities. In both cases, the option of choosing how to present yourself on surveys and in everyday social interactions exists for Whites because of social changes and societal conditions that have created a great deal of social mobility, immigrant assimilation, and political and economic power for Whites in the United States. Specifically, the option of being able to not claim any ethnic identity exists for Whites of European background in the United States because they are the majority group—in terms of holding political and social power, as well as being a numerical majority. The option of choosing among different ethnicities in their family backgrounds exists because the degree of discrimination and social distance attached to specific European backgrounds has diminished over time. . . .

SYMBOLIC ETHNICITIES FOR WHITE AMERICANS

What do these ethnic identities mean to people and why do they cling to them rather than just abandoning the tie and calling themselves American? My own field research with suburban Whites in California and Pennsylvania found that later-generation descendants of European origin maintain what are called "symbolic ethnicities." Symbolic ethnicity is a term coined by Herbert Gans (1979) to refer to ethnicity that is individualistic in nature and without real social cost for the individual. These symbolic identifications are essentially leisure-time activities, rooted in nuclear family traditions and reinforced by the voluntary enjoyable aspects of being ethnic (Waters 1990). Richard Alba (1990) also found later-generation Whites in Albany, New York, who chose to keep a tie with an ethnic identity because of the enjoyable and voluntary aspects to those identities, along with the feelings of specialness they entailed. An example of symbolic ethnicity is individuals who identify as Irish, for example, on occasions such as Saint Patrick's Day, on family holidays, or for vacations. They do not usually belong to Irish American organizations, live in Irish neighborhoods, work in Irish jobs, or marry other Irish people. The symbolic meaning of being Irish American can be constructed by individuals from mass media images, family traditions, or other intermittent social activities. In other words, for later-generation White ethnics, ethnicity is not something that influences their lives unless they want it to. In the world of work and school and neighborhood, individuals do not have to admit to being ethnic unless they choose to. And for an increasing number of European-origin individuals whose parents and grandparents have intermarried, the ethnicity they claim is largely a matter of personal choice as they sort through all of the possible combinations of groups in their genealogies. . . .

RACE RELATIONS AND SYMBOLIC ETHNICITY

However much symbolic ethnicity is without cost for the individual, there is a cost associated with symbolic ethnicity for the society. That is because symbolic ethnicities of the type described here are confined to White Americans of European origin. Black Americans, Hispanic Americans, Asian Americans, and American Indians do not have the option of a symbolic ethnicity at present in the United States. For all of the ways in which ethnicity does not matter for White Americans, it does matter for non-Whites. Who your ancestors are does affect your choice of spouse, where you live, what job you have, who your friends are, and what your chances are for success in American society, if those ancestors happen not to be from Europe. The reality is that White ethnics have a lot more choice and room for maneuver than they themselves think they do. The situation is very different for members of racial minorities, whose lives are strongly influenced by their race or national origin regardless of how much they may choose not to identify themselves in terms of their ancestries.

When White Americans learn the stories of how their grandparents and great-grandparents triumphed in the United States over adversity, they are usually told in terms of their individual efforts and triumphs. The important role of labor unions and other organized political and economic actors in their social and economic successes are left out of the story in favor of a generational story of individual Americans rising up against communitarian, Old World intolerance, and New World resistance. As a result, the "individualized" voluntary, cultural view of ethnicity for Whites is what is remembered.

One important implication of these identities is that they tend to be very individualistic. There is a tendency to view valuing diversity in a pluralist environment as equating all groups. The symbolic ethnic tends to think that all groups are equal; everyone has a background that is their right to celebrate and pass on to their children. This leads to the conclusion that all identities are equal and all identities in some sense are interchangeable—"I'm Italian American, you're Polish American. I'm Irish American, you're African American." The important thing is to treat people as individuals and all equally. However, this assumption ignores the very big difference between an individualistic symbolic ethnic identity and a socially enforced and imposed racial identity.

My favorite example of how this type of thinking can lead to some severe misunderstandings between people of different backgrounds is from the *Dear Abby* advice column. A few years back a person wrote in who had asked an acquaintance of Asian background where his family was from. His acquaintance answered that this was a rude question and he would not reply. The bewildered White asked Abby why it was rude, since he thought it was a sign of respect to wonder where people were from, and he certainly would not mind anyone asking HIM about where his family was from. Abby asked her readers to write in to say whether it was rude to ask about a person's ethnic background. She reported that she got a large response, that most non-Whites thought it was a sign of disrespect, and Whites thought it was flattering:

> Dear Abby,
>   I am 100 percent American and because I am of Asian ancestry I am often asked "What are you?" It's not the personal nature of this question that bothers me, it's the question itself. This query seems to question my very humanity. "What am I? Why I am a person like everyone else!"
>
>   Signed, A REAL AMERICAN
>
> Dear Abby,
>   Why do people resent being asked what they are? The Irish are so proud of being Irish, they tell you before you even ask. Tip O'Neill has never tried to hide his Irish ancestry.
>
>   Signed, JIMMY.
>   (Reprinted by permission of Universal Press Syndicate)

In this exchange Jimmy cannot understand why Asians are not as happy to be asked about their ethnicity as he is, because he understands his ethnicity and theirs to be separate but equal. Everyone has to come from somewhere—his family from Ireland, another's family from Asia—each has a history and each should be proud of it. But the reason he cannot understand the perspective of the Asian American is that all ethnicities are not equal; all are not symbolic, costless, and voluntary. When White Americans equate their own symbolic ethnicities with the socially enforced identities of non-White Americans, they obscure the fact that the experiences of Whites and non-Whites have been qualitatively different in the United States and that the current identities of individuals partly reflect that unequal history.

In the next section I describe how relations between Black and White students on college campuses reflect some of these asymmetries in the understanding of what a racial or ethnic identity means. While I focus on Black and White students in the following discussion, you should be aware that the myriad other groups in the United States—Mexican Americans, American Indians, Japanese Americans—all have some degree of social and individual influences on their identities, which reflect the group's social and economic history and present circumstance.

RELATIONS ON COLLEGE CAMPUSES

Both Black and White students face the task of developing their race and ethnic identities. Sociologists and psychologists note that at the time people leave home and begin to live independently from their parents, often ages eighteen to twenty-two, they report a heightened sense of racial and ethnic identity as they sort through how much of their beliefs and behaviors are idiosyncratic to their families and how much are shared with other people. It is not until one comes in close contact with many people who are different from oneself that individuals realize the ways in which their backgrounds may influence their individual personality. This involves coming into contact with people who are different in terms of their ethnicity, class, religion, region, and race. For White students, the ethnicity they claim is more often than not a symbolic one—with all of the voluntary, enjoyable, and intermittent characteristics I have described above.

Black students at the university are also developing identities through interactions with others who are different from them. Their identity development is more complicated than that of Whites because of the added element of racial discrimination and racism, along with the "ethnic" developments of finding others who share their background. Thus Black students have the positive attraction of being around other Black students who share some cultural elements, as well as the need to band together with other students in a reactive and oppositional way in the face of racist incidents on campus.

Colleges and universities across the country have been increasing diversity among their student bodies in the last few decades. This has led in many cases to strained relations among students from different racial and ethnic backgrounds. The 1980s and 1990s produced a great number of racial incidents and high racial tensions on campuses. While there were a number of racial incidents that were due to bigotry, unlawful behavior, and violent or vicious attacks, much of what happens among students on campuses involves a low level of tension and awkwardness in social interactions.

Many Black students experience racism personally for the first time on campus. The upper-middle-class students from White suburbs were often isolated enough that their presence was not threatening to racists in their high schools. Also, their class background was known by their residence and this may have prevented attacks being directed at them. Often Black students at the university who begin talking with other students and recognizing racial slights will remember incidents that happened to them earlier that they might not have thought were related to race.

Black college students across the country experience a sizeable number of incidents that are clearly the result of racism. Many of the most blatant ones that occur between students are the result of drinking. Sometimes late at night, drunken groups of White students coming home from parties will yell slurs at single Black students on the street. The other types of incidents that happen include being singled out for special treatment by employees, such as being followed when shopping at the campus bookstore, or going to the art museum with your class and the guard stops you and asks for your I.D. Others involve impersonal encounters on the street—being called a nigger by a truck driver while crossing the street, or seeing old ladies clutch their pocketbooks and shake in terror as you pass them on the street. For the most part these incidents are not specific to the university environment, they are the types of incidents middle-class Blacks face every day throughout American society, and they have been documented by sociologists (Feagin 1991).

In such a climate, however, with students experiencing these types of incidents and talking with each other about them, Black students do experience a tension and a feeling of being singled out. It is unfair that this is part of their college experience and not that of White students. Dealing with incidents like this, or the ever-present threat of such incidents, is an ongoing developmental task for Black students that takes energy, attention, and strength of character. It should be clearly understood that this is an asymmetry in the "college experience" for Black and White students. It is one of the unfair aspects of life that results from living in a society with ongoing racial prejudice and discrimination. It is also very understandable that it makes some students angry at the unfairness of it all, even if there is no one to blame specifically. It is also very troubling because, while most Whites do not create these incidents, some do, and it is never clear until you know someone well whether they are the type of person who could do something like this. So one of the reactions of Black students to these incidents is to band together.

In some sense then, as Blauner (1992) has argued, you can see Black students coming together on campus as both an "ethnic" pull of wanting to be together to share common experiences and community, and a "racial" push of banding together defensively because of perceived rejection and tension from Whites. In this way the ethnic identities of Black students are in some sense similar to, say, Korean students wanting to be together to share experiences.

And it is an ethnicity that is generally much stronger than, say, Italian Americans. But for Koreans who come together there is generally a definition of themselves as "different from" Whites. For Blacks reacting to exclusion, there is a tendency for the coming together to involve both being "different from" but also "opposed to" Whites.

The anthropologist John Ogbu (1990) has documented the tendency of minorities in a variety of societies around the world, who have experienced severe blocked mobility for long periods of time, to develop such oppositional identities. An important component of having such an identity is to describe others of your group who do not join in the group solidarity as devaluing and denying their very core identity. This is why it is not common for successful Asians to be accused by others of "acting White" in the United States, but it is quite common for such a term to be used by Blacks and Latinos. The oppositional component of a Black identity also explains how Black people can question whether others are acting "Black enough." On campus, it explains some of the intense pressures felt by Black students who do not make their racial identity central and who choose to hang out primarily with non-Blacks. This pressure from the group, which is partly defining itself by not being White, is exacerbated by the fact that race is a physical marker in American society. No one immediately notices the Jewish students sitting together in the dining hall, or the one Jewish student sitting surrounded by non-Jews, or the Texan sitting with the Californians, but everyone notices the Black student who is or is not at the "Black table" in the cafeteria.

An example of the kinds of misunderstandings that can arise because of different understandings of the meanings and implications of symbolic versus oppositional identities concerns questions students ask one another in the dorms about personal appearances and customs. A very common type of interaction in the dorm concerns questions Whites ask Blacks about their hair. Because Whites tend to know little about Blacks, and Blacks know a lot about Whites, there is a general asymmetry in the level of curiosity people have about one another. Whites, as the numerical majority, have had little contact with Black culture; Blacks, especially those who are in college, have had to develop bicultural skills—knowledge about the social worlds of both Whites and Blacks. Miscommunication and hurt feelings about White students' questions about Black students' hair illustrate this point. One of the things that happens freshman year is that White students are around Black students as they fix their hair. White students are generally quite curious about Black students' hair—they have basic questions such as how often Blacks wash their hair, how they get it straightened or curled, what products they use on their hair, how they comb it, etc. Whites often wonder to themselves whether they should ask these questions. One thought experiment Whites perform is to ask themselves whether a particular question would upset them. Adopting the "do unto others" rule, they ask themselves, "If a Black person was curious about my hair would I get upset?" The answer usually is "No, I would be happy to tell them." Another example is an Italian American student wondering to herself, "Would I be upset if someone asked me about calamari?" The answer is no, so she asks her Black roommate about collard greens, and the roommate explodes with an angry response such as, "Do you think all Black people eat watermelon too?" Note that if this Italian American knew her friend was Trinidadian American and asked about peas and rice the situation would be more similar and would not necessarily ignite underlying tensions.

Like the debate in *Dear Abby*, these innocent questions are likely to lead to resentment. The issue of stereotypes about Black Americans and the assumption that all Blacks are alike and have the same stereotypical cultural traits has more power to hurt or offend a Black person than vice versa. The innocent questions about Black hair also bring up a number of asymmetries between the Black and White experience. Because Blacks tend to have more knowledge about Whites than vice versa, there is not an even exchange going on, the Black freshman is likely to have fewer basic questions about his White roommate than his White roommate has about him. Because of the differences historically in the group experiences of Blacks and Whites there are some connotations to Black hair that don't exist about White hair. (For instance, is straightening your hair a form of assimilation, do some people distinguish between women having "good hair" and "bad hair" in terms of beauty and how is that related to looking "White"?) Finally, even a Black freshman who cheerfully disregards or is unaware that there are these asymmetries will soon slam into another asymmetry if she willingly answers every innocent question asked of her. In a situation where Blacks make up only 10 percent of the student body, if every non-Black needs to be educated about hair, she will have to explain it to nine other students. As one Black student explained to me, after you've been asked a couple of times about something so personal you begin to feel like you are an attraction in a zoo, that you are at the university for the education of the White students.

INSTITUTIONAL RESPONSES

Our society asks a lot of young people. We ask young people to do something that no one else does as successfully on such a wide scale—that is to live together with people from very different backgrounds, to respect one another, to appreciate one another, and to enjoy and learn from one another. The successes that occur every day in this endeavor are many, and they are too often overlooked. However, the problems and tensions are also real, and they will not vanish on their own. We tend to see pluralism working in the United States in much the same way some people expect capitalism to work. If you put together people with various interests and abilities and resources, the "invisible hand" of capitalism is supposed to make all the parts work together in an economy for the common good.

There is much to be said for such a model—the invisible hand of the market can solve complicated problems of production and distribution better than any "visible hand" of a state plan. However, we have learned that unequal power relations among the actors in the capitalist marketplace, as well as "externalities" that the market cannot account for, such as long-term pollution, or collusion between corporations, or the exploitation of child labor, means that state regulation is often needed. Pluralism and the relations between groups are very similar. There is a lot to be said for the idea that bringing people who belong to different ethnic or racial groups together in institutions with no interference will have good consequences. Students from different backgrounds will make friends if they share a dorm room or corridor, and there is no need for the institution to do any more than provide the locale. But like capitalism, the invisible hand of pluralism does not do well when power relations and externalities are ignored. When you bring together individuals from groups that are differentially valued in the wider society and provide no guidance, there will be problems. In these cases the "invisible hand" of pluralist relations does not work, and tensions and disagreements can arise without any particular individual or group of individuals being "to blame." On college campuses in the 1990s some of the tensions between students are of this sort. They arise from honest misunderstandings, lack of a common background, and very different experiences of what race and ethnicity mean to the individual.

The implications of symbolic ethnicities for thinking about race relations are subtle but consequential. If your understanding of your own ethnicity and its relationship to society and politics is one of individual choice, it becomes harder to understand the need for programs like affirmative action, which recognize the ongoing need for group struggle and group recognition, in order to bring about social change. It also is hard for a White college student to understand the need that minority students feel to band together against discrimination. It also is easy, on the individual level, to expect everyone else to be able to turn their ethnicity on and off at will, the way you are able to, without understanding that ongoing discrimination and societal attention to minority status makes that impossible for individuals from minority groups to do. The paradox of symbolic ethnicity is that it depends upon the ultimate goal of a pluralist society, and at the same time makes it more difficult to achieve that ultimate goal. It is dependent upon the concept that all ethnicities mean the same thing, that enjoying the traditions of one's heritage is an option available to a group or an individual, but that such a heritage should not have any social costs associated with it.

As the Asian Americans who wrote to *Dear Abby* make clear, there are many societal issues and involuntary ascriptions associated with non-White identities. The developments necessary for this to change are not individual but societal in nature. Social mobility and declining racial and ethnic sensitivity are closely associated. The legacy and the present reality of discrimination on the basis of race or ethnicity must be overcome before the ideal of a pluralist society, where all heritages are treated equally and are equally available for individuals to choose or discard at will, is realized.

References

Alba, Richard D. 1985. *Italian Americans: Into the Twilight of Ethnicity.* Englewood Cliffs, NJ: Prentice-Hall.

———. 1990. *Ethnic Identity: The Transformation of White America.* New Haven: Yale University Press.

Barth, Frederick. 1969. *Ethnic Groups and Boundaries.* Boston: Little, Brown.

Blauner, Robert. 1992. "Talking Past Each Other: Black and White Languages of Race." *American Prospect* (Summer): 55–64.

Feagin, Joe R. 1991. "The Continuing Significance of Race: Anti-Black Discrimination in Public Places." *American Sociological Review* 56: 101–17.

Gans, Herbert. 1979. "Symbolic Ethnicity: The Future of Ethnic Groups and Cultures in America." *Ethnic and Racial Studies* 2: 1–20.

Lieberson, Stanley. 1985. *Making It Count: The Improvement of Social Research and Theory.* Berkeley: University of California Press.

Ogbu, John. 1990. "Minority Status and Literacy in Comparative Perspective." *Daedalus* 119: 141–69.

Waters, Mary C. 1990. *Ethnic Options: Choosing Identities in America.* Berkeley: University of California Press.

Weber, Max. [1921]/1968. *Economy and Society: An Outline of Interpretive Sociology.* Eds. Guenther Roth and Claus Wittich, trans. Ephraim Fischoff. New York: Bedminister Press.

"MEXICANNESS" IN NEW YORK

47

Migrants Seek New Place in Old Racial Order

Robert Smith

Over the past decade, one of the more dramatic population developments in the United States has been the burgeoning of the Mexican community all

From *North American Congress on Latin America (NACLA) Report on the Americas,* 35, (September/October) 14–17. Reprinted by permission of the publisher.

along the eastern seaboard. Population experts predict that within ten years Mexicans will be the largest minority on the east coast—from Florida to New England. In many places, especially those with relatively small minority populations, such as some towns in New York's Hudson Valley or on the Delmarva Peninsula just east of Washington, DC, they already are. This growth has been especially striking in New York City. From a population that barely existed in the early 1980s, Mexican New Yorkers now number in the hundreds of thousands. The U.S. Census Bureau enumerated over 260,000 Mexicans in New York State in 2000, of whom more than 180,000 were estimated to be in New York City. Allowing for the undercount—many Mexican New Yorkers are undocumented—this means there are probably 450,000–500,000 Mexicans in the state and 300,000 in the city alone.

The new migrants to New York are predominantly from the Mixteca region of southern Mexico, a region that includes large parts of the states of Puebla, Oaxaca and Guerrero. It is a poor, hot agricultural region, called *"tierra caliente,"* that has sent migrant workers to the United States in small numbers since the 1940s, and from which migration to the United States has been accelerating since Mexico's rural economy went into long-term crisis in the late 1980s.

The new immigrants, willing to put in long, hard days of work, have been welcomed by the city's low-wage employers, though their children have received a more ambiguous welcome in the city's schools and neighborhoods. And as the third-largest Latino group in the city, with an enormous potential for growth, Mexicans are beginning to draw the attention of New York's political and community leaders. How their "Mexicanness" will be defined in New York's complex web of ethnicities remains to be seen. Answers will emerge as they begin to negotiate their way through the city's system of social and racial hierarchies.[1]

Mexicans do not fit "naturally" into any one spot in New York's social and racial hierarchies. They enter New York both as immigrants and as Latinos. As immigrants they face—and many have internalized—expectations that they conform to an "immigrant analogy" according to which they will work hard and sacrifice their personal well-being so that their children and grandchildren can prosper. As Latinos, they confront social networks that are heavily "racialized"—networks in which ethnic groups are represented on a continuum that runs from black to white. They face the standard U.S. representations of race, based on the "original sin" of slavery, with "whiteness" signalling that a group is fit for full membership in society. Their immigrant bargain is, in fact, implicitly juxtaposed to the situation of native-born blacks: the native-born do not sacrifice, and they therefore will not prosper.

But the racial categories in New York City are complicated by two things. On the one hand, there are nuanced degrees of white ethnicity and racialization due in part to the presence of first-generation white immigrants such as Greeks, Russians and Italians. On the other hand, African-Americans share the most stigmatized position at the bottom of the hierarchy with a Latino group, Puerto Ricans. Many Puerto Ricans possess phenotypes that most Americans would consider "black," and experience consonant levels of racial segregation and discrimination; but they also speak Spanish and are immigrants. Hence, many Mexican parents and their New York-born children wonder whether their futures will look more like the hard lives they associate with Puerto Ricans and blacks or more like the upwardly mobile lives they associate with white ethnic immigrants. They wonder—worriedly—whether they will become a marginalized, racialized minority or an incorporated ethnic group?[2]

Writing in the 1930s, W.E.B. Du Bois mused that poor southern whites got a "public, psychological wage" by being white that enabled them to feel superior to blacks despite the many commonalities in their material living conditions. Historian David Roediger has used Du Bois' insight to analyze how the "Irish became white." When Irish immigrants started coming to the United States in significant numbers in the 1830s, they had much in common with African-Americans. Both groups did America's dirty work, both were victimized by systematic racism, and they often lived side by side in the poorest parts of northern cities. Roediger answers the question of how the Irish came to not only distance themselves from blacks but to embrace an anti-black racism by analyzing their racialized mobilization by the Catholic Church and the Democratic Party. What the Irish had learned, in essence, was that to be full members of U.S. society, one had not simply to be "not black" but also in many instances "anti-black."[3] This dynamic of incorporation has lived on through generations in a situation in which immigration to the United States is seen simultaneously as a struggle-and-prosper story—the "immigrant analogy"—and one in which generations of immigrants have succeeded by differentiating themselves from blacks, who are understood to form an underclass.

My research with Mexicans in New York over the last fifteen years has yielded a complex picture of responses to the need to define oneself within racialized hierarchies.[4] The immigrant first generation usually has responded by embracing the immigrant analogy, though in different ways in different contexts. In the labor market, for example, Mexican workers frequently confront complex and contradictory forms of ethnic solidarity.[5] While many Korean, Greek, Italian and native-born employers explicitly or implicitly compare Mexican immigrants to native-born blacks and Puerto Ricans, hence treating them as part of the out-group, they also see them as immigrants like themselves, or like their immigrant predecessors, hence part of their in-group. This leads to an emphasis by employers on their own similarities with their Mexican immigrant employees, and the difference of both from native minorities. I have heard many an employer identify with his Mexican employees. A Greek restaurant owner, for example, extolled to me the virtues of Mexican employees while in the same breath he disparaged African-Americans. With tears in his eyes, he told me about working 12-hour days like his

Mexican workers, saying, "When I came to this country I was a good Mexican." The willingness of Mexican immigrants to work hard makes them different from native-born workers, he said, and he detailed how they moved up step by step from being busboys, to dishwashers, to cooks.[6] In addition to African-Americans and Puerto Ricans, native-born whites are occasionally included in these unfavorable comparisons.

Within this social context, racialization—the logic of moral and civic worth based on one's ethnic or racial affiliation—takes place within the Mexican community itself. For example, while some parents express no preferences regarding who their children date, many parents do not want their children to date African-Americans or in some cases Puerto Ricans, even though they themselves may have black and Puerto Rican friends, or even relatives through marriage. Many parents talk of how Puerto Ricans helped them when they came to New York many years earlier, when there were fewer Spanish speakers, but lament what they see as the Puerto Rican population's miserable social conditions. In particular, many Mexican immigrants and their U.S.-born children point to what they see as the tendency of Puerto Rican young men to engage in crime, especially against Mexicans, the propensity of Puerto Rican young women to get pregnant, and the tendency of many Puerto Ricans to depend on government assistance. This is frequently expressed as evidence of the "Americanization" that robs Puerto Ricans of their culture and leaves them vulnerable to the urban vices that Mexicans can resist. . . .

. . . Americanization among Puerto Ricans is seen as a cause of their having failed, while Mexicans are seen to be insulated by their culture from such erosions of values. To be sure, many in the first or second generation do not buy into the immigrant analogy. A second generation woman notes the tendency of Mexicans to assert that they are different from blacks and Puerto Ricans, but is critical of it: "They don't want to be like them," she says of her young compatriots, "but I believe they are, because I believe they do the same things. Why do they have to have a gang? Supposedly that's only blacks and Puerto Ricans. You know, and why do they have to go around drinking 'forties' . . . the big bottles of beer? Mexicans?"[7]

This woman has identified a potential problem in the collective self-image of much of the Mexican community in New York. If that self-image is premised on a juxtaposition with African-Americans and Puerto Ricans, it is being challenged by rising indicators of social distress resulting from the settlement and incorporation of Mexicans in New York. Despite increasing numbers of upwardly mobile second-generation youth, these indicators show increasing numbers of young people, born in Mexico and in the United States, who leave school or join gangs, or become pregnant at an early age. The formation of youth gangs grows out of the migration process itself and the often hostile reception that Mexican youth get here. Concretely, many second generation youth see that their lives are very similar to the blacks and Puerto Ricans with whom they share neighborhoods, parks and schools. Many see limited employment options, little payback from schooling, and feel that the larger society and its institutions think they will fail.

. . . According to the Census, Mexicans as a group went from having one of the highest per capita incomes of all Latinos in New York in 1980 (when the community was smaller) to having one of the lowest in 1990. By 1990, Mexicans also had the largest percentage of 16–19 year olds not in school and not graduated: 47% versus 22% for Dominicans and Puerto Ricans, 18% for blacks and 8% for whites. Part of the reason for these indicators of social distress is the huge surge in Mexican youth migration during the 1990s; younger people earn less, and teen migrants often do not enter school.

The figures are disquieting for members of the Mexican community. So cial psychologists have observed that a positive ethnic identity is usually associated with greater life success, but for many of these young people, self image is formed around an identity that includes being vulnerable on the street, seeing relatively few people who have succeeded through school, and facing the prospect of years of poorly paid hard work.

. . . Among adolescent Mexicans, there has been no single way of adapting to New York life. In addition to the young Mexicans who are dropping out of school at alarming rates, some upwardly mobile youth have used their Mexican ethnicity to differentiate themselves from other minorities in their schools and neighborhoods, whom they see as not serious about school. "Being Mexican helps us do better" describes how many of these adolescents understand themselves. There are other, "cosmopolitan," adolescents who see Mexicanness as only one identity among many, and there is a small group that dissociates itself from Mexicans in public places and identifies instead, in most cases, with upwardly mobile blacks. They see these successful black students as role models whose success they wish to emulate. They fear that if they hang around with other Mexicans they will be pressured to cut school a lot or eventually drop out.

And the results of adolescent engagements with racialized social relations have sometimes been ironic. For example, many Mexican youth, especially boys, now fear being "stepped up to"—stopped by others to verify if they are in a rival gang or not—by members of Mexican gangs more than they fear confrontation with blacks or Puerto Ricans. Most Mexican gang confrontations actually involve conflicts with other Mexicans, not Puerto Ricans or African-Americans. Other Mexican youth I have interviewed report that they are not stepped up to because they "look Dominican" or "look Ecuadorian," and hence are let pass uninspected.

All this suggests recent immigrants and their allies should step up their fight for an immigrant policy that includes more bilingual education, programs for parents to learn English, summer and after school programs for youth and tighter links between U.S. institutions and the communities from which immigrants come. Such a policy, of course, would have to be linked to the rebirth of anti-poverty efforts on behalf of the entire population.

For now, as Mexicans enter the political, social and economic worlds of New York, they do so within the pervasive and dangerous context of racialization. Immigration is not a straight "struggle and prosper" story. We should take note of how the Irish and other "formerly non-white" immigrants became white, and what this means for how immigrants, past and present, have learned and internalized what it means to be "American."

NOTES

1. See Robert Smith, "Mexicans: Social, Educational, Economic and Political Problems and Prospects in New York," in Nancy Foner, ed., *New Immigrants in New York* (New York: Columbia University Press, 2001).

2. For Mexicans in California, the answer will be different than for those in New York. In California, there has already been a racialization and stigmatization of Chicanos, despite the fact that there has been significant upward mobility in the population there as well, especially among women. See Dowell Myers and Cynthia Cranford, "Temporal Differentiation in the Occupational Mobility of Immigrant and Native-Born Latina Workers," *American Sociological Review*, Vol. 63, No. 1, February, 1998, pp. 68–93. It is also interesting that the comments I have heard usually compare Mexicans to Puerto Ricans and blacks or other minorities, and not with whites. Part of the reason for this is that the lifeworlds of most Mexicans and Mexican Americans, especially youth, do not include too many whites. Many of the people we have interviewed for our projects report not knowing many whites, except for teachers or employers, and that most of their classmates, workmates, or people they see in the parks or in their neighborhoods, are black or Latino.

3. David Roediger, *The Wages of Whiteness* (New York: Verso, 1991), especially chapters 1 and 7. See also Joel Ignatiev, *How the Irish Became White* (New York: Routledge, 1995) and Matthew Jacobson, *Whiteness of a Different Color: European Immigrants and the Alchemy of Race* (Cambridge: Harvard University Press, 1998).

4. This work can be consulted in a variety of venues: Robert Smith, "'Los Ausentes Siempre Presentes': The Imagining, Making and Politics of Community Between Ticuani, Puebla, Mexico and New York City," Doctoral Dissertation, Department of Political Science, Columbia University, 1995; "Mexicans in New York City: Membership and Incorporation of New Immigrant Group," in Sherrie L. Baver and Gabriel Haslip-Viera, eds. *Latinos in New York* (South Bend: University of Notre Dame Press, 1996); "Gender, Race and Schools in Educational and Work Outcomes of Second Generation Mexican Americans in New York," in Marcelo Suarez-Orozco and Mariela Paez, eds., *Latinos in the 21st Century* (Berkeley: University of California Press, 2001); "Mexicans," in Foner, ed. See also Robert Smith, Hector Cordero-Guzman and Ramon Grosfoguel, "Introduction: New Analytical Perspectives on Migration, Race and Transnationalization" in Cordero-Guzman, Smith and Grosfoguel, eds., Migration, Race and Transnationalization in a Changing New York (Philadelphia: Temple University Press, 2001); and Robert Smith, *Migration, Settlement and Transnational Life*, 2002. Research assistance during 1998–2000 by Sara Guerrero Rippberger, Sandra Lara, Agustin Vecino, Carolina Pérez, Griscelda Pérez, Linda Rodriguez and Lisa Peterson is gratefully acknowledged.

5. I call this "doubly bounded solidarity." It is my adaptation of the concept of bounded solidarity as developed by Alejandro Portes and his colleagues. See Portes' chapters in Portes, *The Economic Sociology of Immigration* (New York: Russell Sage Foundation, 1995). Also see Min Zhou, "Segmented Assimilation: Issues, Controversies, and Recent Research on the New Second Generation," in Charles Hirschman, Philip Kasinitz, and Josh DeWind, eds., *Handbook of International Migration: The American Experience* (New York: Russell Sage Foundation, 1999), pp. 196–212.

6. Author's interview, 1993, New York City.

7. Author's interview, 1997, New York City.

MIGRATION AND VIETNAMESE AMERICAN WOMEN

48

Remaking Ethnicity

Nazli Kibria

VIETNAMESE AMERICANS AND THE RISE IN WOMEN'S POWER

My research on the adaptive strategies of a community of Vietnamese refugees in Philadelphia revealed some of the ways in which women and men struggled and clashed with each other in efforts to shape the social organization of family and community life. From 1983 to 1985, I gathered information on family life and gender relations through participant observation in household and community settings, as well as in-depth interviews with women and men in the ethnic community.

The Vietnamese of the study were recent immigrants who had arrived in the United States during the late 1970s and early 1980s. Most were from urban, middle-class backgrounds in southern Vietnam. At the time of the study, over 30 percent of the adult men in the households of study were unemployed. Of the men who were employed, over half worked in low-paying, unskilled jobs in the urban service sector or in factories located in the outlying areas of the city. Women tended to work periodically, occupying jobs in the informal economic sector as well as in the urban service economy. Eight of the twelve households had members who collected public assistance. Both the family economy and informal community exchange networks were important means by which the households dealt with economic scarcities. Family and community were of tremendous economic salience to the group, as they were important resources for survival in the face of a rather inhospitable economic and social environment.

As suggested by the high rate of the men's unemployment, settlement in the United States had generated some shifts in power in favor of the women in the group. Traditional Vietnamese family and gender relations were modeled on Confucian principles, which placed women in subordination to men in every aspect of life. A key aspect of the social and economic oppression of women in traditional Vietnamese life was the patrilineal extended household. Its organization dictated that women married at a young age, following which they entered the household of their husband's father. This structure ensured the concentration of economic resources in the hands of men and men's control of women through the isolation of women from their families of origin.[1]

It is important to note the deep-seated changes in traditional family and gender structures in Vietnam during this century. War and urbanization eroded the structure of the patrilineal extended household. While unemployment was high in the cities, men from middle-class backgrounds were able to take advantage of the expansion of middle-level positions in the government bureaucracy and army. Such occupational opportunities were more limited for women: the women study participants indicated that they engaged in seasonal and informal income-generating activities or worked in low-level jobs in the growing war-generated service sector in the cities. The transition from rural to urban life had generated a shift in the basis of men's control over economic and social resources. However, families relied on men's income to maintain a middle-class standard of living. Thus women remained in a position of economic subordination to men, a situation that served to sustain the ideals of the traditional family system and men's authority in the family. Restrictions on women's sexuality were important for middle-class families who sought to distinguish themselves from the lower social strata. My data suggest that families were especially conscious of the need to distance themselves from poorer "fallen" women who had become associated with the prostitution generated by the American military presence.

Within the Vietnamese American community of study, I found several conditions that were working to undermine the bases on which male authority had rested in Vietnam. Most important, for the Vietnamese men, the move to the United States had involved a profound loss of social economic status. Whereas in pre-1975 Vietnam the men held middle-class occupations, in the United States they had access to largely unskilled, low-status, and low-paying jobs. Also, because of their difficulties with English and their racial-ethnic status, the men found themselves disadvantaged within social arenas of the dominant society. Compounding these problems was the dearth of strong formal ethnic organizations in the community that could have served as a vehicle for the men's political assertion into the dominant society.

As a result of these losses, the comparative access of men and women to the resources of the dominant society had to some extent become equalized. In contrast to the experiences of the men, migration had not significantly altered the position of the women in the economy. As in Vietnam, the women tended to work sporadically, sometimes in family businesses or, more commonly, in temporary jobs in the informal and service sector economies of the city. However, the economic contributions of women to the family budget had risen in proportion to those of the men. I have suggested that in modern, urban South Vietnam the force and legitimacy of male authority had rested heavily on the ability of men to ensure a middle-class status and standard of living for their families. In the United States, the ability of men to fulfull this expectation had been eroded. Among the men, there was widespread concern about the consequences of this situation for their status in the family, as is revealed by the words of a former lieutenant of the South Vietnamese army: "In Vietnam, the man earns and everyone depends on him. In most families, one or two men could provide for the whole family. Here the man finds he can never make enough money to take care of the family. His wife has to work, his children have to work, and so they look at him in a different way. The man isn't strong anymore, like he was in Vietnam."

Such changes had opened up the possibilities for a renegotiation of gender relations, and were the cause of considerable conflict between men and women in the family and community. The shifts in power had also enhanced the ability of women to construct and channel familial and ethnic resources in ways that they chose. Previously I suggested that the changes in the balance of power between men and women generated by migration are crucial to understanding the manner and degree to which immigrant family and community reveal themselves to be gender contested. How, then, did the fairly drastic shift in the gender balance of power among the Vietnamese Americans reflect itself in the ability of the men and women in this group to influence family and community life? In the following section, I describe some of the ways in which gender interests and conflict shaped family and community life for the Vietnamese Americans.

FAMILY AND ETHNICITY AS GENDER CONTESTED

One of the most intriguing and important strategies of Vietnamese American adaptation that I observed was the rebuilding of kinship networks. Family ties had undergone tremendous disruption in the process of escape from Vietnam and resettlement in the United States. Despite this, the households of the group tended to be large and extended. The process by which this occurred was one in which the study participants actively worked to reconstruct family networks by building kin relationships. In order for this to take place, the criteria for inclusion in the family had become extremely flexible. Thus close friends were often incorporated into family groups as fictive kin. Also, relationships with relatives who were distant or vaguely known in Vietnam were elevated in importance. Perhaps most important for women, the somewhat greater significance traditionally accorded the husband's kin had receded in importance.[2] Given the scarcity of relatives in the United States, such distinctions were considered a luxury, and the demands of life made the rebuilding of family a valuable, if not a necessary, step in the process of adaptation to the dominant society.

While important for the group as a whole, the reconstruction of kinship as it took place had some special advantages for women. One consequence of the more varied and inclusive nature of the kinship network was that women were rarely surrounded exclusively by the husband's relatives and/or friends. As a result, they were often able to turn to close fictive kin and perhaps members of their families of origin for support during conflicts with men in the family. Another condition that enhanced the power of married women in the family was that few had to deal with the competing authority of their mother-in-law in the household, because elderly women have not been among those likely to leave Vietnam.

The reconstruction of kinship thus had important advantages for women, particularly as it moved the Vietnamese perhaps even further from the ideal model of the patrilineal extended household that it had been in the past. But women were not simply passive beneficiaries of the family rebuilding process. Rather, they played an active part in family reconstruction, attempting to shape family boundaries in ways that were to their advantage. I found women playing a vital part in creating fictive kin by forging close ties. And women were often important, if not central, "gatekeepers" to the family group and household. Thus the women helped to decide such matters as whether the marriage of a particular family member was a positive event and could be taken as an opportunity to expand kinship networks. At other times the women passed judgment on current or potential family members, as to whether they had demonstrated enough commitment to such important familial obligations as the sharing of economic and social resources with kin.

Although women undoubtedly played an important part in family reconstruction, their control over decisions about family membership was by no means exclusive or absolute. In fact, the question of who was legitimately included in the family group was often a source of tension within families, particularly between men and women. The frequency of disputes over this issue stemmed in part from the fluidity and subsequent uncertainty about family boundaries, as well as the great pressures often placed on individuals to subordinate their needs to those of the family collective. Beyond this, I also suggest that disputes over boundaries arose from the fundamental underlying gender divisions in the family. That is, the different interests of women and men in the family spurred efforts to shape the family in ways that were of particular advantage to them. For the reasons I have previously discussed, the Vietnamese American women had greater influence and opportunity in the shaping of family in the United States than they had in the past. The women tended to use this influence to construct family groups that extended their power in the family.

In one case that I observed, considerable tension developed between a couple named Nguyet and Phong concerning the sponsorship[3] of Nguyet's nephew and his family from a refugee camp in Southeast Asia. Nguyet and Phong had been together with their three children (two from Nguyet's previous marriage) for about seven years, since they had met in a refugee camp in Thailand. Phong remained married to a woman who was still living in Vietnam with his children, a fact that was the source of some stress for Nguyet and Phong. The issue of the nephew's sponsorship seemed to exacerbate tensions in the relationship. Phong did not want to undertake the sponsorship because of the potentially heavy financial obligations it entailed. He also confessed that he was worried that Nguyet would leave him after the nephew's arrival, a threat often made by Nguyet during their quarrels. Finally, he talked of how Nguyet's relationship with the nephew was too distant to justify the sponsorship. Nguyet had never even met the nephew, who was the son of a first cousin rather than of a sibling.

Confirming some of Phong's fears, Nguyet saw the presence of the nephew and his family as a potentially important source of support for herself. She spoke of how she had none of "my family" in the country, in comparison with Phong, whose sister lived in the city. She agreed that she did not know much about her nephew, but nonetheless felt that his presence would ease her sense of isolation and also would provide a source of aid if her relationship with Phong deteriorated. Eventually she proceeded with the sponsorship, but only after a lengthy dispute with Phong.

While the issue of sponsorship posed questions about kinship in an especially sharp manner, there were other circumstances under which women and men clashed over family boundaries. When kin connections could not be questioned (for example, in the case of a sibling), what came under dispute was the commitment of the particular person involved to familial norms and obligations. One of my woman respondents fought bitterly with her older brother about whether their male cousin should live with them. Her brother objected to the cousin's presence in the household on the grounds that he had not responded to their request for a loan of money two years ago. The woman respondent wanted to overlook this breach of conduct because of her extremely close relationship with the cousin, who had been her "best friend" in Vietnam.

Regardless of the particular circumstances, gender conflict seemed an important part of the family reconstruction process. Women and men shared an interest in creating and maintaining a family group that was large and cohesive enough to provide economic and social support. However, their responses to the family reconstruction process were framed by their differing interests, as men and women, within the family. Men and women attempted to channel family membership in ways that were to their advantage, such that their control over the resources of the family group was enhanced.

Gender divisions and conflicts also entered into the community life of the group. The social networks of the Vietnamese American women were central to the dynamics and organization of the ethnic community. They served to organize and regulate exchange between households. While "hanging out" at informal social gatherings, I observed women exchanging information, money, goods, food, and tasks such as child care and cooking. Given the precarious economic situation of the group, these exchanges played an important role in ensuring the economic survival and stability of the households. The women's centrality to these social networks gave them the power not only to regulate household exchange but also to act as agents of social control in the community in a more general sense. I found that women, through the censure of gossip and the threat of ostracism, played an important part in defining community norms. In short, the relative rise in power that had accrued to the Vietnamese American women as a result of migration expressed itself in their considerable influence over the organization and dynamics of the ethnic community. Like kinship, community life was a negotiated arena, one over which women and men struggled to gain control.

The gender-contested quality of ethnic forms was also apparent in the efforts of women to reinterpret traditional Vietnamese familial ideologies on their own terms. In general, the Vietnamese American women continued to espouse and support traditional ideologies of gender relations as important ideals. For example, when asked during interviews to describe the "best" or ideal roles of men and women in the family, most of my respondents talked of a clear division of roles in which women assumed primary responsibility for maintaining the home and taking care of the children, and men for the economic support of the family. Most felt that household decisions should be made jointly, although the opinion of the man was seen to carry more weight. About half of those interviewed felt that a wife should almost always obey her husband. Even more widespread were beliefs in the importance of restrictions on female (but not male) sexuality before marriage.

While women often professed such beliefs, their relationship to traditional ideologies was active rather than passive and inflexible. In other words, the women tended to emphasize certain aspects of the traditional familial ideology over others. In particular, they emphasized parental authority and the obligation of men to sacrifice individual needs and concerns in order to fulfill the needs of the family, traditional precepts they valued and hoped to preserve in the United States. The women's selective approach to Vietnamese "tradition" emerged most clearly in situations of conflict between men and women in the family. In such disputes, women selectively used the traditional ideologies to protect themselves and to legitimate their actions and demands (Kibria 1990). Thus, husbands who were beating their wives were attacked by other women in the community on the grounds that they (the husbands) were inadequate breadwinners. The women focused not on the husband's treatment of his wife but on his failure to fulfill his family caretaker role. Through this selective emphasis, the women managed to condemn the delinquent husband without appearing to depart from "tradition." In short, for the Vietnamese American women, migration had resulted in a greater ability to shape family and community life.

CONCLUSION

For immigrant women, ethnic ties and institutions may be both a source of resistance and support, and of patriarchal oppression. Through an acknowledgment of this duality we can arrive at a fuller understanding of immigrant women's lives: one that captures the multifaceted constraints as well as the resistances that are offered by immigrant women to the oppressive forces in their lives. In patterns similar to those noted by studies of other racial-ethnic groups (Stack 1974; Baca Zinn 1975), the Vietnamese Americans presented in this [article] relied on family and community for survival and resistance. Their marginal status made the preservation of these institutions an important priority.

Like other racial-ethnic women, the ability of the Vietnamese American women to shape ethnicity was constrained by their social-structural location in the dominant society. These women saw the traditional family system as key to their cultural autonomy and economic security in American society. Migration may have equalized the economic resources of the men and women, but it had not expanded the economic opportunities of the women enough to make independence from men an attractive economic reality. The Vietnamese American women, as is true for other women of color, were especially constrained in their efforts to "negotiate" family and community in that they faced triple disadvantages (the combination of social class, racial-ethnic, and gender statuses) in their dealings with the dominant society.

Recognition of the role of ethnic institutions in facilitating immigrant adaptation and resistance is essential. However, it is equally important to not lose sight of gender divisions and conflicts, and the ways in which these influence the construction of ethnic institutions. Feminist scholars have begun to explore the diverse ways in which immigrant women manipulate family and community to enhance their own power, albeit in ways that are deeply constrained by the web of multiple oppressions that surround them (Andezian 1986; Bhachu 1986; Kibria 1990). Such work begins to suggest the complexity of immigrant women's relationship to ethnic structures, which is informed by both strength and oppression.

NOTES

1. Some scholars stress the fact that the reality of women's lives was far different from that suggested by these Confucian ideals. Women in traditional Vietnam also had a relatively favorable economic position in comparison with Chinese women due to Vietnamese women's rights of inheritance as well as their involvement in commercial activities (see Hickey 1964; Keyes 1977). Despite these qualifications, there is little to suggest that the economic and social subordination of women was not a fundamental reality in Vietnam.

2. Hy Van Luong (1984) has noted the importance of two models of kinship in Vietnamese life, one that is patrilineal-oriented and another in which bilateral kin are of significance. Thus the flexible, encompassing conceptions of family that I found among the group were not entirely new, but had their roots in Vietnamese life; however, they had acquired greater significance in the context of the United States.

3. Refugee resettlement in the United States involves a system of sponsorship by family members or other interested parties who agree to assume part of the responsibility for taking care of those sponsored for a period of time after their arrival.

References

Andezian, Sossie. 1986. "Women's Roles in Organizing Symbolic Life: Algerian Female Immigrants in France." Pp. 254–266 in *International Migration: The Female Experience*, edited by R. J. Simon and C. B. Brettell. Totowa, N.J.: Rowman and Allenheld.

Baca Zinn, Maxine. 1975. "Political Familism: Toward Sex Role Equality in Chicano Families." *Aztlan* 6, no. 1: 13–26.

Bhachu, Parminder K. 1986. "Work, Dowry and Marriage Among East African Sikh Women in the U.K." Pp. 241–254 in *International Migration: The Female Experience*, edited by R. J. Simon and C. B. Brettell. Totowa, N.J.: Rowman and Allenheld.

Hickey, Gerald C. 1964. *Village in Vietnam*. New Haven: Yale University Press.

Keyes, Charles F. 1977. *The Golden Peninsula*. New York: Macmillan.

Kibria, Nazli. 1990. "Power, Patriarchy and Gender Conflict in the Vietnamese Immigrant Community." *Gender & Society* 4, no. 1 (March): 9–24.

Luong, Hy Van. 1984. "'Brother' and 'Uncle': An Analysis of Rules, Structural Contradictions and Meaning in Vietnamese Kinship." *American Anthropologist* 86, no. 2: 290–313.

Stack, Carol. 1974. *All Our Kin*. New York: Harper & Row.

CHAPPALS AND GYM SHORTS   **49**

An Indian Muslim Woman in the Land of Oz

Almas Sayeed

It was finals week during the spring semester of my sophomore year at the University of Kansas, and I was buried under mounds of papers and exams. The stress was exacerbated by long nights, too much coffee and a chronic, building pain in my permanently splintered shins (left over from an old sports injury). Between attempting to understand the nuances of Kant's *Critique of Pure Reason* and applying the latest game-theory models to the 1979 Iranian revolution, I was regretting my decision to pursue majors in philosophy, women's studies *and* international studies.

My schedule was not exactly permitting much down time. With a full-time school schedule, a part-time job at Lawrence's domestic violence shelter and preparations to leave the country in three weeks, I was grasping to hold onto what little sanity I had left. Wasn't living in Kansas supposed to be more laid-back than this? After all, Kansas was the portal to the magical land of Oz, where wicked people melt when doused with mop water and bright red, sparkly shoes could substitute for the services of American Airlines, providing a quick getaway. Storybook tales aside, the physical reality of this period was that my deadlines were inescapable. Moreover, the most pressing of these deadlines was completely non-school related: my dad, on his way home to Wichita, was coming for a brief visit. This would be his first stay by himself, without Mom to accompany him or act as a buffer.

Dad visited me the night before my most difficult exam. Having just returned from spending time with his family—a group of people with whom he historically had an antagonistic relationship—Dad seemed particularly relaxed in his stocky six-foot-four frame. Wearing one of the more subtle of his nineteen cowboy hats, he arrived at my door, hungry, greeting me in Urdu, our mother tongue, and laden with gifts from Estée Lauder for his only daughter. Never mind that I rarely wore makeup and would have preferred to see the money spent on my electric bill or a stack of feminist theory books from my favorite used bookstore. If Dad's visit was going to include a conversation about how little I use beauty products, I was not going to be particularly receptive.

"Almas," began my father from across the dinner table, speaking in his British-Indian accent infused with his love of Midwestern colloquialisms, "You know that you won't be a spring chicken forever. While I was in Philadelphia, I realized how important it is for you to begin thinking about our culture, religion and your future marriage plans. I think it is time we began a two-year marriage plan so you can find a husband and start a family. I think twenty-two will be a good age for you. You should be married by twenty-two."

I needed to begin thinking about the "importance of tradition" and be married by twenty-two? This, from the only Indian man I knew who had Alabama's first album on vinyl and loved to spend long weekends in his rickety, old camper near Cheney Lake, bass fishing and listening to traditional Islamic Quavali music? My father, in fact, was in his youth crowned "Mr. Madras," weightlifting champion of 1965, and had left India to practice medicine and be an American cowboy in his spare time. But he wanted *me* to aspire to be a "spring chicken," maintaining some unseen hearth and home to reflect my commitment to tradition and culture.

Dad continued, "I have met a boy that I like for you very much. Masoud's son, Mahmood. He is a good Muslim boy, tells great jokes in Urdu and is a promising engineer. We should be able to arrange something. I think you will be very happy with him!" Dad concluded with a satisfied grin.

Masoud, Dad's cousin? This would make me and Mahmood distant relatives of some sort. And Dad wants to "arrange something"? I had brief visions of being paraded around a room, serving tea to strangers in a sari or a shalwar kameez (a traditional South Asian outfit for women) wearing a long braid and chappals (flat Indian slippers), while Dad boasted of my domestic capabilities to increase my attractiveness to potential suitors. I quickly flipped through my mental Rolodex of rhetorical devices, acquired during years of women's studies classes and found the card blank. No doubt, even feminist scholar Catherine MacKinnon would have been rendered speechless sitting across the table in a Chinese restaurant speaking to my overzealous father.

It is not that I hadn't already dealt with the issue. In fact, we had been here before, ever since the marriage proposals began (the first one came when I was fourteen). Of course, when they first began, it was a family joke, as my parents understood that I was to continue my education. The jokes, however, were always at my expense: "You received a proposal from a nice boy living in our mosque. He is studying medicine," my father would come and tell me with a huge, playful grin. "I told him that you weren't interested because you are too busy with school. And anyway you can't cook or clean." My father found these jokes particularly funny, given my dislike of household chores. In this way, the eventuality of figuring out how to deal with these difficult issues was postponed with humor.

Dad's marriage propositions also resembled conversations that we had already had about my relationship to Islamic practices specific to women, some negotiated in my favor and others simply shelved for the time being. Just a year ago, Dad had come to me while I was home for the winter holidays, asking me to begin wearing *hijab*, the traditional headscarf worn by Muslim women. I categorically refused, maintaining respect for those women who chose to do so. I understood that for numerous women, as well as for Dad, hijab symbolized something much more than covering a woman's body or hair; it symbolized a way to adhere to religious and cultural traditions in order to prevent complete Western immersion. But even my sympathy for this concern didn't change my feeling that hijab constructed me as a woman first and a human being second. Veiling seemed to reinforce the fact that inequality between the sexes was a natural, inexplicable phenomenon that is impossible to overcome, and that women should cover themselves, accommodating an unequal hierarchy, for the purposes of modesty and self-protection. I couldn't reconcile these issues and refused my father's request to don the veil. Although there was tension—Dad claimed I had yet to have my religious awakening—he chose to respect my decision.

Negotiating certain issues had always been part of the dynamic between my parents and me. It wasn't that I disagreed with them about everything. In fact, I had internalized much of the Islamic perspective of the female body while simultaneously admitting to its problematic nature (To this day, I would rather wear a wool sweater than a bathing suit in public, no matter how sweltering the weather). Moreover, Islam became an important part of differentiating myself from other American kids who did not have to find a balance between two opposing cultures. Perhaps Mom and Dad recognized the need to concede certain aspects of traditional Islamic norms, because for all intents and purposes, I had been raised in the breadbasket of America.

By the time I hit adolescence, I had already established myself outside of the social norm of the women in my community. I was an athletic teenager, a competitive tennis player and a budding weightlifter. After a lot of reasoning with my parents, I was permitted to wear shorts to compete in tennis tournaments, but I was not allowed to show my legs or arms (no tank tops) outside of sports. It was a big deal for my parents to have agreed to allow me to wear shorts in the first place. The small community of South Asian Muslim girls my age, growing up in Wichita, became symbols of the future of our community in the United States. Our bodies became the sites to play out cultural and religious debates. Much in the same way that Lady Liberty had come to symbolize idealized stability in the *terra patria* of America, young South Asian girls in my community were expected to embody the values of a preexisting social structure. We were scrutinized for what we said, what we wore, being seen with boys in public and for lacking grace and piety. Needless to say, because of disproportionate muscle mass, crooked teeth, huge Lucy glasses, and a disposition to walk pigeon-toed, I was not among the favored.

To add insult to injury, Mom nicknamed me "Amazon Woman," lamenting the fact that she—a beautiful, petite lady—had produced such a graceless, unfeminine creature. She was horrified by how freely I got into physical fights with my younger brother and armwrestled boys at school. She was particularly frustrated by the fact that I could not wear her beautiful Indian jewelry, especially her bangles and bracelets, because my wrists were too big. Special occasions, when I had to slather my wrists with tons of lotion in order to squeeze my hands into her tiny bangles, often bending the soft gold out of shape, caused us both infinite amounts of grief. I was the snot-nosed, younger sibling of the Bollywood (India's Hollywood) princess that my mother had in mind as a more appropriate representation of an Indian daughter. Rather, I loved sports, sports figures and books. I hated painful makeup rituals and tight jewelry.

It wasn't that I had a feminist awakening at an early age. I was just an obnoxious kid who did not understand the politics raging around my body. I did not possess the tools to analyze or understand my reaction to this process of social conditioning and normalization until many years later, well after I had left my parents' house and the Muslim community in Wichita. By positioning me as a subject of both humiliation and negotiation, Mom and Dad had inadvertently laid the foundations for me to understand and scrutinize the process of conditioning women to fulfill particular social obligations.

What was different about my dinner conversation with Dad that night was a sense of immediacy and detail. Somehow discussion about a "two-year marriage plan" seemed to encroach on my personal space much more than had previous jokes about my inability to complete my household chores or pressure to begin wearing hijab. I was meant to understand that that when it came to marriage, I was up against an invisible clock (read: social norms) that would dictate how much time I had left: how much time I had left to remain desirable, attractive and marriageable. Dad was convinced that it was his duty to ensure my long-term security in a manner that reaffirmed traditional Muslim culture in the face of an often hostile foreign community. I recognized that the threat was not as extreme as being shipped off to India in order to marry someone I had never met. The challenge was far more subtle than this. I was being asked to choose my community; capitulation through arranged marriage would show my commitment to being Indian, to being a good Muslim woman and to my parents by providing that they had raised me with a sense of duty and the willingness to sacrifice for my culture, religion and family.

There was no way to tell Dad about my complicated reality. Certain characteristics of my current life already indicated failure by such standards. I was involved in a long-term relationship with a white man, whose father was a prison guard on death row, an occupation that would have mortified my upper-middle-class, status-conscious parents. I was also struggling with an insurmountable crush on an *actress* in the Theater and Film Department. I was debating my sexuality in terms of cultural compatibility as well as gender. Moreover, there was no way to tell Dad that my social circle was supportive of these nontraditional romantic explorations. My friends in college had radically altered my perceptions of marriage and family. Many of my closest friends, including my roommates, were coming to terms with their own life-choices, having recently come out of the closet but unable to tell their families about their decisions. I felt inextricably linked to this group of women, who, like me, often had to lead double lives. The immediacy of fighting for issues such as queer rights, given the strength and beauty of my friends' romantic relationships, held far more appeal for me than the topics of marriage and security that my father broached over our Chinese dinner. There was no way to explain to my loving, charismatic, steadfastly religious father, who was inclined to the occasional violent outburst, that a traditional arranged marriage not only conflicted with the feminist ideology I had come to embrace, but it seemed almost petty in the face of larger, more pressing issues.

Although I had no tools to answer my father that night at dinner, feminist theory had provided me with the tools to understand *why* my father and I were engaged in the conversation in the first place. I understood that in his mind, Dad was fulfilling his social obligation as father and protector. He worried about my economic stability and, in a roundabout way, my happiness. Feminism and community activism had enabled me to understand these things as part of a proscribed role for women. At the same time, growing up in Kansas and coming to feminism here meant that I had to reconcile a number of different issues. I am a Muslim, first-generation Indian, feminist woman studying in a largely homogeneous white, Christian community in Midwestern America. What sacrifices are necessary for me to retain my familial relationships as well as a sense of personal autonomy informed by Western feminism?

The feminist agenda in my community is centered on ending violence against women, fighting for queer rights and maintaining women's reproductive choices. As such, the way that I initially became involved with this community was through community projects such as "Womyn Take Back the Night," attending pride rallies and working at the local domestic violence shelter. I am often the only woman of color in feminist organizations and at feminist events. Despite having grown up in the Bible belt, it is difficult for me to relate to stories told by my closest friends of being raised on cattle ranches and farms, growing up Christian by default and experiencing the strict social norms of small, religious communities in rural Kansas. Given the context of this community—a predominantly white, middle-class, college town—I have difficulty explaining that my feminism has to address issues like, "I should be able to wear *both* hijab *and* shorts if I chose to." The enormity of our agenda leaves little room to debate issues equally important but applicable only to me, such as the meaning of veiling, arranged marriages versus dating and how the north-south divide uniquely disadvantages women in the developing world.

It isn't that the women in my community ever turned to me and said, "Hey you, brown girl, stop diluting our priorities." To the contrary, the majority of active feminists in my community are eager to listen and understand my sometimes divergent perspective. We have all learned to share our experiences as women, students, mothers, partners and feminists. We easily relate to issues of male privilege, violence against women and figuring out how to better appreciate the sacrifices made by our mothers. From these commonalities we have learned to work together, creating informal social networks to complete community projects.

The difficulty arises when trying to put this theory and discussion into practice. Like last year, when our organization, the Womyn's Empowerment Action Coalition, began plans for the Womyn Take Back the Night march and rally, a number of organizers were eager to include the contribution of a petite, white belly dancer in the pre-march festivities. When I voiced my concern that historically belly dancing had been used as a way to objectify women's bodies in the Middle East, one of my fellow organizers (and a very good friend) laughed and called me a prude: "We're in Kansas, Almas," she said. "It doesn't mean the same thing in our culture. It is different here than over *there*." I understood what she meant, but having just returned from seven months in the West Bank, Palestine two months before, for me over there *was* over here. In the end, the dance was included while I wondered about our responsibility to women outside of the United States and our obligation to address the larger social, cultural issues of the dance itself.

To reconcile the differences between my own priorities and those of the women I work with, I am learning to bridge the gap between the Western white women (with the occasional African-American or Chicana) feminist canon and my own experience as a first-generation Indian Muslim woman living in the Midwest. I struggle with issues like cultural differences, colonialism, Islam and feminism and how they relate to one another. The most difficult part has been to get past my myopic vision of simply laying feminist theory written by Indian, Muslim or postcolonial theorists on top of American-Western feminism. With the help of feminist theory and other feminists, I am learning to dissect Western models of feminism, trying to figure out what aspects of these models can be applied to certain contexts. To this end, I have had the privilege of participating in projects abroad, in pursuit of understanding feminism in other contexts.

For example, while living with my extended family in India, I worked for a micro-credit affiliate that advised women on how to get loans and start their own businesses. During this time I learned about the potential of micro-enterprise as a weapon against the feminization of poverty. Last year, I spent a semester in the West Bank, Palestine studying the link between women and economics in transitional states and beginning to understand the importance of women's efforts during revolution. These experiences have been invaluable to me as a student of feminism and women's mobilization efforts. They have also shaped my personal development, helping me understand where the theoretical falls short of solving for the practical. In Lawrence, I maintain my participation in local feminist projects. Working in three different contexts has highlighted the amazing and unique ways in which feminism develops in various cultural settings yet still maintains certain commonalities.

There are few guidebooks for women like me who are trying to negotiate the paradigm of feminism in two different worlds. There is a delicate dance here that I must master—a dance of negotiating identity within interlinking cultural spheres. When faced with the movement's expectations of my commitment to local issues, it becomes important for me to emphasize that differences in culture and religion are also "local issues." This has forced me to change my frame of reference, developing from a rebellious tomboy who resisted parental imposition to a budding social critic, learning how to be a committed feminist and still keep my cultural, religious and community ties. As for family, we still negotiate despite the fact that Dad's two-year marriage plan has yet to come to fruition in this, my twenty-second year.

Sexuality

THE GENDER OF SEXUALITY   50

Pepper Schwartz and Virginia Rutter

The gender of the person you desire is a serious matter seemingly fundamental to the whole business of romance. And it isn't simply a matter of whether someone is male or female; how well the person fulfills a lover's expectations of masculinity or femininity is of great consequence. . . .

. . . Although sex is experienced as one of the most basic and biological of activities, in human beings it is profoundly affected by things other than the body's urges. Who we're attracted to and what we find sexually satisfying is not just a matter of the genital equipment we're born with. . . .

On one level, sex can be regarded as having both a biological and a social context. The biological (and physiological) refers to how people use their genital equipment to reproduce. In addition, as simple as it seems, bodies make the experience of sexual pleasure available—whether the pleasure involves other bodies or just one's own body and mind. It should be obvious, however, that people engage in sex even when they do not intend to reproduce. They have sex for fun, as a way to communicate their feelings to each other, as a way to satisfy their ego, and for any number of other reasons relating to the way they see themselves and interact with others.

Another dimension of sex involves both what we do and how we think about it. Sexual behavior refers to the sexual acts that people engage in. These acts involve not only petting and intercourse but also seduction and courtship. Sexual behavior also involves the things people do alone for pleasure and stimulation and the things they do with other people. Sexual desire, on the other

---

hand, is the motivation to engage in sexual acts. It relates to what turns people on. A person's sexuality consists of both behavior and desire.

The most significant dimension of sexuality is gender. Gender relates both to the biological and social contexts of sexual behavior and desire. People tend to believe they know whether someone is a man or a woman not because we do a physical examination and determine that the person is biologically male or biologically female. Instead, we notice whether a person is masculine or feminine. Gender is a social characteristic of individuals in our society that is only sometimes consistent with biological sex. Thus, animals, like people, tend to be identified as male and female in accordance with the reproductive function, but only people are described by their gender, as a man or woman.

When we say something is gendered we mean that social processes have determined what is appropriately masculine and feminine and that gender has thereby become integral to the definition of the phenomenon. For example, marriage is a gendered institution: The definition of marriage involves a masculine part (husband) and a feminine part (wife). Gendered phenomena, like marriage, tend to appear "naturally" so. But, as recent debates about same-sex marriage underscore, the role of gender in marriage is the product of social processes and beliefs about men, women, and marriage. In examining how gender influences sexuality, moreover, you will see that gender rarely operates alone: Class, culture, race, and individual differences also combine to influence sexuality. . . .

DESIRE: ATTRACTION AND AROUSAL

The most salient fact about sex is that nearly everybody is interested in it. Most people like to have sex, and they talk about it, hear about it, and think about it. But some people are obsessed with sex and willing to have sex with anyone or anything. Others are aroused only by particular conditions and hold exacting criteria. For example, some people will have sex only if they are positive that they are in love, that their partner loves them, and that the act is sanctified by marriage. Others view sex as not much different from eating a sandwich. They neither love nor hate the sandwich; they are merely hungry, and they want something to satisfy that hunger. What we are talking about here are differences in desire. As you have undoubtedly noticed, people differ in what they find attractive, and they are also physically aroused by different things. . . .

Many observers argue that when it comes to sex, men and women have fundamentally different biological wiring. Others use the evidence to argue that culture has produced marked sexual differences among men and women. We believe, however, that it is hard to tease apart biological differences and social differences. As soon as a baby enters the world, it receives messages about gender and sexuality. In the United States, for example, disposable diapers come adorned in pink for girls and blue for boys. In case people aren't sure whether to treat the baby as masculine or feminine in its first years of life, the diaper signals them. The assumption is that girl babies really are different from boy babies and the difference ought to be displayed. This different treatment continues throughout life, and therefore a sex difference at birth becomes amplified into gender difference as people mature.

Gendered experiences have a great deal of influence on sexual desire. As a boy enters adolescence, he hears jokes about boys' uncontainable desire. Girls are told the same thing and told that their job is to resist. These gender messages have power not only over attitudes and behavior (such as whether a person grows up to prefer sex with a lover rather than a stranger) but also over physical and biological experience. For example, a girl may be discouraged from vigorous competitive activity, which will subsequently influence how she develops physically, how she feels about her body, and even how she relates to the adrenaline rush associated with physical competition. . . .

THE BIOLOGY OF DESIRE: NATURE'S EXPLANATION

Biology is admittedly a critical factor in sexuality. Few human beings fall in love with fish or sexualize trees. Humans are designed to respond to other humans. And human activity is, to some extent, organized by the physical equipment humans are born with. Imagine if people had fins instead of arms or laid eggs instead of fertilizing them during intercourse. Romance would look quite different.

Although biology seems to be a constant (i.e., a component of sex that is fixed and unchanging), the social world tends to mold biology as much as biology shapes humans' sexuality. Each society has its own rules for sex. Therefore, how people experience their biology varies widely. In some societies, women act intensely aroused and active during sex; in others, they have no concept of orgasm. In fact, women in some settings, when told about orgasm, do not even believe it exists, as anthropologists discovered in some parts of Nepal. Clearly, culture—not biology—is at work, because we know that orgasm is physically possible, barring damage to or destruction of the sex organs. Even ejaculation is culturally dictated. In some countries, it is considered healthy to ejaculate early and often; in others, men are told to conserve semen and ejaculate as rarely as possible. The biological capacity may not be so different, but the way bodies behave during sex varies according to social beliefs.

Sometimes the dictates of culture are so rigid and powerful that the so-called laws of nature can be overridden. Infertility treatment provides an example: For couples who cannot produce children "naturally," a several billion dollar industry has provided technology that can, in a small proportion of cases, overcome this biological problem (Rutter 1996). . . .

The Social Origins of Desire

Your own experience should indicate that biology and genetics alone do not shape human sexuality. From the moment you entered the world, cues from the environment were telling you which desires and behaviors were "normal" and which were not. The result is that people who grow up in different circumstances tend to have different sexualities. Who has not had their sexual behavior influenced by their parents' or guardians' explicit or implicit rules? You may break the rules or follow them, but you can't forget them. On a societal level, in Sweden, for example, premarital sex is accepted, and people are expected to be sexually knowledgeable and experienced. Swedes are likely to associate sex with pleasure in this "sex positive" society. In Ireland, however, Catholics are supposed to heed the Church's strict prohibitions against sex outside of marriage, birth control, and the expression of lust. In Ireland the experience of sexuality is different from the experience of sexuality in Sweden because the rules are different. Certainly, biology in Sweden is no different from biology in Ireland, nor is the physical capacity to experience pleasure different. But in Ireland, nonmarital sex is clandestine and shameful. Perhaps the taboo adds excitement to the experience. In Sweden, nonmarital sex is acceptable. In the absence of social constraint, it may even feel a bit mundane. These culturally specific sexual rules and experiences arise from different norms, the well-known, unwritten rules of society.

Another sign that social influences play a bigger role in shaping sexuality than does biology is the changing notions historically of male and female differences in desire. Throughout history, varied explanations of male and female desire have been popular. At times, woman was portrayed as the stormy temptress and man the reluctant participant, as in the Bible story of Adam and Eve. At other times, women were seen as pure in thought and deed while men were voracious sexual beasts, as the Victorians would have it.

These shifting ideas about gender are the social "clothing" for sexuality. The concept of gender typically relies on a dichotomy of male versus female sexual categories, just as the tradition of women wearing dresses and men wearing pants has in the past made the shape of men and women appear quite different. Consider high heels, an on-again-off-again Western fashion. Shoes have no innate sexual function, but high heels have often been understood to be "sexy" for women, even though (or perhaps because) they render women less physically agile. (Of course, women cope. As Ginger Rogers, the 1940s movie star and dancing partner to Fred Astaire, is said to have quipped, "I did everything Fred did, only backwards and in high heels.") Social norms of femininity have at times rendered high heels fashionable. So feminine are high heels understood to be that a man in high heels, in some sort of visual comedy gag, guarantees a laugh from the audience. Alternatively, high heels are a required emblem of femininity for cross-dressing men.

Such distinctions are an important tool of society; they provide guidance to human beings about how to be a "culturally correct" male or female. Theoretically, society could "clothe" its members with explicit norms of sexuality that de-emphasize difference and emphasize similarity or even multiplicity. Picture unisex hairstyles and men and women both free to wear skirts or pants, norms that prevail from time to time in some subcultures. What is remarkable about dichotomies is that even when distinctions, like male and female norms of fashion, are reduced, new ways to assert an ostensibly essential difference between men and women arise. Societies' rules, like clothes, are changeable. But societies' entrenched taste for constructing differences between men and women persists.

The Social Construction of Sexuality

. . . In a heterogeneous and individualistic culture like North America, sexual socialization is complex. A society creates an "ideal" sexuality, but different families and subcultures have their own values. For example, even though contemporary society at large may now accept premarital sexuality, a given family may lay down the law: Sex before marriage is against the family's religion and an offense against God's teaching. A teenager who grows up in such a household may suppress feelings of sexual arousal or channel them into outlets that are more acceptable to the family. Or the teenager may react against her or his background, reject parental and community opinion, and search for what she or he perceives to be a more "authentic" self. Variables like birth order or observations of a sibling's social and sexual expression can also influence a person's development.

As important as family and social background are, so are individual differences in response to that background. In the abstract, people raised to celebrate their sexuality must surely have a different approach to enjoying their bodies than those who are taught that their bodies will betray them and are a venal part of human nature. Yet whether or not a person is raised to be at ease with physicality does not always help predict adult sexual behavior. Sexual sybarites and libertines may have grown up in sexually repressive environments, as did pop culture icon and Catholic-raised Madonna. Sometimes individuals whose families promoted sex education and free personal expression are content with minimal sexual expression.

Even with the nearly infinite variety of sexuality that individual experience produces, social circumstances shape sexual patterns. For example, research shows that people who have had more premarital sexual intercourse are likely to have more extramarital intercourse, or sex with someone other than their spouse (Blumstein and Schwartz 1983). Perhaps early experience creates a desire for sexual variety and makes it harder for a person to be monogamous. On the other hand, higher levels of sexual desire may generate both the premarital and extramarital propensities. Or perhaps nonmonogamous, sexually active individuals are "rule breakers" in other areas also, and resist not only the traditional rules of sex but also other social norms they encounter. Sexual history is useful for predicting sexual future, but it does not provide a complete explanation. . . .

Social Control of Sexuality

So powerful are norms as they are transmitted through both social structures and everyday life that it is impossible to imagine the absence of norms that control sexuality. In fact, most images of "liberated" sexuality involve breaking a social norm—say, having sex in public rather than in private. The social norm is always the reference point. Because people are influenced from birth by the social and physical contexts of sexuality, their desires are shaped by those norms. There is no such thing as a truly free sexuality. For the past two centuries in North America, people have sought "true love" through personal choice in dating and mating (Freedman and D'Emilio 1988). Although this form of sexual liberation has generated a small increase in the number of mixed pairs—interracial, interethnic, interfaith pairs—the rule of homogamy, or marrying within one's class, religion, and ethnicity, still constitutes one of the robust social facts of romantic life. Freedom to choose the person one loves turns out not to be as free as one might suppose.

Despite the norm of true love currently accepted in our culture, personal choice and indiscriminate sexuality have often been construed across cultures and across history as socially disruptive. Disruptions to the social order include liaisons between poor and rich; between people of different races, ethnicities, or faiths; and between members of the same sex. Traditional norms of marriage and sexuality have maintained social order by keeping people in familiar and "appropriate" categories. Offenders have been punished by ostracism, curtailed civil rights, or in some societies, death. Conformists are rewarded with social approval and material advantages. Although it hardly seems possible today, mixed-race marriage was against the law in the United States until 1967. Committed same-sex couples continue to be denied legal marriages, . . . income tax breaks, and health insurance benefits; heterosexual couples take these social benefits for granted.

Some social theorists observe that societies control sexuality through construction of a dichotomized or gendered (male-female) sexuality (Foucault 1978). Society's rules about pleasure seeking and procreating are enforced by norms about appropriate male and female behavior. For example, saying that masculinity is enhanced by sexual experimentation while femininity is demeaned by it gives men sexual privilege (and pleasure) and denies it to women. Furthermore, according to Foucault, sexual desire is fueled by the experience of privilege and taboo regarding sexual pleasure. That is, the very rules that control sexual desire shape it and even enhance it. The social world could just as plausibly concentrate on how much alike are the ways that men and women experience sex and emphasize how broadly dispersed sexual conduct is across genders. However, social control turns pleasure into a scarce resource and endows leaders who regulate the pleasure of others with power. . . .

Sexual Identity and Orientation

. . . Sexual identity and sexual orientation . . . are used to mean a variety of things. We use these terms to refer to how people tend to classify themselves sexually—either as gay, lesbian, bisexual, or straight. Sexual behavior and sexual desire may or may not be consistent with sexual identity. That is, people may identify themselves as heterosexual, but desire people of the same sex—or vice versa.

It is hard to argue with the observation that human desire is, after all, organized. Humans do not generally desire cows or horses (with, perhaps, the exception of Catherine the Great, the Russian czarina who purportedly came to her demise while copulating with a stallion). More to the point, humans are usually quite specific about which sex is desirable to them and even whether the object of their desire is short or tall, dark or light, hairy or sleek.

In the United States, people tend to be identified as either homosexual or heterosexual. Other cultures (and prior eras in the United States) have not distinguished between these two sexual orientations. However, our culture embraces the perspective that, whether gay or straight, one has an essential, inborn desire, and it cannot change. Many people seem convinced that homosexuality is an essence rather than a sexual act. . . . People tend to assume that the object of desire is a matter of the gender of the object. That is, they think even homosexual men desire someone who is feminine and that homosexual women desire someone who is masculine. In other words, even among gay men and lesbians, it is assumed that they will desire opposite-gendered people, even if they are of the same sex.

Historians have chronicled in Western culture the evolution of homosexuality from a behavior into an identity (e.g., Freedman and D'Emilio 1988). In the past, people might engage in same-gender sexuality, but only in the twentieth century has it become a well-defined (and diverse) lifestyle and self-definition. Nevertheless, other evidence shows that homosexual identity has existed for a long time. The distinguished historian John Boswell (1994) believes that homosexuals as a group and homosexuality as an identity have existed from the very earliest of recorded history. He used evidence of early Christian same-sex "marriage" to support his thesis. Social scientist

Fred Whitman (1983) has looked at homosexuality across cultures and declared that the evidence of a social type, including men who use certain effeminate gestures and have diverse sexual tastes, goes far beyond any one culture. Geneticist Dean Hamer provides evidence that sexual attraction may be genetically programmed, suggesting that it has persisted over time and been passed down through generations.

On the other side of the debate is the idea that sexuality has always been invented and that sexual orientations are socially created. A gay man's or lesbian's sexual orientation has been created by a social context. Although this creation takes place in a society that prefers dichotomous, polarized categories, the social constructionist vision of sexuality at least poses the possibility that sexuality could involve a continuum of behavior that is matched by a continuum of fantasy, ability to love, and sense of self. . . .

References

Blumstein, P. and P. Schwartz. 1983. *American Couples: Money, Work, and Sex.* New York: William Morrow.

Boswell, J. 1994. *Same-Sex Unions in Pre-Modern Europe.* New York: Villard.

Freedman, E. and J. D. D'Emilio. 1988. *Intimate Matters: A History of Sexuality in America.* New York: Harper & Row.

Foucault, M. 1978. *A History of Sexuality: Vol. 1. An Introduction.* New York: Pantheon.

Rutter, V. 1996. "Who Stole Fertility?" *Psychology Today*, March/April, pp. 44–70.

Whitman, F. 1983. "Culturally Invariable Properties of Male Homosexualities: Tentative Conclusions from Cross-Cultural Research." *Archives of Sexual Behavior* 12: 207–26.

BLACK SEXUALITY

51

The Taboo Subject

Cornel West

Americans are obsessed with sex and fearful of black sexuality. The obsession has to do with a search for stimulation and meaning in a fast-paced, market-driven culture; the fear is rooted in visceral feelings about black bodies fueled by sexual myths of black women and men. The dominant myths draw black women and men either as threatening creatures who have the potential for sexual power over whites, or as harmless, desexed underlings of a white culture. There is Jezebel (the seductive temptress), Sapphire (the evil, manipulative bitch), or Aunt Jemima (the sexless, long-suffering nurturer). There is Bigger Thomas (the mad and mean predatory craver of white women), Jack Johnson, the super performer—be it in athletics, entertainment, or sex—who excels others naturally and prefers women of a lighter hue), or Uncle Tom (the spineless, sexless—or is it impotent?—sidekick of whites). The myths offer distorted, dehumanized creatures whose bodies—color of skin, shape of nose and lips, type of hair, size of hips—are already distinguished from the white norm of beauty and whose feared sexual activities are deemed disgusting, dirty, or funky and considered less acceptable.

Yet the paradox of the sexual politics of race in America is that, behind closed doors, the dirty, disgusting, and funky sex associated with black people is often perceived to be more intriguing and interesting, while in public spaces talk about black sexuality is virtually taboo. Everyone knows it is virtually impossible to talk candidly about race without talking about sex. Yet most social scientists who examine race relations do so with little or no reference to how sexual perceptions influence racial matters. My thesis is that black sexuality is a taboo subject in white and black America and that a candid dialogue about black sexuality between and within these communities is requisite for healthy race relations in America.

The major cultural impact of the 1960s was not to demystify black sexuality but rather to make black bodies more accessible to white bodies *on an equal basis.* The history of such access up to that time was primarily one of brutal white rape and ugly white abuse. The Afro-Americanization of white youth—given the disproportionate black role in popular music and athletics—has put white kids in closer contact with their own bodies and facilitated more humane interaction with black people. Listening to Motown records in the sixties or dancing to hip hop music in the nineties may not lead one to question the sexual myths of black women and men, but when white and black kids buy the same Billboard hits and laud the same athletic heroes the result is often a shared cultural space where some humane interaction takes place.

This subterranean cultural current of interracial interaction increased during the 1970s and 1980s even as racial polarization deepened on the political front. We miss much of what goes on in the complex development of race relations in America if we focus solely on the racial card played by the Republican Party and overlook the profound multicultural mix of popular culture that has occurred in the past two decades. In fact, one of the reasons Nixon, Reagan, and Bush had to play a racial card, that is, had to code their language about race, rather than simply call a spade a spade, is due to the changed *cultural* climate of race and sex in America. The classic scene of Senator Strom Thurmond—staunch segregationist and longtime opponent of interracial sex and marriage—strongly defending Judge Clarence Thomas—married to a white woman and an alleged avid consumer of white pornography—shows how this change in climate affects even reactionary politicians in America.

Needless to say, many white Americans still view black sexuality with disgust. And some continue to view their own sexuality with disgust. Victorian morality and racist perceptions die hard. But more and more white Americans are willing to interact sexually with black Americans *on an equal basis*—even if the myths still persist. I view this as neither cause for celebration nor reason for lament. Anytime two human beings find genuine pleasure, joy, and love, the stars smile and the universe is enriched. Yet as long as that pleasure, joy, and love is still predicated on myths of black sexuality, the more fundamental challenge of humane interaction remains unmet. Instead, what we have is white access to black bodies on an equal basis—but not yet the demythologizing of black sexuality.

This demythologizing of black sexuality is crucial for black America because much of black self-hatred and self-contempt has to do with the refusal of many black Americans to love their own black bodies—especially their black noses, hips, lips, and hair. Just as many white Americans view black sexuality with disgust, so do many black Americans—but for very different reasons and with very different results. White supremacist ideology is based first and foremost on the degradation of black bodies in order to control them. One of the best ways to instill fear in people is to terrorize them. Yet this fear is best sustained by convincing them that their bodies are ugly, their intellect is inherently underdeveloped, their culture is less civilized, and their future warrants less concern than that of other peoples. Two hundred and forty-four years of slavery and nearly a century of institutionalized terrorism in the form of segregation, lynchings, and second-class citizenship in America were aimed at precisely this devaluation of black people. This white supremacist venture was, in the end, a relative failure—thanks to the courage and creativity of millions of black people and hundreds of exceptional white folk like John Brown, Elijah Lovejoy, Myles Horton, Russell Banks, Anne Braden, and others. Yet this white dehumanizing endeavor has left its toll in the psychic scars and personal wounds now inscribed in the souls of black folk. These scars and wounds are clearly etched on the canvass of black sexuality.

How does one come to accept and affirm a body so despised by one's fellow citizens? What are the ways in which one can rejoice in the intimate moments of black sexuality in a culture that questions the aesthetic beauty of one's body? Can genuine human relationships flourish for black people in a society that assaults black intelligence, black moral character, and black possibility?

These crucial questions were addressed in those black social spaces that affirmed black humanity and warded off white contempt—especially in black families, churches, mosques, schools, fraternities, and sororities. These precious black institutions forged a mighty struggle against the white supremacist bombardment of black people. They empowered black children to learn against the odds and supported damaged black egos so they could keep fighting; they preserved black sanity in an absurd society in which racism ruled unabated; and they provided opportunities for black love to stay alive. But these grand yet flawed black institutions refused to engage one fundamental issue: *black sexuality.* Instead, they ran from it like the plague. And they obsessively condemned those places where black sexuality was flaunted: the streets, the clubs, and the dance-halls.

Why was this so? Primarily because these black institutions put a premium on black survival in America. And black survival required accommodation with and acceptance from white America. Accommodation avoids any sustained association with the subversive and transgressive—be it communism or miscegenation. Did not the courageous yet tragic lives of Paul Robeson and Jack Johnson bear witness to this truth? And acceptance meant that only "good" negroes would thrive—especially those who left black sexuality at the door when they "entered" and "arrived." In short, struggling black institutions made a Faustian pact with white America: avoid any substantive engagement with black sexuality and your survival on the margins of American society is, at least, possible.

White fear of black sexuality is a basic ingredient of white racism. And for whites to admit this deep fear even as they try to instill and sustain fear in blacks is to acknowledge a weakness—a weakness that goes down to the bone. Social scientists have long acknowledged that interracial sex and marriage is the most *perceived* source of white fear of black people—just as the repeated castrations of lynched black men cries out for serious psycho-cultural explanation.

Black sexuality is a taboo subject in America principally because it is a form of black power over which whites have little control—yet its visible manifestations evoke the most visceral of white responses, be it one of seductive obsession or downright disgust. On the one hand, black sexuality among blacks simply does not include whites, nor does it make them a central point of reference. It proceeds as if whites do not exist, as if whites are invisible and simply don't matter. This form of black sexuality puts black agency center stage with no white presence at all. This can be uncomfortable for white people accustomed to being the custodians of power.

On the other hand, black sexuality between blacks and whites proceeds based on underground desires that Americans deny or ignore in public and over which laws have no effective control. In fact, the dominant sexual myths of black women and men portray whites as being "out of control"—seduced, tempted, overcome, overpowered by black bodies. This form of black sexuality makes white passivity the norm—hardly an acceptable self-image for a white-run society.

Of course, neither scenario fully accounts for the complex elements that determine how any particular relationship involving black sexuality *actually* takes place. Yet they do accent the crucial link between black sexuality and black power in America. In this way, to make black sexuality a taboo subject is to silence talk about a particular kind of power black people are perceived to have over whites. On the surface, this "golden" side is one in which black people simply have an upper hand sexually over whites given the dominant myths in our society.

. . . Black male sexuality differs from black female sexuality because black men have different self-images and strategies of acquiring power in the patriarchal structures of white America and black communities. Similarly, black male heterosexuality differs from black male homosexuality owing to the self-perceptions and means of gaining power in the homophobic institutions of white America and black communities. The dominant myth of black male sexual prowess makes black men desirable sexual partners in a culture obsessed with sex. In addition, the Afro-Americanization of white youth has been more a male than a female affair given the prominence of male athletes and the cultural weight of male pop artists. This process results in white youth—male and female—imitating and emulating black male styles of walking, talking, dressing, and gesticulating in relation to others. One irony of our present moment is that just as young black men are murdered, maimed, and imprisoned in record numbers, their styles have become disproportionately influential in shaping popular culture. For most young black men, power is acquired by stylizing their bodies over space and time in such a way that their bodies reflect their uniqueness and provoke fear in others. To be "bad" is good not simply because it subverts the language of the dominant white culture but also because it imposes a unique kind of order for young black men on their own distinctive chaos and solicits an attention that makes others pull back with some trepidation. This young black male style is a form of self-identification and resistance in a hostile culture; it also is an instance of machismo identity ready for violent encounters. Yet in a patriarchal society, machismo identity is expected and even exalted—as with Rambo and Reagan. Yet a black machismo style solicits primarily sexual encounters with women and violent encounters with other black men or aggressive police. In this way, the black male search for power often reinforces the myth of black male sexual prowess—a myth that tends to subordinate black and white women as objects of sexual pleasure. This search for power also usually results in a direct confrontation with the order-imposing authorities of the status quo, that is, the police or criminal justice system. The prevailing cultural crisis of many black men is the limited stylistic options of self-image and resistance in a culture obsessed with sex yet fearful of black sexuality.

This situation is even bleaker for most black gay men who reject the major stylistic option of black machismo identity, yet who are marginalized in white America and penalized in black America for doing so. In their efforts to be themselves, they are told they are not really "black men," not machismo-identified. Black gay men are often the brunt of talented black comics like Arsenio Hall and Damon Wayans. Yet behind the laughs lurks a black tragedy of major proportions: the refusal of white and black America to entertain seriously new stylistic options for black men caught in the deadly endeavor of rejecting black machismo identities.

The case of black women is quite different, partly because the dynamics of white and black patriarchy affect them differently and partly because the degradation of black female heterosexuality in America makes black female lesbian sexuality a less frightful jump to make. This does not mean that black lesbians suffer less than black gays—in fact, they suffer more, principally owing to their lower economic status. But this does mean that the subculture of black lesbians is fluid and the boundaries are less policed precisely because black female sexuality in general is more devalued, hence more marginal in white and black America.

The dominant myth of black female sexual prowess constitutes black women as desirable sexual partners—yet the central role of the ideology of white female beauty attenuates the expected conclusion. Instead of black women being the most sought after "objects of sexual pleasure"—as in the case of black men—white women tend to occupy this "upgraded," that is, degraded, position primarily because white beauty plays a weightier role in sexual desirability for women in racist patriarchal America. The ideal of female beauty in this country puts a premium on lightness and softness mythically associated with white women and downplays the rich stylistic manners associated with black women. This operation is not simply more racist to black women than that at work in relation to black men; it also is more devaluing of women in general than that at work in relation to men in general. This means that black women are subject to more multilayered bombardments of racist assaults than black men in addition to the sexist assaults they receive from black men. Needless to say, most black men—especially professional ones—simply recycle this vulgar operation along the axis of lighter hues that results in darker black women bearing more of the brunt than their already devalued lighter sisters. The psychic bouts with self-confidence, the existential agony over genuine desirability, and the social burden of bearing and usually nurturing black children under these circumstances breeds a spiritual strength of black women unbeknownst to most black men and nearly all other Americans.

As long as black sexuality remains a taboo subject, we cannot acknowledge, examine, or engage these tragic psychocultural facts of American life. Furthermore, our refusal to do so limits our ability to confront the overwhelming realities of the AIDS epidemic in America in general and in black

America in particular. Although the dynamics of black male sexuality differ from those of black female sexuality, new stylistic options of self-image and resistance can be forged only when black women and men do so together. This is so not because all black people should be heterosexual or with black partners, but rather because all black people—including black children of so-called "mixed" couples—are affected deeply by the prevailing myths of black sexuality. These myths are part of a wider network of white supremacist lies whose authority and legitimacy must be undermined. In the long run, there is simply no way out for all of us other than living out the truths we proclaim about genuine humane interaction in our psychic and sexual lives. Only by living against the grain can we keep alive the possibility that the visceral feelings about black bodies fed by racist myths and promoted by market-driven quests for stimulation do not forever render us obsessed with sexuality and fearful of each other's humanity.

WHERE HAS GAY LIBERATION GONE?

An Interview with Barbara Smith

52

Amy Gluckman and Betsy Reed

The links between homophobia and sexism have been analyzed a lot, but the linkages between homophobia and other forms of oppression, especially class and race oppression, haven't been thought about or analyzed as much. What connections do you see between, first of all, gay oppression and our economic system, and issues of class?

I think that's really a very complicated issue. I often say that, unlike racial oppression, lesbian and gay oppression is not economically linked, and is not structurally and historically linked, to the founding of this country and of capitalism in the United States. This country was not founded on homophobia; it was founded on slavery and racism and, before that, prior to the importation of slaves, it was founded on the genocide of the indigenous people who lived here, which also had profound racial consequences and rationales.

From Amy Gluckman and Betsy Reed, eds. *Homo Economics: Capitalism, Community, and Lesbian and Gay Life* (New York: Routledge, 1997), pp. 195–207. Reprinted by permission.

I think that sometimes, certain lesbians and gay men get very upset when I and others—I'm not the only person who would say it—put that kind of analysis out because they think it means that I'm saying that lesbian and gay oppression is not serious. That's not what I'm saying at all.

Do you think that extends into the future? In other words, would you say that capitalism could go along its merry way and assimilate gay people and provide gay rights, but otherwise remain the same?

. . . As with all groups, I think that our economic system has the most implications for lesbians and gay men when their class position makes them vulnerable to that economic system. So in other words, it's not that in general being lesbian or gay puts you into a critical relationship to capitalism, it's that a large proportion of lesbians and gay men are poor or working class, but of course they're completely invisible the way the movement's politics are defined now.

So much of this society is about consumerism. As long as lesbians and gay men are characterized as people who have huge amounts of disposable income and who are kind of fun and trendy—nice entertainment type people, k. d. langs and Martinas, just a little on the edge but not really that threatening—as long as they're characterized in that way, probably capitalism can incorporate them. If they begin to think about how extending lesbian and gay rights fully might shake up the patriarchal nuclear family and the economic arrangements that are tied to that, then that might be the point at which capitalists would say, "Well no, I don't think we can include that." As long as it's about k. d. lang and Cindy Crawford on the cover of *Vanity Fair*, capitalism doesn't have any problem with that because that's nothing but an image. . . .

. . . *Right-wing organizations have used inflated income information about gays and lesbians to try to drum up homophobia, which is almost a direct appeal to people's economic frustration. As an organizer, have you encountered that? To what extent do you think that the backlash against lesbian and gay rights has to do with the degree to which people in this country under capitalism are economically exploited and might seek a hateful outlet for that?*

What they say is that these people do not qualify as a disenfranchised group because look at their income levels, as if the only way you could be disenfranchised is by income or lack of access to it. When they tell the untruth that all of us are economically privileged, of course that fans it. I can't say that I have personally seen that myth being picked up. When I come into contact with straight people, it's usually Black people. And it's usually around Black issues.

I don't know that the economic thing is such an issue for heterosexual Black communities; I think it's the moral thing. It's not so much that those white gays are rich, it's that those white gays are sinners, they're going against

God. And they're also white, so they must be racist. And if you are a Black lesbian or gay man, then you must be a racial traitor. Those are the kinds of things that I hear, not so much that they're so rich. But I think there's an assumption on the part of most people of color that *most* white people are better off than we are anyway. It's an assumption that in some cases is accurate and in some cases is a myth that is perpetuated to keep people apart who should be in solidarity with each other because of class. It's a myth that keeps people away from each other, because there are some real commonalities between white and poor and being of color and poor. . . .

Can you say more about the links that you see between racism and gay oppression?

All of the different kinds of oppression are tied to each other, particularly when it comes to the kinds of repression and oppression that are practiced against different groups of people. When you look at a profile of how people who are oppressed experience their oppression, you see such similar components: demonization, scapegoating, police brutality, housing segregation, lack of access to certain jobs and employment, even the taking away of children—custody. They have done that to poor women of color and to poor women from time out of mind. This is not a new phenomenon that women who had children couldn't keep them because the state intervened. It's just that this group of people—lesbians and some gay men—are now experiencing the same thing. There are many similarities in what we experience. And also the same people who are hounding the usual scapegoat target groups, they're hounding the lesbian and gay community too. Our enemies are the same. That to me is the major thing that should be pulling us together.

The militant right wing in this country has targeted the lesbian and gay community, but they have targeted other groups, too. And they're really being quite successful during this time period. Does the Oklahoma City bombing have anything to do with the fate of the lesbian and gay community? Absolutely it does! The Oklahoma bombing epitomizes what dire straits the country is in as a whole, and if we had a responsible lesbian/gay/bisexual/transgendered people movement, it would be asking questions like, "Okay, what is our movement supposed to be doing now, post–Oklahoma City?" That really should have been a wake-up call.

We need to look at why those people are so antigovernment. The press never explains; all it gave us after the bombing was, "They're antigovernment and they're very upset over Waco." The reason those people are so antigovernment is that they think this government is a Zionist conspiracy that privileges Black savages. That's what they think. It's never explained that the reason they're so antigovernment is because they're white supremacists, they're anti-Semitic, they're homophobic, and they're definitely opposed to women's freedom too.

You asked about race and lesbian and gay oppression. One of the things that I wanted to say early on is that the clearest responses that I have to that are out of my own experience as a person who has those identities linked. I think that the clearest answers come from those of us who simultaneously experience these oppressions; however, identity politics has been so maligned during this period that I almost hesitate to bring it up, because I don't want people to think I'm saying that the only reason it makes a difference is because it's bothering *me*. Of course, I am concerned about these issues because in my own life and experience I know what struggle is about, I know what oppression is about, and I have seen and experienced suffering myself.

That's not the only reason I am an activist, though. I care about everyone who is under siege. Some people think that because I'm so positively pro-Black and because I speak out against racism at every turn, that I don't really care if white people are suffering. Quite to the contrary. I care about all people who are not getting a fair shake, who are not getting an equal chance to fulfill their maximum potential and to live without fear and to have the basic things that every human needs: shelter, clothing, quality health care, meaningful and fairly compensated work, love and caring and freedom to express and to create, all those things that make life worth living. How limited would my politics be if I was only concerned about people like me! Given who I am as a Black woman and a lesbian and a person from a working-class home who is a socialist opposed to the exploitation of capitalism, well maybe that isn't so narrow. But what if I only cared about other Black lesbians? I would be sitting on the head of a pin.

The best heroes and the best heroines that we have throughout history have been those people who have gone beyond the narrow expectations that their demographic profile would lead you to expect. . . .

Those privileged, white gay men you mentioned that you view as setting the agenda currently—suppose homophobia could be eased in some ways. What stake do they have in other forms of oppression being ended?

The systems of oppression really do tie together. The plans and the strategies for oppressing and repressing our various groups are startlingly similar, and a society that is unjust, it's like cancer or a bleeding ulcer or something. You can't contain it. I guess it was Martin Luther King who said that when you have injustice anywhere, you really have it everywhere. It poisons the body politic of the society as a whole, and therefore you can't have singular solutions. Those white gay men who have disposable income and who think that all they need to do is get rid of the most blatant homophobia in corporate, government, and military settings, in the legal system and on TV, and everything else will fall into place, that they'll have a nice life in their little enclaves—they're dreaming. Let's say they got rid of homophobia in those places that I mentioned—which is not really getting rid of homophobia, it's getting rid of it in places that make their lives difficult—but suppose they were able to do that. If they were living in or near a city that has imploded, like Los Angeles, because of racial and economic exploitation and oppression, then how free are they going to be? I don't know why people can't understand how interconnected our fates are as creatures on the planet. . . .

One of the arguments people make for single-issue activism is that the gay commu-nity is too diverse to agree on an entire multi-issue agenda. They say, "We'll never get anywhere if we have to agree on everything else, so we're just going to work on one thing that we do agree about."

. . . If gay rights were put in place tomorrow, my behind would still be on fire. I would still be in ultimate danger here, because racism is still in the saddle. And so is class oppression, and so is sexual oppression. So getting a middle-of-the-road, mainstream gay rights agenda passed, how's that going to stop me from being raped? How's that going to help me not get breast cancer? How is it going to help the environment not get poisoned? How is it going to help the children of my community to have a chance for a decent life?

GLOBALIZING SEX WORKERS' RIGHTS   **53**

Kamala Kempadoo

When I first heard about prostitutes organizing for their rights in Suriname in 1993, I was both excited and puzzled by the news. Was it a singular incident spurred by an outsider, or did it reflect a local movement? I wanted to know. Also, in this part of the world, were women serious about staying in the sex in-dustry or anxious to have prostitution abolished? What were the aims of such an organization, and who were the activists? Was this an isolated group, and what was the response to this initiative from the rest of the women's move-ment in this corner of South America? Questions outweighed any answer I

From Kamala Kempadoo and J. Doezema, *Global Sex Workers: Rights, Resistence, and Redefinition* (New York: Routledge, 1998), pp. 1–28. Reproduced by permission of Routledge, Inc., part of the Taylor & Francis Group.

could find in libraries or books—I decided to travel to Suriname to find out more. . . .

SEX WORKER, PROSTITUTE OR WHORE?

Identity, rights, working conditions, decriminalization, and legitimacy have been central issues collectively addressed by prostitutes for many years. Through these struggles the notion of the sex worker has emerged as a counterpoint to traditionally derogatory names, under the broad banner of a prostitutes' rights movement, with some parts recovering and valorizing the name and identity of "whore." In this [work] we have chosen the term "sex worker" to reflect the current use throughout the world, although . . . "sex worker" and "prostitute" are used interchangeably. It is a term that suggests we view prostitution not as an identity—a social or a psychological characteristic of women, often indicated by "whore"—but as an income-generating activity or form of labor for women and men. The definition stresses the social location of those engaged in sex industries as working people.

The idea of the sex worker is inextricably related to struggles for the recognition of women's work, for basic human rights and for decent working conditions. The definition emphasizes flexibility and variability of sexual labor as well as its similarities with other dimensions of working people's lives. Sex work is experienced as an integral part of many women's and men's lives around the world, and not necessarily as the sole defining activity around which their sense of self or identity is shaped. Moreover, commercial sex work . . . is not always a steady activity, but may occur simultaneously with other forms of income-generating work such as domestic service, informal commercial trading, market-vending, shoeshining or office work. Sex work can also be quite short-lived or be a part of an annual cycle of work—in few cases are women and men engaged full-time or as professionals. Consequently, in one person's lifetime, sex work is commonly just one of the multiple activities employed for generating income, and very few stay in prostitution for their entire adulthood. In most cases, sex work is not for individual wealth but for family well-being or survival; for working class women to clothe, feed and educate their children; and for young women and men to sustain themselves when the family income is inadequate. For many, sex work means migration away from their hometown or country. For others, it is associated with drug use, indentureship or debt-bondage. For the majority, participation in sex work entails a life in the margins.

The concept of sex work emerged in the 1970s through the prostitutes' rights movement in the United States and Western Europe. . . . Defining human activity or work as the way in which basic needs are met and human life produced and reproduced, Than-Dam Truong argues that activities involving purely sexual elements of the body and sexual energy should also be considered vital to the fulfillment of basic human needs: for both procreation and bodily pleasure.[1] Truong thus introduces the concept of sexual labor to capture the notion of the utilization of sexual elements of the body and as a way of understanding a productive life force that is employed by women and men. In this respect she proposes that sexual labor be considered similar to other forms of labor that humankind performs to sustain itself—such as mental and manual labor, all of which involve specific parts of the body and particular types of energy and skills. Furthermore, she points out, the social organization of sexual labor has taken a variety of forms in different historical contexts and political economies, whereby there is no universal form or appearance of either prostitution or sex work. Instead, she proposes, analyses of prostitution need to address and take into account the specific ways in which sexual subjectivity, sexual needs and desires are constructed in specific contexts. Wet-nursing, temple prostitution, "breeding" under slavery, surrogate child-bearing, donor sex, commercial sex and biological reproduction can thus be seen as illustrations of historical and contemporary ways in which sexual labor has been organized for the re-creation and replenishment of human and social life.

Perhaps one of the most confounding dimensions in the conceptualization of prostitution as labor concerns the relation that exists in many people's minds between sexual acts and "love," and with prevailing ideas that without love, sexual acts are harmful and abusive. After all, isn't sex supposed to be about consensual sharing of our "most personal, private, erotic, sensitive parts of our physical and psychic being," as some would argue?[2] And aren't women in particular harmed or violated by sexual acts that are not intimate? In such perspectives, the sale of one's sexual energies is confused with a particular morality about sexual relations and essentialist cultural interpretations are imposed upon the subject. This conflation of sex with the highest form of intimacy presupposes a universal meaning of sex, and ignores changing perceptions and values as well as the variety of meanings that women and men hold about their sexual lives. . . .

. . . There is a persistent pattern through much of history that positions the social gendered category "women" as the sellers or providers of sexual labor and "men" as the group deriving profits and power from the interactions. The subordination of the female and the feminine is the overriding factor for this arrangement in a variety of cultural, national and economic contexts, producing stigmas and social condemnation of persons who defy the socially defined boundaries of womanhood. Categories of "good" and "bad" women (virgin/whore, madonna/prostitute, chaste/licentious women) exist in most patriarchal societies, where the "bad" girl becomes the trope for female sexuality that threatens male control and domination. Female sexual acts that serve women's sexual or economic interests are, within the context of masculinist hegemony, dangerous, immoral, perverted, irresponsible and indecent.

Construed in this fashion, the image of the whore disciplines and divides women, forcing some to conform to virginity, domesticity and monogamy and demonizing those who transgress these boundaries. Sex work positions women in dominant discourse as social deviants and outcasts. Today the majority of the world's sex workers are women, working within male-dominated businesses and industries, yet while the social definition of the provider of sexual labor is often closely associated with specific cultural constructions of femininity, and "the prostitute" rendered virtually synonymous with "woman," these gendered relations are clearly also being contested and redefined in different ways throughout the world. Various trends acutely challenge the tendency to essentialize the sex worker with biological notions of gender. In the Caribbean for example, so-called romance tourism is based on the sale by men of "love" to North American and European women, and "rent-a-dreads" and beach boys dominate the sex trade in the tourism industry in some islands.[3] ... Thai sex workers in Japan report to sometimes buy sex for their own pleasure from male strippers, Brazilian "miches"—young male hustlers—get by through selling sex to other men, and in Europe and Malaysia male-to-female transgender sex workers also service men. Across the globe, "genetic" men and boys engage in sex work, selling sex to both men and women in homosexual and heterosexual relations, as feminine and masculine subjects. Nevertheless, even with the increasing visibility of genetic men and boys in sex work, gender inequality and discrimination remain evident. . . .

Children, both boys and girls, are also increasingly evident in prostitution, particularly in Third World settings, making the picture of gendered relations even more complex. However, child participation in sex industries invariably raises other questions and problems than those to do with gender, and it is within the international debates on "child prostitution" that a discourse of sexual labor and sex work is also apparent. While some attribute the rise of adolescents and pre-pubescent children in prostitution to the insatiable sexual appetites of depraved western men, or to cultural preferences in Third World countries for sex with virgins, a highly compelling explanation involves an analysis of the global political economy and processes of development and underdevelopment. Studies by the International Labor Organization (ILO) show that the proliferation of earning activity by children is associated with the development process "with its intrinsic features of population and social mobility, urbanization, and progressive monetization of all forms of human activity," and the growth of the modern tourist industry based on the accumulation of wealth and disposable incomes in the industrialized world.[4] The disruption that such development brings to the organization of production in developing countries, draws children into marginal and servile occupations sometimes requiring parents to deploy the income-generating capacity of their children in order to ensure that the household survives. The research suggests we include child prostitution in the context of the global exploitation of child labor in order to effectively address the problem. Such understandings of child labor undergird various child worker movements around the world. . . .

Sex work, as we understand it here, is not a universal or ahistorical category, but is subject to change and redefinition. It is clearly not limited to prostitution or to women, but certainly encompasses what is generally understood to fall into these two categories. However, even though human sexual and emotional resources have been organized and managed in different ways and acquired different meanings, capital accumulation, liberal free market politics and the commodification of waged labor has transformed various social arrangements in a consistent fashion. . . .

If sexual labor is seen to be subject to exploitation, as with any other labor, it can also be considered as a basis for mobilization in struggles for working conditions, rights and benefits and for broader resistances against the oppression of working peoples, paralleling situations in other informal and unregulated sectors. And by recognizing sexual labor in this fashion, it is possible to identify broader strategies for change. . . . The conceptualization of prostitutes, whores, strippers, lap dancers, escorts, exotic dancers, etc., as "sex workers" insists that working women's common interests can be articulated within the context of broader (feminist) struggles against the devaluation of "women's" work and gender exploitation within capitalism. . . .

SEX WORK AND RACISM

Besides the location of women in the sex trade as workers, migrants and agents we address the specificity of racism in positioning Third World sex workers in international relations. . . .

. . . Images of "the exotic" are entwined with ideologies of racial and ethnic difference: the "prostitute" is defined as "other" in comparison to the racial or ethnic origin of the client. Such boundaries, between which women are defined as "good" and "bad," or woman and whore, reinforce sexual relations intended for marriage and family and sets limits on national and ethnic membership. The brown or black woman is regarded as a desirable, tantalizing, erotic subject, suitable for temporary or non-marital sexual intercourse—the ideal "outside" woman—and rarely seen as a candidate for a long-term commitment, an equal partner, or as a future mother. She thus represents the unknown or forbidden yet is positioned in dominant discourse as the subordinated "other." . . . It is not simply grinding poverty that underpins a woman's involvement in prostitution. Race and ethnicity are equally important factors for any understanding of contemporary sex industries. . . . Prostitution is a realm of contradictions. Thus, even with the heightened exoticization of the sexuality of Third World women and men, they are positioned within the global sex industry second to white women. White sex workers invariably work in safer, higher paid and more comfortable environments; brown women—Mulatas, Asians, Latinas—form a middle class; and Black women are still conspicuously overrepresented in the poorest and most dangerous sectors of the trade, particularly street work. Whiteness continues to represent the hegemonic ideal of physical and sexual attractiveness and desirability, and white sexual labor is most valued within the global sex industry. . . .

. . . . Some prostitutes' rights advocates assume that western development, capitalist modernization and industrialization will enable women in developing countries to exercise choice and attain "freedom." Seen to be trapped in underdeveloped states, Third World prostitutes continue to be positioned in this discourse as incapable of making decisions about their own lives, forced by overwhelming external powers completely beyond their control into submission and slavery. Western women's experience is thus made synonymous with assumptions about the inherent superiority of industrialized capitalist development and Third World women placed in categories of pre-technological "backwardness," inferiority, dependency and ignorance. . . .

Third World and anti-racist feminisms have over the past two decades intensely critiqued the universalism and totalizing effect of unnuanced western (feminist) theorizing—modernist and postmodernist—arguing that many of the concepts and theories produced about women's oppression are, and have been for many years, grounded in struggles of middle-class white women and may be quite antithetical to other women's experiences, if not representative of imperialist feminist thought. Nevertheless the need for feminist theory to engage with racialized sexual subjectivities in tandem with the historical weight of imperialism, colonialism and racist constructions of power has only been raised recently in the context of this feminist theorizing on prostitution. In view of histories of the oversexualization of non-western women in western cultures and the colonial legacies of the rape and sexual abuse of indigenous, and other Third World women, a hesitancy to explore topics of Third World women's sexual agency and subjectivity in prostitution is quite understandable. Yet in an era when women can no longer be defined exclusively as victims, where Third World women speak for themselves in various forums, where increasingly analyses have shifted focus from simple hierarchies and dichotomies to the problematization of multiple spaces, seemingly contradictory social locations and plural sites of power, it would seem that experiences, identities and struggles of women in the global sex industry cannot be neglected. . . .

TRANSNATIONAL SEX WORK AND THE GLOBAL ECONOMY

. . . Since the 1970s a global restructuring of capitalist production and investment has taken place and this can be seen to have wide-scale gendered implications and, by association, an impact on sex industries and sex work internationally. New corporate strategies to increase profit have developed, involving the movement of capital from industrial centers to countries with cheap labor, the circumvention of unionized labor, and so-called flexible employment policies. Unemployment and temporary work plagues the industrialized centers as well as "developing" countries. . . .

The emerging global economic order has already wreaked havoc on women's lives. Recent studies document an increasing need of women to contribute to the household economy through waged labor, yet having to deal with declining real wages, lower wage structures than men and longer working hours. Seasonal or flexible employment is the norm for women all over the world. Skilled and unskilled female workers constitute the main labor force in the new export-oriented industries—for shoe, toy, textile and garment production, in agribusinesses and electronic factories—where they are faced with poor working conditions, are continually threatened with unemployment due to automation and experience mass dismissals due to relocations of whole sectors of the industry. In many instances, minimum wage, health and safety laws are overridden by the transnational corporations in these new production zones, leaving women workers in particularly hazardous situations. Furthermore, with disruptions to traditional household and family structures, women are increasingly becoming heads of households, providing and nurturing the family. With dwindling family resources and the western emphasis on the independent nuclear family, women must also increasingly rely on the state for provisions such as maternity leave and child-care, yet fewer funds are allocated by governments for social welfare and programs. Informal sector work and "moonlighting" is growing and engagement in the booming sex industries fills a gap created by globalization.

Migration is a road many take to seek other opportunities and to break away from oppressive local conditions caused by globalization. A 1996 ILO report describes the "feminization" of international labor migration as "one of the most striking economic and social phenomena of recent times." This "phenomenon" according to the authors of the report, is most pronounced in Asian countries where women are migrating as "autonomous, economic agents" in their own right, "trying to seize economic opportunities overseas."[5] The Philippines has put more women onto the overseas labor market than any other country in the world.[6] Within all this dislocation and movement, some migrant women become involved in sex work.

However, laws prohibiting or regulating prostitution and migration, particularly from the South, combine to create highly complex and oppressive situations for women if they become involved in sex work once abroad. The illegal movement of persons for work elsewhere, commonly known as "trafficking" also becomes a very real issue for those who are being squeezed on all sides and have few options other than work in underground and informal sectors. Traffickers take advantage of the illegality of commercial sex work and migration, and are able to exert an undue amount of power and control over those seeking political or economic refuge or security. In such cases, it is the laws that prevent legal commercial sex work and immigration that form the major obstacles.

A related dimension to globalization with the expansion of sex industries, a heightened necessity for transnational migration for work, and increasing immiseration of women worldwide, is the spread of AIDS. Paul Farmer links the pandemic in sub-Saharan Africa to the social realities of the migrant labor system, rapid urbanization, high levels of war with military mobilization, landlessness and poverty that have been exacerbated by an economic crises caused by "poor terms of trade, the contradictions of post-colonial economies which generate class disparities and burdensome debt service" since the mid 1970s.[7] These factors, he contends, are intricately intertwined with pervasive gender inequality and specific socially constructed meanings of gender and sex, creating a very complex situation regarding the epidemiology and, consequently, the prevention of AIDS. . . .

. . . Sex workers are continually blamed for the spread of the disease, with Eurocentric racist notions of cultural difference compounding the effect for Third World populations. Consequently, inappropriate methods of intervention have been introduced and sex workers burdened with having to take responsibility for the prevention and control of the disease. . . .

THE GLOBAL MOVEMENT

. . . Sex workers in Third World and other non-western countries have been busy, taking action, demonstrating against injustices they face, and demanding human, civil, political and social rights. Thus not only was an Ecuadorian association formed in 1982, but they held a sex workers' strike in 1988. In Brazil, a national prostitutes conference took place in 1987, giving rise to the establishment of the National Network of Prostitutes, Da Vida. In Montevideo, Uruguay, AMEPU inaugurated its childcare center and new headquarters after making its first public appearance in the annual May Day march in 1988. The Network of Sex Work Projects, founded in 1991, began to make links with sex workers' rights and health care projects in the Asian and Pacific region, slowly creating a truly international network that today includes at least forty different projects and groups in as many different countries around the world. 1992 witnessed the founding of the Venezuelan Association of Women for Welfare and Mutual Support (AMBAR), with the Chilean group Association for the Rights of Women, "Angela Lina" (APRODEM) and the Mexican Unión Única following suit in 1993. Two national congresses were held by the Ecuadorian sex workers' rights association in 1993 and 1994. The Maxi Linder Association in Suriname, the Indian Mahila Samanwaya Committee, and the Colombian Association of Women (Cormujer), were also all established by 1994. In the same year, around 400 prostitutes staged a protest against the closing of a brothel in Lima, Peru, with the slogan "We Want to Work, We Want to Work" and in Paramaribo, Suriname, sex workers made a first mass public appearance on AIDS Day, marching through the city with the banner "No Condom, No Pussy," drawing attention to their demands for safe sex. Also, 1994 witnessed the founding of The Sex Worker Education and Advocacy Taskforce (SWEAT) in South Africa. In 1996, groups in Japan and the Dominican Republic—Sex Workers! Encourage, Empower, Trust and Love Yourselves! (SWEETLY) and Movement of United Women (MODEMU)—were formed, and in the same year the Indian organization held its first congress in Calcutta, as well as organizing several protests and demonstrations against harassment and brutality. In 1997, with the help of AMBAR in Venezuela, the Association for Women in Solidarity (AMAS) became the first Nicaraguan group, comprised mainly of street workers. Other sex worker organizations have been reported to exist in Indonesia, Tasmania, Taiwan, and Turkey. . . .

While this list of organizations is not exhaustive, and keeps growing, we must keep in mind that each group has a history that pre-dates its formal founding date. Sex workers as individuals and in informal groups have battled against stigmas and discriminatory laws, denounced social and political injustices, and fought for their basic human rights in non-western settings for many years and there are often several years of organized activity before a formal organization appears on the map. . . .

. . . While clearly there is a need for autonomous organizing and consolidation of each group's position, within its own political, economic and cultural context, it is evident that sex workers do not view the struggle as isolated from that of other members of society. As prostitutes, migrant workers, transgenders, family breadwinners, single parents, HIV-positives, or teenagers, many recognize the multiple arenas in which their lives play, and consequently the multiple facets of social life that must be addressed. Gay, lesbian, bisexual, and transgender organizations, legal and human rights activists, health care workers, labor unions, and other sex industry workers are potential allies in the struggle to transform sexual labor into work that is associated with dignity, respect and decent working conditions. The coalitions that are taking shape through everyday resistances of sex workers also brings new meanings to the women's movement and feminism.

NOTES

1. Than-Dam Truong, *Sex, Money and Morality: Prostitution and Tourism in South East Asia* (London: Zed Books, 1990).

2. A position argued by Kathleen Barry, *The Prostitution of Sexuality: The Global Exploitation of Women* (New York: New York University Press, 1995).

3. Clayton M. Press, Jr. "Reputation and Respectability Considered: Hustling in a Tourist Setting." *Caribbean Issues* 4 (1978): 109–19; and Deborah Pruitt and Suzanne LaFont, "For Love and Money: Romance Tourism in Jamaica," *Annals of Tourism Research* 22, no. 2 (1995), 422–40.

4. Maggie Black, *In the Twilight Zone: Child Workers in the Hotel, Tourism and Catering Industry* (Geneva: International Labour Office, 1995).

5. International Labor Organization. Web site: www.ilo.org.

6. Ninotchka Rosca, "The Philippines' Shameful Export," *The Nation* (April 17, 1995).

7. Paul Farmer, "Women, Poverty, and AIDS," in *Women, Poverty and AIDS: Sex, Drugs and Structural Violence*, ed. Paul Farmer, Margaret Connors, and Janie Simmons (Monroe, Me: Common Courage Press, 1996), p. 71.

GETTING OFF ON FEMINISM

54

Jason Schultz

Minutes after my best friend told me he was getting married, I casually offered to throw a bachelor party in his honor. Even though such parties are notorious for their degradation of women, I didn't think this party would be much of a problem. Both the bride and groom considered themselves feminists, and I figured that most of the men attending would agree that sexism had no place in the celebration of this union. In fact, I thought the bachelor party would be a great opportunity to get a group of men together for a social event that didn't degenerate into the typical antiwomen, homophobic male-bonding thing. Still, ending one of the most sexist traditions in history—even for one night—was a lot tougher than I envisioned.

I have to admit that I'm not a *complete* iconoclast: I wanted to make the party a success by including at least some of the usual elements, such as good food and drink, great music, and cool things to do. At the same time, I was determined not to fall prey to traditional sexist party gimmicks such as prostitutes, strippers jumping out of cakes, or straight porn. But after nixing

From Rebecca Walker, ed., *To Be Real: Telling the Truth and Changing the Face of Feminism* (New York: Anchor Books, 1995), pp. 107–26. Reprinted by permission.

all the traditional lore, even *I* thought it sounded boring. What were we going to do except sit around and think about women?

"What about a belly dancer?" one of the ushers suggested when I confided my concerns to him. "That's not as bad as a stripper." I sighed. This was supposed to be an occasion for the groom and his male friends to get together, celebrate the upcoming marriage, and affirm their friendship and connection with each other as men. "What . . . does hiring a female sex worker have to do with any of that?" I shouted into the phone. I quickly regained my calm, but his suggestion still stung. We had to find some other way.

I wanted my party to be as "sexy" as the rest of them, but I had no idea how to do that in the absence of female sex workers. There was no powerful alternative image in our culture from which I could draw. I thought about renting some gay porn, or making it a cross-dressing party, but many of the guests were conservative, and I didn't want to scare anyone off. Besides, what would it say about a bunch of straight men if all we could do to be sexy was act queer for a night?

Over coffee on a Sunday morning, I asked some of the other guys what they thought was so "sexy" about having a stripper at a bachelor party.

"Well," David said, "it's just a gag. It's something kinda funny and sexy at the same time."

"Yeah," A.J. agreed. "It's not all that serious, but it's something special to do that makes the party cool."

"But *why* is it sexy and funny?" I asked. "Why can't we, as a bunch of guys, be sexy and funny ourselves?"

"'Cause it's easier to be a guy with other guys when there's a chick around. It gives you all something in common to relate to."

"Hmm. I think I know what you mean," I said. "When I see a stripper, I get turned on, but not in the same way I would if I was with a lover. It's more like going to a show or watching a flick together. It's enjoyable, stimulating, but it's not overwhelming or intimate in the same way that sex is. Having the stripper provides a common emotional context for us to feel turned on. But we don't have to do anything about it like we would if we were with a girlfriend, right?"

"Well, my girlfriend would kill me if she saw me checking out this stripper," Greg replied. "But because it's kind of a male-bonding thing, it's not as threatening to our relationship. It's not because it's the stripper over her, it's because it's just us guys hanging out. It doesn't go past that."

Others agreed. "Yeah. You get turned on, but not in a serious way. It makes you feel sexy and sexual, and you can enjoy feeling that way with your friends. Otherwise, a lot of times, just hanging out with the guys is pretty boring. Especially at a bachelor party. I mean, that's the whole point, isn't it—to celebrate the fact that we're bachelors, and he"—referring to Robert, the groom—"isn't!"

Through these conversations, I realized that having a female sex worker at the party would give the men permission to connect with one another without becoming vulnerable. When men discuss sex in terms of actions—who they "did," and how and where they did it—they can gain recognition and validation of their sexuality from other men without having to expose their *feelings* about sex.

"What other kinds of things make you feel sexy like the stripper does?" I asked several of the guys.

"Watching porn sometimes, or a sexy movie."

A. J. said, "Just getting a look from a girl at a club. I mean, she doesn't even have to talk to you, but you still feel sexy and you can still hang out with your friends."

Greg added, "Sometimes just knowing that my girlfriend thinks I'm sexy, and then talking about her with friends, makes me feel like I'm the man. Or I'll hear some other guy talk about his girlfriend in a way that reminds me of mine, and I'll still get that same feeling. But that doesn't happen very often, and usually only when talking with one other guy.

This gave me an idea. "I've noticed that same thing, both here and at school with my other close guy friends. Why doesn't it happen with a bunch of guys, say at a party?"

"I don't know. It's hard to share a lot of personal stuff with guys," said Adam, "especially about someone you're seeing, if you don't feel comfortable. Well, not comfortable, because I know most of the guys who'll be at the party, but it's more like I don't want them to hassle me, or I might say something that freaks them out."

"Or you're just used to guys talking shit about girls," someone else added. "Like at a party or hanging out together. They rag on them, or pick out who's the cutest or who wants to do who. That's not the same thing as really talking about what makes you feel sexy."

"Hmm," I said. "So it's kind of like if I were to say that I liked to be tied down to the bed, no one would take me seriously. You guys would probably crack up laughing, make a joke or two, but I'd never expect you to actually join in and talk about being tied up in a serious way. It certainly wouldn't feel "sexy," would it? At least not as much as the stripper."

"Exactly. You talking about being tied down here is fine, 'cause we're into the subject of sex on a serious kick and all. But at a party, people are bullshitting each other and gabbing, and horsing around. The last thing most of us want is to trip over someone's personal taste or start thinking someone's a little queer."

"You mean queer as in homosexual?" I asked.

"Well, not really, 'cause I think everyone here is straight. But more of queer in the sense of perverted or different. I mean, you grow up in high school thinking that all guys are basically the same. You all want the same thing from girls in the same way. And when someone like you says you like to be tied down, it's kinda weird—almost like a challenge. It makes me have to respond in a way that either shows me agreeing that I also like to be tied down or not. And if someone's a typical guy and he says that, it makes you think he's different—not the same guy you knew in high school. And if he's not the same guy, then it challenges you to relate to him on a different level."

"Yeah, I guess in some ways it's like relating to someone who's gay," Greg said. "He can be cool and all, and you can get along totally great. But there's this barrier that's hard to cross over. It kinda keeps you apart. And that's not what you want to feel toward your friends, especially at a party like this one, where you're all coming together to chill."

As the bachelor party approached, I found myself wondering whether my friends and I could "come together to chill"—and affirm our status as sexual straight men—without buying into homophobic or sexist expressions. At the same time, I was doing a lot of soul-searching on how we could challenge the dominant culture's vision of male heterosexuality, not only by deciding against having a stripper at our party, but also by examining and redefining our own relationships with women.

SEX AND THE SENSITIVE MAN

According to the prevailing cultural view, "desirable" hetero men are inherently dominant, aggressive, and, in many subtle and overt ways, abusive to women. To be sexy and powerful, straight men are expected to control and contrive a sexuality that reinforces their authority. Opposing these notions of power subjects a straight guy to being branded "sensitive," submissive, or passive—banished to the nether regions of excitement and pleasure, the unmasculine, asexual, "vanilla" purgatory of antieroticism. Just as hetero women are often forced to choose between the images of the virgin and the whore, modern straight men are caught in a cultural tug-of-war between the Marlboro Man and the Wimp.

So where does that leave straight men who want to reexamine what a man is and change it? Can a good man be sexy? Can a sexy man be good? What is good sex, egalitarian sex? More fundamentally, can feminist women and men coexist comfortably, even happily, within the same theoretical framework—or the same bedroom?

Relationships with men remain one of the most controversial topics among feminists today. Having sex, negotiating emotional dependency, and/or raising children force many hetero couples to balance their desire to be together with the oppressive dynamics of sexism. In few other movements are the oppressor group and the oppressed group so intimately linked.

But what about men who support feminism? Shouldn't it be okay for straight feminist women to have sex with them? Straight men aren't always oppressive in their sexuality, are they?

You may laugh at these questions, but they hold serious implications for straight feminist sex. I've seen many relationships between opposite-sex activists self-destruct because critical assumptions about power dynamics and desires were made in the mind, but not in the bed. I've even been told that straight male feminists can't get laid without (A) feeling guilty; (B) reinforcing patriarchy; or (C) maintaining complete passivity during sexual activity. Each of these three options represents common assumptions about the sexuality of straight men who support feminism. Choice A, "feeling guilty," reflects the belief that straight male desire inherently contradicts the goals of feminism and fails to contribute to the empowerment of women. It holds that any man who enjoys sex with a woman must be benefiting from sexist male privilege, not fighting against it. In other words, hot sex becomes a zero-sum game where if men gain, feminism loses.

Choice B represents the assumption that hetero male sex is inherently patriarchal. Beyond merely being of no help, as in Choice A, straight male sexuality is seen as part of the problem. Within this theory, one often hears statements such as "all heterosexual sex is rape." Even though these statements are usually taken out of context, the ideas behind them are problematic. In essence, they say that you can never have a male/female interaction that isn't caught up in oppressive dynamics. Men and women can never be together, especially in such a vulnerable exchange as sexuality, without being subject to the misdistribution of power in society.

The third choice, "maintaining complete passivity," attempts a logical answer to the above predicament. In order to come even close to achieving equality in heterosexuality (and still get laid), men must "give up" all their power through inactivity. A truly feminist man should take no aggressive or dominant position. He should, in fact, not act at all; he should merely lie back and allow the woman to subvert male supremacy through her complete control of the situation. In other words, for a man and a woman to share sexuality on a "level playing field," the man must remove all symptoms of his power through passivity, even though the causes of that inequality (including his penis!) still exist. . . .

Does it have to be this way? Must male heterosexuality always pose a threat to feminism? What about the sensitive guy? Wasn't that the male cry (whimper) of the nineties? Sorry, but all the media hype about sensitivity never added up to significant changes in behavior. Straight male sexuality still remains one of the most underchallenged areas of masculinity in America. . . .

Why did sensitivity fail? Were straight women, even feminists, lying to men about what they wanted? The answer is "yes" and "no." I don't think sensitivity was the culprit. I think the problem was men's passivity specifically, men's lack of assertiveness and power.

In much of our understanding, power is equated with oppression of white supremacists dominating people of color, men dominating and the rich dominating the poor underline the histories of many cultu__ and societies. But power need not always oppress others. One can, I believe, be powerful in a nonoppressive way.

In order to find this sort of alternative, we need to examine men's experience with power and sexuality further. Fortunately, queer men and women have given us a leg up on the process by reenergizing the debate about what is good sex and what is fair sex. Gay male culture has a long history of exploring nontraditional aspects of male sexuality, such as cross-dressing, bondage and dominance, and role playing. These dynamics force gay men to break out of a singular experience of male sexual desire and to examine the diversity within male sexuality in the absence of gender oppression. Though gay men's culture still struggles with issues such as the fetishizing of men of color and body fascism, it does invite greater exploration of diversity than straight male culture. Gay culture has broader and more inclusive attitudes about what is sexy and a conception of desire that accommodates many types of sex for many types of gay men. For straight men in our culture, there is such a rigid definition of "sexy" that it leaves us few options besides being oppressive, overbearing, or violent.

Part of the success that gay male culture enjoys in breaking out of monolithic notions of male sexuality lies in the acceptance it receives from its partners and peers. Camp, butch, leather, drag-queen culture is constantly affirming the powerful presence of alternative sexualities. Straight male culture, on the other hand, experiences a lack—a void of acceptance—whenever it tries to assert some image other than the sexist hetero male. Both publicly and in many cases privately, alternative straight male sexualities fail to compete for attention and acceptance among hetero men and women. . . .

We need to assert a new feminist sexuality for men, one that competes with the traditional paradigm but offers a more inviting notion of how hetero men can be sexual while tearing apart the oppressive and problematic ways in which so many of us have experienced sexuality in the past. We need to find new, strong values and ideas of male heterosexuality instead of passive identities that try to distance us from sexist men. We need to stop trying to avoid powerful straight sexuality and work to redefine what our power means and does. We need to find strength and desire outside of macho, antiwomen ways of being masculine. . . .

Gay men and lesbians have engaged in a cultural dialogue around sexuality over the last twenty-five years; straight women are becoming more and more vocal. But straight men have been almost completely silent. This silence,

I think, stems in large part from fear: our cultures tell us that being a "real" man means not being feminine, not being gay, and not being weak. They warn us that anyone who dares to stand up to these ideas becomes a sitting target to have his manhood shot down in flames.

BREAKING THE SILENCE

Not becoming a sitting target to have *my* manhood shot down was high on my mind when the evening of my best friend's bachelor party finally arrived. But I was determined not to be silent about how I felt about the party and about new visions for straight men within our society.

We decided to throw the party two nights before the wedding. We all gathered at my house, each of us bringing a present to add to the night's activities. After all the men had arrived, we began cooking dinner, breaking open beer and champagne, and catching up on where we had left off since we last saw each other.

During the evening, we continued to talk off and on about why we didn't have a stripper or prostitute for the party. After several rounds of margaritas and a few hands of poker, tension started to build around the direction I was pushing the conversation.

"So what don't you like about strippers?" David asked me.

This was an interesting question. I was surprised not only by the guts of David to ask it, but also by my own mixed feelings in coming up with an answer. "It's not that I don't like being excited, or turned on, per se," I responded. "In fact, to be honest, watching a female stripper is an exciting and erotic experience for me. But at the same time, it's a very uncomfortable one. I get a sense when I watch her that I'm participating in a misuse of pleasure, if that makes sense."

I looked around at my friends. I couldn't tell whether the confused looks on their faces were due to the alcohol, the poker game, or my answer, so I continued. "Ideally, I would love to sit back and enjoy watching someone express herself sexually through dance, seduction, flirtation—all the positive elements I associate with stripping," I said. "But at the same time, because so many strippers are poor and forced to perform in order to survive economically, I feel like the turn-on I get is false. I feel like I get off easy, sitting back as the man, paying for the show. No one ever expects me to get up on stage.

"And in that way, it's selling myself short sexually. It's not only saying very little about the sexual worth of the woman on stage, but the sexual worth of me as the viewer as well. By *only* being a viewer—just getting off as a member of the audience—the striptease becomes a very limiting thing, an imbalanced dynamic. If the purpose is for me to feel sexy and excited, but not to act on those feelings, I'd rather find a more honest and direct way to do it.

So personally, while I would enjoy watching a stripper on one level, the real issues of economics, the treatment of women, and the limitation of my own sexual personae push me to reject the whole stripper thing in favor of something else."

"But what else do you do to feel sexy?" A.J. asked.

"That's a tough question," I said. "Feeling sexy often depends on the way other people act toward you. For me, right now, you guys are a huge way for me to feel sexy. [Some of the men cringe.] I'm not saying that we have to challenge our sexual identities, although that's one way. But we can cut through a lot of this locker-room macho crap and start talking with each other about how we feel sexually, what we think, what we like, etc. Watching the stripper makes us feel sexy because we get turned on through the dynamic between her performance and our voyeurism. We can find that same erotic connection with each other by re-creating that context between us. In such a case, we're still heterosexual—we're no more having sex with each other than we are with the stripper. But we're not relying on the imbalanced dynamic of sex work to feel pleasure as straight men." . . .

The guys were silent for a few seconds, but soon afterwards, the ice seemed to break. . . . They agreed that, as heterosexual men, we should be able to share with each other what we find exciting and shouldn't *need* a female stripper to feel sexy. In some ways it may have been the desire to define their own sexuality that changed their minds; in others it may have been a traditionally masculine desire to reject any dependency on women. In any case, other men began to speak of their own experiences with pleasure and desire, and we continued to talk throughout the night, exploring the joys of hot sex, one-night stands, and even our preferences for certain brands of condoms. We discussed the ups and downs of monogamy versus "open" dating and the pains of long-distance relationships.

Some men continued to talk openly about their desire for straight pornography or women who fit the traditional stereotype of feminity. But others contradicted this, expressing their wish to move beyond that image of women in their lives. The wedding, which started out as the circumstance for our gathering, soon fell into the background of our thoughts as we focused away from institutional ideas of breeder sexuality and began to find common ground through our real-life experiences and feelings as straight men. In the end, we all toasted the groom, sharing stories, jokes, and parts of our lives that many of us had never told. Most importantly, we were able to express ourselves sexually without hiding who we were from each other.

Thinking back on the party, I realized that the hard part was figuring out what we all wanted and how to construct a different way of finding that experience. The other men there wanted it just as much as I did. The problem was that we had no ideas of what a different kind of bachelor party might look like. Merely eliminating the old ways of relating (i.e., the female sex workers) left a gap, an empty space which in many ways *felt* worse than the sexist connection that existed there before; we felt passive and powerless. Yet we found a new way of interacting—one that embraced new ideas and shared the risk of experiencing them.

Was the party sexy? Did we challenge the dominance of oppressive male sexuality? Not completely, but it was a start. I doubt anyone found my party as "sexy" as traditional ones might be, but the dialogue has to start somewhere. It's going to take a while to generate the language and collective tension to balance the cultural image of heterosexual male sexuality with true sexual diversity. Still, one of my friends from high school—who's generally on the conservative end of most issues—told me as he was leaving that of all the bachelor parties he had been to, this was by far the best one. "I had a great time," he said. "Even without the stripper." . . .

When it comes to sex, feminist straight men must become participants in the discourse about our own sexuality. We have to fight the oppressive images of men as biological breeders and leering animals. We must find ways in which to understand our diverse backgrounds, articulate desires that are not oppressive, and acknowledge the power we hold. We must take center stage when it comes to articulating our views in a powerful voice. I'm not trying to prescribe any particular form of sexuality or specify what straight men should want. But until we begin to generate our own demands and desires in an honest and equitable way for feminist straight women to hear, I don't think we can expect to be both good *and* sexy any time soon.

Violence

THE HARM THAT HAS NO NAME

55

Street Harassment, Embodiment,

and African American Women

Deirdre E. Davis

In her article *Street Harassment and the Informal Ghettoization of Women*, law professor Cynthia Grant Bowman explores street harassment as a harm and the necessity of legally recognizing street harassment's oppressive effects.[1] This chapter explores the idea that "[w]e cannot hope to understand the meaning of a person's experiences, including her experiences of oppression, without first thinking of her as embodied, and second thinking about the particular meanings assigned to that embodiment"[2] in the context of street harassment and African American women. Street harassment silences women. Similarly, racism has silenced and continues to silence African American women. Writing about street harassment and African American women legitimizes and recognizes the existence and importance of both.

THE MECHANICS OF STREET HARASSMENT

There are three ways to define and understand street harassment. Specifically, particular acts constitute street harassment. Normatively, the following characteristics identify particular acts of street harassment: the locale; the gender of and the relationship between the harasser and the target; the unacceptability of "thank you" as a response; and the reference to body parts. Systematically,

From *UCLA Women's Law Journal* 4 (Spring 1994): 133–78.

street harassment can be understood as an element of a larger system of sexual terrorism.

Specific Acts of Street Harassment

Cheris Kramarae, professor of speech communication and sociology, and Elizabeth Kissling describe street harassment as "verbal and nonverbal markers . . . wolf-whistles, leers, winks, grabs, pinches, catcalls and street remarks."[3] Specific remarks commonly include "Hey, pretty," "Hey, whore," "What ya doin" tonight?" "Look at them legs," "Wanna fuck?" "Are you working?" "Great legs," "Hey, cunt," "Smile," "Smile for me, baby," "Smile, bitch," "Come here, girl," and "I'll be back when you get a little older, baby." When these acts occur on a public street, street harassment takes place.[4]

Street Harassment's Role in Sexual Terrorism

Recognizing street harassment's role in sexual terrorism is crucial to understanding its potential to harm. Carole Sheffield defines sexual terrorism as men's systematic control and domination of women through actual and implied violence.[5] She views sexual terrorism as both the objective condition of women's existence and the theoretical framework that creates and maintains social orders.[6] Sexual terrorism and violence play crucial roles in the ongoing process of female subordination.[7] Violence is not one particular act, nor is it static; rather, it is a continuum of behavior in which street harassment must be placed if we are to understand the depth and pervasiveness of sexual terrorism.[8]

Street harassment "frightens women and reinforces fears of rape and other acts of terrorism."[9] Rape is generally viewed as a violent act of power occurring in a context limited to particular individuals or situations. However, rape may begin with an act of street harassment. Potential rapists can test the accessibility of a victim by making derogatory sexual comments to determine whether she can be intimidated.[10] As a result, street harassment plays a definite role in the objective condition of women fearing bodily harm on a day-to-day basis.

Women also experience the connection between rape and street harassment on a subjective level. Regardless of whether there is the possibility of actual rape, when women endure street harassment, they fear the possibility of rape.[11] That one of every eight adult women has been raped makes rape a constant possibility on a subjective level.[12] The sexual content of street harassment "reminds women of their vulnerability to violent attack in American urban centers, and to sexual violence in general" and intensifies the fear of the possibility of rape.[13] As a precursor to rape and an escalator of the fear of rape, street harassment entraps women in a sexually terroristic environment.

Within the framework of sexual terrorism, the specific acts and normative characteristics of street harassment identify the range of behavior that constitutes street harassment. Once it is realized that "street harassment is not a product of a sexually terroristic culture, but an active factor in creating such a culture," then the ability of an act of street harassment to cause harm becomes clearer.[14]

GENDERIZATION OF THE STREET: THE EFFECTS AND CONTEXT OF STREET HARASSMENT

Street harassment genderizes the street by distributing power in such a way that perpetuates male supremacy and female subordination.[15] Consequently, street harassment transforms the street into yet another forum that perpetuates and reinforces the gender hierarchy.

In order for the social effects of street harassment to occur, a preexisting context must exist that enables street harassment. Psychological oppression serves as the context that allows street harassment to genderize the street.

Despite street harassment's clear socially and psychologically oppressive effects, street harassment remains invisible as a harm. Because men do not suffer street harassment to the extent women do, street harassment is characterized as something other than harassment. Acts that are legally cognizable harms gain recognition "as an injury of the systematic abuse of power in hierarchies [when it is an exercise of] power men recognize."[16] Men view street harassment as innocuous, trivial, "boys will be boys" type of behavior and blame women for attaching negative meanings to their acts. Street harassment remains invisible because it is not a harm men suffer, and therefore it is not a harm men, or society as a whole, recognize.

While some women view street harassment as a trivial part of their everyday lives, they can still suffer extreme consequences. First, because street harassment has been trivialized, women do not talk about it and are thus silenced. This reinforces the invisibility of street harassment and its effects. Moreover, when a woman thinks about ending the silence, she may have a lot of doubt, given that street harassment—a pervasive part of everyday life—is so trivialized. Ignoring street harassment causes women to become complicit supporters of a system of sexual terrorism.[17] Finally, the failure to perceive street harassment as a harm causes women to "transform the pain into something else, such as, for example, punishment, or flattery, or transcendence, or unconscious pleasure."[18]

Giving a harm a name is the first step in making the harm visible. Given that "an injury uniquely sustained by a disempowered group will *lack a name*, a history, and in general *a linguistic reality*," it is crucial for the targets of street harassment to name the harm.[19] Naming is not a random or neutral process, but is biased. One need only look at workplace sexual harassment, date rape, domestic violence, and marital rape to understand the importance of naming a harm. While these harms have been a part of society for a long time, once they had a name, their visibility, both as acts and as harms, increased and led to the possibility of redress.

It is important for women to name the harms they suffer, because "[b]y taking the power of naming for themselves [and gaining cultural autonomy], women can determine with what bias street harassment will be encoded."[20] Discussing and naming street harassment are crucial steps toward erasing a constant source of women's pain and making street harassment visible as a harm.

INCLUDING AFRICAN AMERICAN WOMEN IN THE DISCOURSE ON STREET HARASSMENT: GENDERIZATION AND RACIALIZATION OF THE STREET

African American Women and Street Harassment: Recurring Images of Slavery

By refusing to acknowledge difference, street harassment discourse has excluded African American women's experiences. Many have argued that street harassment just "is," and that race, class, and sexual orientation are irrelevant: "In fact, women will sometimes comment that they think that women of all races, classes, and ages are subject to attacks from men—of all races, classes, and ages."[21] This statement relies on the idea that street harassment is primarily based on gender domination. Consequently, this statement implies that, because all women experience street harassment, it has no significance beyond its gender meaning: "[T]he women who do find street remarks disturbing, disgusting, or dangerous evidently hear them as more sexist than racist or classist."[22] The race of the harassers has also been disregarded: "You can say what you like about class and race. Those differences are real. But in this everyday scenario, any man on earth, no matter what his color or class is, has the power to make any woman who is exposed to him hate herself and her body."[23] This nuanced treatment of race ignores the relevant inquiry: the issue is not the act's independence from these differences, but the fact that the act occurs in spite of the differences. Abstracting the categories of identity limits understanding of the dynamics of street harassment.

All women are subjected to street harassment and, consequently, street harassment is a form of gender subordination. However, when African American women are subjected to street harassment, street harassment is, at the very least, genderized and racialized. This is not to say that street harassment has one meaning for African American women and a different meaning for all other women. Given the various histories of women of differing races and ethnicities, including white women, street harassment is both genderized and racialized for every woman; but the racial aspect is set in the particular historical context to which the particular woman belongs.

During and as a result of slavery, African American women have experienced the preexisting context that enables street harassment to be a factor in our sexually terroristic environment. Consequently, the psychological oppression of street harassment has a different—not a double—impact on African American women given their embodiment as indivisible beings. Street harassment forces African American women to realize that the ideologies of slavery still exist.

Although slavery has been legally eradicated, the racist ideology perpetuated during the slave era still exists with a different fate. While the "formal barriers and symbolic manifestations of subordination"[24] have disappeared, "[t]he white norm . . . has not disappeared; it has only been submerged in popular consciousness."[25] White men struggle to maintain their hierarchical position in a social structure that is constantly being challenged, questioned, and chiseled away.

Street harassment is a forum that allows white men, in the absence of slavery, to maintain the boundaries of their relationship with black women and to perpetuate the image of African American women as "blackwomen." The legal and cultural invisibility of street harassment gives white men a way of oppressing African American women that replaces the historical slave/master structure.

The Cult of True Womanhood: The White Woman as Paradigm

Street harassment oppresses women because it denies women an authentic choice of self and mandates conformance to gender stereotypes. Such oppression also formed the basis of the slave era's dominant gender ideology. In her work *Reconstructing Womanhood*, English and African American studies professor Hazel Carby explores women slaves' relationship to the predominant ideology of the "cult of true [white] womanhood."[26] Based on notions of motherhood and womanhood, the cardinal tenets of piety, purity (sexual and nonsexual), submissiveness, and domesticity characterized the cult of true womanhood.[27]

The ideology had two cultural effects: "[I]t was dominant, in the sense of being the most subscribed to convention governing female behavior, but it was also clearly recognizable as a dominating image, describing the parameters within which women were measured and declared to be, or not to be, women."[28] White men used the cult of true womanhood to establish the normative ideal for white women and to establish the boundaries outside of which slave women were placed.[29] Despite the opposing definitions of motherhood

and womanhood for white women and slave women, the definitions were dependent on one another.[30]

The cult of true womanhood also illustrates how stereotypes obscure
women's reality by focusing on men's interpretations. During slavery, slave
owners and buyers perceived some characteristics as negative in white women,
yet as positive, economic assets in slave women. For example, "[s]trength and
ability to bear fatigue, argued to be so distasteful a presence in a white woman,
were positive features to be emphasized in the promotion and selling of a
black female field hand at a slave auction."[31]

African American slave women internalized the underlying ideological
beliefs; "[b]y completely accepting the female role as defined by patriarchy,
enslaved blackwomen embraced and upheld an oppressive sexist social order
and became (along with their white sisters) both accomplices in the crimes
perpetuated against women and the victims of those crimes."[32] Consequently,
cultural domination has led African American women to believe that street harassment is an acceptable, natural part of everyday life, given the slave culture.

The Controlling Image of Jezebel: African American Woman as (White) Man's Temptress

During slavery, white men developed a racist ideology particular to slave
women, which consisted of four "interrelated, socially constructed controlling
images."[33] Created by white men to justify, maintain, and perpetuate the subordination of African American women, the most powerful of these controlling images is that of the female slave as a Jezebel. The Jezebel image—the
slave woman as "whore, sexually aggressive wet nurse," and "sexual temptress"—served two functions. First, it justified white men's sexual abuse of
slave women. Second, it justified the inapplicability of the cult of true womanhood to slave women—if a slave woman was seen as a sexual animal, then
she was not a real woman. The hypersexual Jezebel image dehumanized black
women and justified their exploitation in the fields. White men used the controlling image of Jezebel, in conjunction with other images, to create and
maintain the existing slave/master social and economic structure.

*Multiple Subordination: The Intersection of White Men's and African American
Men's Objectification of African American Women*   While "[r]acism has always
been a divisive force separating black men and white men, [it has been] sexism
[that] has been a force that unites the two groups."[34] Like white men, African
American men have been socialized to exercise their male status to oppress
women: "As Americans, they [African American men] had not been taught to
really believe that social equality was an inherent right all people possess, but
they had been socialized to believe that it is the nature of males to desire and
have access to power and privilege."[35] One of the entitlements of being a

member of the male gender is the ability to exercise dominance over women. During the slavery era, the African American man, "though obviously deprived of the social status that would enable him to protect and provide for himself and others, had a higher status than the black female slave based solely on his being male."[36] Again, the binary framework of racism and sexism ignores the fact that "[r]acism does not prevent black men from absorbing the same sexist socialization white men are inundated with."[37] As a result of the intersection of race and gender, while racism "cause[s] white men to make blackwomen targets,"[38] "sexism . . . causes all men to think they can verbally or physically assault women sexually with impunity."[39]

When African American Men Seek a Position of Whiteness: The Experience of Intraracial Harassment Given the pervasiveness of the controlling images of African American womanhood created and perpetuated by white men, African American men also view African American women "as nothing more than mammies, matriarchs, or Jezebels." As bell hooks points out, "[a]s sexist ideology has been accepted by black people, these negative myths and stereotypes have effectively transcended class and race boundaries and affected the way black women [are] perceived by members of their own race."[40]

African American men exercise the power implicit in sexism from a "position of whiteness." A position of whiteness consists of the "historically derived constellation of privileges associated with white [male] racial domination."[41] A person acting from a position of whiteness creates a racial hierarchy and produces and reinforces stereotypical images. This position of whiteness is not limited to white men, but can be seen as a position of authority attended by the privileges associated with authority.

Whereas white men assign and invoke the Jezebel image of African American women in order to maintain dominance, African American men use the same image in order to try to obtain that which white men have—the power to define the position of whiteness. As bell hooks has noted, "[t]heir [black men's] expressions of rage and anger are less a critique of the white male patriarchal social order and more a reaction against the fact that *they have not been allowed full participation in the power game.*"[42] Engaging in any form of oppression when you are a member of a marginalized group may make a person feel more powerful and less oppressed. Nevertheless, the established social order, though possibly capable of change, is still relative. Consequently, "men of color are not able to reap the material and social rewards for their participation in patriarchy. In fact they often suffer from blindly and passively acting out a myth of masculinity that is life-threatening. Sexist thinking blinds them to this reality. They become victims of the patriarchy."[43]

Some have argued that street harassment does not harm African American women because "[i]n many African American communities, men and women engage in sexually oriented banter in public."[44] Even if this rapping does exist between African American men and women, the speech rights are asymmetrical because "although many African American women respond assertively to rapping, they typically do not initiate it."[45] Furthermore, the fact that some African American women may engage in rapping does not negate the fact that

> [b]lack leaders, male and female, have been unwilling to acknowledge black male sexist oppression of black women because they do not want to acknowledge that racism is not the only oppressive force in our lives. *Nor do they wish to complicate efforts to resist racism by acknowledging that black men can be victimized by racism but at the same time act as sexist oppressors of black women.*[46]

Finally, characterizing street harassment as an African American cultural phenomenon ignores the intersection of race and gender by "overlook[ing] the way in which this sexual discourse reflects a differential power relationship between men and women."[47] This characterization also allows society to avoid examining "the different means by which these [African American cultural] practices are maintained and legitimated," thereby perpetuating the subordination of African American women.[48]

Multiple Consciousness and Street Harassment: Incorporating African American Women's Experience

Including African American women's experiences in the street harassment discourse enlightens women to their multiple consciousness and provides all women with a tool that enables them to cope with the social and psychologically oppressive effects of street harassment. By embracing the multiplicitous self, African American women, as the descendants of slaves, have learned how to handle the multiple forms of oppression, including gender and racial oppression. Recognizing a multiple consciousness helps women deconstruct and accept their experiences with street harassment. By embracing the multiple parts of self and moving away from a binary construction of "self," we also recognize the possibility for internal contradiction.

African American women have historically and consistently existed in a zone of dissonance. African American women, as society's "other," are disenfranchised and excluded from society in many ways. At the same time, African American women are "essential for [society's] survival because those individuals who stand at the margins of society clarify its boundaries. African American women, by not belonging, emphasize the significance of belonging."[49]

The multiple consciousness is yet another site of identity, where both the subjective self and the objective self coexist. Recognizing this site of identity will allow women to shift their energies from deconstructing and understanding women's response to street harassment, to eradicating street harassment.

A New Definition for the Effects of Street Harassment: Spirit-Murder

Including African American women's experiences in street harassment discourse provides a fuller understanding of how all women may experience street harassment. This inclusion also provides access to a broader term that may more fully reflect street harassment's invidious role in terrorizing all women. Law professor Patricia Williams states that "[a] fundamental part of ourselves and of our dignity is dependent upon the uncontrollable, powerful, external observers who constitute society."[50] Engaging in racist behavior, which is the overt expression of the internalized "system of formalized distortions of thought,"[51] leads to the "disregard for others whose lives qualitatively depend on our regard."[52] Williams terms this disregard "spirit-murder," a phenomenon that creates and perpetuates social structures that are defined by hate and fear, and give unexpressed feeling an outlet. While Williams's discussion of spirit-murder encompasses only race, Adrien Wing incorporates sexism into the concept of spirit-murder.[53] While spirit-murder is the cumulative effect, it is made up of micro aggressions, "[h]undreds, if not thousands of spirit injuries and assaults—some major, some minor—the cumulative effect of which is the slow death of the psyche, the soul and the persona."[54] In the context of street harassment, it is easy to understand spirit-murder as being subjected to many incidences of street harassment each day. To gain a fuller understanding of street harassment and its impact on African American women, it is necessary to place street harassment in the continuum of behavior that includes spirit-murder. Using these terms, one can understand the full extent to which all women are terrorized.

Redefining street harassment as spirit-murder benefits not only African American women, but all women. It allows for recognition of the "other" harms that women suffer, which may or may not be due to their gender. Furthermore, defining street harassment as "spirit-murder" both helps to give street harassment a name and identifies the harasser's wrong instead of focusing on the target. An objective definition of street harassment focuses on the harasser's actions as a form of intrusion instead of "looking to" or blaming the female target. When African American women's experiences are incorporated into the street harassment discourse, women are empowered with a terminology that fully portrays the depth of women's experiences with street harassment.

The first step in recognizing an act as a harm is the accurate construction of that act. Once street harassment is constructed and understood to be a harm that plays a role in the sexual terrorism that governs women's lives by genderizing the street in order to perpetuate female subordination, street harassment becomes visible as a harm. In order to address, deconstruct, and eradicate a harm, we must give the harm a name. Employing the term "street harassment" to describe the type of behavior is one step toward breaking the silence and misconceptions that surround street harassment.

Including African American women's experiences within street harassment and recognizing the different ways African American women experience street harassment due to their experiences with slavery broaden street harassment discourse. This inclusion provides both access to a term, multiple consciousness, that defines the site in which the harm occurs, and a broader meaning of the effects of street harassment, spirit-murder. These terms give women a fuller understanding of the harm street harassment causes.

Naming the harm gives all women the tools with which street harassment can be dismantled and gives them the strength to speak out, up, and loud in response to street harassment.

NOTES

1. Cynthia G. Bowman, *Street Harassment and the Informal Ghettoization of Women*, 106 Harv. L. Rev. 517 (1993).

2. Elizabeth V. Spelman, Inessential Woman: Problems of Exclusion in Feminist Thought 129–30 (1988).

3. Elizabeth A. Kissling & Cheris Kramarae, *Stranger Compliments: The Interpretation of Street Remarks*, 14 Women's Stud. Comm. 75, 75–76 (1991) (reporting results from a computer notes file discussion on street harassment).

4. I focus on the street/sidewalk as the situs. I exclude places like buses, bus stations, taxis, stores, and other public accommodations to highlight the arbitrariness of street harassment. Although I choose to focus on the street, harassment can and does occur in other places.

5. Carole J. Sheffield, *Sexual Terrorism: The Social Control of Women, in* Analyzing Gender 171, 171 (Beth B. Hess & Myra Marx Ferree eds., 1987).

6. *Id.* at 172. The manifestations of sexual terrorism include wife-battering, sexual harassment in the workplace, incest, sexual slavery, prostitution, and rape. *Id.* at 171.

7. *Id.* at 172.

8. There is an "unstated relationship [between] compliments, verbal hostility and physical attack." Kissling & Kramarae, *supra* note 3, at 78; *see also* Cristina Del Sesto, *Our Mean Streets: D.C.'s Women Walk Through Verbal Combat Zones*, Wash. Post, Mar. 18, 1990, at B1 ("I'm afraid everyday that a verbal assault is going to turn into a physical one.").

9. Elizabeth A. Kissling, *Street Harassment: The Language of Sexual Terrorism*, 2 Discourse & Soc'y 451, 456 (1991).

10. *See* Bowman, *supra* note 1, at 536.

11. Kissling & Kramarae, *supra* note 3, at 84–85.

12. Bowman, *supra* note 1, at 536 n.86 (quoting *Study: Rapes Far Underestimated*, Chi. Trib., Apr. 24, 2991 § 1, at 3).

13. Kissling & Kramarae, *supra* note 3, at 76.

14. *See* Kissling, *supra* note 9, at 456.

15. *See* Catharine A. MacKinnon, *Difference and Dominance: On Discrimination, in* Feminism Unmodified: Discourses on Life and Law 32, 40 (1987).

16. Catharine A. MacKinnon, *Sexual Harassment: Its First Decade in Court, in* Feminism Unmodified, *supra* note 15, at 103, 107.

17. Kissling, *supra* note 9, at 456.

18. Robin West, *The Difference in Women's Hedonic Lives: A Phenomenological Critique of Feminist Legal Theory*, 3 Wis. Women's L.J. 81, 85 (1987). This is also something the street harasser does.

19. *Id.* (emphasis added).

20. *See* Kissling, *supra* note 9, at 457.

21. Kissling & Kramarae, *supra* note 3, at 90.

22. *Id.* It would be helpful to know the race, sexual orientation, and economic status of the women commentators.

23. Meredith Tax, *Woman and Her Mind: The Story of Everyday Life, in* Radical Feminism 23, 28 (Anne Koedt et al., eds., 1973). While this may be true, there is still a power differential among men based on race.

24. Kimberlé W. Crenshaw, *Race, Reform, and Retrenchment: Transformation and Legitimation in Antidiscrimination Law*, 101 Harv. L. Rev. 1331, 1378 (1988).

25. *Id.* at 1379.

26. Hazel V. Carby, Reconstructing Womanhood: The Emergence of the Afro-American Woman Novelist 23 (1987).

27. *Id.* at 20–39.

28. *Id.*

29. Hazel V. Carby, Lecture in Class on Black Women Writers at Wesleyan University (Sept. 9, 1987).

30. For example, although all women "had" to reproduce, white women were responsible for producing heirs and slave women were responsible for producing property for the heirs to inherit. Carby, *supra* note 29.

31. Carby, *supra* note 26, at 25.

32. bell hooks, Ain't I a Woman? Black Women and Feminism 49 (1981).

33. Patricia Hill Collins, Black Feminist Thought: Knowledge, Consciousness, and the Politics of Empowerment 71 (1991). The other controlling images of African American womanhood are the "mammy," the "matriarch," and the "welfare mother." *See Id.* at 71–77.

34. bell hooks, Feminist Theory: From Margin to Center 99 (1984).

35. *Id.* at 98.

36. *Id.* at 88–89.

37. *Id.* at 101–102.

38. *Id.* at 68.

39. *Id.* at 68–69 (emphasis added).

40. hooks, *supra* note 32, at 70.

41. Neil Gotanda, *"Race-ing" Racial Non-Recognition, and Racial Stratification: Re-*

Reading Judge Joyce Karlin's Sentencing Colloquy, in People v. Soon Ja Du 23 (Mar. 12, 1993) (unpublished manuscript).

42. hooks, *supra* note 32, at 94 (emphasis added).

43. bell hooks, *Reflections on Race and Sex, in* Yearning Race, Gender and Cultural Politics 57, 63 (1990).

44. Bowman, *supra* note 1, at 532. This type of banter has been referred to as "rapping."

45. *Id.* at 532.

46. hooks, *supra* note 32, at 88 (emphasis added).

47. Kimberlé W. Crenshaw, *Whose Story Is It Anyway? Feminist and Antiracist Appropriations of Anita Hill, in* Race-ing Justice, Engendering Power: Essays on Anita Hill, Clarence Thomas, and the Construction of Social Reality 402, 429 (Toni Morrison ed., 1992).

48. *Id.* at 431.

49. Collins, *supra* note 33, at 68.

50. Patricia J. Williams, *Spirit-Murdering the Messenger: The Discourse of Fingerpointing as the Law's Response to Racism,* 42 U. Miami L. Rev. 127, 151 (1987) (also chapter 28 in this volume).

51. *Id.*

52. *Id.*

53. Adrien K. Wing, *Brief Reflections toward a Multiplicative Theory and Praxis of Being,* 6 Berkeley Women's L.J. 181, 186 (1990–91) (also chapter 3 in this volume).

54. *Id.*

MORE POWER THAN WE WANT

Masculine Sexuality and Violence

56

Bruce Kokopeli and George Lakey

Masculine sexuality involves the oppression of women, competition among men, and homophobia (fear of homosexuality). Patriarchy, the systematic domination of women by men through unequal opportunities, rewards, punishments, and the internalization of unequal expectations through sex role differentiation, is the institution which organizes these behaviors. Patriarchy

From *Off Their Backs . . . and on Our Own Two Feet* (Philadelphia: New Society Publishers, 1983), pp. 17–24. Reprinted by permission.

is men having more power, both personally and politically, than women of the same rank. This imbalance of power is the core of patriarchy, but definitely not the extent of it.

Sex inequality cannot be routinely enforced through open violence or even blatant discriminatory agreements—patriarchy also needs its values accepted in the minds of people. . . . Patriarchy assigns a list of human characteristics according to gender: women should be nurturant, gentle, in touch with their feelings, etc.; men should be productive, competitive, super-rational, etc. Occupations are valued according to these gender-linked characteristics, so social work, teaching, housework, and nursing are of lower status than business executive, judge, or professional football player.

When men do enter "feminine" professions, they disproportionately rise to the top and become chefs, principals of schools, directors of ballet, and teachers of social work. A man is somewhat excused from his sex role deviation if he at least dominates within the deviation. Domination, after all, is what patriarchy is all about.

Access to powerful positions by women (i.e., those positions formerly limited to men) is contingent on the women adopting some masculine characteristics, such as competitiveness. They feel pressure to give up qualities assigned to females (such as gentleness) because those qualities are considered inherently weak by patriarchal culture. . . .

Patriarchy also shapes men's sexuality so it expresses the theme of domination. Notice the masculine preoccupation with size. The size of a man's body has a lot to say about his clout or his vulnerability, as any junior high boy can tell you. Many of these schoolyard fights are settled by who is bigger than whom, and we experience in our adult lives the echoes of intimidation and deference produced by our habitual "sizing up" of the situation.

Penis size is part of this masculine preoccupation, this time directed toward women. Men want to have large penises because size equals power, the ability to make a woman "really feel it." The imagery of violence is close to the surface here, since women find penis size irrelevant to sexual genital pleasure. "Fucking" is a highly ambiguous word, meaning both intercourse and exploitation/assault.

It is this confusion that we need to untangle and understand. Patriarchy tells men that their need for love and respect can only be met by being masculine, powerful, and ultimately violent. As men come to accept this, their sexuality begins to reflect it. Violence and sexuality combine to support masculinity as a character ideal. To love a woman is to have power over her and to treat her violently if need be. The Beatles' song "Happiness Is a Warm Gun" is but one example of how sexuality gets confused with violence and power. We know one man who was discussing another man who seemed to be highly fertile—he had made several women pregnant. "That guy," he said, "doesn't shoot any blanks."

Rape is the end logic of masculine sexuality. Rape is not so much a sexual act as an act of violence expressed in a sexual way. The rapist's mind-set—that violence and sexuality *can* go together—is actually a product of patriarchal conditioning, for most of us men understand the same, however abhorrent rape may be to us personally.

In war, rape is astonishingly prevalent even among men who "back home" would not do it. In the following description by a marine sergeant who witnessed a gang rape in Vietnam, notice that nearly all the nine-man squad participated:

> They were supposed to go after what they called a Viet Cong whore. They went into her village and instead of capturing her, they raped her—every man raped her. As a matter of fact, one man said to me later that it was the first time he had ever made love to a woman with his boots on. The man who led the platoon, or the squad, was actually a private. The squad leader was a sergeant but he was a useless person and he let the private take over his squad. Later he said he took no part in the raid. It was against his morals. So instead of telling his squad not to do it, because they wouldn't listen to him anyway, the sergeant went into another side of the village and just sat and stared bleakly at the ground, feeling sorry for himself. But at any rate, they raped the girl, and then, the last man to make love to her, shot her in the head. [Vietnam Veterans Against the War, statement by Michael McClusker in *The Winter Soldier Investigation: An Inquiry Into American War Crimes.*]

Psychologist James Prescott adds to this account:

> What is it in the American psyche that permits the use of the word "love" to describe rape? And where the act of love is completed with a bullet in the head! [*Bulletin of the Atomic Scientists,* November 1975, p. 17.]

MASCULINITY AGAINST MEN: THE MILITARIZATION OF EVERYDAY LIFE

Patriarchy benefits men by giving us a class of people (women) to dominate and exploit. Patriarchy also oppresses men, by setting us at odds with each other and shrinking our life space.

The pressure to win starts early and never stops. Working-class gangs fight over turf; rich people's sons are pushed to compete on the sports field. British military officers, it is said, learned to win on the playing fields of Eton.

Competition is conflict held within a framework of rules. When the stakes are really high, the rules may not be obeyed; fighting breaks out. We men mostly relate through competition, but we know what is waiting in the wings. John Wayne is not a cultural hero by accident.

Men compete with each other for status as masculine males. Because masculinity equals power, this means we are competing for power. The ultimate proof of power/masculinity is violence. A man may fail to "measure up" to the macho stereotype in important ways, but if he can fight successfully with the person who challenges him on his deviance, he is still all right. . . .

The close relationship between violence and masculinity does not need much demonstration. War used to be justified partly because it promoted "manly virtue" in a nation. Those millions of people in the woods hunting deer, in the National Rifle Association, and cheering on the bloodiest hockey teams are overwhelmingly men.

The world situation is so much defined by patriarchy that what we see in the wars of today is competition between various patriarchal ruling classes and governments breaking into open conflict. Violence is the accepted masculine form of conflict resolution. Women at this time are not powerful enough in the world situation for us to see mass overt violence being waged on them. But the violence is in fact there; it is hidden through its legitimization by the state and by culture.

In everyday middle-class life, open violence between men is of course rare. The defining characteristics of masculinity, however, are only a few steps removed from violence. Wealth, productivity, or rank in the firm or institution translate into power—the capacity (whether or not exercised) to dominate. . . .

Patriarchy teaches us at very deep levels that we can never be safe with other men (or perhaps with anyone!), for the guard must be kept up lest our vulnerability be exposed and we be taken advantage of. At a recent Quaker conference in Philadelphia, a discussion group considered the value of personal sharing and openness in the Quaker Meeting. In almost every case the women advocated more sharing and the men opposed it. Dividing by gender on that issue was predictable; men are conditioned by our life experience of masculinity to distrust settings where personal exposure will happen, especially if men are present. Most men find emotional intimacy possible only with women; many with only one woman; some men cannot be emotionally intimate with anyone.

Patriarchy creates a character ideal—we call it masculinity—and measures everyone against it. Many men as well as women fail the test and even men who are passing the test today are carrying a heavy load of anxiety about tomorrow. Because masculinity is a form of domination, no one can really rest secure. The striving goes on forever unless you are actually willing to give up and find a more secure basis for identity.

MASCULINITY AGAINST GAY MEN: PATRIARCHY FIGHTS A REAR GUARD ACTION

Homophobia is the measure of masculinity. The degree to which a man is thought to have gay feelings is the degree of his unmanliness. Because patriarchy presents sexuality as men over women (part of the general dominance theme), men are conditioned to have only that in mind as a model of sexual expression. Sex with another man must mean being dominated, which is very scary. A nonpatriarchal model of sexual expression as the mutuality of equals doesn't seem possible; the transfer of the heterosexual model to same-sex relations can at best be "queer," at worst, "perverted." . . .

Notice the importance of violence in defending yourself against the charge of being a "pansy." Referring to your income or academic degrees or size of your car is no defense against such a charge. Only fighting will re-establish your respect as a masculine male. Because "gay" appears to mean "powerless," one needs to go to the masculine source of power—violence—for adequate defense. . . .

Different kinds of homosexual behavior bring out different amounts of hostility, curiously enough. That fact gives us further clues to violence and female oppression. In prisons, for example, men can be respected if they fuck other men, but not if they are themselves fucked. (We use the word "fucked" intentionally for its ambiguity.) Often prison rapes are done by men who identify as heterosexual; one hole substitutes for another in this scene, for sex is in either case an expression of domination for the masculine mystique.

But for a man to be entered sexually, or to use effeminate gestures and actions, is to invite attack in prison and hostility outside. Effeminate gay men are at the bottom of the totem pole because they are *most like women*, which is nothing less than treachery to the Masculine Cause. Even many gay men shudder at drag queens and vigilantly guard against certain mannerisms because they, too, have internalized the masculinist dread of effeminacy. . . .

A ticket of admission to masculinity, then, is sex with women, and bisexuals can at least get that ticket even if they deviate through having gay feelings as well. This may be why bisexuality is not feared as much as exclusive gayness among men. Exclusively gay men let down the Masculine Cause in a very important way—those gays do not participate in the control of women through sexuality. Control through sexuality matters because it is flexible; it usually is mixed with love and dependency so that it becomes quite subtle. (Women often testify to years of confusion and only the faintest uneasiness at their submissive role in traditional heterosexual relationships and the role sex plays in that.)

Now we better understand why women are in general so much more supportive of gay men than nongay men are. Part of it of course is that heterosexual men are often paralyzed by fear. Never very trusting, such men find gayness one more reason to keep up the defenses. But heterosexual women are drawn to active support for the struggles of gay men because there is a common enemy—patriarchy and its definition of sexuality as domination. Both heterosexual women and gay men have experienced first hand the violence of sexism; we all have experienced its less open forms such as put-downs and discrimination, and we all fear its open forms such as rape and assault.

Patriarchy, which links characteristics (gentleness, aggressiveness, etc.) to gender, shapes sexuality as well, in such a way as to maintain male power. The Masculine Cause draws strength from homophobia and resorts habitually to violence in its battles on the field of sexual politics. It provides psychological support for the military state and is in turn stimulated by it.

JUST WALK ON BY   57
A Black Man Ponders His Power
to Alter Public Space

Brent Staples

My first victim was a woman—white, well dressed, probably in her early twenties. I came upon her late one evening on a deserted street in Hyde Park, a relatively affluent neighborhood in an otherwise mean, impoverished section of Chicago. As I swung onto the avenue behind her, there seemed to be a discreet, uninflammatory distance between us. Not so. She cast back a worried glance. To her, the youngish black man—a broad six feet two inches with a beard and billowing hair, both hands shoved into the pockets of a bulky military jacket—seemed menacingly close. After a few more quick glimpses, she picked up her pace and was soon running in earnest. Within seconds she disappeared into a cross street.

That was more than a decade ago. I was 22 years old, a graduate student newly arrived at the University of Chicago. It was in the echo of that terrified woman's footfalls that I first began to know the unwieldy inheritance I'd come into—the ability to alter public space in ugly ways. It was clear that she thought herself the quarry of a mugger, a rapist, or worse. Suffering a bout of insomnia, however, I was stalking sleep, not defenseless wayfarers. As a softy who is scarcely able to take a knife to a raw chicken—let alone hold it to a person's throat—I was surprised, embarrassed, and dismayed all at once. Her flight made me feel like an accomplice in tyranny. It also made it clear that I was indistinguishable from the muggers who occasionally seeped into the area from the surrounding ghetto. That first encounter, and those that followed,

From *Ms.* (September 1986). Reprinted by permission of the author.

signified that a vast unnerving gulf lay between nighttime pedestrians—particularly women—and me. And I soon gathered that being perceived as dangerous is a hazard in itself. I only needed to turn a corner into a dicey situation, or crowd some frightened, armed person in a foyer somewhere, or make an errant move after being pulled over by a policeman. Where fear and weapons meet—and they often do in urban America—there is always the possibility of death.

In that first year, my first away from my hometown, I was to become thoroughly familiar with the language of fear. At dark, shadowy intersections in Chicago, I could cross in front of a car stopped at a traffic light and elicit the *thunk, thunk, thunk, thunk* of the driver—black, white, male, female—hammering down the door locks. On less-traveled streets after dark, I grew accustomed to but never comfortable with people who crossed to the other side of the street rather than pass me. Then there were the standard unpleasantries with police, doormen, bouncers, cab drivers, and others whose business it is to screen out troublesome individuals *before* there is any nastiness.

I moved to New York nearly two years ago and I have remained an avid night walker. In central Manhattan, the near-constant crowd cover minimizes tense one-on-one street encounters. Elsewhere—visiting friends in SoHo, where sidewalks are narrow and tightly spaced buildings shut out the sky—things can get very taut indeed.

Black men have a firm place in New York mugging literature. Norman Podhoretz in his famed (or infamous) 1963 essay, "My Negro Problem—And Ours," recalls growing up in terror of black males; they "were tougher than we were, more ruthless," he writes—and as an adult on the Upper West Side of Manhattan, he continues, he cannot constrain his nervousness when he meets black men on certain streets. Similarly, a decade later, the essayist and novelist Edward Hoagland extols a New York where once "Negro bitterness bore down mainly on other Negroes." Where some see mere panhandlers, Hoagland sees "a mugger who is clearly screwing up his nerve to do more than just *ask* for money." But Hoagland has "the New Yorker's quick-hunch posture for broken-field maneuvering," and the bad guy swerves away.

I often witness that "hunch posture," from women after dark on the warrenlike streets of Brooklyn where I live. They seem to set their faces on neutral and, with their purse straps strung across their chests bandolier style, they forge ahead as though bracing themselves against being tackled. I understand, of course, that the danger they perceive is not a hallucination. Women are particularly vulnerable to street violence, and young black males are drastically overrepresented among the perpetrators of that violence. Yet these truths are no solace against the kind of alienation that comes of being ever the suspect, against being set apart, a fearsome entity with whom pedestrians avoid making eye contact.

It is not altogether clear to me how I reached the ripe old age of 22 without being conscious of the lethality nighttime pedestrians attributed to me.

Perhaps it was because in Chester, Pennsylvania, the small, angry industrial town where I came of age in the 1960s, I was scarcely noticeable against a backdrop of gang warfare, street knifings, and murders. I grew up one of the good boys, had perhaps a half-dozen fist fights. In retrospect, my shyness of combat has clear sources.

Many things go into the making of a young thug. One of those things is the consummation of the male romance with the power to intimidate. An infant discovers that random flailings send the baby bottle flying out of the crib and crashing to the floor. Delighted, the joyful babe repeats those motions again and again, seeking to duplicate the feat. Just so, I recall the points at which some of my boyhood friends were finally seduced by the perception of themselves as tough guys. When a mark cowered and surrendered his money without resistance, myth and reality merged—and paid off. It is, after all, only manly to embrace the power to frighten and intimidate. We, as men, are not supposed to give an inch of our lane on the highway; we are to seize the fighter's edge in work and in play and even in love; we are to be valiant in the face of hostile forces.

Unfortunately, poor and powerless young men seem to take all this nonsense literally. As a boy, I saw countless tough guys locked away; I have since buried several, too. They were babies, really—a teenage cousin, a brother of 22, a childhood friend in his mid-twenties—all gone down in episodes of bravado played out in the streets. I came to doubt the virtues of intimidation early on. I chose, perhaps even unconsciously, to remain a shadow—timid, but a survivor.

The fearsomeness mistakenly attributed to me in public places often has a perilous flavor. The most frightening of these confusions occurred in the late 1970s and early 1980s when I worked as a journalist in Chicago. One day, rushing into the office of a magazine I was writing for with a deadline story in hand, I was mistaken for a burglar. The office manager called security and, with an ad hoc posse, pursued me through the labyrinthine halls, nearly to my editor's door. I had no way of proving who I was. I could only move briskly toward the company of someone who knew me.

Another time I was on assignment for a local paper and killing time before an interview. I entered a jewelry store on the city's affluent Near North Side. The proprietor excused herself and returned with an enormous red Doberman pinscher straining at the end of a leash. She stood, the dog extended toward me, silent to my questions, her eyes bulging nearly out of her head. I took a cursory look around, nodded, and bade her good night. Relatively speaking, however, I never fared as badly as another black male journalist. He went to nearby Waukegan, Illinois, a couple of summers ago to work on a story about a murderer who was born there. Mistaking the reporter for the killer, police hauled him from his car at gunpoint and but for his press credentials would have tried to book him. Such episodes are not uncommon. Black men trade tales like this all the time.

In "My Negro Problem—And Ours," Podhoretz writes that the hatred he feels for blacks makes itself known to him through a variety of avenues— one being his discomfort with that "special brand of paranoid touchiness" to which he says blacks are prone. No doubt he is speaking here of black men. In time, I learned to smother the rage I felt at so often being taken for a criminal. Not to do so would surely have led to madness—via that special "paranoid touchiness" that so annoyed Podhoretz at the time he wrote the essay.

I began to take precautions to make myself less threatening. I move about with care, particularly late in the evening. I give a wide berth to nervous people on the subway platforms during the wee hours, particularly when I have exchanged business clothes for jeans. If I happen to be entering a building behind some people who appear skittish, I may walk by, letting them clear the lobby before I return, so as not to seem to be following them. I have been calm and extremely congenial on those rare occasions when I've been pulled over by the police.

And on late-evening constitutionals along streets less traveled by, I employ what has proved to be an excellent tension-reducing measure: I whistle melodies from Beethoven and Vivaldi and the more popular classical composers. Even steely New Yorkers hunching toward nighttime destinations seem to relax, and occasionally they even join in the tune. Virtually everybody seems to sense that a mugger wouldn't be warbling bright, sunny selections from Vivaldi's *Four Seasons*. It is my equivalent of the cowbell that hikers wear when they know they are in bear country.

WHERE RACE AND GENDER MEET

Racism, Hate Crimes, and Pornography

58

Helen Zia

There is a specific area where racism, hate crimes, and pornography intersect, and where current civil rights law fails: racially motivated, gender-based

From Laura Lederer and Richard Delgado, eds., *The Price We Pay: The Case against Racist Speech, Hate Propaganda and Pornography*. Copyright © 1995 by Laura Lederer and Richard Delgado. Reprinted by permission of Farrar, Strauss and Giroux, LLC.

crimes against women of color. This area of bias-motivated sexual assault has been called "ethnorape"; I refer to it as "hate rape."

I started looking into this issue after years of organizing against hate killings of Asian Americans. After a while, I noticed that all the cases I could name concerned male victims. I wondered why. Perhaps it was because Asian-American men came into contact with perpetrator types more often or because they are more hated and therefore more often attacked by racists. But the subordination and vulnerability of Asian-American women, who are thought to be sexually exotic, subservient, and passive, argued against that interpretation. So where were the Asian-American women hate-crime victims?

Once I began looking, I found them, in random news clippings, in footnotes in books, through word of mouth. Let me share with you some examples I unearthed of bias-motivated attacks and sexual assaults:

- In February 1984, Ly Yung Cheung, a nineteen-year-old Chinese woman who was seven months pregnant, was pushed in front of a New York City subway train and decapitated. Her attacker, a white male high school teacher, claimed he suffered from "a phobia of Asian people" and was overcome with the urge to kill this woman. He successfully pleaded insanity. If this case had been investigated as a hate crime, there might have been more information about his so-called phobia and whether it was part of a pattern of racism. But it was not.
- On December 7, 1984, fifty-two-year-old Japanese American Helen Fukui disappeared in Denver, Colorado. Her decomposed body was found weeks later. Her disappearance on Pearl Harbor Day, when anti-Asian speech and incidents increase dramatically, was considered significant in the community. But the case was not investigated as a hate crime and no suspects were ever apprehended.
- In 1985 an eight-year-old Chinese girl named Jean Har-Kaw Fewel was found raped and lynched in Chapel Hill, North Carolina—two months after *Penthouse* featured pictures of Asian women in various poses of bondage and torture, including hanging bound from trees. Were epithets or pornography used during the attack? No one knows—her rape and killing were not investigated as a possible hate crime.
- Recently a serial rapist was convicted of kidnapping and raping a Japanese exchange student in Oregon. He had also assaulted a Japanese woman in Arizona, and another in San Francisco. He was sentenced to jail for these crimes, but they were never pursued as hate crimes, even though California has a hate statute. Was hate speech or race-specific pornography used? No one knows.
- At Ohio State University, two Asian women were gang raped by fraternity brothers in two separate incidents. One of the rapes was part of a

racially targeted game called the "Ethnic Sex Challenge," in which the fraternity men followed an ethnic checklist indicating what kind of women to gang rape. Because the women feared humiliation and ostracism by their communities, neither reported the rapes. However, campus officials found out about the attacks, but did not take them up as hate crimes, or as anything else.

All of these incidents could have been investigated and prosecuted either as state hate crimes or as federal civil rights cases. But they were not. To have done so would have required one of two things: awareness and interest on the part of police investigators and prosecutors—who generally have a poor track record on race and gender issues—or awareness and support for civil rights charges by the Asian-American community—which is generally lacking on issues surrounding women, gender, sex, and sexual assault. The result is a double-silencing effect on the assaults and deaths of these women, who become invisible because of their gender and their race.

Although my research centers on hate crimes and Asian women, this silence and this failure to provide equal protection have parallels in all of the other classes protected by federal civil rights and hate statutes. That is, all other communities of color have a similar prosecution rate for hate crimes against the women in their communities—namely, zero. This dismal record is almost as bad in lesbian and gay antiviolence projects: the vast preponderance of hate crimes reported, tracked, and prosecuted concern gay men—very few concern lesbians. So where are all the women?

The answer to this question lies in the way our justice system was designed, and the way women are mere shadows in the existing civil rights framework. But in spite of this history, federal and state law do offer legal avenues for women to be heard. Federal civil rights prosecutions, for example, can be excellent platforms for high-visibility community education on the harmful impact of hate speech and behavior. When on June 19, 1982, two white auto workers in Detroit screamed racial epithets at Chinese-American Vincent Chin and said, "It's because of you motherfuckers that we're out of work," a public furor followed, raising the level of national discourse on what constitutes racism toward Asian Americans. Constitutional law professors and members of the American Civil Liberties Union and the National Lawyers Guild had acted as if Asian Americans were not covered by civil rights law. Asian Americans emphatically corrected that misconception.

Hate crimes remedies can be used to force the criminal justice bureaucracy to adopt new attitudes. Patrick Purdy went to an elementary school in Stockton, California, in which 85 percent of the students came from Southeast Asia. When he selected that school as the place to open fire with his automatic weapon and killed five eight-year-olds and wounded thirty other children, the police and the media did not think it was a bias-motivated crime. Their denial reminds me of the response by the Montreal officials to the anti-feminist killings of fourteen women students there. But an outraged Asian-American community forced a state investigation into the Purdy incident and uncovered hate literature in the killer's effects. As a result, the community was validated, and, in addition, the criminal justice system and the media acquired a new level of understanding.

Imagine if a federal civil rights investigation had been launched in the case of the African-American student at St. John's University who was raped and sodomized by white members of the school lacrosse team, who were later acquitted. Investigators could have raised issues of those white men's attitudes toward the victim as a black woman, found out whether hate speech or race-specific pornography was present, investigated the overall racial climate on campus, and brought all of the silenced aspects of the incident to the public eye. Community discourse could have been raised to a high level.

Making these investigations happen will not be an easy road. Hate-crime efforts are generally expended on blatant cases, with high community consensus, not ones that bring up hard issues like gender-based violence. Yet these intersections of race and gender hatred are the very issues we must give voice to.

There is a serious difficulty with pushing for use of federal and state hate remedies. Some state statutes have been used against men of color: specifically, on behalf of white rape victims against African-American men. We know that the system, if left unchecked, will try to use antihate laws to enforce unequal justice. On the other hand, state hate statutes could be used to prosecute men of color who are believed to have assaulted women of color of another race—interminority assaults are increasing. Also, if violence against women generally were made into a hate crime, women of color could seek prosecutions against men in their own community for their gender-based violence—even if this would make it harder to win the support of men in communities of color, and of women in those communities who would not want to be accused of dividing the community

But at least within the Asian-American antiviolence community, this discourse is taking place now. Asian-American feminists in San Francisco have prepared a critique of the Asian movement against hate crimes and the men of that movement are listening. Other communities of color should also examine the nexus between race and gender for women of color, and by extension, for all women.

The legal system must expand the boundaries of existing law to include the most invisible women. There are hundreds of cases involving women of color waiting to be filed. Activists in the violence-against-women movement must reexamine current views on gender-based violence. Not all sexual assaults are the same. Racism in a sexual assault adds another dimension to the pain and harm inflicted. By taking women of color out of the legal shadows, out of invisibility, all women make gains toward full human dignity and human rights.

HOW SAFE IS AMERICA?   59

Desiree Taylor

I saw a picture in a magazine in which a woman is walking away from the collapsed World Trade Center towers covered in orange dust from head to toe. Her face is twisted into a shocked and horrifying expression and she is turned around slightly looking back at the photographer as she walks away from the scene. What strikes me about this picture is that I have never seen another like it, such an elegant and stylish depiction of war. The woman is so immaculately dressed, her hair so stylishly cut, that the whole scene looks like a clever fashion spread from an upscale magazine. This just couldn't be real, but it is. Looking at this picture, I try to really see the woman in it. I try to feel the horror of her experience, but as I do I can also see that she is not from my America. She is from an America that before September 11 was in many ways safer and freer than mine before or after that day.

On September 12, the day after the attack on the United States, I watched the media coverage and a very middle-class looking woman interviewed on the street said she "no longer felt safe in America." I was born and have grown up in this country. As a mixed race, half Black, half white woman born into poverty, I have never felt safe here.

In America, life within one class is nothing at all like it is in another. On September 11 thousands of people died in the collapsed World Trade towers. They were not alone. Every day in this country people die from exploitation that originates right here at home. Some who toil and slave in service to a system of wealth and prestige, who don't even earn a living for their trouble, slit their wrists out of desperation and pain. Some die used up and exhausted, in hospital beds with two dollars and eighty-five cents in their purse, like my own mother. Some people work two or more jobs and try to fit some kind of life in between, maybe an education, which gets harder and harder as they lose out on more and more sleep. The amount of safety one can truly feel in America is directly related to how much money you have.

SAFETY DEPENDS ON MONEY

I think about these people who were and are anything but safe here. I think about those who sweat out unappreciated labor to make the American Dream seem so real, to make the consumer culture function. I wonder if the woman in the picture ever looked as closely at the people who are dying in the class war here at home as I do at her.

It enrages me to hear people saying they no longer feel safe here. The United States that is being attacked is the one the woman in the picture belongs to. It is the prosperous, comfortable United States. It is not my United States. But all of a sudden we are all in it together. The flags are brought out and everybody sings, "I'm proud to be an American."

But it's impossible for me to suddenly forget that the United States empire was built upon and is still maintained by abuses against the poor and minorities. I think about those who will fight and die in this war. I think about all the poor, and often Black, students from my high school who enlisted in the armed services to earn money for college. I think about all the students who wouldn't have been caught dead joining the armed forces, kids for whom the military was not one of their few options to move out of poverty. I think about how many low-income families are worrying right now about the lives of their enlisted children. I wonder what proportion of persons who will fight in this war will be those who are not sharing in the "American Dream." I believe that it will be disproportionally high.

POOR INVISIBLE TO MAINSTREAM

The plight of low-income people in this country is invisible to mainstream America. This invisible other America, the poor, enters through side and back doors of hotels, through the servants' entrances. They live in segregated neighborhoods, and work jobs in which they are unseen even though they are in plain view.

For example, I went to Walgreens the other day and was handed a receipt that read at the top: "I'm Laqueeta [name changed]. I'm here to serve you." The message continued on the next line: "with our seven service basics." A middle-class person in this corporation decided that Laqueeta should hand this message out on every one of her receipts. What made someone earning a good salary think that this was a good idea? We're not even told what those seven service basics are, and they don't really matter. What matters is that the plantation-type American Dream is being acted out here. I, as a consumer, am for a moment in the seat of power with someone to serve me. This is meant to register with me, but the server is not. She doesn't matter. If she did, she would be able to survive doing this kind of work.

I decided to inquire about a cashier's job at Walgreens. The pay rate, I was told, is $6.75 an hour in Boston. When I worked hourly wage jobs in retail stores I learned that employers would keep employees just under 40 hours a week, so that officially they were not full-time and therefore not entitled by law to benefits. I do not know if Walgreens does this, but the practice is widespread. For 39 hours of work at $6.75 an hour, that is $263.25 a week before taxes and $758.16 per month after taxes. After subtracting $35 for a subway pass and at least $460 a month for rent and bills (and for rent this low that means living in a hovel with probably four other people she doesn't know), that leaves Laqueeta just $65.79 a week for food, clothing, savings, entertainment, household expenses, healthcare, and any other expense that might come up, including saving for college and a computer. I hope she doesn't have a child. How safe is Laqueeta in America? Does anybody care? The appearance of America's bounty is maintained by the exploitation of people right here at home who are in positions to be easily misused.

The United States feels very much to me like several countries made up of separate social/economic classes, who don't and perhaps won't take the time to really look at each other; they can't feel each other's pains, cannot relate, and largely live in separate worlds. Now that we're at war with outside forces, we are supposed to come together within the United States. We are all supposed to feel the same hurts and the same threats. We are all supposed to feel each other's pain. We are all supposed to defend justice and freedom.

During the past few weeks we have heard over and over again that these acts of terrorism are not only attacks on the United States, but on everything the country stands for. They tell us these are attacks on freedom and justice itself. But how is this possible when here at home justice, freedom, and the American Dream are denied to so many?

Thinking Further

After reading the articles in this section, you will find it helpful to think about the following:

On some social issues, it is very difficult to find writing that utilizes a race/class/gender framework to discuss and analyze the issue at hand. Studies of disability, for example, sometimes make the link between disability and gender—and sometimes also sexuality—but studies that approach disability using a race, class, *and* gender framework are rare. Why? What would such an analysis look like? We have discovered that it is easier to find writings about race, class, and gender on some topics than others. Why is this? What would an analysis look like on topics not included in this anthology, such as health care, violence against different age groups, rural poverty, or political elections? How are these questions different from those you would ask were you using only race or only gender or only class to understand this topic?

Suggested Readings

Collins, Patricia Hill. 1998. *Fighting Words: Black Women and the Search for Social Justice*. Minneapolis: University of Minnesota Press.

Ferber, Abby. 1998. *White Man Falling: Race, Gender, and White Supremacy*. Lanham, MD: Rowman & Littlefield.

Ragone, Helene, and France Winddance Twine. 2000. *Ideologies and Technologies of Motherhood: Race, Class, Sexuality, Nationalism*. New York: Routledge.

Roberts, Dorothy. 1997. *Killing the Black Body: Race, Reproduction, and the Meaning of Liberty*. New York: Vintage.

Russell, Diana E. H., and Roberta A. Harmes. 2001. *Femicide in Global Perspective*. New York: Teachers College Press.

Ruzek, Sheryl Burt, Virginia L. Olsen, and Adele Clark. 1997. *Women's Health: Complexities and Differences*. Columbus: Ohio State University Press.

 ## InfoTrac College Edition: Search Terms

Use your access to the online library, InfoTrac College Edition, to learn more about subjects covered in this section. The following are some suggested terms for searches:

genocide	racial profiling
hate crimes	rape and warfare
migration/immigration	sexual terrorism

InfoTrac College Edition: Bonus Reading

You can use the InfoTrac College Edition feature to find an additional reading pertinent to this section and can also search on the author's name or keywords to locate the following reading:

Narayan, Uma. 1995. "Mail-Order Brides: Immigrant Women, Domestic Violence and Immigration Law." *Hypatia* 10 (Winter): 104–19.

Uma Narayan documents the heightened vulnerability of immigrant women to isolation, domestic violence, and threats of deportation. Her discussion of "mail-order brides" also reveals that race, gender, and class intermingle to create this phenomenon. Were she writing this article today, given the increased restrictions on immigrants since 9/11, what might she say about how race, class, and gender are now affecting the experiences of immigrant women?

Wadsworth Sociology Resource Center: Virtual Society

For additional links, quizzes, and learning tools related to this section, see the Wadsworth Sociology Resource Center, Virtual Society, at www.wadsworth .com/sociology_d/

V

Making a Difference

What in your life do you care about so much that it would spur you to work for social change? Is it your family, your children, your faith, your neighborhood, a concern for a social issue, or an ethical framework that requires not just talk but action? Most people think that people who work for social change must be somehow extraordinary, like Martin Luther King or other heroic figures. But most people who engage in social activism are ordinary, everyday people who decide to take action about something that touches their lives.

Why don't we know more about these people? Dominant social institutions and the ideologies that defend them obscure the individual and collective political activism of everyday individuals. By making the political activism of everyday people from historically marginalized groups invisible, social institutions suppress the strength of these groups and render them more easily exploited. Investigating forms of power used by historically marginalized groups offers one way of rethinking social change and reconceptualizing how people work for social justice. African Americans, Native Americans, women, Latinos, gays and lesbians, and members of the poor and working class have never been powerless. The question is how to identify and use forms of power that often go unrecognized.

The political activism of individuals from historically privileged groups is hidden as well. Many people who currently benefit within existing race, class, and gender arrangements have little interest in studying—much less showcasing—the ideas and actions of those who not only see the unfairness of race, class, and gender relations in the United States, and globally, but who try to make a difference. Political protest from those harmed by existing race, class, and gender relations is routinely discredited. Stereotyping people of color, women, working-class and poor people, and gays, lesbians, and bisexuals as angry minorities, strident feminists, disgruntled trade unionists, or hypersensitive homosexuals effectively recasts political protest in negative terms and strips their protest of meaning. These tactics remain less effective against those who benefit from existing structures of race, class, gender, and sexuality, but who decide to use their privilege in new ways.

In some ways, individuals who are White, middle class, male, and straight and who reject current social arrangements that privilege them may be more threatening than other groups because they often have the power to make change. Consider how different things would be if White American children learned ways to be "White" that rejected racial privilege, or if men learned models of masculinity that do not demean women. Relegating political protest of this sort to the margins remains vitally important to dominant interests.

Despite their invisibility on both sides of privilege, most of the people who work for social justice are people just like you. Often some sort of catalyst spurs them to action. The needless death of a friend or relative, the placement of a toxic dump site in their neighborhood, losing a job when an employer closes shop, a police beating, or hearing a sermon about Christian brotherhood one Sunday morning in the familiarity of a racially segregated church have all been catalysts for activism. Because people engaging in social activism come from all walks of life, their paths into social activism can be quite unexpected, and the strategies they select for making a difference can be richly diverse.

The articles in Part V, "Making a Difference," explore the various ways that people across race, class, and gender groups participate in social activism. They also provide new visions about what is possible. Collectively, these authors apply an intersectional framework of race, class, and gender to their social activism. In doing so, they reveal the kinds of questions we might ask and actions that we might consider in working for social justice.

WHO ARE THE ACTIVISTS?

When we ask students in our classes to describe an "activist," they often think of a "radical"-looking young person, typically with strange hair and unusual clothing, who gives fiery speeches and engages in losing causes. People who work for social change rarely fit this stereotype and actually reflect a much broader spectrum. Often they remain invisible because we do not label their activities as activism.

There is no "typical" activist. Almost anyone can make a difference in the context of her or his everyday life. Celene Krauss's article "Women of Color on the Front Line" describes how the majority of participants in the environmental justice movement are mothers who became involved out of concern for their children's health. Her article highlights how women often become politically involved—through concerns over family, children, and quality of life in their neighborhoods. Still, women's activism often is overlooked. Although academic credentials, positions of authority, and economic resources can do much to help individuals challenge hierarchies of race, class, and gender, Krauss suggests that the individual and collective actions of ordinary people form the bedrock for many social movements.

Sometimes an activist emerges within the context of traditional or even conservative institutions. The selection "'Whosoever' Is Welcome Here: An Interview with Reverend Edwin C. Sanders II" describes how African American minister Sanders managed to build a church that included everyone yet that still retained ties to Southern Black church colleagues. Reverend Sanders describes how a small group of people came together in Nashville, Tennessee, in search of a community of worship that did not exclude anyone. From its small beginnings, the Metropolitan Interdenominational Church grew and welcomed people of different races, genders, classes, sexualities, and religious traditions. When possessing a vision of equality and an ethical framework that helps people see how inequalities of race, class, gender, and sexuality dehumanize us all, one person can make a difference.

WORKING FOR CHANGE IN MANY PLACES

People decide to work for change from diverse social locations. Workplaces, families, schools, and the government—some of the social institutions we examined in Part III—all represent locations where many people engage in social activism. The media, religious institutions, health care organizations, the penal system, and recreational organizations also constitute sites of social activism.

People often think that, without a group of some sort, they cannot make a difference. Again, this narrow view of activists and activism obscures the tremendous impact that even one individual can have when she or he tries. Whether by choice or by necessity, individuals and groups work outside workplaces, schools, and other formal social institutions. Charon Asetoyer's article, "From the Ground Up," points to how much women and men can accomplish through building their own organizations and social institutions. When Asetoyer found herself moving to the Yankton Sioux reservation in South Dakota, she did not explicitly search out activist activities. Instead, they found her. Through conversations with her neighbors, she learned of problems that women and children were having in her new community. Health issues came up first, and these were the issues that she and others initially addressed. Over time, the informal discussions about women's health issues on the reservation grew to the point where her group was able to purchase a building and organize the Native Women's Reproductive Rights Coalition, involving women from eight states. Asetoyer's article provides an historical overview of how one individual decided to respond in the context of her everyday life, instead of throwing up her hands in defeat.

Sometimes institutional policies and practices become the focus of change. Because existing institutions are structured around race, class, and gender oppression, there are limits to the kinds of change that can be achieved solely from inside them. Furthermore, even when African Americans, Native Americans, women, Asian Americans, the poor, gays and lesbians, Latinos, and other historically marginalized groups gain power within social institutions, change does not automatically follow. Replacing one type of dominant group with another of a different color or gender may improve the lives of some people without addressing the fundamental inequalities that pervade existing social institutions. The familiar boycotts, picketing, public demonstrations, leafleting, and other direct-action strategies long associated with social movements of all types typically constitute actions taken outside an institution designed to pressure it to change. Although activities such as these can be trivialized in the media, it is important to remember that direct action from outsider locations represents one important way to make a difference.

At other times, the goal is to enlist support from these same social institutions for a social agenda. This acknowledges that some social problems cannot be solved by ordinary citizens: Institutional involvement is often needed. Despite the cynicism that many people express regarding the proper role of government, many social issues cannot be addressed without government intervention of some sort. Thus, actions of the federal government were required to desegregate the American South in the 1950s and 1960s.

Although government cannot solve social issues created through current race, class, and gender arrangements, social issues such as homelessness, environmental concerns, violence against children, poverty, racial discrimination, and gender discrimination often require partnership with government to support the actions of everyday citizens who care about these problems. At other times, actions of the government may be part of the problem, and people must find other means of challenging oppressive institutional arrangements.

STRATEGIES: WHAT WORKS?

We are often asked, what strategies make a difference? The range of strategies that individuals and groups use in working for social change is immense. Some engage in individualistic strategies. One well-placed individual with a clear sense of what is possible and why it is important to try can accomplish amazing things. Others engage in collective actions where a small group of like-minded individuals organize around some concern that is important to them. By engaging in strategic resistance, a few people can make a big difference if they use their resources wisely. Still other activists participate in national and global mass movements.

Regardless of the strategies selected, one important issue surfaces repeatedly—the importance of building coalitions. Bringing ordinary people together who have long been separated by race, class, and gender (among other barriers) identifies the necessity of building coalitions across differences and the difficulty of doing so. African American activist Bernice Johnson Reagon (1983) has pointed out how individuals who have been harmed by the damaging effects of race, class, and gender oppression may need to create nurturing spaces and communities where they can try and recover. On some college campuses, for example, African American, Latino, women's, Asian American, gay/lesbian/bisexual/transgendered, and similar identity-based organizations try to create such communities. They want to offer safe spaces where students from historically marginalized groups can retreat from the difficulties many encounter on hostile campuses, learn ways of supporting one another, and identify the positive qualities of their group. These spaces can be valuable, but as Reagon also points out, they can become "barred rooms" that confine and constrict. Identity-based communities and the celebration of group difference that may accompany them can become self-defeating if groups fail to look beyond their own "rooms" and learn to appreciate differences of other groups. Moreover, over time, groups of all sorts—whether identity-based or founded around social issues—that fail to build coalitions with other groups may find themselves increasingly isolated and ineffective.

Coalition-building operates best when individuals and groups share a concern for a common social issue while recognizing the diversity of ways that people experience, think about, and want to deal with that issue. The women on the front line of the environmental justice movement recognized that, although they had varying histories of race and class, as individuals and as part of identity-based groups, they needed to find ways to work together. The members of Metropolitan Interdenominational Church in Nashville realized that the tenets of their faith-based community were tested by their ability to form coalitions with one another that transcended the divisive issues of race, class, and sexuality. These examples illustrate how thinking inclusively, and applying a race/class/gender–inclusive framework to social activism is essential to making a difference.

Coalition-building requires hard work that creates tensions and discomforts. This tension can be further aggravated when groups of different power within hierarchies of race, class, and gender try to build coalitions. As the specific suggestions in John Anner's article, "Having the Tools at Hand: Building Successful Multicultural Social Justice Organizations," remind us, people can learn to work together across all sorts of differences. Through case studies of several grassroots groups who realized that their political effectiveness would be increasingly compromised if they did not learn to deal with diversity, Anner develops a list of models and mechanisms that can be used to build successful multicultural social justice organizations. Despite the difficulties, "Change means growth, and growth can be painful," observes Audre Lorde in her essay included in Part I. "But we sharpen self-definition by exposing the self in work and struggle together with those whom we define as different from ourselves, although sharing the same goals."

TOWARD SOCIAL JUSTICE

Power is typically equated with domination and control over people or things. Social institutions depend on this version of power to reproduce hierarchies of race, class, and gender. Exploring the experiences of African Americans, Latinos, women, Native Americans, Asian Americans, and the poor reveals much about how dominant groups exercise power. At the same time, investigating the experiences of historically marginalized groups also reveals much about resistance, because members of these groups engage in individual acts of resistance and organized political activism to challenge race, class, and gender oppression.

Re-envisioning and exercising power to bring about social change requires a sense of purpose and a vision that encourages us to look beyond what already exists. We must learn to imagine what is possible. For example, what type of environmental policies would result if working-class White women and women of color were central in the environmental movement's decision-making processes? If teachers tried to implement antiracist education from kindergarten through post-graduate schooling, how might we better understand issues of immigration, citizenship, and democracy? If all religious institutions welcomed all people, how might families and communities be changed? How might health care be organized if Native American women's ideas were central to planning? If men and women truly learned to work together, how might economic security be better provided for all?

Trying to make a difference takes time. One can take the long view or focus on the here and now. These are not either/or choices but two ways of looking at the same thing. Many people become cynical because they expect change to be immediate and dramatic rather than long-term and continual. How people deal with this disappointment is revealing. The White working-class women who became concerned about the effects of environmental dumping on their children initially trusted the government to provide solutions. When it did not, they could easily have become cynical and thrown in the towel. Instead, they tried to find another way because too much was at stake. Their initial idealism became tempered by pragmatism.

Taking a long view, seeing the connectedness among all sorts of people, and involving people across race, class, and gender—and beyond—will be necessary for bringing lasting change. Clearly this will not happen tomorrow. But if people do not try to make a difference, it just won't happen at all.

Trying to make a difference also requires looking backward in order to plan for and move into the future. Youth are vitally important for social justice projects. Ironically, many young adults think that all the important social activism happened before their time. Like Shani Jamila in her article, "Can I Get a Witness? Testimony from a Hip Hop Feminist," they think that they have "missed" the real social activism. Because they were too young to attend the marches of the civil rights movement or to participate in the picket lines of major union activism, or have only seen activism on film and DVD, many feel that social activism is a thing of the past. Because so many young adults now benefit from the struggles of their parents, teachers, and ordinary citizens— they attend integrated schools, possess formal reproductive rights, are assured legal equal opportunity in jobs, and have helped create an exciting, global multicultural youth culture—some question the need for social activism.

Shani Jamila reminds us that each generation inherits the task of forging a new path of social activism for themselves and for those who will follow. For Jamila, the task lies in synthesizing the received wisdom of feminism with the new challenges raised by hip-hop culture. In this quest, Jamila found that liberating her definitions of activism from the "constraints and constructs of the sixties" opened up her mind to new possibilities. As she points out, "The most important thing we can do as a generation is to see our new positions as power and weapons to be used strategically in the struggle rather than as spoils of war."

Further Resources

For additional materials relating to this section, see the features on pages 560–61 at the conclusion of this section.

Reference

Reagon, Bernice Johnson. 1983. "Coalition Politics: Turning the Century." In *Home Girls: A Black Feminist Anthology*, ed. Barbara Smith (New York: Kitchen Table Press), 356–68.

Making a Difference

WOMEN OF COLOR ON THE FRONT LINE

60

Celene Krauss

Toxic waste disposal is a central focus of women's grass-roots environmental activism. Toxic waste facilities are predominantly sited in working-class and low-income communities and communities of color, reflecting the disproportionate burden placed on these communities by a political economy of growth that distributes the costs of economic growth unequally. Spurred by the threat that toxic wastes pose to family health and community survival, female grass-roots activists have assumed the leadership of community environmental struggles. As part of a larger movement for environmental justice, they constitute a diverse constituency, including working-class housewives and secretaries, rural African American farmers, urban residents, Mexican American farm workers, and Native Americans.

These activists attempt to differentiate themselves from what they see as the white, male, middle-class leadership of many national environmental organizations. Unlike the more abstract, issue-oriented focus of national groups, women's focus is on environmental issues that grow out of their concrete, immediate experiences. Female blue-collar activists often share a loosely defined ideology of environmental justice and a critique of dominant social institutions and mainstream environmental organizations, which they believe do not address the broader issues of inequality underlying environmental hazards. At the same time, these activists exhibit significant diversity in their conceptualization of toxic waste issues, reflecting different experiences of class, race, and ethnicity.

From Robert D. Bullard, ed., *Unequal Protection: Environmental Justice and Communities of Color* (San Francisco: Sierra Club Books, 1994), pp. 256–71. Reprinted by permission.

This [essay] looks at the ways in which different working-class women formulate ideologies of resistance around toxic waste issues and the process by which they arrive at a concept of environmental justice. Through an analysis of interviews, newsletters, and conference presentations, I show the voices of white, African American, and Native American female activists and the resources that inform and support their protests. What emerges is an environmental discourse that is mediated by subjective experiences and interpretations and rooted in the political truths women construct out of their identities as housewives, mothers, and members of communities and racial and ethnic groups.

THE SUBJECTIVE DIMENSION OF GRASS-ROOTS ACTIVISM

Grass-roots protest activities have often been trivialized, ignored, and viewed as self-interested actions that are particularistic and parochial, failing to go beyond a single-issue focus. This view of community grass-roots protests is held by most policymakers as well as by many analysts of movements for progressive social change.

In contrast, the voices of blue-collar women engaged in protests regarding toxic waste issues tell us that single-issue protests are about more than the single issue. They reveal a larger world of power and resistance, which in some measure ends up challenging the social relations of power. This challenge becomes visible when we shift the analysis of environmental activism to the experiences of working-class women and the subjective meanings they create around toxic waste issues.

In traditional sociological analysis, this subjective dimension of protest has often been ignored or viewed as private and individualistic. Feminist theory, however, helps us to see its importance. For feminists, the critical reflection on the everyday world of experience is an important subjective dimension of social change. Feminists show us that experience is not merely a personal, individualistic concept. It is social. People's experiences reflect where they fit in the social hierarchy. Thus, blue-collar women of differing backgrounds interpret their experiences of toxic waste problems within the context of their particular cultural histories, starting from different assumptions and arriving at concepts of environmental justice that reflect broader experiences of class and race.

Feminist theorists also challenge a dominant ideology that separates the "public" world of policy and power from the "private" and personal world of everyday experience. By definition, this ideology relegates the lives and concerns of women relating to home and family to the private, nonpolitical arena, leading to invisibility of their grass-roots protests about issues such as toxic wastes. As Ann Bookman has noted in her important study of working-class women's community struggles, women's political activism in general, and working-class political life at the community level in particular, remain "peripheral to the historical record . . . where there is a tendency to privilege male political activity and labor activism."[1] The women's movement took as its central task the reconceptualization of the political itself, critiquing this dominant ideology and constructing a new definition of the political located in the everyday world of ordinary women rather than in the world of public policy. Feminists provide a perspective for making visible the importance of particular, single-issue protests regarding toxic wastes by showing how ordinary women subjectively link the particulars of their private lives with a broader analysis of power in the public sphere.

Social historians such as George Rudé have pointed out that it is often difficult to understand the experience and ideologies of resistance because ordinary working people appropriate and reshape traditional beliefs embedded within working-class culture, such as family and community. This point is also relevant for understanding the environmental protests of working-class women. Their protests are framed in terms of the traditions of motherhood and family; as a result, they often appear parochial or even conservative. As we shall see, however, for working-class women, these traditions become the levers that set in motion a political process, shaping the language and oppositional meanings that emerge and providing resources for social change.

Shifting the analysis of toxic waste issues to the subjective experience of ordinary women makes visible a complex relationship between everyday life and the larger structures of public power. It reveals the potential for human agency that is hidden in a more traditional sociological approach and provides us with a means of seeing "the sources of power which subordinated groups have created."[2]

The analysis presented in this [essay] is based on the oral and written voices of women involved in toxic waste protests. Interviews were conducted at environmental conferences such as the First National People of Color Environmental Leadership Summit, Washington, D.C., 1991, and the World Women's Congress for a Healthy Planet, Miami, Florida, 1991, and by telephone. Additional sources include conference presentations, pamphlets, books, and other written materials that have emerged from this movement. This research is part of an ongoing comparative study that will examine the ways in which experiences of race, class, and ethnicity mediate women's environmental activism. Future research includes an analysis of the environmental activism of Mexican American women in addition to that of the women discussed here.

TOXIC WASTE PROTESTS
AND THE RESOURCE OF MOTHERHOOD

Blue-collar women do not use the language of the bureaucrat to talk about environmental issues. They do not spout data or marshal statistics in support of their positions. In fact, interviews with these women rarely generate a lot of discussion about the environmental problem per se. But in telling their stories about their protest against a landfill or incinerator, they ultimately tell larger stories about their discovery or analysis of oppression. Theirs is a political, not a technical, analysis.

Working-class women of diverse racial and ethnic backgrounds identify the toxic waste movement as a women's movement, composed primarily of mothers. Says one woman who fought against an incinerator in Arizona and subsequently worked on other anti-incinerator campaigns throughout the state, "Women are the backbone of the grass-roots groups; they are the ones who stick with it, the ones who won't back off." By and large, it is women, in their traditional role as mothers, who make the link between toxic wastes and their children's ill health. They discover the hazards of toxic contamination: multiple miscarriages, birth defects, cancer deaths, and so on. This is not surprising, as the gender-based division of labor in a capitalist society gives working-class women the responsibility for the health of their children.

These women define their environmental protests as part of the work that mothers do. Cora Tucker, an African American activist who fought against uranium mining in Virginia and who now organizes nationally, says:

> It's not that I don't think that women are smarter, [she laughs] but I think that we are with the kids all day long. . . . If Johnny gets a cough and Mary gets a cough, we try to discover the problem.

Another activist from California sums up this view: "If we don't oppose an incinerator, we're not doing our work as mothers."

For these women, family serves as a spur to action, contradicting popular notions of family as conservative and parochial. Family has a very different meaning for these women than it does for the middle-class nuclear family. Theirs is a less privatized, extended family that is open, permeable, and attached to community. This more extended family creates the networks and resources that enable working-class communities to survive materially given few economic resources. The destruction of working-class neighborhoods by economic growth deprives blue-collar communities of the basic resources of survival; hence the resistance engendered by toxic waste issues. Working-class women's struggles over toxic waste issues are, at root, issues about survival. Ideologies of motherhood, traditionally relegated to the private sphere, become political resources that working-class women use to initiate and justify their resistance. In the process of protest, working-class women come to reject the dominant ideology, which separates the public and private arenas.

Working-class women's extended network of family and community serves as the vehicle for spreading information and concern about toxic waste issues. Extended networks of kinship and friendship become political resources of opposition. For example, in one community in Detroit, women discovered patterns of health problems while attending Tupperware parties. Frequently, a mother may read about a hazard in a newspaper, make a tentative connection between her own child's ill health and the pollutant, and start telephoning friends and family, developing an informal health survey. Such a discovery process is rooted in what Sarah Ruddick has called the everyday practice of mothering.[3] Through their informal networks, they compare notes and experiences and develop an oppositional knowledge used to resist the dominant knowledge of experts and the decisions of government and corporate officials.

These women separate themselves from "mainstream" environmental organizations, which are seen as dominated by white, middle-class men and concerned with remote issues. Says one woman from Rahway, New Jersey: "The mainstream groups deal with safe issues. They want to stop incinerators to save the eagle, or they protect trees for the owl. But we say, what about the people?"

Another activist implicitly criticizes the mainstream environmental groups when she says of the grass-roots Citizens' Clearinghouse for Hazardous Wastes:

> Rather than oceans and lakes, they're concerned about kids dying. Once you've had someone in your family who has been attacked by the environment—I mean who has had cancer or some other disease—you get a keen sense of what's going on.

It is the traditional, "private" women's concerns about home, children, and family that provide the initial impetus for blue-collar women's involvement in issues of toxic waste. The political analyses they develop break down the public-private distinction of dominant ideology and frame a particular toxic waste issue within broader contexts of power relationships.

THE ROLE OF RACE, ETHNICITY, AND CLASS

Interviews with white, African American, and Native American women show that the starting places for and subsequent development of their analyses of toxic waste protests are mediated by issues of class, race, and ethnicity.

White working-class women come from a culture in which traditional women's roles center on the private arena of family. They often marry young; although they may work out of financial necessity, the primary roles from which they derive meaning and satisfaction are those of mothering and taking care of family. They are revered and supported for fulfilling the ideology of a patriarchal family. And these families often reflect a strong belief in the existing political system. The narratives of white working-class women involved in toxic waste issues are filled with the process by which they discover the injustice of their government, their own insecurity about entering the public sphere of politics, and the constraints of the patriarchal family, which, ironically prevent them from becoming fully active in the defense of their family, especially in their protest. Their narratives are marked by a strong initial faith in "their" government, as well as a remarkable transformation as they become disillusioned with the system. They discover "that they never knew what they were capable of doing in defense of their children."

For white working-class women, whose views on public issues are generally expressed only within family or among friends, entering a more public arena to confront toxic waste issues is often extremely stressful. "Even when I went to the PTA," says one activist, "I rarely spoke. I was so nervous." Says another: "My views have always been strong, but I expressed them only in the family. They were not for the public." A strong belief in the existing political system is characteristic of these women's initial response to toxic waste issues. Lois Gibbs, whose involvement in toxic waste issues started at Love Canal, tells us, "I believed if I had a problem I just had to go to the right person in government and he would take care of it."

Initially, white working-class women believe that all they have to do is give the government the facts and their problem will be taken care of. They become progressively disenchanted with what they view as the violation of their rights and the injustice of a system that allows their children and family to die. In the process, they develop a perspective of environmental justice rooted in issues of class, the attempt to make democracy real, and a critique of the corporate state. Says one activist who fought the siting of an incinerator in Sumter County, Alabama: "We need to stop letting economic development be the true God and religion of this country. We have to prevent big money from influencing our government."

A recurring theme in the narratives of these women is the transformation of their beliefs about government and power. Their politicization is rooted in the deep sense of violation, betrayal, and hurt they feel when they find that their government will not protect their families. Lois Gibbs sums up this feeling well:

> I grew up in a blue-collar community. We were very into democracy. There is something about discovering that democracy isn't democracy as we know it. When you lose faith in your government, it's like finding out your mother was fooling around on your father. I was very upset. It almost broke my heart because I really believed in the system. I still believe in the system, only now I believe that democracy is of the people and by the people, that people have to move it, it ain't gonna move by itself.

Echoes of this disillusionment are heard from white blue-collar women throughout the country. One activist relates:

> We decided to tell our elected officials about the problems of incineration because we didn't think they knew. Surely if they knew that there was a toxic waste dump in our county they would stop it. I was politically naive. I was really surprised because I live in an area that's like the Bible Belt of the South. Now I think the God of the United States is really economic development, and that has got to change.

Ultimately, these women become aware of the inequities of power as it is shaped by issues of class and gender. Highly traditional values of democracy and motherhood remain central to their lives. But in the process of politicization through their work on toxic waste issues, these values become transformed into resources of opposition that enable women to enter the public arena and challenge its legitimacy. They justify their resistance as a way to make democracy real and to protect their children.

White blue-collar women's stories are stories of transformations: transformations into more self-confident and assertive women; into political activists who challenge the existing system and feel powerful in that challenge; into wives and mothers who establish new relationships with their spouses (or get divorced) and new, empowering relationships with their children as they provide role models of women capable of fighting for what they believe in.

African American working-class women begin their involvement in toxic waste protests from a different place. They bring to their protests a political awareness that is grounded in race and that shares none of the white blue-collar women's initial trust in democratic institutions. These women view government with mistrust, having been victims of racist policies throughout their lives. Individual toxic waste issues are immediately framed within a broader political context and viewed as environmental racism. Says an African American activist from Rahway, New Jersey:

> When they sited the incinerator for Rahway, I wasn't surprised. All you have to do is look around my community to know that we are a dumping ground for all kinds of urban industrial projects that no one else wants. I knew this was about environmental racism the moment that they proposed the incinerator.

An African American woman who fought the siting of a landfill on the South Side of Chicago reiterates this view: "My community is an all-black community isolated from everyone. They don't care what happens to us." She describes her community as a "toxic doughnut":

> We have seven landfills. We have a sewer treatment plant. We have the Ford Motor Company. We have a paint factory. We have numerous chemical

companies and steel mills. The river is just a few blocks away from us and is carrying water so highly contaminated that they say it would take seventy-five years or more before they can clean it up.

This activist sees her involvement in toxic waste issues as a challenge to traditional stereotypes of African American women. She says, "I'm here to tell the story that all people in the projects are not lazy and dumb!"

Some of these women share experiences of personal empowerment through their involvement in toxic waste issues. Says one African American activist:

Twenty years ago I couldn't do this because I was so shy. . . . I had to really know you to talk with you. Now I talk. Sometimes I think I talk too much. I waited until my fifties to go to jail. But it was well worth it. I never went to no university or college, but I'm going in there and making speeches.

However, this is not a major theme in the narratives of female African American activists, as it is in those of white blue-collar women. African American women's private work as mothers has traditionally extended to a more public role in the local community as protectors of the race. As a decade of African American feminist history has shown, African American women have historically played a central role in community activism and in dealing with issues of race and economic injustice. They receive tremendous status and recognition from their community. Many women participating in toxic waste protests have come out of a history of civil rights activism, and their environmental protests, especially in the South, develop through community organizations born during the civil rights movement. And while the visible leaders are often male, the base of the organizing has been led by African American women, who, as Cheryl Townsend Gilkes has written, have often been called "race women," responsible for the "racial uplift" of their communities.[4]

African American women perceive that traditional environmental groups only peripherally relate to their concerns. As Cora Tucker relates:

This white woman from an environmental group asked me to come down to save a park. She said that they had been trying to get black folks involved and that they won't come. I said, "Honey, it's not that they aren't concerned, but when their babies are dying in their arms they don't give a damn about a park." I said, "They want to save their babies. If you can help them save their babies, then in turn they can help you save your park." And she said, "But this is a real immediate problem." And I said, "Well, these people's kids dying is immediate."

Tucker says that white environmental groups often call her or the head of the NAACP at the last minute to participate in an environmental rally because they want to "include" African Americans. But they exclude African Americans from the process of defining the issues in the first place. What African American communities are doing is changing the agenda.

Because the concrete experience of African Americans' lives is the experience and analysis of racism, social issues are interpreted and struggled with within this context. Cora Tucker's story of attending a town board meeting shows that the issue she deals with is not merely the environment but also the disempowerment she experiences as an African American woman. At the meeting, white women were addressed as Mrs. So-and-So by the all-white, male board. When Mrs. Tucker stood up, however, she was addressed as "Cora":

> One morning I got up and I got pissed off and I said, "What did you call me?" He said, "Cora," and I said, "The name is Mrs. Tucker." And I had the floor until he said "Mrs. Tucker." He waited five minutes before he said "Mrs. Tucker." And I held the floor. I said, "I'm Mrs. Tucker," I said, "Mr. Chairman, I don't call you by your first name and I don't want you to call me by mine. My name is Mrs. Tucker. And when you want me, you call me Mrs. Tucker." It's not that—I mean it's not like you gotta call me Mrs. Tucker, but it was the respect.

In discussing this small act of resistance as an African American woman, Cora Tucker is showing how environmental issues may be about corporate and state power, but they are also about race. For female African American activists, environmental issues are seen as reflecting environmental racism and linked to other social justice issues, such as jobs, housing, and crime. They are viewed as part of a broader picture of social inequity based on race. Hence, the solution articulated in a vision of environmental justice is a civil rights vision— rooted in the everyday experience of racism. Environmental justice comes to mean the need to resolve the broad social inequities of race.

The narratives of Native American women are also filled with the theme of environmental racism. However, their analysis is laced with different images. It is a genocidal analysis rooted in the Native American cultural identification, the experience of colonialism, and the imminent endangerment of their culture. A Native American woman from North Dakota, who opposed a landfill, says:

> Ever since the white man came here, they keep pushing us back, taking our lands, pushing us onto reservations. We are down to 3 percent now, and I see this as just another way for them to take our lands, to completely annihilate our races. We see that as racism.

Like that of the African American women, these women's involvement in toxic waste protests is grounded from the start in race and shares none of the white blue-collar women's initial belief in the state. A Native American woman from southern California who opposed a landfill on the Rosebud Reservation in South Dakota tells us:

> Government did pretty much what we expected them to do. They supported the dump. People here fear the government. They control so many aspects of our life. When I became involved in opposing the garbage landfill, my people told me to be careful. They said they annihilate people like me.

Another woman involved in the protest in South Dakota describes a government official's derision of the tribe's resistance to the siting of a landfill:

> If we wanted to live the life of Mother Earth, we should get a tepee and live on the Great Plains and hunt buffalo.

Native American women come from a culture in which women have had more empowered and public roles than is the case in white working-class culture. Within the Native American community, women are revered as nurturers. From childhood, boys and girls learn that men depend on women for their survival. Women also play a central role in the decision-making process within the tribe. Tribal council membership is often equally divided between men and women; many women are tribal leaders and medicine women. Native American religions embody a respect for women as well as an ecological ethic based on values such as reciprocity and sustainable development: Native Americans pray to Mother Earth, as opposed to the dominant culture's belief in a white, male, Anglicized representation of divinity.

In describing the ways in which their culture integrates notions of environmentalism and womanhood, one woman from New Mexico says:

> We deal with the whole of life and community; we're not separated, we're born into it—you are it. Our connection as women is to the Mother Earth, from the time of our consciousness. We're not environmentalists. We're born into the struggle of protecting and preserving our communities. We don't separate ourselves. Our lifeblood automatically makes us responsible; we are born with it. Our teaching comes from a spiritual base. This is foreign to our culture. There isn't even a word for dioxin in Navajo.

In recent years, Native American lands have become common sites for commercial garbage dumping. Garbage and waste companies have exploited the poverty and lack of jobs in Native American communities and the fact that Native American lands, as sovereign nation territories, are often exempt from local environmental regulations. In discussing their opposition to dumping, Native American women ground their narratives in values about land that are inherent in the Native American community. They see these projects as violating tribal sovereignty and the deep meaning of land, the last resource they have. The issue, says a Native American woman from California, is

> protection of the land for future generations, not really as a mother, but for the health of the people, for survival. Our tribe bases its sovereignty on our land base, and if we lose our land base, then we will be a lost people. We can't afford to take this trash and jeopardize our tribe.
>
> If you don't take care of the land, then the land isn't going to take care of you. Because everything we have around us involves Mother Earth. If we don't take care of the land, what's going to happen to us?

In the process of protest, these women tell us, they are forced to articulate more clearly their cultural values, which become resources of resistance in helping the tribe organize against a landfill. While many tribal members may not articulate an "environmental" critique, they well understand the meaning of land and their religion of Mother Earth, on which their society is built.

CONCLUSION

The narratives of white, African American, and Native American women involved in toxic waste protests reveal the ways in which their subjective, particular experiences lead them to analyses of toxic waste issues that extend beyond the particularistic issue to wider worlds of power. Traditional beliefs about home, family, and community provide the impetus for women's involvement in these issues and become a rich source of empowerment as women reshape traditional language and meanings into an ideology of resistance. These stories challenge traditional views of toxic waste protests as parochial, self-interested, and failing to go beyond a single-issue focus. They show that single-issue protests are ultimately about far more and reveal the experiences of daily life and resources that different groups use to resist. Through environmental protests, these women challenge, in some measure, the social relations of race, class, and gender.

These women's protests have different beginning places, and their analyses of environmental justice are mediated by issues of class and race. For white blue-collar women, the critique of the corporate state and the realization of a more genuine democracy are central to a vision of environmental justice. The definition of environmental justice that they develop becomes rooted in the issue of class. For women of color, it is the link between race and environment, rather than between class and environment, that characterizes definitions of environmental justice. African American women's narratives strongly link environment justice to other social justice concerns, such as jobs, housing, and crime. Environmental justice comes to mean the need to resolve the broad social inequities of race. For Native American women, environmental justice is bound up with the sovereignty of the indigenous peoples.

In these women's stories, their responses to particular toxic waste issues are inextricably tied to the injustice they feel as mothers, as working-class women, as African Americans, and as Native Americans. They do not talk about their protests in terms of single issues. Thus, their political activism has implications far beyond the visible, particularistic concern of a toxic waste dump site or the siting of a hazardous waste incinerator.

NOTES

1. Sandra Morgen, "'It's the Whole Power of the City Against Us!': The Development of Political Consciousness in a Women's Health Care Coalition," in *Women and the Politics of Empowerment*, eds. Ann Bookman and Sandra Morgen (Philadelphia: Temple University Press, 1988), p. 97.

2. Sheila Rowbotham, *Women's Consciousness, Man's World* (New York: Penguin, 1973).

3. See Sara Ruddick, *Maternal Thinking: Towards a Politics of Peace* (New York: Ballantine, 1989).

4. Cheryl Townsend Gilkes, "Building in Many Places: Multiple Commitments and Ideologies in Black Women's Community Work," in *Women and the Politics of Empowerment*, op. cit.

"WHOSOEVER" IS WELCOME HERE

An Interview with Reverend Edwin C. Sanders II

61

Gary David Comstock

Reverend Edwin C. Sanders II is the founding pastor of Metropolitan Interdenominational Church, which is located in a working class neighborhood of small houses in Nashville, Tennessee. Reverend Sanders is African American and the congregation is predominantly African American. Lesbian/bisexual/gay/transgendered people are welcome and encouraged to participate in the life of the church. I attended a packed Sunday morning service in June 1998 and interviewed Rev. Sanders in the afternoon.

From Eric Brandt, ed., *Dangerous Liaisons: Blacks, Gays, and the Struggle for Equality* (New York: The New Press, 1999), pp. 142–57. Reprinted by permission of the New Press.

GARY COMSTOCK: How did the church get started?

REVEREND SANDERS: I had been the Dean of the Chapel at Fisk University. I left in 1980 in a moment of controversy. A new president had come, and we weren't able to mesh. I left and had no where to go. I didn't have a plan. And instantly there were folks who were part of the chapel experience there at Fisk who wanted to organize a new church. I felt no spiritual interest whatsoever in organizing a new church, but about seven months later I felt like I clearly heard the voice saying, "This is something to do." I'm glad it worked out that way, because I think if I had done it directly after Fisk it would have been born out of a reaction and we probably wouldn't have developed the kind of identity, sense of mission, and direction the way we did. There were twelve people who came together and said they wanted to do this.

One of my good friends, Bill Turner, is a sociologist, and we were at Fisk together. Bill has a theory he advanced that institutions—and he built his theory around black institutions—cannot break out of the mold from which they were born. There was something about the way an institution is framed in its beginning, and no matter what you do you don't escape it. I thought that was absurd, but in time I came to think that he had something. The congregation was a mix of white and black men and women from all kinds of denominational backgrounds, and that mix turned out to be significant because from the beginning people identified us as being inclusive at least across racial terms and definitely inclusive and equal in gender terms. . . .

In that original group we had also a young man—one of my very dear friends—who was gay. Don was living a bisexual lifestyle at that point, but mainly to keep up appearances for professional purposes. Another one of my dearest friends went through a major mental breakdown at the time we were starting the church. I felt it very important not to abandon him and to include him. So we had a guy going through major psychological problems, somebody who was gay, and we had the racial mix.

We did not have the class mix at the beginning. Most of the folks were associated in some way with the academic community. Nothing like we have now. Today we have an unbelievable mix of people. I mean there are people who are doctors, lawyers, dentists and business people, and we also have a lot of folks who are right off the street, blue collar workers, in treatment for drugs and alcohol, going through a lot of transition. And that mix has evolved. Although we said in the beginning that's what we wanted to be, in actuality the current mix goes beyond that of the original twelve members.

The presence of Don, the one gay black male in the original congregation, had a lot to do with our current mix of people, and it had a lot to do with how we got so involved with HIV/AIDS ministry. The church began in '81 and he died in '84. It's almost hilarious when I think about it because I'm so involved in HIV/AIDS work now, but when he died I remember he was real sick, we didn't know what was going on, and they told us he died of toxicosis.

I remember saying what in the world is that? I researched it and found out that it had something to do with cat and bird droppings, but Don didn't have cats. It was AIDS. You'd hear people talking about this strange disease because then it was 1984, but what that meant for us was that we got involved before it became a publicly recognized and discussed issue. Don's presence and death immediately pushed us in ministering to and being responsive to folks with AIDS and in dealing with the issue of homosexuality.

The other thing I was going to tell you which is kind of funny is the name. I will never forget when I told one of my friends we were going to name the church Metropolitan Interdenominational. He said are you sure you want to do that. I said what do you mean, you know, I was pretty naive. He said all the gay churches across the country are called Metropolitan churches. I said that's where I feel the Lord leads me, I feel Metropolitan. In my mind what that meant was that Nashville happens to have been the first metropolitan government in the United States. It's the first place where the county and city combined, so everything in and around Nashville is referred to as metropolitan government. But my friend said to me that everybody is going to instantly say you're the gay church. I said so be it and we went on with it. Like I said, there was a hand bigger than mine at work. . . .

COMSTOCK: What was the turnaround point for the class mix? How and when did it happen?

REV. SANDERS: It was real clear. Early on what happened was we got involved in prison ministry—actually directed a ministry called the Southern Prison Ministry for a while. Going in and out of the prisons we started to develop relationships that translated to folks coming out of prison and getting involved in the church. After I had done the prison ministry for a while it became crystal clear to me that 80 percent of the people I was dealing with in prison were there for alcohol and drug related issues. I started reassessing this whole issue of drugs and alcohol and decided that's an area we had to begin to focus our ministries. So, I got involved, did the training, and became certified as a counselor. I would venture to say that 25 percent of the people in this church are folks that I first met in treatment. Thirty to 35 plus percent of this congregation are folks who are in treatment from alcohol and drug use. This is a place where a lot of folks know they can come. We've got all these names, you know, the drug church, the AIDS church, we get those labels. But it's all right because that's what we do. We hit those themes a lot.

I've learned some things over the years that I pretty much hold to and this is one of them. . . . We don't say we are a black church, we don't say that we are a gay church or a straight church, we don't say that we are anything other than a church that celebrates our oneness in Christ. I'm convinced that has turned out to be the real key to being able to hold this diverse group of folks together. I must admit I'm a person that has a negative thing about the word diversities. I don't use it much. We don't celebrate our differences, we celebrate our oneness. That ends up being an avenue for a lot of folks being attracted and feeling comfortable here. New folks say I got here and I just felt like no one was looking at me strange, no one was treating me different, I was just able to be here. . . .

COMSTOCK: I was impressed by the informality of the service today. People seem to relax and fit in. The choir is not so focused on performance that they're not interacting with the congregation. And your own manner is informal. It's a style that lets people in. You aren't just saying all people are welcomed here, you actually do something that let's them feel at home. I was also struck by the openness of the windows—the plants inside and the view into the park. . . .

REV. SANDERS: . . . We have the sense of informality. People are arriving from the time we welcome the guests at the beginning until just after the sermon when there's maximum presence in the audience. People come in slow, and we give folks an opportunity to leave early. We even say it in the bulletin, we just say if you need to leave just leave quietly. And we know folks do. There's a lot of folks who want to hear the choir or they want to hear the sermon, but they don't want all the rest of it. So we do it that way. It works out. One of the real hooks for just about everyone at Metropolitan is the fellowship circle at the end. We actually have a few Jews and a couple of Muslims who worship here regularly. The Muslim family does a very interesting thing which helped me to appreciate the significance of the fellowship circle at the end. We offer communion—the Eucharist, the Lord's supper—every Sunday, and they stay until we get to that part of the service. Then they go outside or into the vestibule, but they come back in. I remember when Omar first started attending, I said to him it's interesting to me that you don't leave at that point. He said, "No, no, no, I wouldn't miss the fellowship circle." And it made me realize that the communion of the fellowship circle is more important than the bread and the wine. Probably the real communion is what we do when we stand there at the end and sing "We've Come Too Far to Turn Back Now." We do that every week. It's our theme. That's a very significant moment. I've heard people tell me when they have to be late that they rush to get here just to catch the end of it. One woman said to me it was enough for her if she just got here for the fellowship circle.

A lot of the informality is very intentional. For instance, the only thing at Metropolitan that's elevated is the altar. Nothing else is above ground level. None of the seating is differentiated. In most churches the ministers have different, higher, bigger chairs than everyone else. We don't do that. We sit in the same seats the other folks do. My choir director is always telling me we'd get better sound if we could elevate the back rows. I say no, everybody's got to be on the same level and got to sit in the same seats. We do a lot of symbolic things like that.

Comstock: Do other clergy give you much flack for working with needle exchange, welcoming gay folks, and working on issues that they may see as too progressive?

Rev. Sanders: They do. But let me tell you something. I have been able to have a level of involvement with ministers in this community that probably has brought credibility to what we do. I would like to think that we have maintained our sense of integrity especially as it relates to our consistency in ministry. Folks tend to respect that, so even when they disagree, they also look more seriously and harder at what it is they're questioning. . . .

Comstock: "Whosoever" is from John 3:16–17: "For God so loved the world that he gave his only Son, that whosoever believes in him should not perish but have eternal life." Has it been an expressed theme from the beginning?

Rev. Sanders: It's been our theme for the last seven or eight years. . . . I think we're growing in this inclusivity all the time. The language issue was big for us. We try to use inclusive language. If somebody else comes into the church that is not into that, it sticks out like a sore thumb. Does our inclusivity also mean that there is a real tolerance for folks who are perhaps not where we think folks should be in terms of issues like inclusive language, issues that relate to sexuality, gay and lesbian issues? I think the answer to that question is yes, you have to make room for those folks too. That's a real struggle. Another one of our little clichés, and we don't have a lot of them, is we say we try to be inclusive of all and alienating to none. Not being alienating is a real trick. It's amazing how easy you can alienate folks without realizing it, in ways that you're just not aware of. . . .

Comstock: What would you do if some new people had trouble with cross dressing, transgendered, lesbian or gay people? How do you get them to stay and deal with it rather than leave?

Rev. Sanders: We tracked that issue a couple of times. Let me tell you what's happened. Most folk are here for a while before some of it settles in, before they start to notice everything. It's amazing to me how people get caught up in Metropolitan. They'll join and get involved and then they'll go through membership class, that's usually when it starts to hit them, that they say to themselves, "Oh oh, I'm seeing some stuff that I'm not sure about here." But what we've discovered is that what seems to help us more than anything else is that folks end up remembering what their initial experience had been when they first came here. We even have one couple, this is my favorite story to tell, who has talked about coming here on one of those Sundays when we were hitting the theme of inclusive hard. And they left here and said what kind of church is that? But then a couple of weeks later they said let's go back over there again. They came back and then they didn't come again for about two or three months. They were visiting other churches. They said when they got down to thinking about all the churches, they had felt warmth and connection here. They said you know that church really did kind of work; and they ended up coming back. When folks are here there's a warmth they feel, there is a connection they feel, there's a comfort zone in which they'll eventually deal with the issues that might be their point of difference. We've seen folks move in their thinking and there's some folks that have not been able to do it. I've got one young man—I really do think he just loves being here—who says, "I just can't fathom this gay thing. Why do you insist on it." I said, "You know, you're the one who's lifting this up."

As I said, we try to make sure that when any issue is brought up, it's in the course of things. It's not like we stop and have gay liberation day, just like we don't celebrate Black History month. We don't focus on or celebrate these identities or difference, but yet if you're around here you can't help but pick up on the church's support for these issues and differences. Today, for instance, we sang "Lift Up Your Voice and Sing," which is known as the black national anthem. But we call it "Our Song of Liberation," and I've tried to help people to understand that. . . .

I have a lot of divinity students who serve as my pastoral assistants, and most of them are women. This place has become a real refuge for black women in the ministry. There are not a lot of clergy opportunities for them, so I've tried to figure out how to incorporate them into the life and ministry of the church. They actually run a lot of our ministries. And most of them are extremely well prepared academically. They're more prepared than I am to do the work and they do it well. When they started coming we suddenly became a magnet, a place where there was this rush of folks out of divinity school to come and be a part of ministry here. At one level I probably seem like I'm extremely freewheeling and loose, but I'm probably a lot more intentional about how things are going than folks realize, especially as they relate to the focus of ministry here at the church. One of my real concerns was not having the time to orient the young ministers, to bring them into a full awareness of what makes this place click. What I have discovered is that if I get the inclusive piece established in the beginning with them, I don't have to worry about the rest of it as much. If they buy into understanding that inclusivity is at the heart of this church, I don't have to worry about keeping my eye on them all the time. . . .

COMSTOCK: You know that the example here of accepting and welcoming lesbians and gay men is an anomaly in the Church. You provide inspiration and hope, an example of something positive that is actually happening. Most gay people feel alienated from their churches, but I've found that African-American more often than white gay people emphasize the importance of religion in their family, community, and history and say that most of the pain and sadness in their lives centers around the Church. They claim that being rejected by the black church is especially devastating because this institution has been, and continues to be, the only place where they can take real refuge from the racism they experience in the society. Clearly that's not happening here at Metropolitan.

REV. SANDERS: We realized when we started doing our HIV/AIDS ministry that there are organizations that were established by gay white men who had done a good job of developing services, but we kept seeing there was something that was not happening for gay and lesbian African Americans. And we realized it was community. No matter how much they tried to get that in other communities, there is the whole thing of cultural comfort. They were looking for the context where there were people who looked like them, who were extensions of their family supporting them. It became clear to me that what we needed to do was to establish a place where people could literally come and where people had a sense of community. Consequently, what we call the Wellness Center here is more than anything else just a place where you can come and feel at home. It has sofas and tables and chairs. Folks sit around. It's a place where there is a community that is an insulating, supporting place to come to. And although we provide some direct services for those folks, more than anything else I think the greatest service we provide is a safe place, a comfort zone, a community, a way to be connected to community. So I know real clearly what you're talking about, and I've heard it spoken too. It's one of the real issues for African Americans who are gay and lesbian. So our simple response has been to try to create a space where folks will have a certain level of comfort.

The problem I often run into, which speaks to what you were asking about, is that I think the greatest trouble we have sometimes is folks in the gay and lesbian community want to celebrate their sense of life and lifestyle more than Metropolitan lends itself to. In the same way, I have folks who don't understand why we don't do more things that are clearly more defined as being Afrocentric. There's a strong Afrocentric movement now within religious circles, and I end up dealing with them the same way I do with the folks who want to have ways in which they celebrate being gay and lesbian. My greatest struggle ends up being with folks who want to lift that up more, and I say the only thing we lift up is the basis of our oneness. I think the Church has been a place where folks have not been able to find community, when they have been rejected. The African American church is a pretty conservative entity, always has been. It's probably be even better at holding up the conventions of American Christianity than other institutions. Consequently folks can draw some pretty hard lines, and that's what we've been dealing with at Metropolitan. But the other side of the Black church is that it is known as being an institution of compassion. So at the same time you have folks giving voice to some conservative ideas and practices, you can also appeal to a tradition and practice of compassion. That's why in our HIV/AIDS ministry we've been able to bring other churches into the loop. We're trying to get thirty churches to be involved in doing education and intervention on HIV/AIDS. We've been able to engage them on the level of compassion. Once we get them involved at that level then, we see folks open their eyes to other issues. African American churches have been very effective in compassion ministry for years around issues of sickness, death and dying. What we're trying to do is get folks to put as much emphasis on living and not just being there to minister to sickness and dying.

Another thing is the contradictions in the Church. The unspoken message that says it's all right for you to be here, just don't say anything, just play your little role. You can be in the choir, you can sit on the piano bench, but don't say you're gay. We had an experience here in Nashville which I'm sure could be evidenced in other places in the world. You know how a few years ago in the community of male figure skaters there was a series of deaths related to HIV/AIDS. The same thing happened a few years ago with musicians in the black churches. At one point here in Nashville there were six musicians who died of AIDS. In every instance it was treated with a hush. Nobody wanted to deal with the fact that all of these men were gay black men, and yet they'd been there leading the music for them. It's that contradiction where folks say yeah you're here but don't say anything about who you really are, don't be honest and open about yourself. I believe that the way in which you get the Church to respond is to continue to force the issue in terms of the teachings of Christ, to be forthright in seeing how the issue is understood in relationship to Jesus Christ. One thing I've learned about dealing with inner city African Americans is that you have to bring it home for them in a way that has some biblical basis. I'm always challenged by this, but I'm always challenging them to find a place where Jesus ever rejected anyone. I don't think anyone can find it. I don't think there's anybody that Jesus did not embrace.

FROM THE GROUND UP

62

Charon Asetoyer

My husband is Dakota Sioux from the Yankton Sioux reservation, located in the South Central part of South Dakota. His father passed away in 1983, and in 1985 we moved to his reservation in order to fulfill our traditional

From *Women's Review of Books* 11 (July 1994): 22. Reprinted by permission.

commitment. It is a tradition to hold a memorial on the anniversary of a loved one's death for the following four years. It was to be a temporary stay, until our commitment was over. We had planned to move to Oklahoma so I could work for my tribe.

After we had moved to South Dakota, I applied for a job with the Yankton Sioux Tribe. The tribe is small and jobs were held by members of the tribe; it was not likely that I would be hired. I was turned down. But word got around that Clarence Rockboy had married a Comanche woman and had returned home with her, and people were curious to meet me.

Women in particular started to drop by, and I took that opportunity to invite others to come and talk. Before long a group of women were meeting regularly in my home to talk about "women's things." We moved the meetings to an empty bedroom in our basement. Out conversations centered around problems that women and children were having in the community and what we might do to address them.

By 1986 we were incorporated as a nonprofit organization. My small basement bedroom turned into an office. Women would stop in to have coffee and talk about their problems at home. Early one morning a woman came by to ask me what I knew about Fetal Alcohol Syndrome (FAS). She told me that she had gone on drinking during her pregnancy and that her child, now a teenager, was having many learning problems. She wanted to know where she could get the child tested for FAS. She was frustrated and had nowhere to turn for support or help. Before long we were meeting on a regular basis; she began bringing other mothers to join in. Soon we had a community task force on FAS, the beginning of our first project, "Women and Children in Alcohol."

It was natural that health issues should come up first: in reservation communities across the U.S., health problems are devastating, and some of the worst are on the Yankton Sioux Reservation. The infant mortality rate has been as high as 23.8 per thousand. Diabetes affects 70 percent of the people over the age of 40, and the rates of secondary health conditions caused by diabetes, such as kidney failure, heart attacks, stroke, amputations and blindness, are alarmingly high. Alcoholism is out of control and contributes to many of these health problems, including domestic abuse and fetal alcohol births—one out of four children has FAS. Many health problems are related to poor nutrition, caused by years and even generations of eating government commodity foods that are high in salt, sugars and fat and low in fiber.

Native activists have brought the Indian Health Service under fire for years for their neglect and mistreatment of Native women. In the 1960s sterilization abuses surfaced, as they did again in the eighties when it was revealed that the Indian Health Service had used Depo-Provera on Native women who were developmentally disabled before it was FDA approved. Today the Indian Health Service is still abusing Native women's reproductive rights by limiting our contraception choices. They are promoting contraceptives like

Depo Provera and Norplant, as well as sterilization. When the Hyde Amendment took away federal funding for abortion, Native women lost their access to abortion altogether: the Indian Health Service, our primary health care provider, is a division of the U.S. Public Service—that is, the Federal government.

The direction our programs have taken has been determined by the women who have come through our doors. I was sitting at my desk one afternoon working on an article for our newsletter when the phone rang. "Is it true that a tubal ligation can be temporary and that I can have it reversed when I want to have a child?" the caller asked. When I asked her why she wanted the information, she said an Indian Health Service doctor had recommended that she have her tubes tied. The procedure, he hold her, could be reversed later if she wanted. Incidents like this spurred us to create a reproductive health program.

Other issues evolved after we started doing workshops in the community. One Saturday morning a young woman came to my front door, banging for someone to let her in. I opened the door and she ran in. I recognized the look of fright and helplessness on her face, and locked the door behind her. During my first marriage I too had been a victim of domestic violence. There I was, standing in the hallway of my home faced with something that I had wanted to leave behind. Only this time I was not the one in need of someone to take me in.

In the course of time we created a program to assist women like this one and their children. We started out with a safe house program, sheltering women in local motels. In 1991, after a long court battle over zoning, the Native American Community Board opened a shelter for women and children fleeing from domestic violence and sexual assault. It now houses eighteen women and children and has provided services for over 900 women and children over the past two and a half years.

As more concerns were brought forward, we needed additional programming and additional staff. The office had grown too small for the organization. Several months earlier I had visited the National Black Women's Health Project in Atlanta, Georgia. The Project was located in a large house which seemed ideal: there was room to grow in, space for program work, a kitchen, a yard, and a strong sense of pride among the staff in their unique working environment. It all seemed so right: why not for us too? We wanted to do nutrition work to address diabetics, so we needed a place to prepare food; we wanted to work with children, so we needed space for them as well as for office work. And if we could buy a house we would no longer have to pay rent to a landlord.

Owning our own place seemed like a good idea but no more than a dream at that point, so I put it in the back of my mind. Who would fund a group of women to buy a house anyway, especially a group of Native American women living on a reservation in South Dakota? I had no idea what the cost of real estate might be in Lake Andes, South Dakota. But I kept thinking about the idea. A friend mentioned a small house she knew of, a fixer-upper, around $5,000 to $7,000. I went to look at it: it did seem to suit our needs.

Shortly after, I was invited to a conference sponsored by the National Women's Health Network, a health advocacy organization based in Washington, D.C. At a networking reception the first evening, I met another woman of color, Luz Alveolus Martinets, the director of the National Latina Health Organization. We talked about the work we were doing, and I eventually told her about our dream of buying a house for our project. When I told her what it would cost, she thought I meant the down payment. But $6,000 was the price for the entire house.

Luz told me that most of the women attending the conference were working women and could afford to make a contribution toward the house. All I had to do was to get up during lunch the next day, when we all had to introduce ourselves, and make an appeal for donations. At lunch I tried to avoid Luz: it seemed too hard to stand up in a crowded room of women I really did not know and ask for money. But Luz found me. Still, when it was my turn to introduce myself, I started to sit down without making my request for donations. Luz kept encouraging me to go ahead and make my request. Before long, all the women in the room wanted to know what we were arguing about. So I went ahead and asked for their help. Then I saw Byllye Avery of the National Black Women's Health Project, across the room, reach into her pocketbook and pull out her checkbook. Byllye held up the first check and announced to the room, "I donate $100." The pocketbooks opened and the checks piled up. It was like a dream: the house was soon to become a reality. I left the conference with about half of the money we needed.

Soon after we purchased the house, people from the community came to help clean, paint and panel the walls. A few weeks of hard work transformed the little brick house across from my home into a warm environment where women and children could come together to share in health education activities and to organize around issues confronting our community. In February 1988, the doors opened on the first Native American Women's Health Education Resource Center based on a reservation, located in Lake Andes, South Dakota, on the Yankton Sioux Reservation.

Not long after the Resource Center opened a young Yankton Sioux woman walked in, introduced herself and asked if she could do her college internship with us. She was interested in working with children. It was time, we decided, for us to examine how we could provide services for children with special needs. The Child Development Program was our answer.

The Program brought children with special needs together to work, play and learn in an after-school program. Fetal alcohol-affected children, latchkey children, gifted children, and children from highly stressed families participate in the program. Today the program has evolved into a Dakota Cultural program in which elders of the Yankton Sioux tribe work with children, teaching them Dakota language, values and spirituality.

An Adult Education Program was not something we had even thought about developing—until one day a call came in from one of the local high school counselors, asking if a sixteen-year-old pregnant student could continue her schooling at the Resource Center with tutoring from our staff. "She's getting too big to sit in class all day long," he said. We said "Sure," and the next week we had a student. Soon after that a mother of one of the children in our Child Development Program asked if we could show her how to use a computer. She said that she already knew how to type, and she could get a job if she knew how to use a computer. This was the beginning of a program that offered job readiness skills and GED completion.

As more women heard about the work and the services that the Resource Center was providing, a feeling of trust was established. Women turned to us for help concerning sterilization abuse, and for information about sexually transmitted diseases, AIDS, family planning and abortion. A high school student came into the Resource Center one afternoon wanting information about pregnancy, AIDS and condoms. She asked if we could get her a home pregnancy test kit, since she thought that she might be pregnant, but if she went down to the local drugstore for it, the clerk would recognize her and call her mother.

It was not long before we were going into the schools with AIDS and contraception information, and conducting AIDS workshops in the community. We hired four high school students, two girls and two boys, and trained them to be AIDS peer counselors, so they could go back into the community and schools to give out accurate information to other young people.

It became obvious to us that no one else in our community, and no other Native women's groups on other reservations, were doing this kind of work. As women turned to us for information about abortion and issues of reproductive health, we knew that we needed to organize our reproductive health work more formally. Our next move was to bring Native women together in a project that would address our unmet health care needs and the reproductive abuses committed by the Indian Health Service. In 1990, 36 Native women met in Pierre, South Dakota, to form the first Native Women's Reproductive Rights Coalition. The Coalition has developed a reproductive rights agenda that defines the issues of reproductive rights and health as Native women ourselves understand them.

Over the past three years the Native Women's Reproductive Rights Coalition has grown to involve over 150 women from eight states, and it continues to grow each year. It organizes Native women to address such concerns as Norplant and Depo-Provera abuse, environmental issues, abortion, and the oppressive effects of organized religion in reservation communities.

Another central part of the Coalition's work has been to include the teachings of our elders. The passing down of traditional methods of birthing, child spacing, abortion, parenting and "becoming a woman" ceremonies is an important issue for Native women and has become a major mission of the Coalition. Through its work, combined with the direct services of our Resource Center, we can bring into the national and international arena practices and policies that we ourselves have shaped and that truly represent the issues confronting us.

HAVING THE TOOLS AT HAND   63
Building Successful Multicultural
Social Justice Organizations

John Anner

The long-awaited age of the true international city is fully upon us, although not quite in the cheerful Disney World format expected by the advocates of multiculturalism. Urban America has become a staggeringly diverse melange of cultures and languages; it's no longer hyperbole to talk about the "global village," not with millions of the global villagers themselves living just down the street.

The changing demographic composition of many areas of the country is coupled with other shifts in the global political economy that are having severe negative effects on both urban and rural America. In brief, the unrestrained ability of international corporations to shift production and jobs just about anywhere in the world has meant disrupted communities, declining wages, and lower standards of living for most Americans—and increasing social decay and conflict.

Social justice organizations and movements have to contend with new constituents and new conditions. They have to figure out how to mobilize and organize people who may share geographic proximity and hard times but little else, in a political climate that has shifted the blame for worsening economic conditions onto the immigrants, the poor, and the powerless. And they have to

---

From John Anner, ed., *Beyond Identity Politics: Emerging Social Justice Movements in Communities of Color* (Boston: South End Press, 1996), pp. 153–66. Reprinted by permission.

figure out how to develop effective strategies for defending their constituents in an era in which progressive government-financed social legislation is close to politically impossible.

On the whole, the traditional defenders of low-income and working-class Americans have done a very poor job of building a sophisticated multicultural response to the changes outlined above, which is one reason things are so terrible. Unions, long the main bulwark against attacks on people's livelihoods, are barely awakening from a long period of political sleep. Progressive academics, professionals, leaders of national civil rights organizations, advocates, lobbyists, and politicians appear to be helpless, timid, and out of ideas, endlessly fighting symbolic battles. And for whatever reason, there is no credible leftist political party in America.

It's more or less an article of faith among progressive political thinkers that we need a new mass-based struggle for social justice featuring leadership from people of color and a diverse membership. I believe . . . that there is an emerging "second generation" of grass-roots community and labor organizations that have developed over the past five to ten years. . . . The energy and mutual respect characteristic of the politics of identity have been used to revitalize and strengthen class-based organizing and to build strong relations between communities normally divided by race, language, and culture.

NO MYSTERY TO MULTICULTURALISM

There is no mystery to building effective multicultural social justice organizations, say many activists and organizers. A variety of organizations provide training and consultations, and, as this [essay] indicates, there are specific methods available to activists and organizers. "What matters most of all," says diversity trainer Guadalupe Guajardo, of Technical Assistance for Community Services (TACS) in Portland, Oregon, "is being able to listen and learn, and having the tools at hand to make changes."

This [essay] focuses on the specific tactics and strategies that social justice organizers use to build multicultural organizations. . . . Organizers can begin to confront the larger economic and political forces devastating low- and middle-income communities by building social justice organizations that can adapt to rapid changes in the demographic and cultural environment. While these organizations may not lead a national social justice movement, if and when it develops, they will train its political leaders, provide models of effective organizing methods, educate and develop memberships, keep people involved on a regular basis in community work, and test the local power structure for weak points.

A movement can draw on this base and use it to inspire large numbers of people to take action. This phenomenon occurred in California's anti-Proposition 187 movement. Established community, student, labor, civil rights, professional, and other groups sounded the alarm and laid the ground-work, but in the end a movement exploded outside the boundaries these or-ganizations created. Indeed, one of the biggest protest marches, which took place in Los Angeles, occurred despite attempts by some "Stop 187" organi-zations to prevent it. A movement, when one develops, pulls in political, com-munity, and labor organizations, but also inspires large numbers of previously uninvolved people.

At the same time, the efficacy and longevity of a movement rests on the foundation created through the daily back-breaking work of local grass-roots organizing. This work must include constant attention to developing multi-cultural memberships and alliances.

The models and mechanisms being used to build successful multicultural social justice organizations include:

- Building personal relationships between members from different back-grounds.
- Actively engaging in solidarity campaigns, actions, and activities with so-cial justice organizations in other communities.
- Challenging bigoted statements and attitudes when they arise.
- Holding regular discussions, forums, "educationals," and workshops to enhance people's understandings of other communities and individuals.
- Working to change the culture of the organization so that members see themselves as "members of the community" first instead of members of a particular part of the community.
- Developing issues, tactics, and campaigns that are relevant to different communities and that reveal fundamental areas of common interest.
- Conducting antiracism training to get people to confront and deal with their biases.
- Examining and changing the organization's practices in order to hire, promote, and develop people of color.
- Confronting white privilege and nationalism.
- Hiring, recruiting, and training more people of color for leadership positions.

IT DOESN'T JUST HAPPEN

Elsa Barboza is an organizer with South Los Angeles–based Action for Grassroots Empowerment and Neighborhood Development Alternatives (AGENDA). AGENDA has generally focused on the African-American com-munity, and most of the staff and organizers are African-American. In 1994, says Barboza, "we decided to move towards organizing in the Latino commu-nity for the simple reason that we have a lot of new immigrants from Central

America in the neighborhoods. We wanted to [make AGENDA] an authentic multicultural organization, but we learned an important lesson: it doesn't just happen."

AGENDA organizers quickly found that bringing monolingual Spanish-speaking members to the general membership and committee meetings was not effective in involving the new Latino members. AGENDA staff decided instead to form a separate organization known as the Latino Organizing Committee. The plan is to build leadership in the Committee, take on a few organizing campaigns around issues of particular concern to those members, and "agitate and educate" the South Los Angeles Latino community. "In the beginning we used the same strategies and tactics for everyone," says Barboza, but it simply didn't work in terms of attracting more Latinos to the group.

One solution is to define issues and campaigns so as to make them relevant to all members of the organization. For example, People United for a Better Oakland (PUEBLO) conducted a campaign in 1994 to force Highland Hospital to hire more professional translators. At first glance, this would not seem to be an issue of major concern to the native English speakers in the group. However, PUEBLO organizers and leaders successfully argued that nobody can get adequate medical care if some doctors and nurses are being called away from their duties in order to translate for a patient who doesn't speak English.

AGENDA has taken a different tack, working instead to research and develop campaigns that are of particular concern to the Spanish-speaking members. At some point in the future, when the Latino Organizing Committee has had enough experience organizing, the two organizations will be merged. In the meantime, AGENDA is conducting what they call "educationals" with all their African-American members to "demystify" the changing demographics of Los Angeles, and show how low-income communities of color face similar challenges and problems.

In terms of specific mechanisms to overcome prejudice-based resistance on the part of current members, "we don't have any 'diversity training' going on just yet," says Barboza, "but we'll probably need it in the future." She says that people get past any initial resistance pretty quickly, however, and see through to the larger self-interest that unites people of color. All the members come together for general membership meetings and selected planning meetings. Translation is conducted by an interpreter who sits with the Spanish-speaking members.

. . . [A]nother community group that started out all African-American and gradually changed as the demographics of the neighborhood changed is Direct Action for Rights and Equality (DARE) in Providence, Rhode Island. DARE is a powerful ten-year-old organization well known at City Hall; part of that power, say staff members, comes from the group's representative membership. In the beginning, however, the few Latino and Asian members seldom participated in DARE's activities. Looking back, says former organizer Libero Della Piana, three things stand out as "holding back" other members: language, culture, and the issues DARE worked on.

In contrast to AGENDA, DARE set out from the beginning to avoid creating two different organizations. "We did have a Comité Latino where meetings are conducted only in Spanish," says Della Piana, "but we didn't want it to become a separate group because we didn't think we'd be able to later turn it into a cohesive organization."

Della Piana says that organizers who want to build multicultural organizations have to ask themselves if they are making everyone feel welcome and wanted. Translating all written materials is critical, he says, because it not only makes all members feel part of the organization but is a way of visibly showing that the group is serious about being multiracial. At this point, Providence is changing again, experiencing a big influx of immigrants from South Asia, and DARE "desperately needs" an organizer who speaks one or more of the languages of the Indian subcontinent.

IN THE HEAT OF THE STRUGGLE

Perhaps the most common way that multicultural groups deal with diversity in their memberships is through a political ideology that emphasizes how the struggle at hand transcends differences of race, age, gender, or sexuality. The Committee Against Anti-Asian Violence (CAAAV) is a New York-based community group that has been active and visible on issues ranging from hate crimes to police brutality and economic exploitation. The group's 2,500 members are diverse, but all Asian. "We are a pan-Asian organization," says staff organizer Saleem Osman, "so in our group there is no hostility, only solidarity, because we are all working together for the same things." Although it is certainly true that there is a good deal of mutual dislike and often active hostility among the large variety of immigrant communities that have found themselves living side by side in urban America, it does seem to be the case that these differences can be set aside in the heat of the struggle.

An additional factor might be that the members of multicultural groups tend to be self-selecting. If someone is really not happy being part of a diverse membership organization, they tend to simply not join in the first place, or leave as the group changes. Finally, differences that seemed of intense importance in the region of origin (Chinese versus Japanese, Salvadoran versus Honduran, North Indian versus South Indian) start to lose their meaning once the reality of life in the United States sinks in and the different groups find that more unites them than separates them.

On the other hand, says Osman, working in coalition with other groups requires a lot of internal education. "Our members, who are taxi drivers, garment workers, and vendors mostly, need to be challenged to look at [their prejudices]." Prejudice against Latinos and African Americans among CAAAV members was clearly revealed during city council hearings on a bill that would have granted taxi drivers the right to refuse to pick up any individual based on their appearance. CAAAV organizers knew that drivers were supportive of the bill because it gave them the right to refuse rides to African-American men.

"We had to do a lot of work with the members on that one," says Osman. "That was a law aimed directly at African-Americans. So we said 'no, we won't support it because it's racist.' But first we put together a video and used it to educate drivers" in day-long training sessions.

CAAAV then took the issue one step further, appearing at public hearings to argue against the law, sometimes surprising African-American civil rights groups. "The best way to overcome prejudices between [communities of color]," says Osman, "is to work together in solidarity with each other to build unity." For this reason, says Osman, CAAAV has actively sought to build a working relationship with Puerto Rican and African-American groups active in the fight against police brutality.

Organizations such as CAAAV, AGENDA, and PUEBLO frequently are called upon by other local groups to attend public hearings or demonstrations or turn out their members for community events. Wendall Chin, an organizer for a multiracial group called San Francisco Anglers for Environmental Rights (SAFER) says that an important way to overcome racial prejudices is to get the members involved in soliciting secondary support from other groups and in joining those groups' actions. SAFER is an environmental justice organization sponsored by Communities for a Better Environment that organizes low-income anglers who fish in the San Francisco Bay for food as opposed to sport. SAFER has joined forces with Texas-based Fuerza Unida in their campaign against the Levi Strauss Company, which is headquartered in San Francisco. Similarly, they have been actively involved with PUEBLO and Asian Immigrant Women Advocates (AIWA), both of which are located in Oakland, not too far from SAFER's main area of operations.

The relationship between external and internal politics is not always as easy and obvious as staff and leaders would like; this is especially true when the senior staff is white. "One of the reasons we had such trouble with the NTC [National Toxics Coalition]," says Sonia Peña, "is that the people in charge figured that because they were part of the struggle and did good political work, they therefore could not be considered racist or sexist or authoritarian or any of those things and didn't have to deal with the issues when they came up in the organization." Peña was on the NTC board at the time of its demise due to intractable internal problems in 1993; she is the lead organizer for the multiracial community group Denver Action for a Better Community.

The anti-AIDS activist group ACT UP and the gay visibility network Queer Nation also started to come apart at the seams in the mid-1990s, in part because of their inability to cope with demands by members of color that the particular needs of their communities receive greater attention. Similarly, at the 1995 National People's Action (NPA) conference in Washington, D.C., a group of Latino delegates stormed the stage and took over the microphone from executive director Gale Cinotta. Led by Juan Mireles, they demanded that the NPA start translating meetings and conference plenaries into Spanish and that a minimum of 25 percent of the delegates to the next year's convention be Latino. Mireles told freelance writer Daniel Cordes that part of the problem is that the NPA—as a predominantly white and African-American organization—doesn't see the need to organize around issues of particular concern to Latinos or to figure out mechanisms to bring in a more diverse membership.

For both NPA and NTC, shared concerns among the membership about declining neighborhoods, redlining, and economic justice were not enough to paper over conflicts among staff and/or members of different races. Some formal mechanism is needed for "surfacing" these conflicts, letting them come out into the open, and resolving them as a group.

This can be as simple as not letting bigoted comments pass without comment. When prejudices are aired openly in a meeting or other event, say organizers, it can poison relations unless it is dealt with openly. "I remember one time at a meeting where we had whites, Blacks, and three or four Mexican members," says SAFER organizer Wendell Chin," and an outspoken white leader commented that she was glad that [the anti-immigrant California ballot initiative] Proposition 187 had passed because of the problems too many immigrants were causing. I looked over at the Mexican [members] and they weren't saying anything. So I had to step in and intervene."

"It's always better if the challenge comes from inside the membership of the group," says PUEBLO organizer Danny HoSang, "because then the members who are being challenged don't feel like they are being singled out by the staff. But if nobody speaks up, the staff organizer has to do it. It's not only about consciousness-raising. All the members of the group need to feel like they are welcome and valued."

Sometimes a more involved process is required. When tensions surfaced between Asian and Black participants in a year-long program called A New Collaboration for Hands-On Relationships (ANCHOR), program director Rinku Sen decided to skip the actual program for that week and hold a series of discussions about the differences between the two communities. "Just because we call ourselves people of color doesn't mean we have the same backgrounds," says Sen. "Someone who was born in Cambodia and moved into a Black neighborhood in Oakland might experience racism as a daily fact of life. But that experience itself might be different than it would be for a Black person, and of course it is filtered through their history and current expectations."

SPEAKING MY LANGUAGE

Multicultural organizations are usually multilingual. PUEBLO, DARE, and other community groups have invested in a number of simultaneous translating machines. "They are a costly but highly effective tool," says PUEBLO lead organizer Rosi Reyes. The machines allow members to be seated anywhere in the room, instead of being segregated in one area, while the translator speaks quietly into a transmitter worn over the head. Receivers are smaller than a pack of cigarettes, with tiny earphones.

Some groups break meetings into segments, with some parts translated and others held in one language. Monica Russo, an organizer with a Florida local of the UNITE, uses this technique with a membership that speaks English, Creole, and Spanish. Day-long trainings, for example, feature monolingual sessions combined with collective meals and informal periods.

Most organizers warn, however, that translation demands more than simply literally transcribing what is being said. In order to transmit the real meaning and invite participation, says Peter Cervantes-Gautschi, director of the Portland, Oregon Workers Organizing Committee, "the critical thing is that the [translator] has to be into the movement. Because if [that person] doesn't really understand what we're trying to accomplish, then they are not going to get it right."

FROM WHITE TO RAINBOW

"There are two main obstacles to building multiracial social justice movements," says Libero Della Piana, who edits *Race File* at the Applied Research Center, "nationalism and white racism." Although many barriers divide people of color from each other, diversifying white organizations presents a different and more difficult set of challenges. "People of color generally understand racism as institutional," says Guadalupe Guajardo, "while to white people white privilege is virtually invisible, and they see prejudice as being something personal.

"Sometimes the hardest thing is to get people to face the reality that racism does exist even though polite people don't make racist comments anymore. But the only way to reach people and get them to examine how they benefit [from white privilege] is to start from the assumption that people are basically good. If you call folks racist, you will just make them defensive and won't get anywhere."

When working with an all-white group that wants to become multicultural, TACS trainers lay out five concrete areas to examine and consider changing: (1) the recruitment process, including where it is done and what the qualification requirements are; (2) planning, i.e., who is at the table when plans are being made; (3) decisionmaking, i.e., who is involved when decisions are made, both in formal and informal settings; (4) resource allocation, including money, access, and power; (5) promotion and leadership. Promoting people of color to leadership positions is vital, but only if these individuals have a base of support, are going to be given power and resources, and are held accountable.

"People need solutions," says Guajardo. "We help them find a new model based on an alliance-building or partnership idea, instead of seeing diversity as a win/lost kind of thing."

A number of other organizations around the country conduct similar trainings, ranging from "racial awareness training" to "diversity management," "prejudice reduction," and "coalition building."

CHANGING THE MIX

A criticism frequently directed at national community organizing networks is that the staff organizers and directors are mostly white, while the people they organize are predominantly people of color. This is certainly true for NPA, the Association of Community Organizations for Reform Now (ACORN), and to a lesser extent the Industrial Areas Foundation (IAF). The same criticism is leveled at labor organizations and the progressive press. There are more people of color working at the average metropolitan daily newspaper than at all the left-wing magazines in the country combined.

This situation—at least on the community organizing side of things—is responsible for at least three significant trends in grass-roots social justice organizing that appeared in the 1980s and developed throughout the 1990s. First is the rapid growth of independent organizations in communities of color not connected to any of the community organizing networks or unions. . . .

Second is the formation of organizer training programs specifically designed to train organizers of color to work in community and labor organizations comprised of people of color. Of these, by far the most prominent is the Center for Third World Organizing (CTWO), which has trained several hundred organizers of color through ten years of programs such as the Minority Activist Apprenticeship Program (MAAP) and the Community Partnership Program. Along with the organizer training programs, CTWO has also developed a model of community organizing that relies on organizers of color.

Finally, many emerging social justice organizations have made an explicit commitment to leadership diversity. The New Party constitution, for example, states that 50 percent or more of the leaders of local chapters must be people of color and 50 percent must be women. Greater diversity in the organizing staff can bring immediate rewards for community and labor organizations; it is now pretty much accepted by labor leaders that white men are the least effective organizers.

Teamsters Local 175 in Seattle, Washington, wanted to organize Asian women working in private postal facilities. The notion that these women could not be organized, says organizer Michael Laslett, was partly due to the fact that nobody had ever tried, and partly due to the expected barriers of culture, race, and language. He brought on two Asian-American interns from the AFL-CIO Organizing Institute to go into one particular shop, which resulted in a victorious union drive. "Having Asian organizers is what made organizing this company possible," he said.

In 1994, the United Electrical Workers (UE) were able to organize Mexican workers at a SteelTech factory in Milwaukee by importing an organizer from a sister union in Mexico. Labor organizers of color can be pretty scarce. A perhaps apocryphal story has it that the Oil, Chemical, and Atomic Workers Union at one time employed the only Vietnamese labor organizer in the country, and used to lend him out to other unions that needed a Vietnamese-speaking organizer.

FLUID IDENTITIES

The strategies for diversifying that work with this generation may not work with the next one, however. According to Libero Della Piana, when push comes to shove, the color of the organizer matters less than the person's commitment, especially when working with youth. "Of course it helps to have a diverse organizing staff," he comments, "but in the end that's not what it's about. Members respect the staff that's willing to do the work. It's a delicate balance."

In fact, say some youth organizers, most of what has been outlined above is based on strict definitions that don't fit with the mixed identities and intensely multicultural lives of urban youth. These young people don't need to be taught how to get along with other cultures, nor do they necessarily identify with the racial categories that guide the previous generation.

"It's pretty wild," says Next Generation co-director Mike Perez, who used to work at the Oakland-based youth program Encampment for Citizenship. "Race and class intersect in different ways for young people than they do for their parents." Many young people believe they can choose their racial identity based on how they feel. "I didn't have enough attitude to be a Black girl," one white high school student told *YO!* editor Nell Bernstein, explaining why she dressed like a *cholita*, or Latina gangsta girl.

"You have white kids coming in saying they are just as much a 'nigga' as any Black kid since they come from the same 'hood," says Perez. "You have Asian kids dressing hip-hop and talking [African-American dialect]. Other Black kids talk and act 'white,' in the eyes of their friends. Identities are very fluid, but it's what being young is about now."

Della Piana agrees: "During the campaign to stop Prop 187, the main organizations active in Oakland tried to define it as a Latino issue. The high school and junior high students who organized and marched in the streets refused to see it that way. Their notion of what the fight was about was based on who was in and who was out.

"Kids base their identity on the basis of a complicated formula of territory, race, class, and aspirations," he continues. "There's a connection between kids that nobody understands; they can relate to each other [across racial boundaries] in some way adults have a hard time with."

It is probably true, as some youth organizers argue, that the intense problems that race, gender, and sexuality caused for older political groups and movements will not be as much in evidence as younger people start to move into leadership positions in social justice organizations. Perhaps "the fire next time" will burn brightly in rainbow patterns. But if we want social justice organizing to move beyond the limitations of identity struggles into an enlightened next phase, with justice, community, democracy, and true solidarity on the top of the agenda, we would do well to remember what Elsa Barboza said: It doesn't just happen.

CAN I GET A WITNESS?

64

Testimony from a Hip Hop Feminist

Shani Jamila

I used to think I had missed my time. Thought I was meant to have come of age in the sixties when I could've been a Panther freedom fighter, challenging the pigs alongside Assata, Angela and Kathleen. Oh, but I went deeper than

From Daisy Hernández and Bushra Rehman, eds., *Colonize This! Young Women of Color on Today's Feminism* (New York: Seal Press, 2002), pp. 382–99. Copyright © 2002 Daisy Hernández and Bushra Rehman. Reprinted by permission of the publisher, Seal Press.

that. I saw myself reading my poetry with Sonia, Ntozake, Nikki and June . . . being a peer of Audre, Alice and Paula Giddings . . . kickin' it with revolutionary brothers like Huey, Haki, and Rap . . . all while rocking the shit out of my black beret. When she needed advice, I would've *been* there for Patricia Hill Collins as she bounced her preliminary ideas about *Black Feminist Thought* off me. Now can't you see the beauty in this? I would have been building and bonding with a community of artists and activists that had this whole vibrancy radiating from its core, and so many of my role models would've just been crew.

I know right now some of y'all are probably like, "OK, this child's on crack . . . that decade was not all that!" But don't front. When you heard the stories of your parents, aunts, uncles and family friends—or even if you just watched some TV special talking about the mystique of the sixties—didn't it ever make you wonder what happened with our generation? Who were our revolutionaries? What sparked our passions so high that we were willing to risk our lives to fight for it? Where was our national Black Arts Movement?

It seemed natural that we should have one—after all, we had flavor for days . . . high-tops and Hammer pants, jellies and Jheri Curls bear witness to that! And as an African-American child coming of age in the first generation to endure the United States post-integration, I can *definitely* testify that we had our own struggles: AIDS, apartheid, affirmative action, the prison industrial complex and underdeveloped inner cities are only a few examples. Growing up, these were the things that would run through my mind as I'd cut out collages from *Right On!*, and wonder what happened to us. Seemed like coping with issues like these would've hyped us up enough to create our own culture of resistance. The glossy pages trying to stick to the walls of my room competed futilely with the vibrations emanating from my spastic MTV imitations as Power 99FM's bass blasted. I danced around the images of Public Enemy and Queen Latifah that now decorated my floor, rapping all the lyrics to "Fight the Power" as I mourned our inactivity.

As I got older, I realized what I'd missed in my youth. Largely due to globalization and growing technology, in addition to some banging beats and off-the-chain lyrics, we'd had an *international* Black Arts Movement shaping our generation. As a kid, I didn't recognize hip hop as a vibrant and valuable sociocultural force—I just thought the culture was cool. I loved the music but never conceived of it as a revolutionary outcry. After all, I'd learned in school that activism was a concept confined to the sixties . . . so even though I felt empowered by the Afrocentric vibes and in-your-face lyrics, it seemed like the culture came a few decades too late for a critical context. In fact, individuals and organizations whose work challenged me to think critically about Black people rarely even entered the public school curriculum. In my high school's halls we were taught a very narrow and revisionist view of world history that boiled down to this: white was right, Africa was an afterthought. In addition to the massive amounts of potentially empowering information that was erased by those messages, a holistic history just was not taught.

Not only were my people not reflected in the syllabi, but I didn't see a proportionate reflection in the faces of my classmates either. As one of only three Black faces in the honors program, all of whom were middle-class females, I often questioned why our representation was so disproportionate. The subtext shouted that the reason wasn't a deficiency in the newly integrated school system but rather the failings of people of color. We were tacitly taught that our token presence proved racism and sexism were over, so the problem must have been our peers' inability to achieve. I knew this wasn't true but as a child I was often frustrated because I didn't know how to prove it, as it was often demanded that I do.

See, my generation came of age with the expectancy that we could live, eat and attend school among whites. Race and gender were no longer inscribed in the law as automatic barriers to achievement, making the injustices we encountered less obvious than those our predecessors had faced. But the issues didn't go away. Instead, we found that an adverse consequence of integration and the "gains" of the sixties was even more heavily convoluted notions of race and gender oppression. Economic class stratification has also continued to evolve as a serious complication.

The paradox of the Black middle class as I experienced it is that we are simultaneously affirmed and erased: tokenized and celebrated as one of the few "achievers" of our race but set apart from other Black folks by our economic success. It is the classic divide-and-conquer technique regularly employed in oppressive structures; in this case saying that Black people are pathological—but you somehow escaped the genetic curse, so you must be "different." These lessons were regularly reinforced with camouflaged compliments such as, "Wow, you're so pretty/smart . . . are *both* of your parents Black?" Other times the insults would blaze brazenly, like the comments made by the white girls in my Girl Scouts carpool when we drove by a group of Black children playing in their yard: "Ooooh, Mom! Look at the little niglets playing on the street!" Their snickers echoed in a familiar way that suggested they'd shared this joke before. As the pain and rage began to well up within me, their dismissive comments also gathered force: "God, like, don't be so sensitive, Shani. We're not talking about *you*. You're different." I waited expectantly for the adult in the car to tell her children they were out of line and to apologize. She said nothing. I began to wonder if I was overreacting. Maybe it wasn't such a big deal. Maybe I *was* different. Maybe I thought too much.

Over the years I learned how to censor myself and adapt to different surroundings, automatically tailoring my tongue to fit the ear of whatever crew I was with. Depending on the composition of the crowd, the way I'd speak and even the things I'd talk about were subject to change. White people automatically got a very precise speech, because I knew every word out of my mouth was being measured and quantified as an example of the capabilities of the entire Black race. Around Black people I slipped into the vocabulary I felt more comfortable with but remained aware that I was still being judged. This time it was to see how capably I could fall back into "our talk" without sounding like a foreigner, if I could prove my suburban upbringing and elitist education had not robbed me of my authenticity as a Black woman.

Passing this litmus test meant the most to me. Because if the daily trials weren't enough, when the flood of college acceptance and rejection letters began pouring in and I got into schools my white "friends" weren't admitted to, all of a sudden the color they didn't see before came back fierce. My GPA, test scores, extracurricular activities, and recommendations were rendered irrelevant when they viciously told me the only conceivable reason I was getting in over them had to be affirmative action. I realized the racial logic being used against me was something that pervaded all class spheres. Whether you were a beneficiary of affirmative action or you were seen in the imagery of welfare queens (whose depiction as poor Black women defies actual numerical stats), we were all categorized as niggers trying to get over on the system.

While most of my memories from childhood are happy ones, I also remember a constant struggle to find a sense of balance. For every "reward" token status bestowed, it simultaneously increased the isolation I felt. I didn't think there were many people who could understand how and why I was struggling when by societal standards I was succeeding. I worried that I was being ungrateful because I knew so many who had come before me had given their lives in the hopes that one day their children could have the opportunities I'd grown up with. Despite the public accolades I received for my accomplishments, until I went to college I felt shunned by whites and suspected by Blacks. I was looking for a place to belong.

In 1993 I took my first steps on the campus of Spelman College, a Black woman's space in the middle of the largest conglomeration of historically Black colleges and universities in the world. This is not your typical institution. One of only two colleges of its kind surviving in the States—at Spelman Black women walk proud. Our first address from "Sista Prez" Johnnetta B. Cole told us so. As is characteristic of speeches to incoming first-year students, she instructed us to look to our right and look to our left. We dutifully gazed upon each other's brown faces. She spoke: "Other schools will tell you one of these students will not be here in four years when you are graduating. At Spelman we say we will all see to it: your sister *better* be at your side when you *all* graduate in four years!" Loud cheers erupted—we were our sisters' keepers.

At Spelman I learned new ways of learning, thinking and challenging. It was in this place that I was first introduced to a way of teaching that was unapologetically rooted in Black women's perspectives, that addressed the reality of what it means to be at the center of intersecting discriminations like race, class and gender. My formal education about my people began to expand beyond Malcolm and Martin. I learned about activists like Sojourner Truth and Maria W. Stewart, journalists and crusaders like Ida B. Wells, preachers like Jarena Lee, freedom fighters and abolitionists like Harriet Tubman, scholars like Anna Julia Cooper, poets like Frances Ellen Watkins Harper and community leaders like Mary Church Terrell. Here our core courses were entitled "African Diaspora and the World" and "Images of Women in the Media." The required reading on the syllabi included books like Paulo Freire's *Pedagogy of the Oppressed,* Frantz Fanon's *The Wretched of the Earth*, and Patricia Hill Collins's *Black Feminist Thought.* In these books and classes I found the answers to questions I didn't even have the language to ask with the education I'd received in high school.

This is what made attending a historically Black college such a turning point in my life. I don't want to romanticize my collegiate experience to the point where it was like I opened up a book and suddenly became some sort of guru, but what being exposed to this community of scholars and activists did do was give me a framework for my feelings. The value of having my thoughts nurtured, legitimized and placed into a historical context, in addition to the power of being surrounded by sisters and brothers who were walking refutations to the stereotypes I'd grown up with, gave me a space to blossom in ways I couldn't have imagined. I felt validated and affirmed by the idea that I no longer had to explain why Black folks were different from the purported standard. Instead of being made to justify what mainstream society perceived as deviance, I was supported in the effort to critically challenge how societal norms even came to be. I loved that when we would discuss slavery, an integral part of the conversation was slave revolts—Black resistance had finally entered the curriculum. It was the first time I saw people reflective of myself and my experiences both inside and outside of the classroom. Living and learning like this was revolutionary for me. It changed my life.

Of course, being on an all women's campus, gender was also a regular topic of conversation. I was part of some beautiful dialogues where brothers would share the struggles they endured excelling academically that they didn't face when they'd shine in the more "acceptable" realms of sports or music. Sisters would relate back with testimonies of feeling forced to choose between our Blackness and our womanhood—a choice as impossible, a professor pointed out, as choosing between our left and right sides. In stark contrast to the race debates, however, these moments of raw honesty took place on a slippery slope. Gendered analyses were not granted the same sense of universal urgency attributed to race. Rather, they were received with suspicion. Many people perceived the debate over gender dynamics as a way to pit Black folks against each other. In heated conversations my peers would choose camps, placing race, gender and class in a hierarchy and declaring loyalty to one over the other. Protests would be peppered with frequent warnings that Spelman was notorious for inculcating crazy mentalities in its students. We were told we better watch our backs before we turned into one of those (gasp!) feminists too.

Yup, the dreaded F-word continues to be so weighed down by negative connotations that few people are willing to voluntarily associate with it. Hurled out like an accusation, it is enough to make many sistas start backpedaling faster than the rising stats on violence against women. The reluctance to be identified with something perceived as an internally divisive force inside historically oppressed communities is understandable. Many feminists of color felt it too, which is largely why Black feminist theory and womanism emerged. Unfortunately, much of what Black feminism really stands for has been stereotyped or obscured by school systems that don't devote time to Black women's intellectual traditions. A sad consequence has been that in addition to having something designed to advance our people become a tool of division, millions of people have been kept in the dark (so to speak) from a wealth of really important information and support networks.

Because of all the drama surrounding the word "feminism," there are mad heads who identify with feminist principles but feel conflicted about embracing the term. But let's examine what it really means. At root, Black feminism is a struggle against the pervasive oppression that defines Western culture. Whether taking aim at gender equity, homophobia or images of women, it functions to resist disempowering ideologies and devaluing institutions. It merges theory and action to reaffirm Black women's legitimacy as producers of intellectual work and reject assertions that attack our ability to contribute to these traditions. In stark contrast to the popular misconception that Black feminism is a divisive force that pits sisters against brothers, or even feminists against feminists, I view it as an essential part of a larger struggle for all of our liberation. Our fight for freedom has to be inclusive.

Of course, my understanding of Black feminism is rooted in the theoretical texts written decades before I was first introduced to them in college. Many of these theories remain relevant, at the very least as an essential historical base. However, for any movement to maximize its effectiveness, it has to be applicable to the times. It is incumbent upon us as hip hop feminists not to become complacent in the work that has come before us. We have to write our own stories that address the issues that are specific to our time. For example, some people think it's an oxymoron when I juxtapose a term like feminism alongside a genre of music that has been assailed for its misogyny. It seems obvious to me, however, that just as the shape of what we're fighting has changed, we need to examine how we as a community of activists have changed as well. Hip hop is the dominant influence on our generation.

Since my birth in 1975 four years before the first rap single achieved mainstream success, I have watched the hip hop movement, culture and music evolve. I mark important events in my life by the hip hop songs that were popular at the time, linking my high-school graduation with the Souls of Mischief's album *'93 til Infinity* and my first school dance with the song "It Takes Two" by Rob Bass. My ideas of fashion have often been misled by hip hop artists like Kwamé, whose signature style resulted in the proliferation of polka dots in American schools around 1989. My taste in men has also been molded by hip hop aesthetics. I entered my love life interested in brothers who were rocking gumbies like my first boyfriend. As I got older, I discovered my own poetic voice, and I cannot begin to place a value on the amount of inspiration I got from this musical movement and the culture it birthed. I am a child of the hip hop generation, grounded in the understanding that we enter the world from a hip hop paradigm.

Those of us who embrace feminism can't act like hip hop hasn't been an influence on our lives, or vice versa, simply because claiming them both might seem to pose a contradiction. They are two of the basic things that mold us. However, we must not confuse having love for either one with blind defense. We have to love them enough to critique both of them and challenge them to grow—beyond the materialism and misogyny that has come to characterize too much of hip hop, beyond the extremism that feminism sometimes engages in. As women of the hip hop generation we need a feminist consciousness that allows us to examine how representations and images can be simultaneously empowering and problematic.

We have to engage with the rap lyrics about women and the accompanying images found in video scenes. A friend, Adziko Simba, once told me, "It seems bizarre to me that we African women have reached such an 'enlightened' point that we are defending our right to portray ourselves in ways that contribute to our degradation. . . . Did Sojourner Truth walk all those miles and bear her breast in the name of equality so that her heirs could have the right to jiggle their breasts on BET?" I completely feel her, though I don't think the role of feminism is to construct "proper" femininity, or to place limits on how women are able to define and present themselves. I think doing so is actually antithetical to the movement. Teaching women not to be sensual and erotic beings, or not to show that we are, is diminishing and subverts the locus of our own uniqueness as females. Why shouldn't we be able to celebrate our beauty, sensuality, sexuality, creative ability or our eroticism? They are all unique sites of women's power that we should not be taught to hide, or only display when someone else says it's appropriate. On the flip side we shouldn't support each other to the point of stupidity. We have to demand accountability from each other, no doubt. We need to be cognizant of the power in this music and of how we are representing ourselves on a global scale and on the historical record. These examples demonstrate how wide open the field is for sisters of the hip hop generation to address the constantly shifting space women occupy. But these areas of concern should not be solely relegated to the Black feminist body of work. Hip hop activists, intellectuals and artists all need to take a leading role in confronting the fragmenting issues our generation deals with.

So, yeah, I used to think I had missed my time. I thought the flame lighting the hearts of activists had been snuffed, *Survivor*-style. But liberating my definitions of activism from the constraints and constructs of the sixties opened up my mind to a whole new world of work and progressive thought. Now I draw strength from the knowledge that people have been actively combating sexism, racism and other intersecting discriminations for a long time. Many of those icons I respect are still on the scene actively doing their thing for us. That knowledge is my ammunition as I join with them and my peers to continue fighting those battles and the other fronts unique to our time. We can't get complacent. The most important thing we can do as a generation is to see our new positions as power and weapons to be used strategically in the struggle rather than as spoils of war. Because this shit is far from finished.

Thinking Further

After reading the articles in this section, you will find it helpful to think about the following:

In Part V, "Making a Difference," we focus on forms of activism in the American context. This is important, yet we also recognize that much activism concerning racism, class exploitation, sexism, homophobia, religious intolerance, and other inequities is occurring in global arenas. For example, the United Nations has sponsored conferences on the status of women and on race and racism that are attended by people from around the world. We wonder how the forms of activism that we detail here might participate in these larger global initiatives. How might local activists in the United States whose work is informed by a race, class, and gender analysis connect with similar work in a global context? What form might local-global coalitions take that are grounded in intersectional frameworks of the type detailed here? In the aftermath of 9/11, these questions are critical. We invite you to think further about this important issue.

Suggested Readings

Brandt, Eric, ed. 1999. *Dangerous Liaisons: Blacks, Gays, and the Struggle for Equality.* New York: The New Press.

Bullard, Robert D. 1994. *Unequal Protection: Environmental Justice and Communities of Color.* San Francisco: Sierra Club Books.

Espiritu, Yen Le. 1992. *Asian-American Pan-Ethnicity: Bridging Institutions and Identities.* Philadelphia: Temple University Press.

Freeman, Jo, and Victoria Johnson. 1999. *Waves of Protest: Social Movements Since the Sixties.* Totowa, NJ: Rowman & Littlefield.

Hernández, Daisy, and Bushra Rehman, eds. 2002. *Colonize This! Young Women of Color on Today's Feminism.* New York: Seal Press.

Torres, Andrés, and José E. Velázquez, eds. 1998. *The Puerto Rican Movement: Voices from the Diaspora.* Philadelphia: Temple University Press.

 ## InfoTrac College Edition: Search Terms

Use your access to the online library, InfoTrac College Edition, to learn more about some of the subjects covered in this section. The following are some suggested terms for searches:

Black power movement
civil rights movement

environmental racism
feminist movement
grassroots organizations
La Raza
liberation theology
National Welfare Rights Organization (NWRO)

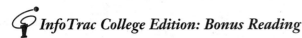

InfoTrac College Edition: Bonus Reading

You can use the InfoTrac College Edition feature to find an additional reading pertinent to this section and can also search on the author's name or keywords to locate the following reading:

Featherstone, Liza. 2000. "The Student Movement Comes of Age." *The Nation* (October 16): 23.

Liza Featherstone reports that contemporary student movements have managed to achieve an impressive measure of power on college campuses. Students have organized to protest the sweatshop labor conditions that are used to produce their college sweatshirts, t-shirts, and other college insignia. More attuned to issues of social justice, student movements have generated changes on many college campuses. Featherstone also raises the important question of how students might use this newfound power. How will students connect multiple movements, such as those against unfair labor practices, those for women's rights, and those for economic and racial justice? How might a race, class, and gender framework inform the work of contemporary student movements?

Wadsworth Sociology Resource Center: Virtual Society

For additional links, quizzes, and learning tools related to this section, see the Wadsworth Sociology Resource Center, Virtual Society, at www.wadsworth .com/sociology_d/

Index

Movies, and the English language, 319
Moynihan, Daniel Patrick, 292, 312
Multiculturalism, 5–6, 51–62
  in activist organizations, 542–52
  in child care, 300–301
  in education, 225, 348–53
Muslim women, 214, 441–47
Myrdal, Gunnar, 363
Mythical norm, 66
Myths
  about African Americans, 455–56
  about Latinas, 337–42
  See also stereotyping

NAACP (National Association for the Advancement of Colored People), 363, 526
NAFTA (North American Free Trade Agreement), 261
Narratives
  personal, 21–22
  slaves, 271
National Advocates for Pregnant Women, 387
National Association for the Advancement of Colored People (NAACP), 363, 526
National Basketball Association (NBA), 76–77, 345
National Black Women's Health Project, 539, 540
National Congress of American Indians, 358
National Council of Negro Women, 122
National Latina Health Organization, 540
National Lawyers Guild, 504
National Network of Prostitutes, 472
National People's Action (NPA), 548, 550
National Rifle Association (NRA), 497
National security, and immigrants, 389–390
National Toxics Coalition (NTC), 547
National Women's Health Network, 540
Native American Community Board, 539
Native Americans
  assimilation of, 325–28, 356–57
  and child care, 296
  colonization of, 83
  and cultural institutions, 224
  domestic violence among, 539
  elderly, 281
  genocide of, 226
  incorporation of, 355–59
  media images of, 47, 327–28
  population of, 1, 45
  and poverty, 113, 155–56
  racism in athletics, 321–24
  racist terminology, 316–17
  relocation of, 358
  removal of, 355–56
  self-determination of, 359
  and symbolic ethnicity, 420
  tribal government among, 359–60
  in U.S. history, 57
  urban, 360
  and white privilege, 143

women
  reproductive health of, 207, 538–39
  reproductive rights of, 388
  reproductive sterilization of, 207, 232, 538–39
  and survival, 44–48
  and toxic waste activism, 527–29
Native American Women's Health Education Resource Center, 540
Native Women's Reproductive Rights Coalition, 514, 541
Naturalization Law of 1790, 56
Nazi Germany, 324–25, 328
NBA (National Basketball Association), 76–77, 345
Network of Sex Work Projects, 472
New Deal, 127
Newman, Katherine S., 219, 248
New Party, 550–51
Next Generation, 551
Nez Perce Indians, 356
Nightingale, Florence, 39
Nixon, Richard Milhouse, 359, 456
North American Free Trade Agreement (NAFTA), 261
NPA (National People's Action), 548, 550
NRA (National Rifle Association), 497
NTC (National Toxics Coalition), 547
Nuremberg Trials, 324–25, 326
Nurturant social skills, 239
Nuyoricans, 253. *See also* Puerto Ricans

Office of Management and Budget (OMB), 380
Oil, Chemical, and Atomic Workers Union, 551
OMB (Office of Management and Budget), 380
O'Neill, Thomas P. ("Tip"), 421
Onondaga Indians, 359
Operation Rescue, 205–6
Opium Wars, 56
Oppression
  black-white model, 111–12
  and class, 31, 32, 34
  double binds, 49–50
  expectations of the oppressed, 64–65
  of homosexuals, 461–62, 463–64
  internalized, 100
  of Latinos, 115
  of men, 49
  and otherness, 33
  racism, 99
  sexual, 41
  vs. suffering, 22
  and toxic waste activism, 521–23
  of women, 31, 32–33, 48–51
    lesbians, 30, 33
    street harassment, 486–90
    working-class, 146–47
Organized sports, 190–201, 290. *See also* athletics